THE ALBANY

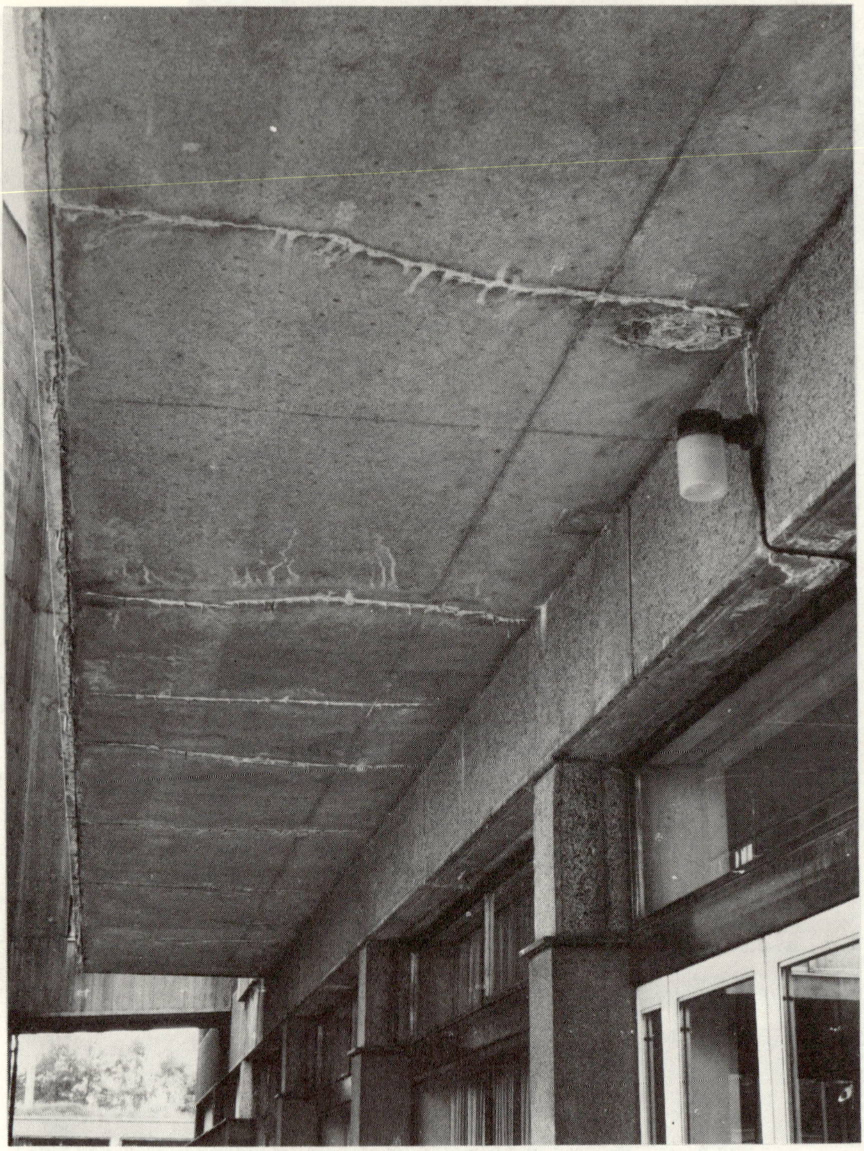

Limit state theory
for reinforced concrete design

Limit state theory for reinforced concrete design

SI units
(Second edition)

B P Hughes
DSc, PhD, C Eng, MICE, FI Struct E
Professor of Civil Engineering
University of Birmingham

Pitman

First published 1971

Second edition 1976
Reprinted with corrections 1977

Pitman Publishing Ltd
Pitman House, Parker Street, Kingsway, London WC2B 5PB
PO Box 46038, Banda Street, Nairobi, Kenya

Pitman Publishing Pty Ltd
Pitman House, 158 Bouverie Street, Carlton, Victoria 3053, Australia

Pitman Publishing Corporation
6 East 43 Street, New York, NY 10017, USA

Sir Isaac Pitman (Canada) Ltd
495 Wellington Street West, Toronto, Canada M5V 1G1

The Copp Clark Publishing Company
517 Wellington Street West, Toronto, Canada M5V 1G1

Paperback edition

ISBN 0 273 00304 6
(1345:73)

Cased edition

ISBN 0 273 01023 9
(1346:73)

Reproduced and printed by photolithography and bound in Great Britain at The Pitman Press, Bath

Preface to Second Edition

Since the publication of the first edition in 1971 the Unified Code based on limit state theory, CP 110:1972,The structural use of concrete, has been issued and it has now been joined by the limit state version of CP 2007 – BS 5337, The structural use of concrete for retaining aqueous liquids. The chairmen of the respective BSI committees, Dr D.D. Matthews and Mr A.W. Hill, are to be congratulated on steering these radical documents to successful conclusions. Not least of the changes required in this second edition has been the adoption of the internationally agreed principles for notation produced by the European Concrete Committee and the American Concrete Institute and approved at the 1971 CEB meeting. The logic of the preferences of the international agreements for both notation and SI units has been respected and this edition is therefore generally in accordance with both CP 110 and BS 5337. Very exceptionally, where CP 110 and BS 5337 violate the international preferences and clarity would have suffered, the international preferences have been followed. (The only other very slight difference with CP 110 is also indicated in the note following the list of symbols.)

Appendix C – the concise summary of CP 110 for rapid location of the essential design requirements for reinforced concrete – has been completely revised and subdivided. New appendixes have been added for plain concrete walls, water-retaining structures (to BS 5337), joints (to CP 110 and BS 5337), fire resistance (to 1975 Joint Report) and wind loading (to CP 3 chapter V, part 2). A new chapter has also been added on mix design and quality control to extend the coverage even further.

It is hoped that this new edition will again assist students to an understanding of the fundamentals of concrete theory and practice, and the rational concepts implicit in the limit state codes, and provide practising engineers with a rapid appreciation of the essential requirements of CP 110 and BS 5337.

It is suggested that most of chapters 1 to 5, section 6.1, chapter 8, sections 9.1 to 9.4 and chapter 15 could be included in a second-year undergraduate course, but this can vary considerably between universities. At Birmingham the concrete structures and technology course is complementary to the general structures course. Questions from recent examination papers are reproduced by permission of the universities of Birmingham and London (King's College). The university course and year, and the time alloted, are shown in brackets where appropriate, e.g. '(UL2 1969, 36 min)' indicates a 36-minute question for second-year internal students of King's College in 1969.

I am indebted in this revision to Mr R.D. Anchor (Appendix M), Dr J.E. Ash and Dr R.M. Willcock (Appendix V), Mr N.I. Fattuhi (Appendix E) and Dr M.R. Prince (Appendix I). I also wish to thank all those who have commented on the first edition, especially Dr J.E. Ash, Mr J. Behan, Dr W.B. Cranston, Mr J.E.C. Farebrother, Mr J. Guest, Prof. L.L. Jones, Mr J.K. Nesbit, Mr A. Smith and Dr P. Swannell.

Extracts from CP 110:1972 and BS 5337:1976 (known as the Unified Code and the Water-retaining Code respectively); CP 3 Codes of basic data for the design of buildings, chapter V, loading part 1: 1967 Dead and imposed loads, part 2:1972 Wind loads; BS 4449:1969 Specification for hot rolled steel bars for the reinforcement of concrete; and BS 4466:1969 Bending dimensions and scheduling of bars for the reinforcement of concrete, are reproduced by permission of the British Standards Institution, 2 Park Street, London W1Y 4AA, from whom copies of the complete documents may be obtained. Extracts from the reports on The detailing of reinforced concrete; A standardized procedure for the presentation of calculations for reinforced and prestressed concrete structures; and Fire resistance of concrete structures, are reproduced by courtesy of the Councils of the Institution of Structural Engineers and the Concrete Society. Individual acknowledgements are also given in the text to other authors and associations whose publications have been quoted and those who have given permission for extracts to be reproduced, including the Cement and Concrete Association and the British Ready Mixed Concrete Association.

Finally I wish to thank my wife and family for their renewed patience, encouragement and editorial assistance with this second edition, and Mrs. J. Soley for typing the manuscript revisions.

Contents

Appendices

Frontispiece

To Anne

Introduction

The introduction of ultimate load theory for the design of reinforced concrete at first tended to discredit the traditional elastic theory approach. Steadily increasing knowledge, however, has brought each theory into perspective and has shown that both have their application within the wider concepts provided by limit state theory. The serviceability limit states generally both utilize elastic theory while the ultimate limit states of collapse incorporate ultimate load theory.

The rational approach and philosophy of limit state theory does more than simply adopt existing elastic and ultimate load theories. It introduces the concepts of partial safety factors, characteristic loads and characteristic strengths based on a semi-probability approach. It provides the framework which will enable design to be more accurate and therefore more economical. The former use of a single safety factor included a large allowance for ignorance which can now be reduced. Limit state theory has enabled design methods to be presented in CP 110 and BS 5337 in such a way that the results of research work can be incorporated as they become available. It provides guide lines for future research by focusing attention on those areas where knowledge is lacking.

The elastic theory as given in CP 114:1957 used a (stress) factor of safety—i.e. ultimate-strength/working-stress ratio—of 3 for concrete and 2 for steel. Ultimate load theory as originally given in CP 114:1957 used a load factor (of safety)—i.e. ultimate-load/working-load ratio—of 2, together with a factor of safety for concrete strength only. Limit state theory uses (at least) two partial safety factors, in association with characteristic loads and strengths, to allow for predicted variations in a rational manner. This enables the design to be carried out with greater confidence and greater economy.

List of Symbols

A_c	area of concrete (=bh)
A_{cc}	actual (or net) area of concrete
A_s	area of steel reinforcement
A_{sa}	cross-sectional area of additional tension steel to balance compression steel (for beams)
A_{sc}	area of compression reinforcement (total area of longitudinal reinforcement in columns)
A'_{sc_1}	area of compression reinforcement at level 1 in the more highly compressed face
A_{sc_2}	area of compression reinforcement at level 2 in the less highly compressed face (can be in tension)
A_{s_1}	area of longitudinal reinforcement provided for torsion
A_{st}	area of tension reinforcement
A_{sv}	area of the two legs of a link
a	deflection; volume fraction of coarse aggregate
a_{cr}	distance of point (crack) considered to surface of nearest longitudinal bar
b	width of section; volume fraction of fine aggregate
b_c	breadth of compression face
b_t	breadth of section at level of tension reinforcement
b_w	breadth of web or rib
C	torsional constant; statistical constant $(= n/\sigma\surd(2\pi))$
${}_NC_n$	number of combinations of n items from N items
c	concrete cover; volume fraction of cement
D	deviation $(= x_i - \mu)$
D'_i	deviation of class i from class o in terms of class width, w

d	effective depth of tension reinforcement (formerly d_1 in CP 114)
d'	depth to compression reinforcement
d_n	neutral axis depth (elastic analysis only: see also x)
d_2	depth to reinforcement
E_c	static secant modulus of elasticity of concrete
E_s	modulus of elasticity of steel
e	base of Napierian logarithms
e	eccentricity
F	ultimate (design) load; force (with subscripts as for area, A); cumulative frequency
F_d	permissible load
F_k	characteristic load
F_{kv}	characteristic variable load
F_{im}	impact load
F_t	basic tie force
F_{var}	variable design load
$_*f$	design strength
f_b	average (or anchorage) bond stress
f_{bs}	local (or flexural) bond stress
f_c	concrete stress (compression +ve)
$_*f_c$	design strength of concrete in compression
f'_c	cylinder crushing strength
f_{cb}	concrete stress in bending at extreme fibre (compression +ve)
f_{cc}	concrete stress in direct compression (compression +ve)
f_{ct}	concrete stress in tension (tension +ve)
f_{cu}	characteristic concrete cube strength
f_{dcc}	permissible concrete stress in direct compression
f_{dcb}	permissible concrete stress in bending
f_{dsc}	permissible steel stress in compression

f_{dst} permissible steel stress in tension

f_i frequency of values in class i

f_k characteristic strength or resistance

f_{ku} characteristic uniaxial compressive strength of concrete

f_m mean strength or resistance

f_{mcu} mean cube strength

f_s steel stress (tension +ve)

f_{sc} compression steel stress (compression +ve)

f'_{sc_1} compression steel in highly compressed face (compression +ve)

f_{sc_2} compression steel in other face (compression +ve)

f_{st} tension steel stress (tension +ve)

f_{sv} steel stress (in tension) in shear reinforcement

f_y characteristic strength of reinforcement

f_{yl} characteristic strength of longitudinal reinforcement

f_{yv} characteristic strength of link reinforcement

$*f_{yc}$ design yield strength of compression steel

$*f_{yt}$ design yield strength of tension steel

G shear modulus

G_k characteristic dead load

g distributed dead load

g_k characteristic dead load per unit area

h overall depth of section in plane of bending; statistical constant ($=(\sigma\sqrt{2})^{-1}$)

i integer

h_c diameter of column head

h_f thickness of flange

h_n depth of neutral axis for uncracked section (elastic theory)

I_c second moment of area of (entire) concrete cross-section

I_e second moment of area of transformed (equivalent cracked) section

I_o	second moment of area of gross (equivalent uncracked) section
K,k	various constants or coefficients
K_b, K_l, K_u	stiffness of beam, lower column and upper column respectively
l_e, ℓ_e	effective span
l_{ex}, ℓ_{ex}	effective height for bending about major axis
l_{ey}, ℓ_{ey}	effective height for bending about minor axis
l_m, ℓ_m	average of l_1 and l_2
l_o, ℓ_o	clear height of column between end restraints
l_x, ℓ_x	length of the shorter side (of rectangular slab)
l_y, ℓ_y	length of the longer side (of rectangular slab)
l_1, ℓ_1	length of slab panel in direction of span measured from centres of columns
l_2, ℓ_2	width of slab panel measured from centres of columns
M	bending moment; bending moment due to ultimate loads
M_a	increased moment in column
M_{add}	maximum additional moment
M_d	design moment of resistance (elastic theory)
M_{dc}	design moment of resistance in compression
M_{dct}	design moment of resistance of concrete in tension
M_{dt}	design moment of resistance in tension
M_e	fixed end moment
M_i	maximum initial moment in a column due to ultimate loads (but $\nless$ 0.05 Nh)
M_{ix}	initial moment about the major axis of a slender column due to ultimate loads
M_{iy}	initial moment about the minor axis of a slender column due to ultimate loads
M_{sx}, M_{sy}	bending moments at midspan on strips of unit width and spans l_x and l_y respectively
M_t	total moment in a column due to ultimate loads

M_{tx} total moment about the major axis of a slender column due to ultimate loads

M_{ty} total moment about the minor axis of a slender column due to ultimate loads

M_u ultimate resistance moment

M_{ux} maximum moment capacity in a short column assuming ultimate axial load and bending about the major axis only

M_{uy} maximum moment capacity in a short column assuming ultimate axial load and bending about the minor axis only

M_x, M_y moments about the major and minor axes of a short column due to ultimate loads

M_o moment necessary to produce zero stress

M_1 smaller initial end moment due to ultimate loads (assumed negative if the column is bent in double curvature)

M_2 larger initial end moment due to ultimate loads (assumed positive)

m resistance moment per unit length along axis of moment

m_x resistance moment (per unit length) of x-direction reinforcement

m_y resistance moment (per unit length) of y-direction reinforcement

N ultimate axial load at section considered; population size

N_{axial} axial load capacity of column with moment of 0.05 Nh

N_{bal} axial load on column corresponding to the balanced condition (see §3.5.7.4)

N_{uz} axial load capacity of column ignoring all bending

n total ultimate load per unit area; number in a sample

n_s number of storeys

n_w axial load per unit length of wall

${}_NP_n$ number of permutations of n items from N items

Q_k characteristic imposed load

q distributed live load

q_k	characteristic live load per unit area
r	internal radius of bend
r_b^{-1}	curvature at midspan (beams) or at support section (cantilevers)
r_{bi}^{-1}	initial curvature
r_{blg}^{-1}	long-term curvature due to permanent loading
r_{cs}^{-1}	curvature due to shrinkage
r_x^{-1}	curvature of a beam at point x
s	spacing of secondary reinforcement generally; crack spacing; (estimated) standard deviation
s_b	spacing of bars
s_v	spacing of links along the member
T	torsional moment; torsional moment due to ultimate loads
t	time
u	perimeter
u_{crit}	length of critical perimeter
V	shear force due to ultimate loads
V_c	ultimate shear resistance of concrete
v	shear stress
v_c	ultimate shear stress in concrete
v_t	torsional shear stress
v_{tu}	ultimate torsional shear stress
W	loads generally
w	crack width; distributed loads generally; distributed wind load; volume fraction of water; class interval (statistics)
W_k	characteristic wind load
W_{var}	variable (design) load (= 0.4 G_k + 1.6 Q_k for collapse)
x	neutral axis depth (inelastic analysis only); co-ordinate
x_i	ith value of an indpendent variable
x_i'	value of an independent variable for midpoint of class i

x'_o value of an independent variable for midpoint of lowest class o

$\overline{x}$ estimate of population mean or sample mean

x_1 smaller dimension of a link

y co-ordinate; co-ordinate distance from neutral axis (the positive direction being downwards for beams)

$\overline{y}$ depth to centroid of stress block

y_1 larger dimension of a link

Z section modulus (= I/y)

z lever arm, co-ordinate; statistics coefficient (= D/σ)

α angle; coefficient

α_c ratio of sum of column stiffnesses to sum of beam stiffnesses

α_{c_1} value of α_c at lower end of column

α_{c_2} value of α_c at upper end of column

α_{cmin} minimum value of α_{c_1} and α_{c_2}

α_e modular ratio

α_n coefficient as a function of column axial loading

α_{sx}, α_{sy} bending moment coefficients for slabs with no provision to resist torsion at the corners or to prevent the corners from lifting

β angle; coefficient

β_b ratio of beam moments with respect to service stress in beams (see §3.3.8.1)

β_{red} ratio of the reduction in resistance moment

β_{sx}, β_{sy} bending moment coefficients for slabs with provision to resist torsion and to prevent corners from lifting

β_w coefficient for plain walls dependent on b/h ratio

β_1 ratio of longer to shorter base sides

γ_f partial safety factor for load

γ_m partial safety factor for strength

ΔM_e difference in fixed end moments

δ coefficient of variation

δ_n neutral axis depth ratio (d_n/d) for elastic theory

ε_{cs} shrinkage strain

ε_m average strain

ε_1 strain at the level considered

ζ coefficient (see ξ,ψ,ζ)

θ rotation

θ_s angle between the compression face and tension reinforcement

λ coefficient

λ_w coefficient for walls dependent upon dimensions and concrete used

μ coefficient of friction; true mean

ν Poisson's ratio

ξ,ψ,ζ coefficients given by design guides

ξ coefficient

ξ_n h_n/h for uncracked section (elastic theory)

ξ_r d_n/h for cracked section (elastic theory)

ξ_s depth of slab factor

ρ tension steel ratio ($= A_{st}/bd$)

ρ' compression steel ratio ($= A_{sc}/bd$)

ρ_c total steel - total concrete ratio ($= A_s/bh$)

ρ_e equal-strength (tensile) steel ratio (elastic theory)

ρ_o coefficient which depends upon the percentage of tension and compression steel in the section

Φ bar size (or diameter)

ϕ angle

$\sum_{i=1}^{n}$ sum from i=1 to i=n

Σu_s	sum of the effective perimeters of the tension reinforcement
σ	true standard deviation
σ_1, σ_2	principal stresses
$\sigma_{\overline{x}}$	standard error of the mean
ψ	angle; coefficient (see ξ, ψ, ζ)

Note: The above notation is in accordance with international recommendations and generally with CP 110:1972 (but see Appendix Y). Also, to emphasize the difference between service load and ultimate load conditions x has been reserved for neutral axis depth at ultimate only.

1 Introduction to limit state theory

Limit state theory is conveniently introduced by considering the fundamental purpose of design so that structural design can be seen in its correct perspective. Rowe et al. [1.1] have defined the purpose of design as 'the provision of a structure complying with the client's (and users') requirements'. The client is often uncertain of his requirements when he first approaches an engineer (or architect), so the engineer should assist him to define them more precisely.

Although the client automatically assumes that the design will satisfy such essential requirements as structural safety and durability, many joint decisions must be made before the engineer has a satisfactory design brief. For example, what are the client's relative priorities regarding early completion, low initial cost and low maintenance? What is the proposed design life of the structure and does the client envisage a change of use within this period that could increase the applied loadings? If so, should the structure (or parts of the structure such as the foundations) be designed for the increased loadings or will the client meet the cost of strengthening the structure as and when necessary? Accurate selection of design loadings cannot be expected to include an allowance for changes in use which are not intended at the design stage.

Once the basic requirements have been defined, the engineer can develop and assess various outline schemes, each with their own use of materials and method of construction, before taking the vital decision for any particular structural concept. A complete structural analysis can then be carried out and, if satisfied for both service and erection conditions, he can proceed with the detailed design. Specifications and detail drawings must be produced to complete the contract documents by which he instructs the contractor, usually through his resident (site) engineer.

The object of structural design is to obtain a structural solution for safety and serviceability throughout the design life, which gives the greatest overall economy for first cost and for maintenance costs. It is clear that structural costs rise rapidly if other requirements become too onerous—if, for example, very large clear spans are specified in a multi-storey garage. It is equally clear, however, that structural costs cannot be considered in isolation. A multi-storey car park with closely spaced columns would not perform its primary function very efficiently. Similarly, the successful enclosure of space provided by a building requires the creation of an environment which is pleasant to all the senses of the occupant. It should be adequately illuminated, heated and ventilated, have an attractive appearance and be satisfactory acoustically, as well as being structurally sound. The cost of the mechanical and electrical services in a hospital, for instance, can be as high as 40% of the total.

1.1 PHILOSOPHY OF STRUCTURAL DESIGN

Satisfactory design must ensure the achievement of an acceptable probability that the specified life of a structure is not curtailed prematurely due to the attainment of an unsatisfactory condition, or limit state. The acceptable probability should be chosen in such a way that a satisfactory balance is achieved between the cost of construction and the combined costs of insurance premiums (which are related to the probability of structural or functional failure) and of maintenance (the costs for design, normal operation and use, and insurance for flooding etc., are normally independent of structural safety and serviceability). Limit state design is therefore based, where possible, on the application of the methods of statistics to the variations that occur in practice.

Limit states

Limit states are concerned with structural safety and serviceability and cover all forms of failure. Unfitness for use can occur in many ways and it is convenient to group limit states together into three major categories as follows.

(a) Ultimate limit states: collapse of the structure due to normal or exceptional loadings or the occurrence of exceptional events.

(b) Serviceability limit states: deflection, cracking, vibration or (if relevant) any special design function.

(c) Other considerations or limit states: fatigue, durability, fire resistance, lightning, etc.

Usually some limit states are inapplicable and others can be satisfied by suitable detailing. The usual approach is to design on the basis of the most critical limit state (ultimate for RC) and then check that the criteria governing any remaining limit states are satisfied. Adequate durability, for example, is ensured for most structures by complying with requirements for crack widths, concrete cover to reinforcement, cement contents and water/cement ratios.

Ultimate limit states

Collapse of the structure, or part of the structure, may arise from the rupture of one or more critical sections, from buckling due to elastic or plastic instability (including the effects of sway where appropriate), from the loss of static equilibrium (e.g. due to transformation into mechanism, due to loss of stability by overall tilting etc.). CP 110 refers to the 'ultimate limit state' (singular) and gives three normal loading cases, another for defined exceptional loadings and another for continued stability after suffering an exceptional event. The first four cases apply to the structure as constructed, whereas the last case concerns the structure after it has sustained localized damage. For clarity, the term limit states of collapse will be used for all loadings which act on the structure as designed, whereas the term limit state of impact will refer to loadings on the damaged structure (see Appendix B).

Limit states of cracking and local damage

Cracking or spalling of the concrete, is excessive if it impairs the appearance, efficiency or durability of the structure or causes damage to finishes, partitions etc. For example, in water-retaining structures the limit state of crack formation in tension on the water-retaining face may be critical. However, cracks can generally be tolerated provided that the width of the individual cracks is insufficient to allow the ingress of moisture, which would corrode the steel; hence it is necessary to check the appropriate limit state of crack width. Crack formation in compression (e.g. initial microcracking) or local damage for steel (i.e. excessive strains) can be controlled by specifying permissible (or allowable) stresses.

Limit states of deflection and vibration

Deflection or deformation becomes excessive when it impairs the appearance or efficiency of the structure, affects non-load-bearing members or finishes adversely, or causes discomfort or alarm to the users of the structure. Excessive vibration (e.g. near resonance) can also cause severe damage and interference to the proper function of the structure. Acceptable limits to the level of vibration vary according to usage. For example, pedestrians on a bridge can be very alarmed by vibrations which are not even noticed by the occupants of a vehicle. It may be necessary to isolate the source of vibration or, in certain circumstances—for example, if sensitive measuring instruments are to be installed—to isolate a part or even the whole of the structure.

Limit state of fatigue

The effects of fatigue should be considered if the imposed load on a structure, or part of a structure, is predominantly cyclic in character.

Limit states of fire resistance and lightning

The main limit states in fire concern stability in fire (i.e. resistance to collapse) and integrity in fire (i.e. resistance to passage of flames or hot gases and resistance to heat transmission) [1.2,1.3]. See Appendix M for details. Reinforcement may be used as part of a lightnin protective system in accordance with CP 326.

Limit state of impact

Structures can be subjected to severe impacts due to misuse or accident, e.g. explosions, impact of heavy bodies, etc. If tie forces are provided as in CP 110, one approach is to assume that the member can absorb the energy of impact provided that these tie forces are also capable of enabling the beam or slab to act as a catenary, localize the damage and prevent cumulative and catastrophic failure [1.4]. If specific static loadings need to be considered, these should be applied to the damaged members (e.g. the catenaries).

1.2 AVOIDANCE OF LIMIT STATES

An acceptable probability that the structure will not reach a limit state throughout its specified life can only be provided if variations in strengths of the constructional materials, and variations in loads on the structure, can be estimated. Ideally, all the probable variations for each limit state should be predicted from statistical data. Although only some of the relevant data are at present available, it is still possible to implement the main principles of the limit state philosophy. Each limit state is considered individually and the variations (or uncertainties) can be due to one or more of the following causes:

Variations in strength of the structure

Variations which can affect the material strength in the structure are:

(i) Those caused by inherent variability of material strength due to the material's method of manufacture and quality of constituents. These variations can be allowed for by a characteristic strength which is calculated by using statistical principles.

(ii) Variations due to other causes, such as construction procedure (treatment, curing); state of stress (axial, biaxial, triaxial); speed, duration and repetition of loading; and environmental conditions in the actual structure. Statistical principles cannot be applied, either because the data are inadequate or because the usual assumptions made in statistical analysis are invalid. These variations are therefore allowed for by a partial safety factor for material strength.

Variations in loads on the structure

Variations which can increase the loading on a structure are:

(i) Those due to the inherent variability of loads applied to a structure of a particular type throughout its design life. These can be allowed for by a characteristic load.

(ii) Variations due to other causes. For example, the load on a member may be excessive due to errors in the design assumptions. These variations are covered by a partial safety factor for loads.

Gross defects

Gross defects in design or construction, such as positioning of the reinforcement in the wrong face of a member, must be avoided entirely by checking and inspection. Although the absence of gross defects and exceptional events (see below) are assumed, it may also be assumed that a major catastrophic failure could not, in general, be caused by a single such defect or event, but only by a combination of at least two major errors from either (i) design (ii) construction, and/or (iii) use. A major error by the user, such as gross overloading, or installation of explosive equipment not in accordance with the designer's brief, should not cause complete catastrophic failure unless combined with a major design or construction error.

Exceptional events

Truly exceptional events, such as major national disasters or enemy action, can only be covered completely by insurance, if at all, and one such event alone would be sufficient to cause catastrophic failure. However, events that are somewhat less exceptional, such as explosions, earthquakes or accidents due to fire or internal gas explosions, should be considered in the design even though extensive damage to the structur can occur if such events take place. Damage to the structure can be localized, and possible loss of life reduced, by taking sensible precautions to provide a certain minimum resistance to fire and impact. A particularly disturbing feature of designs based on CP 114, CP 116, etc., was that hazards could be ignored. A disregard of the limit state of impact, combined with an exceptional event, could and did result in progressive and catastrophic collapse 1.5 . Exceptional events with an even higher probability of occurrence, such as earthquakes in regions where they are known to occur, should be considered specifically in the calculations for collapse and serviceability.

1.3 SOME BASIC STATISTICAL PRINCIPLES

Limit state design is a probabilistic design method in which the values of certain basic design parameters are considered as random. It therefore uses statistical techniques to analyse the variations which occur in practice, either in the strengths of constructional materials due to their method of manufacture) or in the loads on structures of a given type. A brief outline of some elementary statistical principles (which are covered more fully elsewhere, e.g. references [1.6 and 1.7]) is therefore very relevant since it is assumed that the above variations effectively follow a normal (gaussian) distribution for practical purposes.

Errors

There are two main types of error:

(i) Systematic errors. These are due to effects which act consistently under the given circumstances (e.g. an incorrect zero on a compression testing machine, or a measuring cylinder calibrated at one temperature and used at another). Such errors should be avoided by suitable experimental techniques.

(ii) Random errors. These are due to a large number of small effects such as imprecisions in interpolating between graduations on a scale, or imprecisions in allowing for temperature variations when using equipment which has been calibrated at a standard temperature. Notice that these effects have an equal probability of being positive or negative; hence the probability of a given error being positive is one half and the probability of its being negative is also a half.

Gauss function

If it is assumed for convenience that all the small contributory random errors are of equal absolute magnitude ΔD, and for any particular case there is a surplus (or deficiency) of positive errors equal in number to $2r$, then the resultant positive error D equals $2r\Delta D$. If the total number of errors is $2n$ then the total error can range in individual cases from minus $2n\Delta D$ (all negative) to plus $2n\Delta D$ (all positive).

These are the errors that cause variations between observations even under the most carefully controlled conditions. It can be shown from mathematical considerations that the distribution of these errors is given by the so-called Gauss function:

$$y = Ce^{-h^2D^2} \qquad (1.3.1)$$

where $y\Delta D$ is the frequency with which the error lies in the range D to $D + \Delta D$, C is a constant equal to the maximum value of y, and h is a constant related to the spread of the errors. These constants are related to the total number of observations, n, and the standard deviation, σ(which is defined later), as follows:

$$C = n/\sigma\sqrt{(2\pi)}$$
$$h = 1/\sigma\sqrt{2}$$

We shall reconsider the Gauss function later on.

Normal distribution

The distribution given by the Gauss function is commonly known as the normal distribution. This distribution is shown in Fig. 1.3.1 by

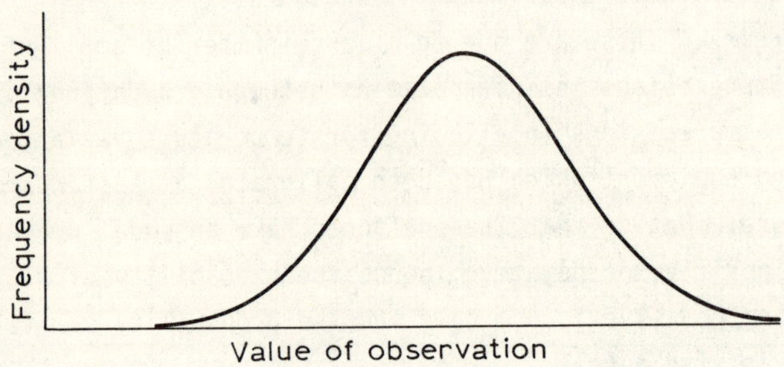

Fig.1.3.1 Frequency density plotted against value of observation for normal probability distribution

plotting frequency density against value of observation. Notice that while the individual values are equally distributed about the mean value, the extreme values can be at $\pm\infty$. Hence although the probability of the occurrence of an exceptionally high or low value is extremely small, it is never zero.

Standard deviation

The deviation of an individual observation x_i is given by $(x_i - \mu)$ where μ is the true mean. The standard deviation is the root-mean-squar deviation and is therefore always positive. The true standard deviation is given by

$$\sigma = \sqrt{\left[\frac{1}{n}\sum_{i=1}^{n}(x_i - \mu)^2\right]} \tag{1.3.2}$$

It is important to notice that μ is the true mean and not the mean of the actual number, n, of observations which have been taken. The true mean is usually unknown and it is necessary to calculate the standard deviation using the sample mean, $\bar{x}$. The sample mean will generally differ from the true mean, and the direct substitution of $\bar{x}$ for μ in the above expression gives a minimum value of $\sum(x_i - \mu)^2$. It is therefore necessary to apply a correction when using deviations from the sample mean instead of deviations from the actual mean, such that the estimated standard deviation is given by

$$s = \sqrt{\left[\frac{1}{n-1}\sum_{i=1}^{n}(x_i - \bar{x})^2\right]} \tag{1.3.3}$$

Standard deviation calculations

The computatation of the standard deviation, s, as given by Eqn.(1.3.3), is tedious for all but the smallest values of n. This computation is expedited by the following alternative expression for $\sum(x - \bar{x})^2$ which can be utilized if all the individual results are grouped together into classes, or ranges of values.

Let w = class interval (i.e. range of values for each class)

x'_i = class midpoint (rather than an individual value)

x'_0 = midpoint of lowest class

f_i = frequency of values within class i

D'_i = deviation of class i from class 0 in terms of w (i.e. deviation from an arbitrary origin in terms of class width w).

Also, let $\sum_{i=1}^{n} x_i$ be written in abbreviated form as $\sum x_i$ for convenience.

It is clear that $n = \sum f_i$ and $\sum x_i = \sum f_i x'_i$

Also $\bar{x} = (1/n)\sum x_i = (1/n)\sum f_i x'_i$

Therefore $\sum(x_i - \bar{x})^2 = \sum f_i(x_i' - \bar{x})^2$

$$= \sum f_i x_i'^2 - 2\bar{x}\sum f_i x_i' + \bar{x}^2 \sum f_i$$

$$= \sum f_i x_i'^2 - (2/n)(\sum f_i x_i')^2 + (\sum f_i x_i')^2(1/n)$$

i.e. $$\sum f_i(x_i' - \bar{x})^2 = \sum f_i x_i'^2 - (1/n)(\sum f_i x_i')^2 \tag{1.3.4}$$

Now $\bar{x} = (1/n)\sum f_i x_i' = x_o' + (w/n)\sum f_i D_i'$

and $x_i' = x_o' + wD_i'$

Therefore $\sum f_i(x_i' - \bar{x})^2 = w^2 \sum f_i(D_i' - n^{-1}\sum f_i D_i')^2$

since x_o' cancels out. Thus x_i' can be replaced by D_i' (and w) in r.h.s. of Eqn (1.3.4) also. Hence, from Eqn (1.3.3),

$$s = w\sqrt{\frac{\sum f_i D_i'^2 - (1/n)(\sum f_i D_i')^2}{n-1}} = \sqrt{\frac{\Sigma x^2 - (\Sigma x)^2/n}{n-1}} \tag{1.3.5}$$

Notice that the expression for σ equivalent to Eqn (1.3.5) is

$$\sigma = \sqrt{\left(\frac{\sum x^2}{n} - \mu^2\right)} \tag{1.3.6}$$

Various 'mean' values

The mean value should always refer to the arithmetic mean. Less common types of average or mean values are as follows. The median is the middle observation when the observations are arranged in order of magnitude. The median therefore divides a histogram and a frequency polygon into two equal areas. The mode is the value which occurs most frequently if the variable is discrete, or the class interval which has the highest frequency if the distribution is continuous. The mode therefore represents a peak value in a frequency distribution. Consider, for example, the values 6, 6, 7, 8, 8, 10, 18. The mean is 9, median 8, and mode 6 or 8.

Coefficient of variation

The coefficient of variation is the ratio of the standard deviation to the mean, expressed as a percentage, i.e.:

$\delta = 100\sigma/\bar{x}$

Fractional cumulative frequency

The number of results higher or lower than a particular value is given by the cumulative frequency. The cumulative frequency divided by the total number of results gives the fractional cumulative frequency. A grading curve showing the percentage passing a particular sieve size is a typical example of a fractional cumulative frequency curve.

Histogram

A histogram is one way of representing a grouped frequency distribution diagrammatically. The class intervals are set out on the horizontal axis and the frequency in a given class interval is shown by the rectangular area above the horizontal axis at the position representing the given class interval. The width of all class intervals should generally be made the same, in which case the ordinate is proportional to the area of each rectangle. A typical histogram will be shown in connection with Example 1.3.1, below.

Frequency distribution curve

If the class width is decreased indefinitely and the total number of results correspondingly increased, the histogram is transformed, in the limit, into a frequency distribution curve.

Frequency polygon

If the frequency within a class interval is plotted as a point whose abscissa is that of the class midpoint, adjacent points can be connected to form a frequency polygon (see Example 1.3.1). Notice that the line between the points has no significance, hence intermediate values must not be read off the line.

Example 1.3.1

The cube strength results (in MN/m^2) for the first 40 separate batches of concrete produced under actual operating conditions on a construction project were as follows:

30.5	32.0	42.0	39.0	29.0	21.0	24.5	30.5
27.0	25.0	30.5	31.5	30.0	36.5	34.0	33.0
33.0	36.5	31.5	34.0	36.0	33.0	30.5	34.5
43.0	33.0	33.0	32.5	36.0	37.0	34.5	28.0
26.5	23.5	34.0	30.0	35.5	37.5	38.0	31.5

Group the strengths into a frequency distribution table with a class interval of 2.5 MN/m² and draw a histogram and a frequency polygon. Hence calculate the mean, median, mode, standard deviation, coefficient of variation and range of the given data.

Solution

Let all stresses be in units of MN/m². The results are tabulated as follows:

Class interval	Class mid-point x'_i	Frequency f_i	Cumulative frequency $\sum_1^i f_i$	Fractional cumulative frequency $\frac{1}{n}\sum_1^i f_i$	Deviation from origin in terms of class width D'_i	$f_i D'_i$	$f_i D'^2_i$
20.0-22.4	21.2	1	1	2.5	0	0	0
22.5-24.9	23.7	2	3	7.5	1	2	2
25.0-27.4	26.2	3	6	15.0	2	6	12
27.5-29.9	28.7	2	8	20.0	3	6	18
30.0-32.4	31.2	10	18	45.0	4	40	160
32.5-34.9	33.7	11	29	72.5	5	55	275
35.0-37.4	36.2	6	35	87.5	6	36	216
37.5-39.9	38.7	3	38	95.0	7	21	147
40.0-42.4	41.2	1	39	97.5	8	8	64
42.5-44.9	43.7	1	40	100.0	9	9	81
		40				183	975

The histogram and frequency polygon can be drawn using the values in the frequency and class interval columns, as shown in Fig. 1.3.2.

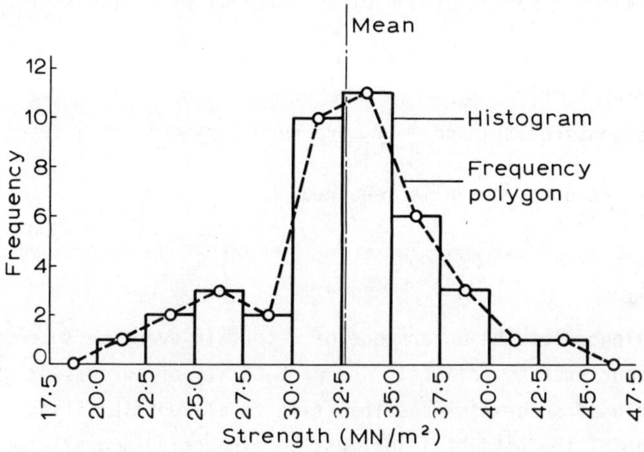

Fig.1.3.2 Example 1.3.1.

The totals for the appropriate columns of the above table give the following values:

$\sum f_i = 40 \quad \sum f_i D_i' = 183$ and $\sum f_i D_i'^2 = 975$

Hence the mean is

$\bar{x} = x_o + w\sum f_i D_i' / \sum f_i = 21.2 + 183 \times 2.5/40$

$= 32.6$ (MN/m²) (Answer)

The standard deviation is

$$s = w\sqrt{\frac{\sum f_i D_i'^2 - (\sum f_i D_i')^2/\sum f_i}{n-1}}$$

$= 2.5(975 - 183^2/40)^{\frac{1}{2}}(39)^{-\frac{1}{2}}$

$= 4.70$ (MN/m²) (Answer)

The coefficient of variation is

$\delta = 100s/\bar{x} = 100 \times 4.70/32.64 = 14.4\%$ (Answer)

The range is $43.0 - 21.0 = 22.0$ (MN/m²) (Answer)

The median is the centroid of the histogram, i.e. the value for $i = 20$, or

$$32.5 + 2w/11 = 32.95 \text{ (MN/m}^2\text{)} \qquad \text{(Answer)}$$

The mode is the midpoint of the class interval with highest frequency, i.e.:

$$\text{mode} = 33.7 \text{ (MN/m}^2\text{)} \qquad \text{(Answer)}$$

Note: columns for $\sum_1^i f_i$ and $\frac{1}{n}\sum_1^i f_i$ are necessary only if a fractional cumulative frequency curve is required.

Probability

For convenience let the occurrence of a certain event be success and its non-occurrence be failure. The probability of success is given by the ratio of successful possibilities to all possibilities. The examination of the possibilities must be exhaustive and all the possibilities must be equally likely and mutually exclusive. If there are n possibilities and success occurs in p cases, the probability of success is p/n and the probability of failure is 1 — (p/n). Hence probability is expressed as a number not greater than 1. Unity denotes certainty and zero denotes impossibility. Two important rules of probability should be noted.

(i) The probability of simultaneous occurrence of a number of independent events is the product of the separate probabilities. For example, if the probability of an excessive load on a structure is 1/20 and the probability of the material in a structur being substandard is also 1/20, the probability of an excessive load being applied to substandard material in a structure is (1/20) × (1/20), or 1/400.

(ii) The probability of occurrence of any one of two or more events, out of a number of independent and mutually exclusive events, is simply the sum of the probabilities of the separate events. For example, if a box contains 2 red balls, 4 blue balls and 6 green balls, the probability of drawing either a red or a blue ball from the box is 2/12 + 4/12 = 1/2. Notice that the sum of the probabilities of all the possible events is always unity.

Normal probability density function

The normal frequency distribution curve can be plotted from the Gauss function (Eqn 1.3.1) and the total number of observations equals the area under the curve:

$$\int_{-\infty}^{+\infty} y\,dD = n$$

To obtain the normal probability curve from the above, it is necessary to make the area under the curve equal unity, since the sum of all probabilities must equal unity.
That is,

$$\int_{-\infty}^{+\infty} y_1\,dD = 1$$

where $y_1 = y/n$. In addition, it is also convenient to express D in terms of standard deviations, i.e. to replace D in Eqn (1.3.1) by $z\sigma$, where z is a coefficient. The new ordinate y_2 must therefore be given by $\sigma y_1 = \sigma y/n$ if the area under the y_2-z curve is to be unity.

Hence from Eqn (1.3.1):

$$y_2 = (\sigma/n)C\,\exp(-h^2\sigma^2z^2)$$

or

$$y = \frac{1}{\surd(2\pi)}\exp(-\tfrac{1}{2}z^2) \qquad (1.3.7)$$

where the suffix 2 for y has been omitted. This is the normal probability density function which is usually referred to in standard statistical tables and which defines the normal distribution. Notice that the mean, μ, is constant for any one distribution, its effect being to determine the position of the normal curve along the x-axis. The shape of the curve (i.e. the horizontal spread) depends on the value of the standard deviation only. The normal probability distribution curve is shown in Fig. 1.3.3 together with the areas under the curve between the limits $0 \le z \le \frac{1}{2}$, $\frac{1}{2} \le z \le 1$, etc. See also Appendix N for tabulated values.

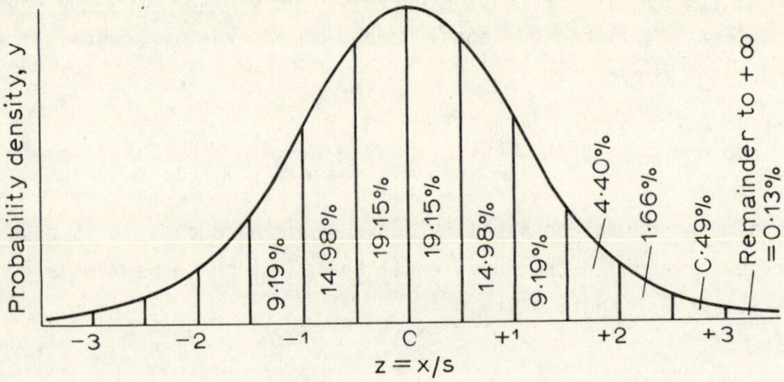

Fig.1.3.3 Areas under the normal probability curve

Level of significance and confidence limits

It is often convenient to consider the probability that any single value will not lie between certain limits, and this probability is known as the level of significance. Thus the 5% level of significance, for example, is given by the limits $\mu \pm 1.96\sigma$. i.e. the remote areas under the two extremities of the curve are each 2½%.
The limits which provide a given probability that any single value will lie between them are known as the confidence limits. Thus the 95% confidence limits are also given by $\mu \pm 1.96\sigma$.

Example 1.3.2

What percentage of all results would lie below the value of $\bar{x} - 2.5s$ for a normal distribution? What is the stress which provides this number of failures for the data in Example 1.3.1? What are the stress limits at the 1.24% level of significance? What are the confidence limits for the band $\bar{x} \pm 2.5s$?

Solution

From Fig. 1.3.3 or Table N1, the percentage of all results that lie below $\bar{x} - 2.5s$ is

$(0.49 + 0.13)\% = 0.62\%$ (Answer)

Hence for the previous data there are 0.62% failures for a stress (MN/m^2 units) of

$$32.6 - 2.5 \times 4.70 = 20.85 \ (MN/m^2) \qquad \text{(Answer)}$$

An equal number of results lies above the value $\bar{x} + 2.5s$. Hence at the 1.24% (= 2 × 0.62%) level of significance we have

stress limits: 20.85 and 44.35 (MN/m^2) (Answer)

The confidence limits for the band $\bar{x} \pm 2.5s$ are

$$(100 - 1.24)\% = 98.76\% \qquad \text{(Answer)}$$

1.4* ACCURACY OF THE SAMPLE MEAN

Sampling is necessary because it is usually impracticable to test the entire population; this is especially true of destructive tests. It is essential, therefore, that each and every sample is selected in a random manner so that it is representative of the population from which it is drawn. This means that every member of the population has an equal chance of being drawn in every trial. The word 'sample' is used in a very broad sense to mean an observation of any type. A value arrived at by sampling is absolutely accurate only for the sample itself. For the population which it represents, a sample can only give an estimate whose accuracy is expressed in terms of probability. The greater the size of the sample (i.e. number of observations), the greater the confidence that can be placed on the estimate for the population.

Population

The term 'population' refers to all the observations that could be made. Samples are tested in order to make inferences about the properties of the population and it is vital to be clear which population is of interest. For example, when considering cube results, are we interested in a particular batch of concrete, or a complete day's placing, or all the concrete in a particular structure?

* This section concerning further statistical principles should be omitted on first reading.

Sample means

If we consider a single batch of concrete as the population, each individual cube test can be considered as the sample. However, if we consider the total output of concrete for a particular day as the population, each group of cube tests taken for a single batch could be considered as the sample. Similarly, if we consider all the concrete in a given structure as the population, all the cube tests taken for a particular day's concreting could be considered as the sample. The question which then arises is: how many observations should be made for a single batch of concrete, or a single day's output, or a week's output? Alternatively, consider a unique quantity (e.g. the mean cube strength of a batch of concrete) and assume that the mean of the observations (e.g. the cube test results) is the appropriate single measure. How is the accuracy (e.g. of the assessment of the mean cube strength of the batch of concrete) increased as the number of observations is increased? These questions can be answered by considering the distribution of the sample means, and the relation between the standard deviation of the means and the standard deviation of the original population.

Permutations and combinations

The number of permutations of n from N is the number of ways of filling n different places with N different items, i.e. it is the number of different arrangements by which these places can be filled. This can be determined as follows.

The 1st place can be filled in N ways
" 2nd " " " " " $N - 1$ ways
" nth " " " " " $N - n + 1$ ways

Therefore the number of permutations of n from N, written as ${}_NP_n$, is given by

$${}_NP_n = N(N - 1) \ldots (N - n + 1) = N!/(N - n)!$$

where N! means factorial N, i.e.:

$N(N - 1)(N - 2) \ldots \times 2 \times 1$

and 0! is one, by definition.

For a combination, the order in which the places are filled (i.e. the order of selection) does not matter, since in this case one is only concerned with the actual items which are selected. Now n items can be arranged in n! ways, hence the number of combinations of n items from N, ${}_N C_n$, is n! times smaller than the number of permutations of n items from N. Hence

$$ {}_N C_n = \frac{{}_N P_n}{n!} = \frac{N!}{n!(N-n)!} \tag{1.4.1} $$

Standard error of the mean

The number of different samples, each of n observations, that can be drawn from a population of N is the number of combinations of n items from a total of N, that is, ${}_N C_n$. Substituting in Eqn (1.3.2) gives the standard deviation for the means of samples of size n drawn from a population of N, $\sigma_{\bar{x}}$, as follows:

$$ \sigma_{\bar{x}} = \sqrt{\left(\frac{\sum(\bar{x}-\mu)^2}{{}_N C_n}\right)} \tag{1.4.2} $$

where the summation is from 1 to ${}_N C_n$.

Substitution into the alternative expression for σ, Eqn (1.3.6), gives

$$ \sigma_{\bar{x}} = \sqrt{\left(\frac{\sum\bar{x}^2}{{}_N C_n} - \mu^2\right)} \tag{1.4.3} $$

The standard deviation of the mean, $\sigma_{\bar{x}}$, is also called the standard error of the mean. Once this expression can be evaluated we can answer the question: how is the accuracy increased as the number of observations in a sample is increased?

Distribution of sample means

To evaluate $\sigma_{\bar{x}}$ in Eqn (1.4.3), notice that

$$ \sum\bar{x}^2 = \sum(n^{-1}\sum x)^2 = n^{-2}\sum(x_1 + \dots + x_n)^2. $$

This means that it is possible to write

$$ \begin{aligned} \sum\bar{x}^2 &= A\sum x^2 + B\sum x_1 x_2 \\ &= A(x_1{}^2 + x_2{}^2 + \dots + x_N{}^2) + B(x_1 x_2 + x_1 x_3 \dots + x_{N-1}x_N) \end{aligned} $$

where x_1, x_2, ... are the individual observations and $\bar{x}$ stands for each sample mean (for n observations only). The number of samples (of n observations) containing any single observation, x_1, is given by ${}_{N-1}C_{n-1}$, i.e. the number of combinations of n − 1 observations from a total of N − 1. Similarly, the number of samples containing both x_1 and x_2 is given by ${}_{N-2}C_{n-2}$. Hence

$$A = \frac{1}{n^2} \times {}_{N-1}C_{n-1} = \frac{1}{n^2} \times \frac{(N-1)!}{(n-1)!(N-n)!}$$

and

$$B = \frac{2}{n^2} \times {}_{N-2}C_{n-2} = \frac{2}{n^2} \times \frac{(N-2)!}{(n-2)!(N-n)!}$$

Therefore

$$A/{}_NC_n = (1/n^2) \times (n/N) = 1/nN$$

and

$$B/{}_NC_n = 2/n^2 \times \left[n(n-1)/N(N-1)\right] = \left[2(n-1)/nN(N-1)\right]$$

Now

$$\mu^2 = (1/N^2)(\sum x^2 + 2\sum x_1x_2)$$

Substituting in Eqn (1.4.3) and squaring both sides gives

$$\sigma_{\bar{x}}^2 = \left(\frac{1}{nN} - \frac{1}{N^2}\right)\sum x^2 + \left[\frac{2(n-1)}{nN(N-1)} - \frac{2}{N^2}\right]\sum x_1x_2$$

$$= \frac{N-n}{nN^2}\left(\sum x^2 - \frac{2\sum x_1x_2}{N-1}\right)$$

Putting n = N in Eqn (1.3.6) and squaring both sides gives

$$\sigma^2 = \sum x^2/N - \mu^2 = (1/N - 1/N^2)\sum x^2 - (2/N^2)\sum x_1x_2$$

$$= \frac{N-1}{N^2}\left(\sum x^2 - \frac{2\sum x_1x_2}{N-1}\right)$$

Hence it is now possible to compare the variance of the sample means with the variance of the original population, i.e.:

$$\sigma_{\bar{x}}^2 = \frac{N-n}{nN^2} \times \frac{N^2}{N-1}\sigma^2$$

or

$$\sigma_{\bar{x}} = \left[\frac{N - n}{n(N - 1)}\right]^{\frac{1}{2}} \sigma \qquad (1.4.4)$$

If N is very large (i.e. $N \gg n$) then

$$\sigma_{\bar{x}} = \sigma/\sqrt{n} \qquad (1.4.5)$$

In general, σ is not known and must be estimated from the sample. Substituting s for σ gives the following estimate for the standard deviation of the mean (or standard error of the mean)

$$s_{\bar{x}} = s/\sqrt{n} \qquad (1.4.6)$$

Notice that a sample of one gives no information about the accuracy of the estimated mean since s cannot be estimated from a single observation. A sample of 2 can provide useful information but the accuracy is poor. This accuracy is doubled if the sample size is increased to 8 since the standard error of the mean varies inversely as the square root of the sample size. To double the accuracy again it would be necessary to increase the sample size to 32: an additional 24 observations. It is clear that the gain in precision for large samples is relatively small compared with the additional effort required. A sample size of 4 or 5 observations will often provide maximum information for a given effort.

Fig. 1.4.1 shows the (probability) distribution of individual observations and the (probability) distribution of sample means drawn

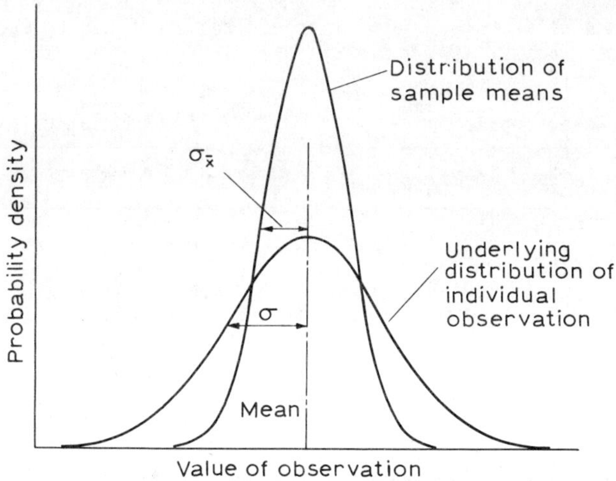

Fig.1.4.1 Probability distribution of sample means compared with the underlying distribution

from the underlying distribution. Notice the much narrower distribution of the sample means and the fact that the peak for this curve is therefore higher, since the areas under both curves are the same, equal to the sum of all probabilities (i.e. unity).

Example 1.4.1

Calculate the sample mean and the standard error of the mean for the following five results from Example 1.3.1: 30.5, 27.0, 33.0, 43.0 and 26.5 MN/m². Hence determine the 68.26% confidence limits, as estimated from these five cubes, for the following:

(i) All the concrete represented by the tests

(ii) The 40 actual cube tests.

<u>Solution</u> This example emphasizes the need for distinguishing between different populations. The population of 40 cube results in (ii) is quite distinct from the total population from which the 40 cubes were selected. Let all stresses be in MN/m² units. The computation of the appropriate standard deviations is most conveniently done from Eqns (1.3.2) and (1.3.3). The alternative form given in Eqn (1.3.5) is only useful if the number of results is sufficiently large to warrant grouping into classes, since it gives a less accurate result, especially when using a slide rule. Tabulation of the values can be made as follows:

(i) Total population

x_i	$x_i-\bar{x}$	$(x_i-\bar{x})^2$
30.5	-1.5	2.25
27.0	-5.0	25.0
33.0	1.0	1.0
43.0	11.0	121.0
26.5	-5.5	30.25
160.0	0	179.5

(ii) Population of 40

x_i	$x_i-\mu$	$(x_i-\mu)^2$
30.5	-2.1	4.4
27.0	-5.6	31.4
33.0	0.4	0.2
43.0	10.4	108.2
26.5	6.1	37.2
160.0		181.4

(i) The sample mean is

$\bar{x} = 160/5 = 32.0$ (MN/m²) (Answer)

From the standard deviation

$$s = \sqrt{\frac{\sum(x_i - \bar{x}^2)}{n - 1}} = \sqrt{\frac{179.5}{4}} = 6.7$$

the standard error of the mean can be calculated:

$s_{\bar{x}} = s/\sqrt{n} = 6.7/\sqrt{5} = 3.0(\text{MN/m}^2)$ (Answer)

The 68.26% confidence limits for the mean of all the concrete, estimated from the 5 cube tests only, are

$\bar{x} \pm s_{\bar{x}} = 32.0 \pm 3.0 = 29.0$ and 35.0 (MN/m^2) (Answer)

(ii) In this case the population is the 40 actual cube tests for which the mean value of 32.6 has been calculated in Example 1.3.1. However, assuming that neither $\mu = 32.6$ nor $\sigma = 4.70$ is known then table (ii) above (which gives $\sigma = 6.0$ and $\sigma_{\bar{x}} = 2.55$) cannot be determined, and it is necessary to substitute $\mu = \bar{x}$ and $\sigma \simeq s = 6.7$ as calculated in (i) above. Hence

$$s_{\bar{x}} = s\sqrt{\frac{N - n}{n(N - 1)}} = 6.7\sqrt{\frac{40 - 5}{5 \times 39}} = 2.8$$

The 68.26% confidence limits for the mean of 40 results estimated from the 5 cube tests only are

$\bar{x} \pm s_{\bar{x}} = 32.0 \pm 2.8 = 29.2$ and 34.8 (MN/m^2) (Answer)

(Note: compare with $\mu \pm \sigma_{\bar{x}} = 32.6 \pm 1.99$ obtained below.)

Example 1.4.2

Calculate the standard error of the mean for the data in Example 1.3.1 and explain its significance. Estimate the probability that a mean of a sample of 5 cubes from the population of 40 cubes will be less than 30 MN/m^2. What are the limits for the mean of a sample of 5 cubes at the 5% level of significance? Assume 9.7% and 2.5% failures for $z = 1.3$ and 1.96 respectively.

Solution Let all stresses be in MN/m^2 units. The standard error of the mean is

$s/\sqrt{n} = 4.70/\sqrt{40} = 0.74$

This means, for example, that there is a 68.26% probability that the mean of the total population (i.e. all the concrete represented by the test cubes) lies within the range 32.6 $\pm$ 0.74.

(Note: compare these limits, of 31.9 and 33.3, with the wider limits of 29.0 and 35.0 obtained in Example 1.4.1 for the first 5 results only.)

The total population of 40 cubes is known exactly so the exact form of the standard error of the mean can be used, i.e.:

$$\sigma_{\bar{x}} = \sigma \sqrt{\frac{N - n}{n(N - 1)}} = 4.70 \sqrt{\frac{40 - 5}{5 \times 39}} = 1.99$$

The deviation is

$D = \mu - 30 = 2.6$

Hence

$z = D/\sigma_{\bar{x}} = 2.6/1.99 = 1.3$

When z = 1.3 there are 9.7% failures (given). Hence the probability that a mean of 5 cubes is less than 30 MN/m^2 is 0.097 (Answer)

The level of significance is the probability of the value falling outside the range $\mu \pm z\sigma$ on either limit, and at 5% level of significance (i.e. 2½% failures), z is 1.96. Hence

$D = 1.96 \times 1.99 = 3.9$

and the limits are

$32.6 \pm 3.9 = 28.7$ and 36.5 (MN/m^2) (Answer)

1.5 CHARACTERISTIC STRENGTHS AND LOADS

In this section the derivation of characteristic strengths and loads, using statistical principles, will be outlined. One important consequence of statistical theory should already be apparent: although the probability of a particular limit state being reached during the working life of a structure can be reduced (by increasing design loads and reducing design stresses), this probability cannot be zero. It is impossible to guarantee that a failure will never occur in any structure.

Mean strength, or resistance

This is the arithmetic mean of a series of test results made to determine the resistances of samples of a material which are nominally similar, or supposedly conform to the same specification. The term 'resistance' commonly implies strength but can also include such properties as resistance to cracking etc.

Characteristic resistance

This is the resistance which takes account of the statistical variation of the resistance of a material. It corresponds to the 'specified' strength or 'minimum' strength referred to in earlier codes. It is defined by the equation

$$f_k = f_m - zs \qquad (1.5.1)$$

where f_k = characteristic resistance
f_m = mean resistance
s = standard deviation
z = coefficient, depending on acceptable probability of test results being below the value f_k.

CEB recommended, and CP 110 has adopted, a value of 1.64 for z, which implies that 5% of the test results will be below the characteristic value.

Characteristic steel strength

The guaranteed minimum yield strength (either yield or 0.2% proof stress) formerly given by steel manufacturers often corresponded to the characteristic yield strength provided that the probability of having results less than the guaranteed value was not more than that envisaged for the present value of 1.64 for z.

The yield point is much reduced, or, virtually eliminated, for hot rolled or cold worked bars respectively, hence codes of practice refer to the 0.2% proof stress—the stress at which the stress-strain curve deviates from the initial tangent modulus by 0.2% strain. See section 2.1 for descriptions of different reinforcements and Example 2.1.2 for statistical interpretation of clauses in BS 4449.

Characteristic cube strength

The characteristic cube strength, f_{cu}, corresponds to the minimum works cube strength formerly given in codes of practice. Hence from Eqn (1.5.1)

$$f_{cu} = f_{mcu} - 1.64s \qquad (1.5.2)$$

The standard deviation depends upon the degree of control exercised during concrete production and should be calculated from recent test results on as many separate batches of concrete produced by the same plant under the same supervision as possible. The characteristic cube strength normally implies the 28-day strength but Eqn (1.5.2) can be applied to cube results of any age.

See Appendix A for concrete grades, current margins and testing plan alternatives and chapter 15 for applications of statistics to mix design and quality control.

Example 1.5.1

Check whether the distribution of the data in Example 1.3.1 is approximately normal by plotting $\frac{1}{n}\sum_{1}^{i} f_i$ against strength on normal probability paper. Assuming a normal distribution, determine the characteristic cube strength for the concrete. How many of the actual results fall below this value? Check your values for mean and characteristic strengths by inspection of your probability plot.

Solution Let all stresses be in MN/m^2 units. Values of the fractional cumulative frequency, $\frac{1}{n}\sum_{1}^{i} f_i$ have been tabulated in Example 1.3.1, and $\frac{1}{n}\sum_{1}^{i} f_i$ is shown plotted against strength on normal probability paper in Fig. 1.5.1. These values are fairly well distributed about a straight line and indicate an approximately linear plot. The data therefore possess an approximately normal distribution.

We know from Example 1.3.1 that $\bar{x}$ is 32.6 and s is 4.70. Hence the characteristic cube strength is

$32.6 - 1.64 \times 4.70 = 24.9$ (MN/m^2) (Answer)

The number of results falling below this value (by inspection) is three. (Answer)

f_{mcu} and f_{cu} are given by 50% and 5% failures respectively. Hence from Fig. 1.5.1 the mean and characteristic cube strengths are approximately 32.6 and 24.9 (MN/m^2) respectively.

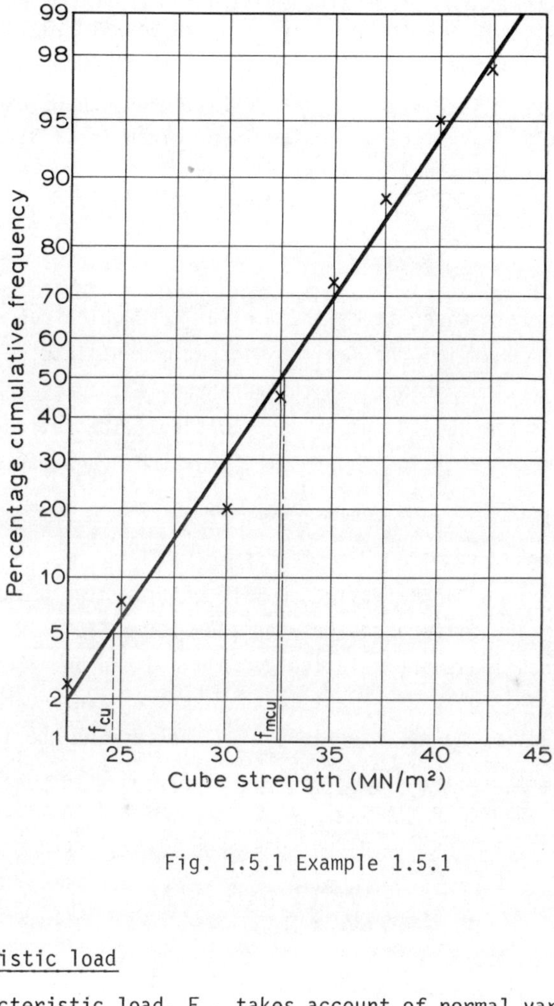

Fig. 1.5.1 Example 1.5.1

Characteristic load

The characteristic load, F_k, takes account of normal variations in load and, ideally, should be determined as a function of the mean load and its standard deviation in the same way as strength. For example, consider the live loading on the floors of a certain type of building such as a block of flats. Let $Q_1, Q_2, \ldots Q_n$ be the maximum values of live loading that occur on the floors of n blocks of flats during their design life. It is assumed that $Q_1, Q_2, \ldots Q_n$ are normally distributed about a mean value Q_m having a standard deviation s. The characteristic load is given by

$$Q_k = Q_m + zs \tag{1.5.3}$$

Hence if z is taken as 1.64 the probability that the maximum load will exceed Q_k is 0.05. (Notice the positive sign in Eqn (1.5.3), written because it is the maximum load in conjunction with minimum strength which is usually of interest.) However, until loads can be expressed in statistical terms, the minimum design loads given in CP 3; chapter V [1.8] can be used for the characteristic loads.

Minimum loads

Notice that in certain cases it is the minimum load which can give the worst effect. For example, the maximum midspan moment in a continuous beam is given by maximum loading on that span while the adjacent spans on each side are subjected to the minimum loading. Theoretically, therefore, the characteristic load for the adjacent spans would be given by

$$F_k' = F_m - zs \tag{1.5.4}$$

However, the dead load can be estimated with much greater precision than the live load, so any differences between the minimum and maximum characteristic dead loads are relatively small and can be allowed for, if necessary, by a slight adjustment in the partial safety factor for load (as explained in the next section). In the case of the live load, the minimum characteristic load is zero, so that the possibility of using a lower value does not arise. It may be assumed in this case, therefore, that for all practical purposes Eqn (1.5.4) need never be used.

Example 1.5.4

If the mean unit weight of normal concrete for a particular locality is 23 kN/m^3 and the standard deviation is found to be 0.5 kN/m^3, what is the difference between the characteristic (upper) load, F_k, and a characteristic lower load, F_k'? Would a value of 0.9 F_k allow for this discrepancy?

Solution 5% failures (and successes) are given by z = 1.64. Hence

$$F'_k = (23 - 1.64 \times 0.5)10^3 = 22.2 \times 10^3$$

and

$$0.9F_k = 0.9(23 + 1.64 \times 0.5)10^3 = 21.5 \times 10^3$$

Thus a 10% reduction in F_k could allow for not considering the correct limit for the dead load in this particular instance.

Note that the factor of 0.9 is necessary, for example, when the overall stability of a structure is being considered, since all the concrete in the structure could be lighter than normal due to the use of a particular aggregate. In the case of adjacent spans of a continuous beam, however, it is reasonable to assume that the aggregate will not have been changed from one span to the next. Hence any variations in unit weight of the concrete will be mainly due to variations in water content only. This is controlled (via strength) and should be small enough for the factor of 0.9 to be unnecessary. The minimum load effects due to variable loads in this case are given by maximum variable loads on adjacent spans and zero variable load on the given span.

Limitations of statistical theory

If observations for two essentially different causes—e.g. loads for domestic dwellings and loads for foundries—are inadvertently mixed together then the resulting frequency density plotted against load curve would have two 'peaks', each centred around the mean values for domestic floor loadings and foundry floor loadings respectively, rather than showing a single peak as indicated in Fig. 1.3.1. When extreme values are being considered, any normally remote extraneous effects can become significant and distort the predictions of the laws of chance based on essentially one variable. For example, the possibility of a client subjecting a floor intended for a caretaker's room to full foundry-floor loading, although remote, is not impossible and could distort the extreme predictions based on a normal gaussian distribution. Similarly, the extreme predictions for the resistance moment of a beam are dramatically affected if the main reinforcement is placed in the top face instead of the bottom!

Another common source of discrepancy is if the actual distribution is skew, rather than normal, because the variable being considered is not linearly related to the variable which is essentially causing the variations. A good example of this is in mix design (see chapter 15) where errors in batching cause variations in c/w ratio which in turn cause variations in strength. However, since strength is not linearly related to c/w ratio for all values, the distribution for strengths cannot also be normal for all values. Very high strengths and negative strengths are both impossible. Thus to reiterate, while strengths and loads can be expected to follow a normal distribution in the vicinity of their mean values, they may not do so at extreme values for a variety of reasons, which include human intervention and other physical constraints and effects, and which are best allowed for by means of partial safety factors as indicated in section 1.6.

1.6 PARTIAL SAFETY FACTORS AND DESIGN VALUES

The characteristic values allow for the inherent variations which occur in loads and material strengths. It is now necessary to consider the other factors and causes of variations which can occur and to make allowances (up and down) for these uncertainties by introducing suitable partial safety factors.

Partial safety factor for resistance

The material in the actual structure, under long-term loading, may be weaker than the characteristic value, for a short-term test, for any of the following additional reasons:

(i) The strength of a material in the actual structure may be reduced as a result of constructional faults or difficulties. For example, the concrete strength can be reduced by inadequate compaction or curing, by bad weather, or by impurities. Similarly, bond strength can be reduced if the surface of the steel is contaminated by dirt, oil, etc.

(ii) The strength of a material under sustained or repeated loading is less than that obtained for a short-term test. The long-term strength of a material may be further reduced by corrosion during the life of the structure even though the requirements for the limit state of durability are complied with. If the imposed loads are predominantly cyclic in character the limit state for fatigue should be considered.

Design resistance

The additional errors given above are taken into account by the partial safety factor for material deficiencies, γ_m. The design strength of the material which is assumed in calculating the strength of the structure is given by:

$$\text{Design resistance, } {}_*f = f_k/\gamma_m \qquad (1.6.1)$$

Partial safety factor for load

The partial safety factors for loads enable the following additional uncertainties to be taken into account for each limit state:

(i) Modification of loads (γ_{f_1}). Possible unfavourable deviations of load beyond those considered in deriving the characteristic loads. For example, there may be an increase in the general level of loading in structures of a particular type, due to the introduction of modern machinery etc., without any change of use.

(ii) Load combinations (γ_{f_2}). The reduced probability that various loadings act together simultaneously at their design (maximum) values. Transient loads, such as those due to wind and earthquakes, rarely reach their design values and when they do it is highly unlikely that all the other loads are also at their maximum values.

(iii) Modification of load effects (γ_{f_3}). Adverse modification of load effects can occur due to errors in design assumptions and calculations or due to constructional discrepancies. These include the possible neglect of thermal, creep and shrinkage stresses, errors in assessing the effects of loading, and the presence of unforeseen stress redistributions within the structure. For example, the rotation capacity of a plastic hinge may be less than assumed. The variations in dimensional accuracy achieved in construction can increase the load on a member due, for example, to excessive thickness of a slab caused by the deflection of formwork; dimensional errors in spans and heights; deviation of columns from the vertical; and errors in setting out arch profiles. The ability of the structure to resist loading can also be reduced by, for example, inadequate or incorrect positioning of reinforcement and formwork, resulting in reduced lever arms and sizes of members or reduced steel cover.

Unduly flexible formwork can affect the relative stiffnesses of members which, in turn, can affect the magnitudes of the forces and bending moments acting throughout a structure. Notice that modifications due to design errors are partly interrelated with constructional errors: hence the reason for grouping them together.

Possible modification of design values

The design values for either materials or loading may require adjustment to take account of the following:

(i) Nature of structure and its behaviour (γ_{c_1}). Structures or parts of structures in which partial or complete collapse can occur without warning, where reduction of internal forces is not possible, or where failure of a single element can lead to overall collapse.

(ii) Seriousness of reaching limit state (γ_{c_2}). The seriousness of attaining a particular limit state with regard to economic consequences, danger to the community, etc. For example, the probability of collapse of a multi-storey block of flats should be much less than that for a small footbridge over a stream in a park.

Russian Code

The Russian Code groups γ_{f_3} with γ_{c_1} and γ_{c_2} under a partial safety factor for service. This third partial safety factor can also allow for other factors which are not included in the calculations but which can either increase or decrease the carrying capacity. However, the justification for considering only two partial safety factors, as in CP 110, is given later.

Design load

The design load for a given type of loading and limit state is obtained from the appropriate characteristic load, F_k, and the partial safety factor for loads, γ_f, as follows:

$$\text{Design load, } F, = F_k\gamma_f \qquad (1.6.2)$$

In general a single coefficient, γ_f, can be used in the calculation where γ_f is a function of γ_{f_1}, γ_{f_2} and γ_{f_3}.

The different types of loads and load effects which can occur are as follows:

(i) Direct loads: permanent and variable imposed loads.

(ii) Imposed deformations: thermal and shrinkage effects, settlement of supports, prestressing etc.

(iii) Impact, inertia or other loads due to exceptional events.

Recommended values for partial safety factors

The values for the partial safety factors for loads and strengths recommended by CP 110 are given in Table B1 in Appendix B. These are usually in approximate agreement with the CEB-FIP recommendations [1.9] which are essentially more general.

The CEB-FIP recommendations for the ultimate limit states are as follows:

(i) Direct loads (maximum)

$$F = 1.5(G_k + F_{kv})$$

where F_{kv} is the set of forces due to the most unfavourable set of characteristic variable loads. These include variable working loads, climatic loads (snow and wind), dynamic or centrifugal forces or vibrating construction loads, exceptional overloads (hurricanes, earthquakes, fires, shocks, etc).

(ii) Direct loads (minimum)

$$F = 0.9G_k + 1.5F_{kv}$$

(iii) Combination of variable loads of different origin

$$F = 1.5 \quad G_k + 0.9F_{kv_1} + 0.8F_{kv_2} + 0.7(F_{kv_3} + F_{kv_4} + \ldots)$$

where $F_{kv_1} \geq F_{kv_2} \geq F_{kv_3} \geq \ldots$

γ_m for the above conditions is 1.15 for steel and either 1.4 for carefully mixed and controlled precast concrete in a factory or 1.5 for concrete generally.

(iv) Exceptional loads of short duration

$$F = G_k + F_{kv(permanent)} + F_{ex}$$

where $F_{kv(permanent)}$ is the fraction of characteristic variable loads of a permanent nature, and F_{ex} is the exceptional load envisaged which falls outside the range of characteristic loads.

For the serviceability limit states $\gamma_m = 1$ in all cases and $\gamma_f = 1$ generally, except that for a combination of variable loads of different origin, the reduction coefficients γ_{f_2} given above for F_{kv} at ultimate should be used (i.e. values of 0.9, 0.8 and 0.7). The partial safety factors for concrete are applied to the standard cylinder (not the standard cube), hence $\gamma_m = 1$ for all serviceability limit states; this difference between CP 110 and International Codes is reconciled for the collapse limit state by differences in assumptions for the concrete stress blocks.

Application of design strengths

It is important to realize that the partial safety factor γ_m allows for material deficiencies only and is used to obtain the design strength, $_*f$. The latter is not necessarily the stress which is intended, or is in fact permissible, for the appropriate limit state. For example, for the limit states of crack width the maximum desirable stress in the steel is $0.8f_y$ (the crack width formula for beams becomes conservative at higher stresses). This value, $0.8f_y$, is considerably less than the design tensile steel strength for crack widths (or local damage) in beams, which is f_y. Similarly, the maximum fibre stress in the concrete under working loads should not usually exceed $0.5f_{cu}$. The design cube strength for local damage from Table B1 is $0.77f_{cu}$ which is appreciably higher. Thus the coefficient 0.5 automatically incorporates the partial safety factor of 1.3. There are other cases, of course, when the design strength does equal the design stress considered for a particular limit state. This is so, for example, for both the steel and concrete stresses for the limit state of collapse, and for the steel stress for the limit states of crack width in a lightly reinforced wall. This observation will become clearer when each of the main limit states is considered in detail later.

1.7 GENERAL CONSIDERATIONS OF DESIGN PHILOSOPHY

The twin concepts of characteristic strength and partial safety factor for material strength require to be kept carefully balanced, if some of the factors assessed by one of them are not to be obscured by an undue emphasis on the other. This and other general considerations which affected the recommendations of the CEB-FIP and CP 110 are briefly discussed in this section.

Safety factors for concrete and steel

Concrete tests on cylinders or cubes cannot give the real strength of the concrete in the structure exactly. The results of tests on steel bar specimens, on the other hand, can correspond approximately to the strengths of the actual bars. In addition, since concrete is basically a brittle material, weakness in the concrete could lead to sudden failure of the structure without warning. For both these reasons the value of γ_m for concrete must be considerably greater than for steel.

Single value of safety factor

The value of γ_m for concrete at collapse, 1.50, is for the control conditions specified in the CEB-FIP Code for in situ concrete. For rigorous control in a factory the value could be reduced, whereas for poor control in situ it would need to be increased. In principle, CP 110 assumes that variations in strength due to constructional faults or difficulties tend to increase with variations in inherent strength. This assumption enables a single value to be adopted for the partial safety factor for concrete strength at collapse.

Characteristic strength v safety factor

A structure should be designed in such a way that the probability of any limit state being attained is substantially constant for all component members and for the structure as a whole. Although the rational determination of structural safety should be based on the theory of probability, it is at present necessary to adopt a semi-probability approach. Statistical theory can be used for determining the characteristic strength, whilst the partial safety factor for material strength allows for those aspects that are much more arbitrary in nature. Hence the guarantee that the actual strength does not fall below the value corresponding to the limit state is achieved in two stages. It is desirable not only to use both concepts, but also to maintain a certain balance between the relative importance of each for the following reasons:

(i) The importance of the concept of a characteristic strength, depending statistically on the variability of the strength, encourages manufacturers of materials to establish methods of control for reducing variability. This desirable effect is diminished if the value of z in Eqn (1.5.1) is reduced.

(ii) The normal statistical distribution of results is invalid for extreme values of strengths, so that the value of z should not be so great that there is an excessive discrepancy between the calculated and actual probabilities of the strength being less than f_k. For example, when considering the design of concrete mixes it is shown that since the strength of fully compacted concrete varies primarily with variations in the c/w ratio, it is preferable to consider variations in this ratio rather than variations in the resulting cube strengths. A normal distribution in c/w ratio values will only result in a normal distribution in corresponding strength values if there is a linear relation between c/w ratio and strength. Since there are two distinct slopes to the strength-c/w-ratio curve, with a transition in between, a skew distribution will generally be obtained for the strength results (see chapter 15). Such limitations in statistical theory indicate that the value of z should not be made too large.

A reasonable compromise between the foregoing effects is achieved if z is between about 1.0 and 2.0 (e.g. 1.64).

Nature of structure

The safety factor γ_f must be suitably increased for structures (or parts of structures) in which redistribution of moments is not possible, and in which partial failure might therefore result in complete collapse, or partial or complete failure occur without warning. The CP 110 approach means that the factor γ_{c_2} must be effectively incorporated or 'built in' where necessary: e.g. as in Table CC1 for critical shear stresses.

Seriousness of failure

The economic consequences of failure include not only the cost of repairing or replacing the structure, but also the cost of providing, while the structure is out of use, alternative means of fulfilling the function for which it was built. Personal considerations include danger to the lives of people in the vicinity of the structure at the time of failure and also the psychological effects of the failure. These considerations are at present covered by the statement that the partial safety factors must correspond to an acceptable probability of the limit state being reached, having regard to the seriousness of failure should it occur. If the seriousness of failure is taken as an indication of the care that must be taken in the calculations and in the execution of the work, it is generally possible to use common values for γ_f such that average accuracy of calculation and average workmanship are adequate where failure is not too serious, but very good accuracy of calculations combined with very good workmanship should be used where failure would be very serious.

Probability of collapse

Collapse occurs if the load F exceeds the resistance R, i.e. if R - F is negative. It has been suggested [1.10,1.11] that R - F approaches a normal distribution more nearly than either of the underlying distributions. Thus an estimate of the probability of collapse can be obtained if the coefficient of variation for the R - F distribution can be determined.

Freudenthal considered a 'central load factor', which will be redefined here as the ratio of the characteristic ultimate resistance (R_k) to the characteristic maximum load (F_k), γ_k

$$\text{i.e.}\quad \gamma_k = \frac{R_k}{F_k} = \frac{R_m(1 - z_R\delta_R)}{F_m(1 + z_F\delta_F)} \tag{1.7.1}$$

where subscripts k, m, F and R refer to characteristic, mean, load and resistance values respectively.

$$\text{Let}\quad \beta^{-1} = \delta_{(R - F)} = \surd(s_R^2 + s_F^2) \,/\, (R_m - F_m)$$

$$\text{i.e.}\quad R_m^2 - 2R_mF_m + F_m^2 = \beta^2(s_R^2 + s_F^2)$$

Dividing throughout by F_m^2 and transposing gives either

$$\beta^2 = \frac{(R_m/F_m)^2 - 2R_m/F_m + 1}{\delta_R^2\,(R_m/F_m)^2 + \delta_F^2} \tag{1.7.2}$$

$$\text{or}\quad (1 - \beta^2\delta_R^2)\,(R_m/F_m)^2 - 2R_m/F_m + (1 - \beta^2\delta_F^2) = 0 \tag{1.7.3}$$

In a typical case for an under-reinforced beam let

$\delta_F = 0.30$ e.g. dead load 10% and live load 50% and equal dead and live loads.

$\delta_R = 0.05$ for reinforcement and 0.15 for concrete.

For under-reinforced sections reduction in concrete strength is not very critical, hence take δ_R as 0.10.

Also, $z_R = z_F = 1.64$ and $\gamma_k = \gamma_f\gamma_m = 1.725$

Hence from Eqns (1.7.1) and (1.7.2),

$$\frac{R_m}{F_m} = \frac{1 + 1.64 \times 0.30}{1 - 1.64 \times 0.10} \times 1.725 = 3.08$$

and

$$\beta^2 = \frac{3.08^2 - 2 \times 3.08 + 1}{(0.1 \times 3.08)^2 + 0.3^2} = (4.8)^2$$

R - F is zero when $(R_m - F_m) - 4.8s_{R-F}$ is zero, hence from Appendix N the probability of R - F being negative is given by 10^{-6} approximately.

It is perhaps of interest to compare the above result with that obtained by assuming normal distribution throughout. The mean values of γ_f and γ_m can be taken as 1.5 and 1.15 as before and the correspondi[illegible] displacements from the mean values can be obtained as follows.

$$(\gamma_f - 1)F_k/s_F = 0.5\ \gamma_f F_m/s_F = 0.5 \times 1.5/0.30 = 2.5$$

and

$$(\gamma_m - 1)R_k/s_m = 0.15R_m/\gamma_m s_m = 0.15/(1.15 \times 0.10) = 1.3$$

Thus from Appendix N these give probabilities of 0.0062 and 0.097 respectively. The probabilities of exceeding the characteristic load or failing the characteristic strength are each 0.05, hence the combined probability of the load exceeding the resistance is given by:

$$0.05 \times 0.05 \times 0.0062 \times 0.097 = 1.5 \times 10^{-6}$$

Notice that this result is not so very different from that obtained previously. Notice also that where the strength of the concrete is more critical, as in columns or in balanced section beams, the probability of collapse would be somewhat different.

1.8 GENERAL ASSUMPTIONS FOR STRUCTURAL ANALYSIS

It is now convenient to consider some of the more general assumptions of CP 110 for the different limit states. In particular, it should be noted that elastic analysis may be used as the basis for determining the force distribution throughout the structure for the limit states of collapse, deflection and local damage.

Analysis of structure

When analysing the structure, or part of the structure, to determine force distributions within the structure, the properties of materials are to be assumed to be those associated with their characteristic strengths irrespective of which limit state is being considered.

Relative stiffnesses

The relative stiffnesses may be based on any one of the following sections:

(i) The concrete section—the entire concrete and reinforcement, cross-section, but ignoring the modular ratio of the reinforcement.

(ii) The gross section (or equivalent uncracked section)— the entire concrete cross-section, including the reinforcement on the basis of modular ratio.

(iii) The transformed (or equivalent cracked section) section— the compression area of the concrete cross-section combined with the reinforcement on the basis of modular ratio.

Analysis of sections

When analysing any section within the structure, the properties of the materials are to be assumed to be those associated with their design strengths appropriate to the limit state being considered. The strength of a section at ultimate should be assessed by inelastic analysis based on short-term stress-strain curves. The deflections of a structure for serviceability should be based on the curvatures at sections allowing for the influences of creep, shrinkage and cracking.

REFERENCES

1.1 Rowe, R.E., Cranston, W.C., and Best, B.C., 'New concepts in the design of structural concrete', Struct. Eng. vol. 43 London, Dec. 1965.

1.2 Anchor, R.D., 'Fire (reinforced concrete)', Concrete Journal vol. 9, January 1975, pp. 45-6.

1.3 Institution of Structural Engineers—Concrete Society, Fire Resistance of Concrete Structures, Joint Committee Report, August 1975, 59 pp.

1.4 Hughes, B.P., 'Limit states of impact and CP 110', Concrete Journal vol. 7, August 1973, pp. 37-40.

1.5 Griffiths, H., Pugsley, A., and Saunders, O., Collapse of flats at Ronan Point, Canning Town, Report of the Inquiry, HMSO, London, 1968, 71 pp.

1.6 Neville, A.M., and Kennedy, J.B., Basic Statistical Methods for Engineers and Scientists, Int. Textbook Co., Scranton Pennsylvanis 1964, 325 pp.

1.7 McIntosh, J.D., Concrete and Statistics, C R Books, London, 1968, 139 pp.

1.8 British Standards Institution (London) CP 3 Code of Basic Data for the Design of Buildings, Chapter V. Loading part 1:1967 Dead and imposed loads part 2: 1972 Wind loads 19 pp.

1.9 CEB-FIP (Comité Européen du Beton-Fédération Internationale de la Précontrainte) Recommendations for the Design and Construction of Concrete Structures. FIP Sixth Congress, Prague, June 1970.
1.10 Freudenthal, A.M., 'Safety and the probability of structural failure', Trans. Amer. Soc. Civil Engrs. vol. 121, 1956, pp. 1337-75.
1.11 Kemp, K.O., 'Concepts of structural safety', Civil Engng and Public Works Rev., vol. 68, February 1973, pp.132-41.

Exercise

1. The cube strength results (MN/m^2) for the first 40 separate batches of grade 60 concrete produced by the batching plant in a precast concrete works were as follows:

66.4	67.8	78.0	74.8	65.2	56.4	60.2	66.6
69.0	61.2	66.6	67.4	66.2	72.8	70.2	69.0
63.2	72.4	67.8	70.0	72.2	69.0	66.4	70.6
64.0	70.8	73.0	72.2	68.6	69.2	69.0	79.2
62.6	59.8	70.2	66.4	71.0	74.0	74.4	67.8

Calculate the mean strength and the standard deviation for the above data. Hence determine the characteristic strength for the concrete. Is this concrete satisfactory? You may use a class interval of 2.5 MN/m^2 for the frequency distribution table. Check that the distribution of the results is approximately normal by plotting the fractional cumulative frequency against strength on normal probability paper which is provided.

(UB3/H 1969,45 min)

2 Introduction to reinforced concrete

Reinforced concrete is by reason of its strength, durability, availability and adaptability an economical material eminently suited for many types of permanent structures. Concrete is relatively strong in compression and can be made sufficiently massive to provide lateral stability. Steel, on the other hand, is strong in tension but often lacks adequate lateral stability in compression due to its slender proportions. A reinforced concrete member in which the concrete resists compression while the steel resists tension is therefore an ideal structural partnership.

Portland cement concrete is sufficiently impervious to moisture, and retains sufficient alkalinity, to inhibit corrosion of the reinforcement even when small cracks are present. Although the idea of reinforced concrete was suggested in 1830 [2.1], the earliest successful use of tensile reinforcement in Portland cement concrete is believed to be due to the French contractor J. L. Lambot. He used a mesh of iron bars to build a reinforced concrete rowing boat for the Paris International Exhibition of 1855. An Englishman, William B. Wilkinson of Newcastle upon Tyne, took out the first patent for reinforced concrete in 1854, which proposed embedding either flat iron rods or second-hand wire ropes in the tension zones of concrete beams. This patent revealed that Wilkinson also understood the need to develop good bond between the concrete and the reinforcement, and the fact that concrete floors possessed good fire resistance. Wilkinson, together with Joseph Monier, a French engineer who later took out several patents are generally regarded as the inventors of reinforced concrete.

2.1 BASIC CONSIDERATIONS FOR REINFORCED CONCRETE

Traditional elastic design is a deterministic design method using a single (stress) factor of safety which is based on elastic theory. That is, the basic parameters are treated as non-random. The stresses occurring under the maximum working loads are compared with fractions of the strengths of the materials, and stresses are assumed to be proportional to strain. Traditional elastic design has been widely used for many years and elastic theory remains an extremely useful concept which provides an excellent introduction to the subject of reinforced concrete as well as an indication of the behaviour of structures under working loads. It therefore serves also to introduce the limit states of serviceability.

Tensile and compressive strengths of concrete

The use of plain concrete in most structures is limited because its ultimate tensile strength is small compared with its ultimate compressive strength. Consider, for example, a rectangular plain concrete beam as

Fig.2.1.1 Failure of (a) plain beam, and (b) reinforced beam. In (a) the bottom fibre is critical, in (b) the top fibre is critical

shown in Fig. 2.1.1. According to elastic theory for a homogeneous beam the extreme fibre stress, f, is given by

$$M = fZ \qquad (2.1.1)$$

where M = bending moment at the section

Z = section modulus.

If the strength of the concrete in tension is 3 MN/m^2 it is clear that the maximum moment which can be resisted by the plain beam is only 3Z MN m, even though the compressive strength of the concrete is of the order of 30 MN/m^2. If reinforcing bars are introduced into the bottom of the beam to take the tensile stresses then the moment which can be resisted becomes more like 30Z MN m, a very substantial increase.

Modulus of rupture

While the foregoing considerations are basically correct, it is essential to realize that elastic theory cannot be expected to predict the stress distribution in a beam at failure. If a plain concrete beam is actually tested in flexure then the moment at failure would be found to be nearer 4Z than the 3Z given above, for a concrete having a uniaxial tensile strength of 3 MN/m^2. This theoretical value is known as the modulus of rupture and must not be confused with a true value of stress. The increased value is due to plastic, rather than elastic, behaviour of concrete at high stresses. The behaviour of reinforced concrete at collapse is considered later. For the present it should be realized that elastic theory can only be expected to apply, in general, up to working load conditions.

Concrete quality

It is essential that all materials used in reinforced concrete conform to a certain minimum standard. Concrete can be produced in different mixes (proportions of cement, sand and aggregate), and different grades (qualities). The quality is usually measured in terms of the compressive strength of a 100-mm or 150-mm cube, f_{cu}, or of a 300-mm long × 150-mm diameter cylinder, in which case the compressive strength is denoted by f'_c.

Reinforcement

There are ten standard sizes of steel reinforcement. The most common are from 6 to 40 mm in diameter but the standard sizes include a 50-mm diameter for use when a larger size is necessary. The characteristic strength, f_y, of the reinforcement is that value of the yield or proof stress below which not more than 5% of the test values fall. The values of f_y to be used in design are the specified characteristic strengths in the appropriate British Standard. These are included in Table CD1.

Types of reinforcement

The main types of steel reinforcement are as follows:

(i) Rolled mild steel bars. Rolled mild steel bars to BS 4449 [2.2] are usually circular although square section bars are also available. These square bars, chamfered or unchamfered, have a cross-sectional area which is the same as that of a standard round bar; for example, a 7.1 mm square bar is actually referred to as an 8-mm RABar (round area bar). The minimum yield stress is specified as 250 MN/m^2. The minimum elongation (on a length of 5.65 × bar diameter) is 22%. There is also a bend test. Mild steel may be made by the open hearth (acid or basic), acid Bessemer, top or bottom blown oxygen ($\ngtr$ 0.008% nitrogen) or electric processes. Mild steel has $\ngtr$ 0.25% carbon, $\ngtr$ 0.06% sulphur and $\ngtr$ 0.06% phosphorous.

(ii) Hot rolled deformed bars. A deformed bar has a surface which has been so shaped that its bond with concrete is increased by at least the amount required for Type 1 to CP 110. The deformation and consequent stretching raises the yield stress but reduces the elongation and the performance in the bend test. The composition is as for mild steel except for sulphur ($\ngtr$ 0.05%) and phosphorous ($\ngtr$ 0.05%). BS 4449 specifies a minimum yield stress of 410 MN/m^2 (taken as 0.2% proof stress or as stress at a total strain of 5 millistrain). The minimum elongation for 'high yield 410' bars, to which these figures apply, is 14%.

(iii) Cold worked deformed bars. Cold working can also raise the yield point (and reduce the elongation). The composition is as for mild steel. BS 4461 [2.2] specifies a minimum yield stress (taken as 0.2% proof stress or as stress at a total strain of 4 millistrain) of 425 and 460 MN/m^2 for bars > 16 mm and $\ngtr$ 16 mm diameter respectively. The minimum elongation is 14% and 12% for bars > 16 mm and $\ngtr$ 16 mm respectively.

(iv) High tensile bars. High tensile steel bars with a 0.2% proof stress of about 900 MN/m^2 or higher are usually used in prestressed concrete only. However, some American tests using reinforcement having yield stresses up to 650 MN/m^2 have indicated that it can be satisfactory for use in reinforced concrete. Bars with a guaranteed minimum yield stress of 550 MN/m^2, and up to 20 mm in diameter, have been accepted for use in internal beams by some local authorities in Britain since 1965. A useful review of the use of steel reinforcement has been given by Lancaster [2.3].

(v) Steel fabric. Steel fabric has either an oblong or a square mesh in which the cross-wires are electrically welded to the main wires at the intersections. Expanded metal (steel) fabric has a diamond-shaped pattern and is formed from a metal sheet. Both types are covered by BS 4483 [2.4]. The minimum 0.2% proof stress is specified in BS 4482 [2.5] as 485 MN/m^2, or in BS 4461 as

460 MN/m². Fabric usually has a wire diameter not exceeding 12 mm and an adequate bond strength even though the wires are not deformed. This is mainly due to the anchorage provided by the welded intersections although the relatively small bar diameters also help to reduce the average bond stress.

Steel bending

Reinforcing bars must usually be bent without heating before being placed in the structure prior to concreting. BS 4449 and BS 4461 therefore specify that sample lengths of bar shall withstand cold bending through 180° round a former of diameter 2Φ for mild steel, 3Φ for high-yield steel with yield strengths up to 460 MN/m² , and 4Φ or 6Φ for stronger steels, where Φ is the nominal bar size or effective diameter. BS 4466 [2.6] specifies that the minimum diameter of bend allowed in service is double that used in the foregoing cold bend test.

BS 4449 and BS 4461 also include an optional re-bend (i.e. bent and partially bent back) test which reveals any loss in strength due to the bending operation. Some deformed bars appear to perform unsatisfactorily in this test due to notch effects at intersections of ribs [2.3]. The test also indicates strain embrittlement which can occur in steels with an excessive nitrogen content.

The bearing stress increases with decreasing radius of bar, so the bearing stress for the concrete may be exceeded (see Appendix CD). It is clear that one disadvantage of high-yield reinforcement is that larger bend radii are necessary and this makes detail design more difficult.

Nominal size of deformed bars

The nominal size of deformed bars can be checked by determining the effective cross-sectional area. BS 4449 and BS 4461 state that if the pattern of deformation is such that the cross-sectional area appears, by visual inspection, to be substantially uniform along the length of the bar, the effective area shall be the gross area, determined by weighing and measuring to a precision of $\pm$ 0.5% a length $\nless$ 0.5 m. If the cross-sectional area varies along its length, it is necessary to weigh before (m) and after (m') removing the transverse ribs. If

$1.03\ m' \nless m$ the mass of the bar should be taken as $1.03\ m'$; otherwise the area is obtained assuming mass m.

Example 2.1.1

The cross-sectional area of a nominal 12-mm diameter hot rolled deformed bar appears to vary along its length. A sample length of 750 mm is weighed before and after removal of the projections and found to have a mass of 0.666 and 0.650 kg respectively. Determine the effective cross-sectional area of the bar.

Solution BS 4449 assumes a density for steel of 7850 kg/m³.

$$1.03\ m' = 1.03 \times 0.650 = 0.669$$

which is > 0.666. Hence the effective cross-sectional area is

$$0.666/(0.75 \times 7850) = 113.1 \times 10^{-6}$$
$$= 113.1\ \text{mm}^2 \qquad \text{(Answer)}$$

Note that rolling tolerances permit variations from the nominal size which should be taken into account. For example, 12-mm and larger hot rolled bars have a tolerance on the nominal mass per metre run of $\pm\ 2\frac{1}{2}\%$ for a batch, or $-$ 4% for an individual bar. The effective area for an individual 12-mm bar can therefore vary by an unspecified positive value and by up to $-$ 4% from the nominal area of 113 mm².

Specified characteristic strength for steel

BS 4449 states the following:

'16.1 The specified characteristic strength of the steel is the value of the yield stress below which shall fall not more than 5% of the test results of the material complying with this standard.

16.2 The steel shall be deemed to comply with the specified characteristic strengths...provided that for each size and type of bar:

(1) not more than two test results of yield stress out of the last 40 consecutive results are less than the specified characteristic strength, and

(2) no test results are less than 0.93 of the specified characteristic strength.'

Example 2.1.2

Determine the significance of the BS 4449 and BS 4461 clauses: 'no test results are less than 0.93 of the specified characteristic strength'.

Solution The basic requirement for characteristic strength is given by

$$f_k = f_m - 1.64s \text{ (5\% failures)}$$

If f is the strength of an individual specimen the additional requirement is given by

$$f \nless 0.93f_k$$

that is

$$f \nless 0.93(f_m - 1.64s) = f_m - (0.07f_m + 1.53s)$$

Now $\delta = s/f_m$, hence this condition becomes

$$f \nless f_m - (0.07/\delta + 1.53)s$$

For example, if $\delta = 10\%$,

$$f \nless f_m - 2.23s \text{ i.e. 1.3\% failures}$$

or, if $\delta = 5\%$,

$$f \nless f_m - 2.93s \text{ i.e. 0.17\% failures.}$$

It can be expected, therefore, that if the coefficient of variation is 10%, over 1% of all batches of steel will be deemed not to comply with the Standard. The second clause therefore provides an additional incentive to improve the quality control. For example, if δ is halved, to 5%, less than 0.2% of all batches can be expected to provide difficulty.

Further tension tests for steel

If a batch of steel does not comply with the Standard, or if a purchaser wishes to verify that a batch of steel is satisfactory, it is specified that $\nless$ 10 bars shall be chosen from the batch and a tensile test specimen cut from each bar. If any one of the bars tested has a yield stress $< 0.93f_k$ then that bar shall be rejected. If another of the bars tested also has a yield stress $< 0.93f_k$, then the whole batch of steel shall be deemed not to comply with the requirements of the Standard. If $\nless$ 9 results are > $0.93f_k$ the steel shall be deemed to comply with the Standard.

Elastic modulus

Up to the limit of proportionality of an elastic material, stress is proportional to strain; the constant of proportionality being the elastic modulus, E, of the material. That is,

Stress = E × Strain

where strain is the extension (change in length) of the material per unit length. E_s and E_c are used to denote the elastic moduli of steel and concrete respectively. Since concrete is not in fact an elastic material, E_c is taken to be the secant modulus at the working stress for a standard rate of loading. Creep effects are included if the rate of loading is extremely slow. CP 114:1957 [2.7] assumed a constant secant modulus of 2×10^6 lbf/in² (13.8 GN/m²) whereas CP 110 assumes that the short-term E_c is given by

$$E_c \simeq 9100 \sqrt[3]{f_{cu}} \text{ (MN/m}^2\text{ units)} \qquad (2.1.2)$$

where f_{cu} is the characteristic cube strength (MN/m² units) of normal density concrete. The long-term E_c is assumed to equal one-half of the above value.

The steel is considered to be perfectly elastic in both tension and compression. The elastic modulus is assumed to be 200 GN/m² for most steels except alloy steels and prestressing steels, where the values are slightly lower.

Modular ratio

Consider two similar bars, one concrete and one steel, which are compressed uniformly so that they deform together by equal amounts. If the overall compressions are the same, then the strains are also equal. Hence by equating these strains

$$f_s/f_c = E_s/E_c = \alpha_e \qquad (2.1.3)$$

where f_s and f_c are the stresses in the steel and concrete respectively. If E_s = 200 GN/m² and E_c (long-term) = 10 GN/m² then the modular ratio, α_e equals 20. CP 114:1957 [2.7] assumed that α_e = 15 for all grades of concrete. The only merits of assuming a single value were simplicity and the fact that the secant modulus of concrete is variable with time as well as with grades of concrete.

Permissible stresses

The permissible stresses are the maximum stresses which are desirable under the action of normal working loads. For example, in beams it is desirable that the tensile stress in the steel should not be too high; otherwise the cracking in the surrounding concrete becomes excessive. Similarly the maximum compressive stress in the concrete should be appreciably less than the design strength (which is given by f_{cu}/γ_m) to ensure that no spalling or microcracking of the concrete can occur under working loads. The permissible stresses for beams in bending for elastic theory as given in CP 114:1957 were $\frac{1}{2}f_y$ for steel and $\frac{1}{3}f_{cu}$ for concrete.

Concrete tension cracks

It follows from the above that if no slipping occurs between a steel bar in the bottom of a loaded concrete beam and the surrounding concrete, the stress in the bar must be α_e times that in the concrete. If a mild steel bar is stressed to $\frac{1}{2} \times 250 = 125$ MN/m² under the working load, the stress in the adjacent concrete for $\alpha_e = 15$ is 125/15 or 8.3 MN/m². Since concrete fractures at a tensile stress far less than this, it must have cracked. Such cracks are always present on the tension side of a reinforced beam or slab, even though they are not visible to the naked eye. The tensile strength of concrete is normally neglected in flexure since the concrete cracks long before the steel is fully stressed under the working load.

2.2 BASIC ASSUMPTIONS FOR ELASTIC THEORY

It is essential for a complete understanding of any theory to appreciate fully the assumptions which are made. The basic assumptions of the elastic theory are therefore summarized so that their full significance may become apparent subsequently as individual aspects of reinforced concrete are considered.

(i) Steel and concrete act in conjunction throughout, there being no relative movement of the bars in the concrete. This is attained by ensuring that adequate bond or frictional stress is developed between the steel and concrete.

(ii) Plane cross-sections of members before stressing remain plane under load.

(iii) Both steel and concrete are perfectly elastic.

(iv) Concrete resists no flexural tension stresses.

(v) Stresses due to thermal or shrinkage movement of the concrete may be neglected.

(vi) The cross-sectional area of the steel is small compared with that of the concrete.

These assumptions will now be discussed individually.

Bond

Adequate bond between concrete and steel is essential in reinforced concrete construction. Only a fraction of the potential strength of an RC member is realized if the concrete and steel do not act in conjunction. The bond is due to (a) adhesion and (b) grip, or frictional resistance. The concrete shrinks upon setting and therefore exerts a pressure upon the steel. This pressure is reduced slightly when the steel and surrounding concrete are stressed in tension, since Poisson's ratio for steel (about 0.29) is greater than that for concrete (0.10 to 0.20). Irregularities on the surface of the reinforcement mechanically lock the steel to the concrete, improving both adhesion and grip; it is this that justifies the indentation of cold-worked bars.

Bernoulli's theorem

The assumption that plane sections remain plane in bending implies that strains are proportional to the distance from the neutral axis.

Elasticity

While steel can be considered an elastic material, concrete is hardly elastic even below the working stress. This assumption is perhaps the least justified since creep in the concrete can cause very considerable stress redistribution with time in continuous beams and columns etc. Nevertheless, this assumption was the main basis used for designs for many years. They generally proved to be satisfactory in practice because due allowance was made in the light of experience for long-term effects such as creep, and because the overall safety margin was usually high.

Tensile stresses

While the tensile strength of concrete is ignored in direct tension and in bending, it is still expected to resist diagonal tension stresses of the order of 1 MN/m^2 due to shear effects.

Secondary stresses

Stresses due to shrinkage and temperature variations are often ignored. Since these effects can lead to tensile stresses in excess of the tensile strength of the concrete, it is necessary to include nominal steel reinforcement in positions and directions where they would otherwise appear unnecessary. Relatively small amounts of reinforcement (see chapter 11) usually enable shrinkage cracks to be distributed as a series of very fine hair cracks at frequent intervals, instead of large cracks at infrequent intervals. Minimum amounts of nominal steel are included in Appendix CD.

Steel area

The cross-sectional areas of the steel bars are sufficiently small for the stress to be assumed constant across the steel section.

2.3 PRACTICAL DESIGN CONSIDERATIONS

Design is based on allowable stresses in the concrete and steel reinforcement (known as 'permissible' stresses for elastic theory and 'design' stresses for limit state theory), as well as on certain guides governing the placing of the steel in the concrete. Elastic design is essentially simple and straightforward, and factors such as shrinkage are not generally considered. Where practice indicated the desirability of taking other factors into account, various 'design guides' became necessary and were incorporated in codes of practice such as CP 114. The practical guides for limit state design and reinforced concrete design (generally as given in CP 110 and BS 5337) are summarized, for convenience in Appendixes B and C respectively. These guides are now much less arbitrary than they used to be for designs based solely on elastic theory. The more elementary practical considerations are given below.

Accuracy of calculation

The assumptions given in section 2.2 emphasize the necessity of avoiding meticulous accuracy in design calculations. The elasticity and compressive strength of concrete differ very considerably from the design assumptions. Many variables are introduced due to inequalities of workmanship, materials and conditions. The compressive strength of concrete varies even in adjacent parts of the same structure, and thicknesses of members as actually placed in practice may well differ slightly from the dimensions given on the drawings, due to undetected errors in fixing formwork. Unpredictable secondary stresses are often set up by shrinkage and expansion of the concrete. These are some of the more important variables which limit state theory attempts to assess but which conventional elastic design virtually ignores except for the use of a single, global, safety factor. Fictitious precision in design calculations is to be avoided; conventional slide-rule accuracy is more than adequate, especially for elastic design.

Reinforcement arrangement

Beams and slabs are provided with bars near the faces where tensile stresses occur. Bars used in a beam run the length of the beam, whereas in a slab either steel bars or steel mesh fabric are used. Slabs must be provided with bars in two directions, even when the slab is intended to be one-way spanning: main bars running the length of the span and smaller transverse bars at right angles to the span. The former take the tensile stress in bending while the latter prevent large cracks forming due to shrinkage or temperature movement of the concrete. The transverse bars are known as distribution reinforcement and are wired to the main bars to form a mesh.

Size of members

A degree of standardization of the size of members within a structure is desirable from economical considerations such as re-use of forms, fewer bar types etc.

Cover

A certain minimum thickness or cover of concrete is necessary around all reinforcement in order to:

(i) Prevent moisture penetration to the steel causing corrosion and flaking off of the concrete around the bar.

(ii) Enable the bar to act in conjunction with the concrete without slip.

Bar sizes and applications

The general sizes and uses of round mild steel bars are as follows:

Links and ties: 6 to 12 mm.

Common sizes for main reinforcement: 10 to 32 mm.

Very heavy construction only: 40 and 50 mm.

Use as few different diameters as possible when designing, since this assists identification on site and avoids unnecessary sorting: there should normally be no need to use more than the ten standard sizes recommended by CP 110.[2.8,9]

Bar lengths

The normal standard length of bars is 12 m, but use shorter lengths for sizes less than 10 mm. Thus longer bars up to 12 m are proportionately cheaper to use than shorter bars if the quantity is sufficient to justify the increased transport and handling costs.

Bar bending

Bars must be bent in order to conform to the geometry of the structure. Some bars have to be hooked at either end to increase their grip or anchorage in the concrete. Bends and hooks or other anchorages are of such forms, dimensions and arrangements as to ensure that the concrete is not over-stressed.

Wire ties

Bars are fixed rigidly in position relative to each other, in the formwork and prior to concreting, by securing some of the intersections of the bars with wire ties.

Effective span

The effective span of a simple beam or slab is taken as the lesser of

(i) The distance between centres of supports, and

(ii) The clear distance between supports plus the effective depth of the beam or slab.

Weight of reinforced concrete

The weight of dense reinforced concrete depends upon the type and proportions of its constituent materials. It also increases with the steel ratio. It usually varies, however, between 22 and 26 kN/m^3 for normal aggregates and is generally assumed for design purposes to be 24 kN/m^3.

REFERENCES

2.1 Farebrother, J.E.C., 'Concrete—past, present and future', Struct. Eng. London, October 1962, pp 337–41.

2.2 British Standards Institution (London), BS 4449: 1969 Hot Rolled Steel Bars for the Reinforcement of Concrete (metric units). Also BS 4461: 1969 Cold Worked Steel Bars for the Reinforcement of Concrete (metric units).

2.3 Lancaster, R.I., 'Steel reinforcement', Structural Concrete vol. 3, London, July–Aug. 1966 pp 159–77, and discussion, pp 178–98.

2.4 British Standards Institution (London), BS 4483:1969 Steel Fabric for the Reinforcement of Concrete (metric units).

2.5 British Standards Institution (London), BS 4482:1969 Hard Drawn Mild Steel Wire for the Reinforcement of Concrete.

2.6 British Standards Institution (London), BS 4466:1969 Bending Dimensions and Scheduling of Bars for the Reinforcement of Concrete.

2.7 British Standards Institution (London), CP 114:1957 Structural Use of Reinforced Concrete in Buildings (Reset and reprinted in 1965).

2.8 Anchor, R.D., 'Current practice sheets IM/09/1 and 2, Steel bar reinforcement and fabric reinforcement', Concrete, vol. 7, January and March 1973.
2.9 Lancaster, R.I., 'Current practice sheet 2P/20/1, Steel reinforcement production and practice', Concrete, vol. 7, June 1973.

3 Elastic analysis of beams

The present chapter considers the elastic *analysis* of members such as beams, slabs and staircases which are essentially subjected to pure flexure. Analysis concerns the determination of certain unknowns *when the section is fully defined* (that is, A_{st}, b and d are known). Two types of problem therefore arise in analysis:

(i) Determination of the stresses for a given loading

(ii) Determination of the maximum (permissible) loading for given permissible stresses.

3.1 STRESS AND STRAIN DISTRIBUTIONS

The stress and strain distributions across a section of a member subjected to pure bending can now be determined from the six assumptions given in section 2.2. The more general considerations outlined in the previous chapter, however, must also be borne in mind.

Neutral axis

Consider a loaded rectangular concrete beam which is reinforced in one layer only as shown in Fig. 3.1.1(a). Notice that the neutral axis, which represents points at zero stress, is not necessarily at half the depth of the beam as it would be for a homogeneous beam.

Effective depth

Since all the concrete below the neutral axis is in tension, and is therefore assumed to take no stress, the depth of the beam that really affects the strength calculations is the depth from the compression face to the centroid of the tensile steel. This is known as the effective depth, d in CP 110 (or d_1 in CP 114).

Strain profile

If no slip occurs between the concrete and the steel, the steel strain equals the strain in the concrete immediately surrounding the steel. The strain at any point is assumed proportional to its distance from the neutral axis, hence the strain profile is linear as shown in Fig. 3.1.1(b). The maximum compressive strain in the concrete, ε_{cb}, occurs at the top edge of the beam. The cross-sectional area of the steel is small, so the tensile strain in the steel is given by a mean value, ε_{st}. It follows from similar triangles that

$$\varepsilon_{st}/\varepsilon_{cb} = (d - d_n)/d_n \tag{3.1.1}$$

Stress profile

The stress profile for the concrete is also linear as shown in Fig. 3.1.1(c), since stress varies as strain. The maximum compressive stress is f_{cb}. The steel stress, f_{st}, is given by:

$$f_{st} = E_s\varepsilon_{st} = \alpha_e \times \text{(adjacent concrete stress)}$$

Conversely, the stress in the concrete around the steel equals f_{st}/α_e. This value is theoretical only since in general the concrete will have cracked, as was shown in chapter 2.

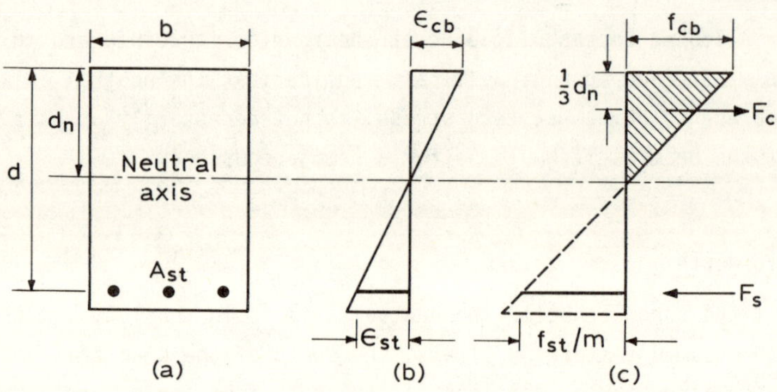

Fig.3.1.1 Beam cross-section showing strain distribution (concrete and steel) and theoretical concrete stress distribution

Maximum stresses

By considering the strain profile and using Hooke's law,

$$f_{st}/f_{cb} = E_s\varepsilon_{st}/E_c\varepsilon_{cb} = \alpha_e(d - d_n)/d_n \qquad (3.1.2)$$

3.2 MOMENT OF RESISTANCE

The moment of resistance of a section is the maximum moment which can be applied to the section without exceeding certain limiting stresses in either the concrete or the steel. For the elastic theory the limiting stresses are the permissible values. The moment of resistance is determined by considering the equilibrium of the internal forces and both permissible stresses.

Resultant compressive force

The compressive stress in the concrete varies uniformly from zero to f_{cb}, hence the average value is $f_{cb}/2$. This stress acts over a width b and depth d_n. The compressive force exerted by the concrete is therefore given by

$$F_c = \tfrac{1}{2}f_{cb}bd_n$$

This force acts at the centre of pressure, (i.e. the centroid of the shaded triangle in Fig. 3.1.1(c)) which is $d_n/3$ below the compressive face for a constant beam breadth.

Resultant tensile force

The total tensile force in the steel is given by

$$F_s = f_{st}A_{st} = f_{st}\rho bd$$

where ρ is the steel ratio and equals A_{st}/bd. The force acts at a depth d from the compressive face.

Equilibrium requirements

If the beam is subjected to pure bending then the resultant compressive and tensile forces must be equal. Hence

$$F_s = F_c \quad \text{and therefore} \quad f_{st}\rho bd = \tfrac{1}{2}f_{cb}bd_n$$

Rearranging,

$$f_{st}/f_{cb} = d_n/2\rho d \qquad (3.2.1)$$

Neutral axis depth

The depth of the neutral axis, d_n, can be determined by eliminating the unknown stress ratio f_{st}/f_{cb} from Eqns (3.1.2) and (3.2.1) as follows:

$$\alpha_e(d-d_n)/d_n = d_n/2\rho d \tag{3.2.3}$$

Rearranging,

$$(d_n/d)^2 + 2\alpha_e\rho d_n/d - 2\alpha_e\rho = 0 \tag{3.2.3}$$

This equation is a quadratic in d_n/d, hence the neutral axis depth is given by

$$d_n/d = - \alpha_e\rho \pm \surd[(\alpha_e\rho)^2 + 2\alpha_e\rho]$$

Only the positive value of d_n/d is appropriate, hence

$$d_n/d = [\surd(1 + 2/\alpha_e\rho) - 1]\alpha_e\rho \tag{3.2.4}$$

The position of the neutral axis for any given values of ρ and α_e can therefore be determined.

Lever arm

The parallel tension and compression forces form a couple which balances the applied bending moment. The bending moment is therefore given by either of the two equal forces multiplied by their distance apart. The arm of the couple is called the lever arm, z, of the beam and is given by

$$z = d - d_n/3 \tag{3.2.5}$$

The stresses in the concrete and steel can therefore be determined from the following equations in terms of the applied bending moment, M, since d_n is known from Eqn (3.2.4).

$$\text{Steel} \quad M = f_{st}\rho bd(d - d_n/3) \tag{3.2.6}$$

$$\text{Concrete} \quad M = \tfrac{1}{2}f_{cb}bd_n(d - d_n/3) \tag{3.2.7}$$

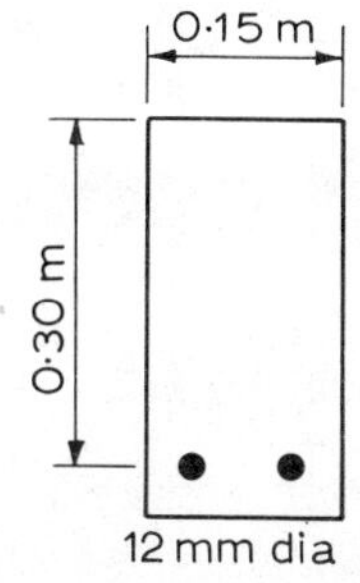

Fig.3.2.1 Example 3.2.1

Example 3.2.1

A rectangular reinforced concrete beam has a breadth of 150 mm and an effective depth of 300 mm. It is reinforced with two 12-mm diameter bars. Calculate the maximum stresses in the concrete and the steel if the beam is subjected to a bending moment of 7.5 kNm (a) for a short time (b) for a prolonged period, if:

E_s = 200 GN/m²

Short-term E_c = 21.4 GN/m²

Long-term E_c = ½ short-term E_c.

Solution

(a) For short-term loading

$$\alpha_e = E_s/E_c = 200/21.4 = 9.35$$

$$\rho = A_{st}/bd = 2 \times (\pi/4) \times 12^2/150 \times 300 = 0.503\%$$

$$100\ \alpha_e\rho = 4.70$$

Hence, from Eqn (3.2.4)*

$$d_n/d = [\surd(1 + 200/4.70) - 1] \times (4.70/100) = 0.263$$

* The student should test his understanding of the text by calculating d_n/d from first principles, as would be necessary in an examination. This applies to all worked examples which, for convenience, quote equations derived in the text.

Therefore

$$z/d = 1 - 0.263/3 = 0.912$$

Hence from Eqns(3.2.6) and (3.2.7)

$$7.5 \times 10^3 = f_{st} \times 0.00503 \times 0.15 \times 0.30^2 \times 0.912$$

$$f_{st} = 121 \times 10^6 = 121 \text{ MN/m}^2 \qquad \text{(Answer)}$$

and

$$7.5 \times 10^3 = \tfrac{1}{2} f_{cb} \times 0.15 \times 0.30^2 \times 0.263 \times 0.912$$

$$f_{cb} = 4.63 \times 10^6 = 4.63 \text{ MN/m}^2 \qquad \text{(Answer)}$$

(b) For long-term loading

$$E_c = \tfrac{1}{2} \times 21.4 \times 10^9 = 10.7 \times 10^9$$

$$\alpha_e = 18.7 \qquad 100\alpha_e\rho = 9.41$$

Therefore

$$d_n/d = [\sqrt{(1 + 200/9.41)} - 1](9.41/100) = 0.351$$

$$z/d = 1 - 0.351/3 = 0.883$$

Therefore

$$7.5 \times 10^3 = f_{st} \times 0.00503 \times 0.15 \times 0.30^2 \times 0.883$$

$$f_{st} = 125 \times 10^6 = 125 \text{ MN/m}^2 \qquad \text{(Answer)}$$

Also

$$7.5 \times 10^3 = \tfrac{1}{2} f_{cb} \times 0.15 \times 0.30^2 \times 0.351 \times 0.883$$

$$f_{cb} = 3.59 \times 10^6 = 3.59 \text{ MN/m}^2 \qquad \text{(Answer)}$$

Notice that the reduced value for E_c has an appreciable effect upon d_n/d and therefore upon f_{cb}, but that it has a relatively small effect upon z/d. The steel stress is therefore only slightly affected by the large change in E_c.

Moment of resistance—steel critical

The moment of resistance of the steel in tension, M_{dt}, is given by the maximum moment which can be resisted by the steel. Substituting M_{dt} for M and f_{dst} for f_{st} in Eqn (3.2.6), therefore, gives

$$M_{dt} = f_{dst}\rho bd(d - d_n/3)$$

Rearranging,

$$M_{dt}/bd^2 = f_{dst}\rho(1 - \tfrac{1}{3} d_n/d) \qquad (3.2.8)$$

d_n/d is known from Eqn (3.2.4), for given values of α_e and ρ, hence

M_{dt}/bd^2 can be determined.

Moment of resistance—concrete critical

Similarly, the maximum moment which can be resisted by the concrete is given by

$$M_{dc}/bd^2 = \tfrac{1}{2}f_{dcb}(d_n/d)(1 - \tfrac{1}{3}d_n/d) \qquad (3.2.9)$$

Example 3.2.2

A concrete beam of breadth 100 mm is reinforced with two 10-mm diameter bars such that the effective depth is 200 mm. Calculate the maximum working load moment that the section can safely carry for a short time, and the maximum steel and concrete stresses induced by that moment, if

$f_{cu} = 22.5$ MN/m² $\quad f_{yt} = 250$ MN/m²

$E_c = 21.4$ GN/m² $\quad E_s = 200$ GN/m²

Assume that

$f_{dcb} = \tfrac{1}{3}f_{cu} \qquad f_{dst} = \tfrac{1}{2}f_{yt}$

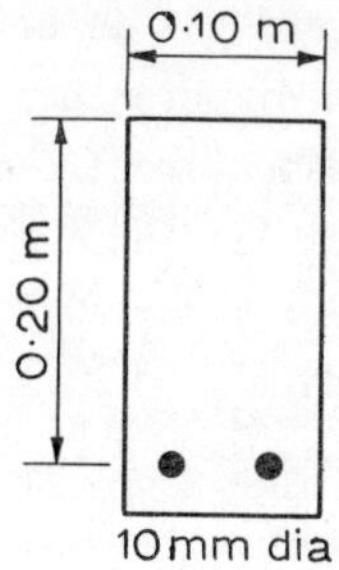

Fig. 3.2.2 Example 3.2.2

Solution For short-term loading, $E_c = 21.4 \times 10^9$ and $\alpha_e = 9.35$, as for Example 3.2.1.

$\rho = 10^{-6} \times 2 \times (\pi/4) \times 10^2/0.1 \times 0.2 = 0.785\%$

$100\alpha_e = 9.35 \times 0.785 = 7.34$

From Eqn (3.2.4):

$$d_n/d = [\sqrt{(1 + 200/7.34)} - 1] \times 7.34/100 = 0.317$$

$$z/d = 1 - 0.317/3 = 0.894$$

$$bd^2 = 0.1 \times 0.2^2 = 4 \times 10^{-3}$$

$$f_{dst} = 125 \times 10^6 \qquad f_{dcb} = 7.5 \times 10^6$$

Hence from Eqns (3.2.8) and (3.2.9),

$$M_{dt} = 125 \times 0.00785 \times 0.894 \times 4 \times 10^{-3+6} = 3.51 \times 10^3$$

$$M_{dc} = \tfrac{1}{2} \times 7.5 \times 0.317 \times 0.894 \times 4 \times 10^{-3+6} = 4.25 \times 10^3$$

Hence the steel is critical and the maximum moment is 3.51 kNm (Answer)

The maximum steel stress is

$f_{st} = f_{dst} = 125$ MN/m^2 (Answer)

The maximum concrete stress, f_{cb}, is given by Eqn (3.2.7). It is also given by

$$(M/M_{dc})f_{dcb} = (3.51/4.25) \times 7.5 \times 10^6$$

$$= 6.19 \text{ MN/m}^2$$ (Answer)

Example 3.2.3

A concrete beam is 100 mm wide and has an effective depth of 215 mm. It is reinforced by three 12-mm diameter bars, and carries a sustained applied moment of 7.543 kNm. Calculate the maximum stresses in the steel and concrete for long-term loading. What is special about this section if:

$f_{cu} = 22.5$ MN/m^2 $f_{yt} = 250$ MN/m^2?

Assume that

$f_{dcb} = \frac{1}{3}f_{cu}$ $\qquad f_{dst} = \frac{1}{2}f_{ty}$

Long-term $E_c = 10.7$ GN/m^2 $\quad E_s = 200$ GN/m^2

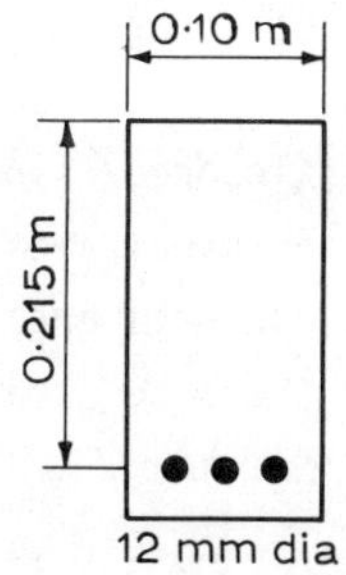

Fig.3.2.3 Example 3.2.3

Solution $\alpha_e = 18.7$, $\rho = 10^{-6} \times 3 \times (\pi/4) \times 12^2/(0.1 \times 0.215) = 1.58\%$

$100\alpha_e\rho = 18.7 \times 1.58 = 29.5$

Therefore, from Eqn (3.2.4),

$$d_n/d = [\sqrt{(1 + 200/29.5)} - 1] \times 29.5/100 = 0.528$$
$$z/d = 1 - 0.528/3 = 0.824$$
$$bd^2 = 0.1 \times 0.215^2 = 4.62 \times 10^{-3}$$

Therefore

$$7.543 \times 10^3 = f_{st} \times 0.0158 \times 4.62 \times 0.824 \times 10^{-3}$$
$$f_{st} = 125 \text{ MN/m}^2 \qquad \text{(Answer)}$$
$$7.543 \times 10^3 = \tfrac{1}{2}f_{cb} \times 4.62 \times 0.528 \times 0.824 \times 10^{-3}$$
$$f_{cb} = 7.5 \text{ MN/m}^2 \qquad \text{(Answer)}$$

This section is very special because the maximum steel and concrete stresses equal the permissible values simultaneously. In general, one of the permissible stresses is critical whilst the other is not.

3.3 THE EQUIVALENT CONCRETE SECTION

The analysis given in section 3.1 and 3.2 derives expressions for the position of the neutral axis by considering first principles— that is, the longitudinal equilibrium of forces, the strain profile geometry and Hooke's law. An alternative and quicker approach is to use the concept of the equivalent concrete section.

Equivalent concrete area

The equivalent concrete area is the actual cross-sectional area represented entirely in terms of concrete having the same flexural rigidity, (i.e. the same values of the product EI). The concrete area below the neutral axis is ignored, since the concrete there is assumed to have cracked, and the steel area is multiplied by α_e to convert it into an equivalent concrete area, since it carries α_e times the concrete stress for the same strain.

Position of neutral axis

The neutral axis passes through the centre of area of the equivalent concrete section. Its position can therefore be obtained by considering the moments of the equivalent concrete areas about the neutral axis. Hence from Fig. 3.3.1, b being constant,

$$b\int_{-d_n}^{0} y\,dy = \alpha_e A_s(d - d_n)$$

Alternatively,

$$\tfrac{1}{2}d_n^{\,2} = \alpha_e \rho d(d - d_n) \qquad (3.3.1)$$

Dividing by d and rearranging gives Eqn (3.2.3), and solving for d_n/d gives Eqn (3.2.4).

Equivalent second moment of area

If f_c and f_s are the concrete and steel stresses respectively at distance y from the neutral axis, and using Appendix Y, sign convention (1), then

$f_c/f_{cb} = y/d_n$ and $f_s/f_{cb} = \alpha_e y/d_n$

(Note: y is positive downwards, hence f_c is positive when y is negative, d_n is always considered positive.)

Considering elemental strips as shown in Fig. 3.3.1:

$\Delta F_c = f_c \Delta A_c = -y\Delta A_c f_{cb}/d_n$

$\Delta F_s = f_s \Delta A_s = \alpha_e y \Delta A_s f_{cb}/d_n$

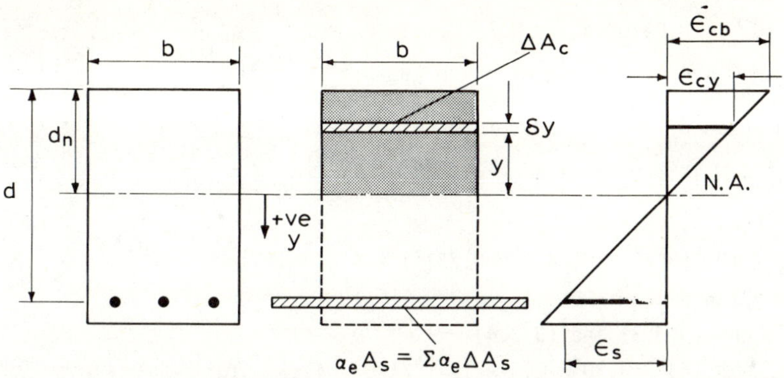

Fig.3.3.1 Cross-section for singly reinforced beam showing equivalent concrete section and strain profile

Taking moments about the neutral axis gives

$$M = [\sum y^2 \Delta A_c + \sum \alpha_e y^2 \Delta A_s] \times (f_{cb}/d_n)$$

The equivalent second moment of area of the section, I_e, is given by

$$I_e = \sum y^2 \Delta A_c + \sum \alpha_e y^2 \Delta A_s \qquad (3.3.2)$$

Hence

$$M/I_e = f_{cb}/d_n \qquad (3.3.3)$$

Similarly

$$M/I_e = f_{st}/\alpha_e(d - d_n) \qquad (3.3.4)$$

Eqns (3.3.4) and (3.3.3) correspond to Eqns (3.2.6) and (3.2.7) respectively. For the rectangular beam shown in Fig. 3.3.1:

$$I_e = bd_n{}^3/3 + \alpha A_{st}(d - d_n)^2$$
$$I_e/bd^3 = \tfrac{1}{3}(d_n/d)^3 + \alpha_e\rho(1 - d_n/d)^2 \qquad (3.3.5)$$

Example 3.3.1

Repeat Example 3.2.1(a) using the concept of the equivalent concrete section.

<u>Solution</u> $E_c = 21.4 \times 10^9$, $\alpha_e = 9.35$, $\rho = 0.503\%$ and $\alpha_e\rho = 0.0470$, as before.

From Eqn (3.3.1)

$$\tfrac{1}{2}d_n{}^2 = 0.0470d(d - d_n)$$

i.e. $\tfrac{1}{2}\delta_n{}^2 + 0.0470\delta_n - 0.0470 = 0$ where $\delta_n = d_n/d$

from which $\delta_n = 0.263$ as before.

$$bd^3 = 0.15 \times 0.30^3 = 4.05 \times 10^{-3} \text{ and } 1 - \delta_n = 0.737$$

From Eqn (3.3.5)

$$I_e = [0.263^3/3 + 0.0470 \times 0.737^2] \times 4.05 \times 10^{-3}$$
$$= 128 \times 10^{-6}$$

From Eqns (3.3.3) and (3.3.4)

$$f_{cb} = 0.263 \times 0.30 \times 7.5 \times 10^{3+6}/128 = 4.63 \times 10^6$$
$$= 4.63 \text{ MN/m}^2 \qquad \text{(Answer)}$$

$$f_{st} = 9.35 \times 0.737 \times 0.30 \times 7.5 \times 10^{3+6}/128 = 121 \times 10^6$$
$$= 121 \text{ MN/m}^2 \qquad \text{(Answer)}$$

Moments of resistance

The moments of resistance of the concrete and of the steel are obtained by substituting the permissible stresses into Eqn (3.3.3) and (3.3.4) and transposing as follows:

$$M_{dc} = f_{dcb}I_e/d_n \qquad (3.3.6)$$

$$M_{dt} = f_{dst}I_e/\alpha_e(d - d_n) \qquad (3.3.7)$$

Example 3.3.2

If the area of steel in the beam in Example 3.2.3 is doubled, calculate the percentage increase in sustained moment which the section could carry. Use the equivalent area method.

Solution $E_c = 10.7 \times 10^9$ and $\alpha_e = 18.7$ as before, but $\rho = 2 \times 1.58\%$

$$100\alpha_e\rho = 18.7 \times 3.16 = 59$$

Therefore from Eqn (3.2.4)

$$d_n/d = [\sqrt{(1 + 200/59)} - 1] \times 0.59 = 0.646$$

$$bd^3 = 0.1 \times 0.215^3 = 0.994 \times 10^{-3}$$

From Eqn (3.3.5)

$$I_e = [0.646^3/3 + 0.59(1 - 0.646)^2] \times 0.994 \times 10^{-3}$$
$$= 163 \times 10^{-6}$$

From Eqns(3.3.6) and (3.3.7)

$M_{dc} = 7.5 \times 163 \times 10^{6-6}/0.646 \times 0.215 = 8800$

$M_{dt} = 125 \times 163 \times 10^{6-6}/18.7 \times 0.354 \times 0.215 = 14320$

Hence M_{dc} is critical and the actual moment of resistance is 8.8×10^3. This represents a percentage increase of

$100 \times (8.8 - 7.543)/7.543 = 17\%$ (Answer)

Notice that the increase in M_{dc} due to the extra steel is relatively small since the tension steel can only have the secondary effect of increasing the neutral axis depth d_n. In these circumstances it is preferable to add compression steel as explained in the next section.

3.4 DOUBLY REINFORCED BEAMS AND SLABS IN FLEXURE

Doubly reinforced beams (and slabs) are reinforced in compression as well as in tension. The use of compression steel should generally be restricted as much as possible in practice since it can cause congestion and tends to increase the cost of the section for a given resistance moment. Its use is often unavoidable, however, and in such cases doubly reinforced beams are desirable or even essential. There is a limit to the amount of tension steel which can profitably be used in a singly reinforced beam.

Applications

Compression steel is justified when the dimensions of a beam are restricted due to requirements of headroom, space, appearance etc. If a high bending moment exists over a relatively short length of the beam only (e.g. over supports), then the cost of additional steel is small compared with the overall saving along the full length of a beam of constant cross-section.

Neutral axis depth

The equivalent concrete area for the doubly reinforced beam shown in Fig. 3.4.1 is given by

$A_e = bd_n + (\alpha_e - 1)A_{sc} + \alpha_e A_{st}$

Taking first moments of area about the neutral axis gives

$$\tfrac{1}{2}bd_n^{\,2} + (\alpha_e - 1)A_{sc}(d_n - d') = \alpha_e A_{st}(d - d_n) \tag{3.4.1}$$

This quadratic can be solved for d_n and corresponds to Eqn (3.3.1) for a singly reinforced beam.

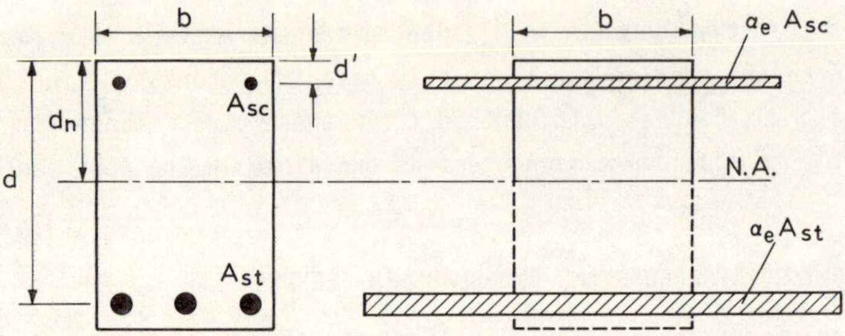

Fig.3.4.1 Cross-section for doubly reinforced beam showing equivalent concrete section

Concrete and steel stresses

The equivalent second moment of area about the neutral axis is given by

$$I_e = \tfrac{1}{3}bd_n^{\,3} + (\alpha_e - 1)A_{sc}(d_n - d')^2 + \alpha_e A_{st}(d - d_n)^2 \tag{3.4.2}$$

The stresses for the doubly reinforced beam are given by

$$f_{cb} = Md_n/I_e \tag{3.4.3}$$

$$f_{sc}\ (= -f_s) = \alpha_e M(d_n - d')/I_e \tag{3.4.4}$$

$$f_{st}\ (= +f_s) = \alpha_e M(d - d_n)/I_e \tag{3.4.5}$$

Notice the sign convention: the symbol f_{sc} specifically implies that the steel stress is compressive and that therefore compression is positive. Similarly, f_{st} specifically implies that the steel stress is tensile. However, if a reversal of stresses could occur then it would be preferable to consider the general case, for which steel stresses are assumed to be tensile. That is, to consider f_s, which can be either positive (tension) or negative (compression). The quantities α_e, d, d', d_n and I_e all have positive values. M is assumed to be a sagging (i.e. positive) moment for convenience. If M is actually a hogging moment (i.e. negative) the section must be inverted, i.e. tension steel at the top of the section and compression steel at the bottom as shown in Fig. 3.4.2.

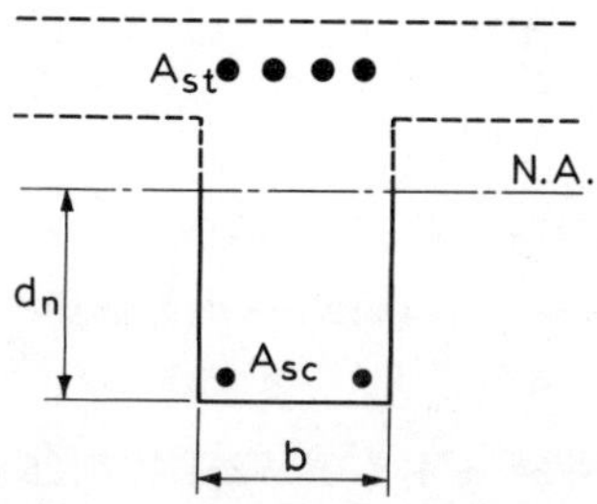

Fig.3.4.2 Doubly reinforced section at support

Note: presence of any slab at support section is immaterial since concrete shown in broken outline is in tension.

Moments of resistance

The moments of resistance can be obtained by transposing Eqns (3.4.3) and (3.4.5) and substituting into them the permissible values as follows:

$$M_{dc} = f_{dcb} I_e / d_n \qquad (3.4.6)$$

$$M_{dt} = f_{dst} I_e / \alpha_e (d - d_n) \qquad (3.4.7)$$

It is not necessary to consider Eqn (3.4.4) since the compression steel is unlikely to reach its permissible value before the concrete fibre stress reaches f_{dcb}. Notice also that Eqns (3.4.6) and (3.4.7) are exactly the same as Eqns (3.3.6) and (3.3.7) since the value for I_e allows for the compression steel when present. The equivalent area method is certainly the most convenient to use when dealing with other than singly reinforced sections.

Example 3.4.1

Calculate the maximum stresses in an RC beam, 200mm wide and of 500mm effective depth, having a tensile steel area of 1960 mm^2 and a compressive steel area of 628 mm^2 at a depth of 75 mm below the compressive face of the beam. The applied bending moment is 100 kNm. Assume that E_c = 10.7 GN/m² and E_s = 200 GN/m².

Solution

$\alpha_e = E_s/E_c = 200/10.7 = 18.7$

From Eqn (3.4.1)

$$\tfrac{1}{2} \times 0.20 d_n^2 + 17.7 \times 628(d_n - 0.075)10^{-6} - 18.7 \times 1960(0.50 - d_n) \times 10^{-6} = 0$$

$$d_n^2 + 0.478 d_n - 0.192 = 0$$

$$d_n = \tfrac{1}{2}[- 0.478 \pm \sqrt{(0.478^2 + 4 \times 0.192)}] = 0.26$$

(since d_n must be + ve). Therefore from Eqn (3.4.2)

$$I = \tfrac{1}{3} \times 0.20 \times 0.26^3 + 17.7 \times 628(0.26 - 0.075)^2 \times 10^{-6} + 18.7 \times 1960(0.50 - 0.26)^2 \times 10^{-6} = 3.67 \times 10^{-3}$$

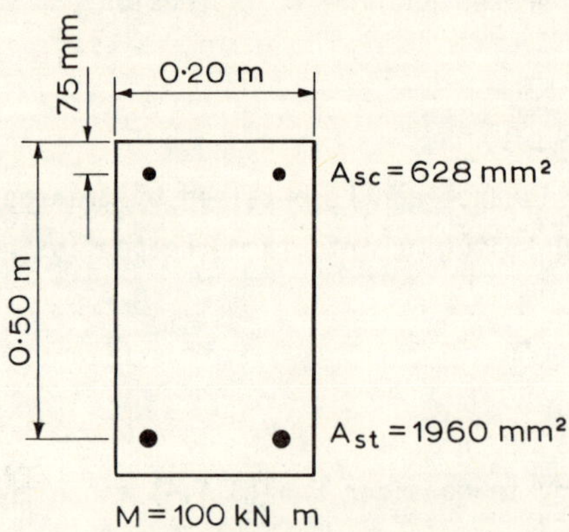

Fig.3.4.3 Example 3.4.1

Therefore

$$f_{cb} = \frac{0.26 \times 100 \times 10^{3+3}}{3.67} = 7.1 \text{ MN/m}^2 \quad \text{(Answer)}$$

$$f_{st} = \frac{18.7 \times 0.24 \times 100 \times 10^{3+3}}{3.67} = 122 \text{ MN/m} \quad \text{(Answer)}$$

$$f_{sc} = \frac{18.7(0.26 - 0.075) \times 100 \times 10^{3+3}}{3.67} = 94.5 \text{ MN/m}^2 \quad \text{(Answer)}$$

Example 3.4.2

Show that the neutral axis depth d_n of a doubly reinforced rectangular concrete beam for which $d' = d_n/3$ is given by

$$\delta_n/\alpha_e\rho = \sqrt{\left[\left(1 + \frac{2\rho'}{3\rho}\right)^2 + \frac{2}{\alpha_e\rho}\right]} - \left(1 + \frac{2\rho'}{3\rho}\right)$$

where $\delta_n = d_n/d$ and it is assumed that $(\alpha_e - 1)$ approximately equals α_e.

<u>Solution</u> Taking first moments of area about the neutral axis,

$$\tfrac{1}{2}bd_n^2 + (\alpha_e - 1)\rho' bd(1 - \tfrac{1}{3})d_n = \alpha_e \rho bd(d - d_n)$$

or, since $\alpha_e - 1 \simeq \alpha_e$ and if $\delta_n = d_n/d$

$$\tfrac{1}{2}\delta_n^2 + \alpha_e(\rho + \tfrac{2}{3}\rho')\delta_n - \alpha_e\rho = 0$$

$$\delta_n = -\alpha_e(\rho + \tfrac{2}{3}\rho') \pm \sqrt{[\alpha_e^2(\rho + \tfrac{2}{3}\rho')^2 + 2\alpha_e\rho]}$$

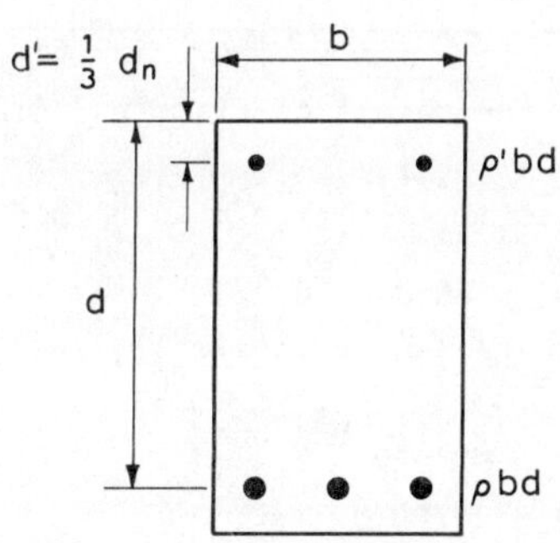

Fig.3.4.4 Example 3.4.2

Neglecting the negative root

$$\delta_n/\alpha_e\rho = \sqrt{\left[\left(1 + \frac{2\rho'}{3\rho}\right)^2 + \frac{2}{\alpha_e\rho}\right]} - \left(1 + \frac{2\rho'}{3\rho}\right) \qquad \text{(QED)}$$

Example 3.4.3

Use the result of Example 3.4.2 to calculate the moment of resistance of a reinforced concrete beam with a total steel area, $A_{st} + A_{sc}$, of

4000 mm² for the following values of ρ'/ρ: 0, 0.3 and 0.6. Assume that

$$f_{dst} = 125 \text{ MN/m}^2 \quad f_{dcb} = 7.5 \text{ MN/m}^2 \quad f_{dsc} = 110 \text{ MN/m}^2$$

$$\alpha_e = 18.7 \quad b = 0.30 \text{ m} \quad d = 0.45 \text{ m}$$

Solution

$$\rho + \rho' = 4000 \times 10^{-6}/0.30 \times 0.45 = 29.6 \times 10^{-3}$$

If $\rho' = k\rho$ then

$$\rho = 29.6 \times 10^{-3}/(1 + k)$$

The values are most conveniently tabulated as follows:

k	$\rho(\%)$	$\rho'(\%)$	$1+\frac{2}{3}k$	$(1+\frac{2}{3}k)^2$	$\alpha_e\rho$	$2/\alpha_e\rho$	$\sqrt{[(1+\frac{2}{3}k)^2+2/\alpha_e\rho]}$
0	2.96	0	1.0	1.00	0.553	3.61	2.15
0.3	2.275	0.685	1.2	1.44	0.425	4.70	2.48
0.6	1.85	1.11	1.4	1.96	0.346	5.78	2.78

δ_n is obtained using the result of Example 3.4.2.

$$I_e/bd^3 = \tfrac{1}{3}\delta_n{}^3 + (\alpha_e - 1)\rho'(\tfrac{2}{3}\delta_n)^2 + \alpha_e\rho(1 - \delta_n)^2$$

$$bd^2 = 0.30 \times 0.45^2 = 60.8 \times 10^{-3}$$

$$M_{dc} = (f_{dcb}/\delta_n)(I_e/bd^3)bd^2 = (456/\delta_n)(I_e/bd^3) \times 10^3$$

$$M_{dt} = [f_{dst}/\alpha_e(1 - \delta_n)](I_e/bd^3)bd^2$$

$$= [407/(1 - \delta_n)]I_e/bd^3) \times 10^3$$

Further tabulation can be made as follows, using units of 10^{-3} except where shown otherwise:

δ_n	$\frac{1}{3}\delta_n{}^3$	$(\alpha_e-1)\rho'(\frac{2}{3}\delta_n)^2$	$\alpha_e\rho(1-\delta_n)^2$	I_e/bd^3	M_{dc}	M_{dt}
636	85.7	0	73.3	159.0	114×10^3	178×10^3
544	53.6	15.9	88.4	157.9	132×10^3	141×10^3
478	36.4	20.0	94.3	150.7	144×10^3	118×10^3

Therefore M_d = 114, 132 and 118 kNm for ρ'/ρ = 0, 0.3 and 0.6 respectively. (Answer)

Notice that there is a certain optimum amount of compression reinforcement. This is considered further in chapter 4.

3.5 T-BEAMS IN FLEXURE

Reinforced concrete beams often carry a floor slab such that the whole construction is monolithic. Each beam can therefore include part of the slab in its structural strength and so produce a T-shaped section. Such beams are clearly very common in practice and even if a reinforced slab is not involved monolithically it is sometimes economical to use a T-beam, despite the additional shuttering costs.

Advantages

The relatively wide horizontal flange of concrete provides the resistance to compression where it is most effective, while the vertical rib gives adequate depth for the lever arm. A relatively narrow rib can contain sufficient steel to provide the tensile resistance. The wide flange can also be useful in providing increased lateral stability.

Compression flange

The breadth b of the flange which effectively acts with the rib and contributes to its strength in bending depends upon the T-beam span, slab thickness, rib breadth, and distance between adjacent T-beams. A certain minimum quantity of transverse reinforcement is also necessary in the flange. A combination of theory and practice has enabled the design rules given in Appendix CB to be formulated for both the flange breadth and the minimum amount of transverse reinforcement.

Example 3.5.1

A continuous slab, 150 mm deep, spans between beams at 750 mm centres as shown in Fig. 3.5.1. Calculate the effective width of the flange if the beams are simply supported on a span of 10 m and have a rib breadth of 180 mm. What is the minimum amount of transverse reinforcement required by CP 110 in the top of the flange? Is the

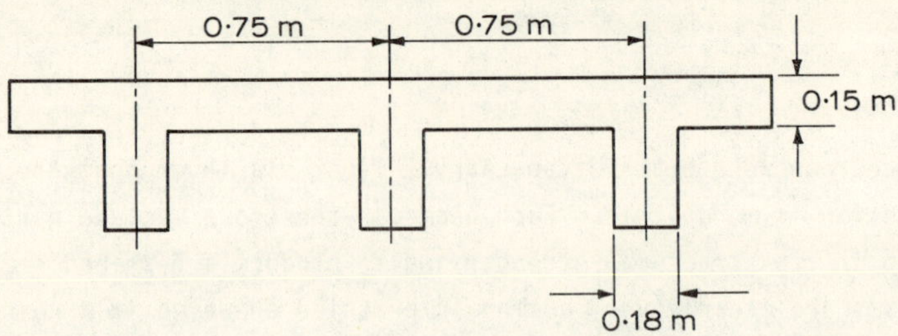

Fig.3.5.1 Example 3.5.1

slenderness of the beam satisfactory if the effective depth is 400 mm?

Solution From Appendix CB

$b \not> 0.750$ (i.e. actual flange width)

and

$b \not> 0.18 + 0.2 \times 10$ (i.e. web width + $\xi\psi$)

Hence

$b = 0.750$ (Answer)

Transverse reinforcement $\not< 0.3 \times 0.15 \times 1.0/100$

$\not< 450 \times 10^{-6}$

The minimum amount of transverse reinforcement is 450 mm² per metre run of beam. (Answer)

$Ld/b_c^2 = 10 \times 0.4/0.75^2 = 7.1$

$\ll$ maximum limit, i.e. 250

$L/b_c = 10/0.75 = 13.3$

$\not> 30$

so no additional reinforcement is required.

Deep flanges

Notice that when the neutral axis lies within the depth of the slab, the section resembles a rectangular beam of breadth b, since the concrete below the neutral axis is irrelevant to the analysis. Only the case where the neutral axis lies below the soffit of the slab need therefore be considered.

Neutral axis depth

The equivalent concrete area for the T-beam shown in Fig. 3.5.2 is given by

$$A_e = bd_n - (b - b_w)(d_n - h_f) + \alpha_e A_{st} \tag{3.5.1}$$

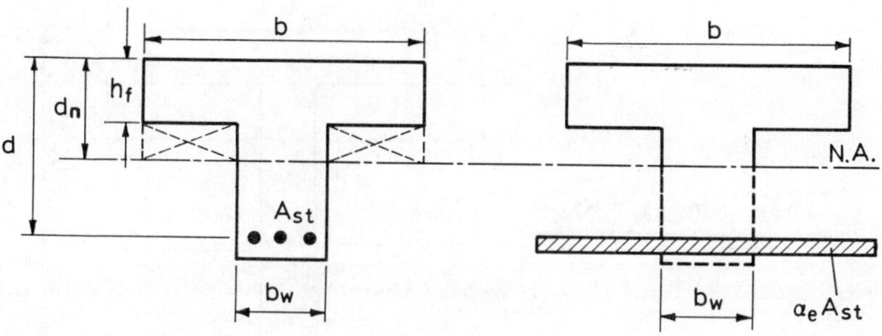

Fig.3.5.2 T-beams, (a) Actual section, (b) Equivalent concrete section

Taking first moments of the equivalent areas about the neutral axis gives

$$\tfrac{1}{2}bd_n{}^2 - \tfrac{1}{2}(b - b_w)(d_n - h_f)^2 = \alpha_e A_{st}(d - d_n)$$

or

$$b_w d_n{}^2 + 2[h_f(b - b_w) + \alpha_e A_{st}]d_n - (b - b_w)h_f{}^2 - 2\alpha_e/A_{st}d = 0 \tag{3.5.2}$$

This quadratic in d_n can be solved in the normal way.

Concrete and steel stresses

The equivalent second moment of area about the neutral axis for the T-beam is given by

$$I_e = \tfrac{1}{3}bd_n{}^3 - \tfrac{1}{3}(b - b_w)(d_n - h_f)^3 + \alpha_e A_{st}(d - d_n)^2 \tag{3.5.3}$$

Hence the concrete and steel stresses are again given by

$$f_{cb} = Md_n/I_e \quad \text{and} \quad f_{st} = \alpha_e M(d - d_n)/I_e$$

Example 3.5.2

A T-beam has a flange width of 750 mm, a rib width of 150 mm, a flange depth of 150 mm, an effective depth of 600 mm and a steel area of 5000 mm². If the beam is simply supported on a span of 10 m, and the total uniformly distributed load is 25 kN/m, calculate the maximum stresses in the beam assuming that $E_c = 10.7\ GN/m^2$ and $E_s = 200\ GN/m^2$.

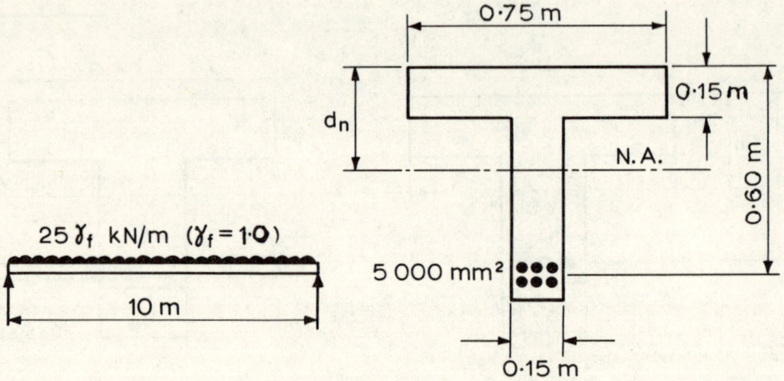

Fig.3.5.3 Example 3.5.2

Solution

$$\alpha_e = 200/10.7 = 18.7$$

The midspan moment is

$$wL^2/8 = (25 \times 10^2/8) \times 10^3 = 313 \times 10^3$$

Assume that $d_n > 0.15$ and take first moments of area about the neutral axis.

$$\tfrac{1}{2} \times 0.75d_n{}^2 - \tfrac{1}{2}(0.75 - 0.15)(d_n - 0.15)^2 = 18.7 \times 5(0.60 - d_n) \times 10^{-3}$$

$$0.15d_n{}^2 + 0.367d_n - 0.126 = 0$$

Hence $d_n = 0.307$ and $d_n > 0.15$ as assumed.

$$I_e = \tfrac{1}{3} \times 0.75d_n{}^3 - \tfrac{1}{3}0.60(d_n - 0.15)^3 + 18.7 \times 5(0.60 - d_n)^2 \times 10^{-3}$$

$$= 14.5 \times 10^{-3}$$

$$f_{cb} = 313 \times 0.307 \times 10^{3+3}/14.5 = 6.62\ MN/m^2 \qquad \text{(Answer)}$$

$$f_{st} = 18.7 \times 313 \times 0.293 \times 10^{3+3}/14.5 = 118\ MN/m^2 \qquad \text{(Answer)}$$

Compression steel

Although it is theoretically possible to have an undersize compression flange, it is preferable to increase the effective depth in such cases since compression steel should generally be avoided for the following reasons.

(i) The stress in the compression steel is low due to a fairly shallow neutral axis depth.

(ii) The compression steel must be tied by stirrups. This blocks the top of the rib more than usual and accentuates the difficulties in placing the concrete.

(iii) Even more compression steel than usual is required at the supports.

L-beams

An L-beam is similar to a T-beam except that the slab, and therefore the flange, is available on one side of the rib only: for example, an end bay. The analysis is similar so that the same formulae apply. The allowable breadth of the compression flange is again covered by arbitrary design rules as given in Appendix CB.

Torsion reinforcement

Since neither an L-beam nor its loading is symmetrical, a bending moment applied in a vertical plane produces torsion. This torsion is resisted by the floor slab and any beams which frame into the L-beam. The torsional resistance is greatly increased by providing top steel along the length of the L-beam, in conjunction with a fairly strong system of stirrups. L-beams are usually located at the outside of a building, so that the top steel is also very effective in resisting the stresses due to the high temperature gradient between the inside and outside of the building. Beams in torsion are considered in detail for the limit state of collapse.

Example 3.5.3

Calculate the central deflection for the T-beam in Example 3.5.2.

Solution One method of determining the elastic deflection is to consider the bending moment for the conjugate beam (i.e. the beam loaded with (1/EI) × bending moment diagram).

$$E_c I_e = 10.7 \times 14.5 \times 10^{3-3} = 155 \times 10^6$$

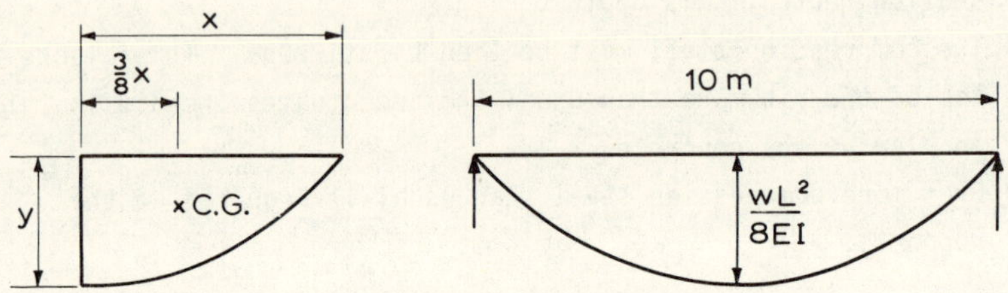

Fig.3.5.4 Example 3.5.3

The BM diagram is parabolic, hence the deflection at midspan is

$$\frac{2}{3} \cdot \tfrac{1}{2}L(wL^2/8EI)(L/2 - 3L/16) = 5wL^4/384EI = 21 \times 10^{-3}$$

i.e. the deflection is 21 mm (Answer)

Although the limit state of deflection is considered in detail later, notice how the concept of the equivalent concrete section can be very conveniently applied.

3.6 STEEL BEAM THEORY

The steel beam theory is a very approximate method of calculation which ignores the strength of the concrete and which can be useful because of its simplicity. It is used mainly as a rough check of the strength of an existing structure, where the reinforcement is known, or to give a rough estimate of the steel required for a job before the detailed design is available.

In the case of doubly reinforced beams having a very high proportion of compression steel it is possible for the steel beam theory to indicate a smaller quantity of steel than the elastic theory for a particular moment and size of section. However, its use other than as a quick guide is not to be recommended since collapse theory, which is considered later, is always preferable for accurate analysis or design.

Theory

The basis of the steel beam theory is that a beam having equal steel top and bottom, as shown in Fig. 3.6.1, may be treated as a steel beam, and that its strength may be calculated on the assumption that the concrete merely holds the steel apart. On this basis, the stress in the compression steel equals that in the tension steel. The strength of the concrete is entirely neglected and the lever arm is simply the distance the bars are apart irrespective of the steel ratio or grade of concrete.

Hence referring to Fig. 3.6.1,

$$M = A_{st}f_{st}(d - d') = A_{sc}f_{sc}(d - d') \tag{3.6.1}$$

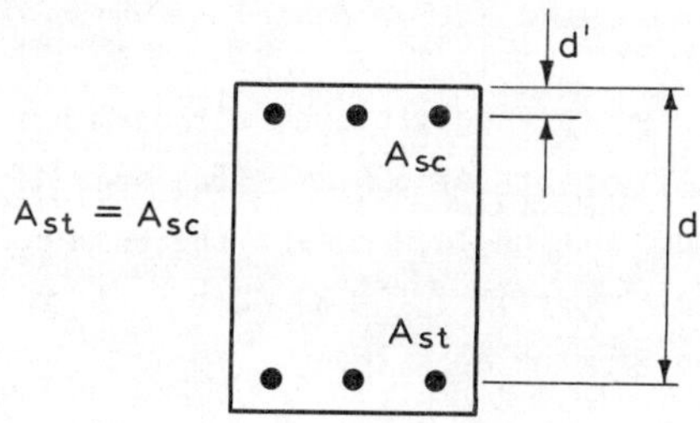

Fig.3.6.1 'Steel beam'

Example 3.6.1

Check the solutions to (a) Example 3.2.2 and (b) Example 3.3.2 assuming that the additional (imaginary) compression steel is 20 mm from the top surface. What are the percentage errors?

Solution

(a) $A_{st} = 2 \times (\pi/4) \times 10^2 \times 10^{-6} = 157 \times 10^{-6}$

$z = 180 \times 10^{-3}$ $f_{dst} = 125 \times 10^6$

Hence for steel-beam

$M_d = f_{dst}A_{st}z = 3.53 \times 10^3$

The percentage error is

$(3.53 - 3.51) \times 100/3.51 = +\ 1\%$ (Answer)

(b) $A_{st} = 6 \times (\pi/4) \times 12^2 \times 10^{-6} = 679 \times 10^{-6}$

Hence for steel-beam

$M_d = 125 \times 679 \times 195 \times 10^{6-6-3}$

$= 16.6 \times 10^3$

The percentage error is

$(16.6 - 8.8) \times 100/8.8 = +\ 89\%$ (Answer)

It is clear that the steel-beam theory can give a good check for well proportioned beams but can be hopelessly wrong where the concrete is critical, because the beam is over-reinforced (in tension).

Exercises

1. A singly reinforced rectangular concrete beam has a breadth of 300 mm and an effective depth of 600 mm. The area of steel is 1800 mm^2. Calculate the maximum working load moment the beam could sustain according to elastic theory if $f_{cu} = 30$ MN/m^2, $f_y = 250$ MN/m^2, $f_{dcb} = f_{cu}/3$, $f_{dst} = \frac{1}{2}f_y$ and $\alpha_e = 16$.

2. A concrete beam is 250 mm wide and has an effective depth of 500 mm. The tensile reinforcement consists of four 25-mm diameter mild steel bars. Calculate the maximum moment that the beam could carry under working load conditions (a) by elastic theory, (b) by collapse theory using the rectangular stress block. Assume that

$f_{cu} = 37.5$ MN/m^2 $f_y = 250$ MN/m^2

Assume that:

For elastic analysis $f_{dcb} = \frac{1}{3}f_{cu}$ $f_{dst} = \frac{1}{2}f_y$ $\alpha_e = 14.5$

For collapse analysis $\gamma_f = 1.6$

Maximum steel stress = $0.87f_y$

Mean concrete stress in the rectangular stress block = $0.40f_{cu}$

(UB2 1969, 40 min)

(Note: attempt part (b) only after chapter 8 has been studied.)

4 Elastic design of beams

Design is concerned with the selection of suitable values for b, d and A_{st}, so that the beam can resist a given bending moment. This is very different to a problem in analysis, where the section is completely defined and the required unknowns are concerned with the stresses in the concrete and steel. An infinite number of different sections can be chosen to resist a given moment, hence good design aims to select only those which are the most economical. This often involves making a further assumption in addition to those already made in the analysis.

The reader must distinguish clearly between analysis and design. A very common mistake among students who are presented with a problem in elastic analysis is to assume that a section has been designed in a special way when this need not be the case. A section is, in general, only at ESR (see below) *if it has been designed accordingly*.

4.1 EQUAL-STRENGTH STEEL RATIO

If the steel ratio is so adjusted that the steel and concrete stresses reach their permissible values simultaneously it is termed the equal-strength steel ratio, or ESR. A design is clearly efficient, theoretically, if it ensures that when the maximum bending moment is applied at a section both the steel and the concrete are stressed to their maximum permissible values. For such a design

$$f_{cb} = f_{dcb} \text{ when } f_{st} = f_{dst}$$

Neutral axis depth at ESR

If the steel and concrete stresses are both made equal to their respective permissible values, substitution in Eqn (3.1.2) gives

$$f_{st}/f_{cb} = f_{dst}/f_{dcb} = \alpha_e(d - d_n)/d_n$$

Rearranging and putting $\delta_n = d_n/d$

$$\delta_n = \frac{\alpha_e}{\alpha_e + f_{dst}/f_{dcb}} \tag{4.1.1}$$

Beam dimensions

The beam dimensions can be so adjusted that the maximum concrete stress just equals the permissible stress. If the beam is too large the concrete is understressed and this may be uneconomic. Conversely, if too small, the concrete is overstressed and this is not permissible. Similarly, if the tensile steel ratio exceeds the ESR value the steel is understressed, whilst if it is less than the ESR value the steel is overstressed.

ESR values

Equating the tensile and compressive forces, and assuming that the stresses in both the concrete and steel equal their respective permissible values,

$$f_{dst}A_{st} = \tfrac{1}{2}f_{dcb}bd_n$$

Rearranging gives the equal-strength (tensile) steel ratio, ρ_e, as follows:

$$\rho_e(= A_{st}/bd) = \tfrac{1}{2}\delta_n f_{dcb}/f_{dst} \tag{4.1.2}$$

Moments of resistance at ESR

The moment of resistance of the concrete for a beam at ESR is given by

$$M_{dc}/bd^2 = \tfrac{1}{2}f_{dcb}bd_n z/bd^2 = \tfrac{1}{2}f_{dcb}\delta_n\delta_z \tag{4.1.3}$$

The moment of resistance of the steel for a beam at ESR is given by

$$M_{dt}/bd^2 = f_{dst}\rho_e\delta_z \tag{4.1.4}$$

Notice that $M_{dc} = M_{dt}$ for an ESR beam (by definition). Notice also that Eqn (3.2.5) still applies and,

$$\delta_z(=z/d) = 1 - \tfrac{1}{3}\delta_n$$

Permissible stresses for elastic design

The elastic theory is concerned with stresses at working loads. It has overall safety factors which are applied to the material strengths in order to obtain the permissible stresses for the steel and the concrete. In terms of limit state theory this means that the partial safety factor for the loads is always unity. The overall (stress) safety factors previously used in conventional elastic design were 2 for the steel yield strength and 3 for the concrete cube strength (for the extreme compressive fibre stress in flexure). The permissible values were therefore given by

$$f_{dst} = \tfrac{1}{2}f_{yt} \text{ and } f_{dcb} = f_{cu}/3$$

Typical coefficients for ESR beams

Consider the typical values of $f_{yt} = 250$ MN/m² and $f_{cu} = 30$ MN/m². If it is assumed that

$$f_{dst} = \tfrac{1}{2}f_{yt} \quad f_{dcb} = f_{cu}/3$$

$$E_s = 200 \text{ GN/m}^2 \text{ and } E_c = 9100 \sqrt[3]{f_{cu}} \text{ (in MN/m}^2\text{ units)}$$

then: $f_{dst} = 125$ MN/m² $\quad f_{dcb} = 10$ MN/m² $\quad f_{dst}/f_{dcb} = 12.5$

$$E_c \simeq 9100 \times 30^{\frac{1}{3}} \text{ MN/m}^2 = 25 \text{ GN/m}^2$$

and $\alpha_e = 200/25 = 8$

These values can then be substituted into Eqns (4.1.1), (3.2.5), (4.1.2) and (4.1.3) or (4.1.4) to give $\delta_n, \delta_z, \rho_e$ and M_d/bd^2 as shown in Table 4.1.1(a). Values have been given for two grades of concrete, $f_{cu} = 20$ and 30 MN/m², as well as for the combinations of hot rolled high yield bars ($f_{yt} = 410$ MN/m²) with concrete grades 30 and 50.

Table 4.1.1 Typical elastic design coefficients for various modular ratios

f_{yt} (MN/m^2)	f_{cu} (MN/m^2)	E_c (GN/m^2)	α_e	f_{dst} (MN/m^2)	f_{dcb} (MN/m^2)	δ_n	δ_z	ρ_e (%)	M_d/bd^2 (MN/m^2)
250	20	25.0	8.00	125	6.7	0.299	0.900	0.80	0.90
250	30	28.0	7.14	125	10.0	0.364	0.879	1.46	1.60
410	30	28.0	7.14	205	10.0	0.258	0.914	0.63	1.18
410	50	34.0	5.88	205	16.7	0.323	0.892	1.31	2.40
(a) Short-term E_c									
250	20	12.5	16.0	125	6.7	0.460	0.847	1.23	1.30
250	30	14.0	14.3	125	10.0	0.534	0.822	2.13	2.19
410	30	14.0	14.3	205	10.0	0.411	0.863	1.00	1.77
410	50	17.0	11.8	205	16.7	0.490	0.837	1.99	3.41
(b) Long-term E_c(=0.5 × short-term E_c)									
250	20	13.3	15.0	125	6.7	0.444	0.852	1.19	1.26
250	30	13.3	15.0	125	10.0	0.545	0.818	2.18	2.23
410	30	13.3	15.0	205	10.0	0.423	0.859	1.03	1.82
410	50	13.3	15.0	205	16.7	0.549	0.817	2.23	3.74
(c) α_e = 15 (as in CP 114:1957)									

Coefficients for long-term loading

Concrete continues to strain with time if the loading is sustained. This means that for a given beam the concrete creeps under the working loads such that its effective 'elastic' modulus decreases and the modular ratio α_e increases. As α_e increases the area of concrete in compression increases, hence δ_n increases and $\delta_z(=1 - \delta_n/3)$ decreases. It follows that the concrete stress, f_{cb}, decreases with time and that the steel stress, f_{st}, increases with time. This is illustrated in Fig. 4.1.1 for the case f_{yt} = 250 MN/m^2 and f_{cu} = 22.5 MN/m^2 and where $f_{dst} = \frac{1}{2}f_{yt}$ and $f_{dcb} = \frac{1}{3}f_{cu}$ are stresses which are assumed to exist initially before any creep has taken place in the concrete. If the creep strain for long-term loading of the concrete is approximately equal in magnitude to the initial elastic strain, then the long-term modulus for concrete equals half the short-term modulus. Hence the modular ratio α_e for long-term loading is doubled. Table 4.1.1(b) shows typical values of the various coefficients assuming that α_e is twice the short-term value given in Table 4.1.1(a).

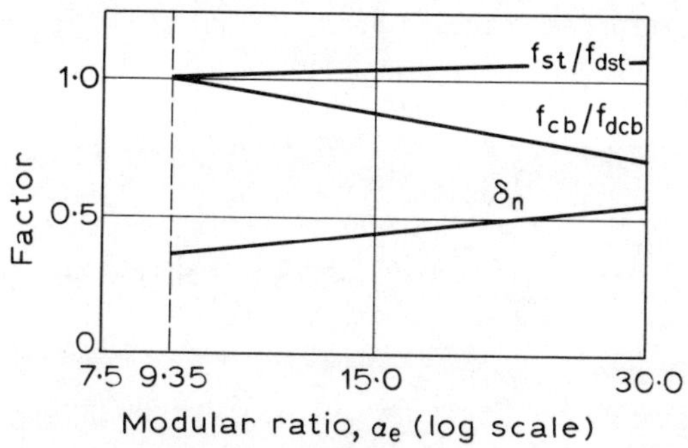

Fig. 4.1.1 Variations in factors for δ_n and the stresses as the modular ratio increases

Modular ratio given in CP 114:1957

The old code of practice assumed a constant value of 15 for the modular ratio, irrespective of the grade of concrete or duration of loading. The values obtained for a constant value of 15 are shown in Table 4.1.1(c). Comparison with Tables 4.1.1(a) and (b) shows that the old code corresponded approximately with the long-term conditions in many cases. This was intended to ensure that neither the (initial) concrete stress, nor the (final) steel stress, was too high. This over-simplification was possible because the overall safety factors were relatively large.

Timber and concrete

It is interesting to compare the moment of resistance of a timber beam with that for reinforced concrete. For a homogeneous rectangular beam

$$M_d = fZ = fbh^2/6$$

The permissible stress in bending for high-quality timber is often about 8 MN/m^2, hence

$$M_d = 1.33bh^2$$

A good-quality timber beam is therefore comparable with a singly reinforced concrete beam at ESR when $f_{dcb} \simeq 7$ MN/m^2.

Example 4.1.1

Design a singly reinforced concrete beam, 200 mm wide, to withstand a bending moment under working loads of 12.6 kNm. Assume that:

$f_{dst} = 125\ \text{MN/m}^2 \quad f_{dcb} = 6.7\ \text{MN/m}^2$ and $\alpha_e = 15$.

<u>Solution</u> From Table 4.1.1 $M_d = 1.26bd^2\ \text{MN/m}^2$. Therefore, for an ESR beam of breadth 0.2, the required d is given by

$$12.6 \times 10^3 = 1.26 \times 0.2d^2 \times 10^6$$

$$d = \sqrt{0.05} = 0.224$$

Assume a bar diameter $\not>$ 12 mm and a nominal cover of 20 mm. Hence let the overall depth be 250 mm and the effective depth be

$$0.250 - 0.020 - 0.012/2 = 0.224$$

The required A_{st} is given by

$$A_{st}(=M/f_{dst}z) = 12.6 \times 10^{3-6}/125 \times 0.852 \times 0.224$$
$$= 528 \times 10^{-6}$$

since $\delta_z = 0.852$ for ESR beam, from Table 4.1.1. Hence use seven 10-mm diameter bars (area = 550 mm²).

Use a 200-mm-wide × 250-mm-deep beam with seven 10-mm diameter bars at an effective depth of 224 mm. (Answer)

It is essential to appreciate both the problem and the method of solution illustrated by this example. The basic problem is to determine suitable values of d and A_{st} to resist a given bending moment, and the elastic design solution is obtained by using the ESR beam.

4.2 EFFECT OF VARYING THE STEEL RATIO

The previous section describes the special case when both the steel and the concrete are stressed to their full permissible values. Although it would appear that the most efficient elastic design is given by the ESR beam, this purely theoretical concept of an efficient section must often be modified in practice. It is not always practicable to design a section at approximately ESR. Simple shuttering necessitates a constant cross-section so that at positions along the beam where the applied moment is low the sections can have less steel. Conversely, the beam dimensions are usually limited by space considerations so that at some sections additional steel is necessary where the moment exceeds M_d for the ESR beam.

General values of the coefficients δ_n, δ_z and M_d/bd^2

It is clear that since the ESR section need not always be employed a knowledge of the effect of varying the steel ratio upon δ_n, δ_z and M_d/bd^2 for a beam of constant cross-section is essential for intelligent design. Table 4.2.1 gives values for the typical case when

f_{dst} = 125 MN/m² f_{dcb} = 7.5 MN/m²

and

E_c = 21.4 GN/m².

It is apparent that as ρ increases, δ_n also increases although δ_z decreases only slightly.

Table 4.2.1 Values of coefficients δ_n, δ_z and M_d/bd^2 for different steel ratios when f_{dcb} = 7.5 MN/m², f_{dst} = 125 MN/m² and α_e = 9.35 (short-term E_c = 21.4 GN/m²)

ρ	δ_n	δ_z	M_{dc}/bd^2	M_{dt}/bd^2	Critical stress
0.05	0.088	0.971	0.32	0.06	Steel
0.1	0.128	0.957	0.46	0.12	
0.2	0.176	0.941	0.62	0.23	
0.5	0.250	0.917	0.88	0.57	
1.075	0.359	0.880	1.19	1.19	Steel and concrete
1.5	0.408	0.864	1.32	1.62	Concrete
2.0	0.452	0.849	1.44	2.12	
5.0	0.588	0.804	1.77	5.03	

Variation in moment of resistance

Fig. 4.2.1 shows M_d/bd^2 plotted against ρ, and clearly indicates that the moment of resistance increases rapidly with increasing steel ratios up to the ESR but increases much more slowly above this value. The ESR therefore represents a critical value. High steel ratios tend to be

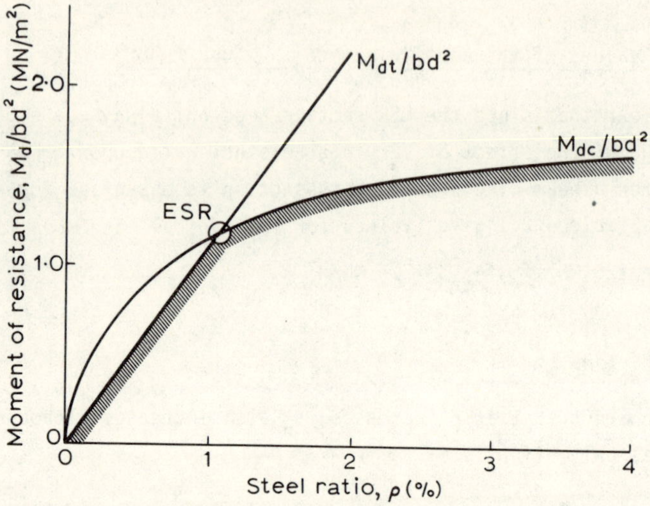

Fig. 4.2.1 Typical variations in moments of resistance with steel ratio (short-term E_c; f_{dcb} = 7.5 MN/m^2, f_{dst} = 125 MN/m^2, α_e = 9.35)

extravagant in steel, low ratios extravagant in concrete. Economical designs would therefore appear to be given by values which do not vary greatly from ESR.

Small variations from ESR

In practical beam design the calculated depths are rounded off to the nearest 10 mm. Similarly, the areas of reinforcement are increased slightly so as to provide a whole number of bars. If ρ varies slightly from the ESR value the effect on δ_z is very small and can be neglected. Even if ρ varies by 50% from the ESR value the change in δ_z is sufficiently small to be neglected for most purposes.

Large variations from ESR

It is sometimes impractical to design near the ESR. Some typical examples are as follows:

(i) Stiffness considerations may necessitate larger beam dimensions (e.g. to avoid excessive deflections).

(ii) At locations near the ends of beams, bars may have either been curtailed or bent up to provide for shear and reversed bending moments.

(iii) Different beams may have to possess the same dimensions, even though they are subjected to different maximum bending moments, for the sake of appearance or standardization of formwork.

It is preferable in such cases to use ρ values less than the ESR, in which case the moment of resistance for the steel is the ruling factor. A greater value is not normally desirable (chiefly because of collapse considerations) but in such cases the concrete becomes critical.

Example 4.2.1

Reconsider the beam which was designed in Example 4.1.1. The quantity of steel has been halved at a particular section nearer the supports since the applied moment is only 50% of the maximum (6.3 kNm). Calculate the actual moment of resistance of the section according to elastic theory and hence comment on the significance of the error for practical design purposes.

<u>Solution</u> 50% of theoretical area of steel = $\frac{1}{2} \times 528 = 264$ mm^2. The actual δ_n is given by Eqn (3.2.4).

$$\rho = 264 \times 10^{-6}/0.2 \times 0.224 = 0.59\% \quad \alpha_e\rho = 88.5 \times 10^{-3}$$

Therefore

$$\delta_n = 88.5 \times 10^{-3} \, [\sqrt{(1 + 2000/88.5)} - 1] = 0.341$$

Hence

$$\delta_z = 1 - 0.341/3 = 0.886$$

$$M_{dt} = 125 \times 264 \times (0.886 \times 0.224) = 6550$$

Hence the error is

$$(6.55 - 6.3)/6.3 = +\,4.0\%$$

The error is only 4% and is on the safe side. Hence the coefficients for the ESR beam can be used for practical design purposes even for a section where the steel ratio is only 50% of the ESR value.

(Answer)

4.3 SINGLY REINFORCED BEAM AND SLAB DESIGN

Floor and wall slabs are usually designed as rectangular beams reinforced in tension only. The following recommended procedure for designing a singly reinforced beam or slab indicates the factors which must be considered:

(a) Assume b, d and h. For a slab, consider a metre-wide strip, i.e. b = 1 m. Assume a maximum span/depth ratio (see Appendix CE).

(b) Calculate the dead and superimposed loading. If the load is uniformly distributed, determine the design load per metre run.

(c) Determine the effective span (see Appendix CB).

(d) Calculate the maximum bending moments.

(e) Determine M_d/bd^2 for the selected grade of concrete (for an ESR beam).

(f) Determine minimum d from $M_d = M$. Adjust assumed d (and hence h and therefore M) if necessary.

(g) Derive z from d.

(h) Calculate A_{st} (from $M = f_{dst}A_{st}z$). Check ξ for span/depth ratio in (a)

(i) Determine a suitable size and number of bars, checking that the bars fit into the assumed breadth of beam.

(j) Decide actual total depth, h, from minimum d, bar size and assumed cover. The depth is rounded off to the nearest 10 or 20 mm greater than the calculated value.

(k) If actual d differs appreciably from the assumed d in (a), repeat steps (a) to (j).

(l) In slabs only, decide size and spacing of distribution bars from $\rho \nless 0.12\%$ or 0.15% (see Appendix CD).

(m) Check for shear and bond and determine shear reinforcement.

Example 4.3.1

Design a floor slab to carry a superimposed loading of 4.0 kN/m^2 on a simple span of 4 m. Use concrete grade 30 MN/m^2, hot rolled deformed steel reinforcement (f_{yt} = 410 MN/m^2) and assume long-term loading. Fire resistance to be 2 hours.

Solution The design procedure is implemented as follows:

(a) Assume span/depth ratio $\ngtr 20 \times 1.2$ (Appendix CE: deflection)
Hence minimum effective depth = 4/24 = 0.167
Assume bar diameter $\ngtr$ 12 mm
Hence nominal cover $\nless$ 15 mm (Appendix CD) but use 20 mm for 2 hours fire resistance
Therefore minimum h = 167 + 20 + 12/2 = 193 mm
Hence try h = 200 mm

(b) Self-weight = $24 \times 0.200 \times 10^3 = 4800$ N/m^2
Therefore total loading for metre-wide strip = (4000 + 4800) = 8800 N/m.

(c) Effective span = 4.0 m

(d) Midspan moment = $1.0 \times 8800 \times 4^2/8 = 17\,600$ Nm (partial safety factor for load = 1.0 for elastic theory)

(e) $M_d/bd^2 = 1.77 \times 10^6$ for $f_{yt} = 410 \times 10^6$, $f_{cu} = 30 \times 10^6$ and $E_c = 14$ GN/m^2(Table 4.1.1)

(f) Hence $1.77 \times 10^6 bd^2 = 17600$, or
$d = 10^{-3}\ \sqrt{(17600/1.77)} = 0.100$
Assumed d of 0.174 is therefore satisfactory

(g) $\delta_z = 0.863$ (from Table 4.1.1) or slightly more since $\rho < \rho_e$
Therefore $z = 0.863 \times 0.174 = 0.150$

(h) $A_{st} = 17.6 \times 10^{3-6}/205 \times 0.150 = 572 \times 10^{-6}$ (or slightly less)

(i) Use 12 mm diameter at 200 mm spacing; area = 566×10^{-6} (Appendix G)

(j) Assumed h = 200 mm and $\xi = 1.2$ are satisfactory

(k) Could use 12 mm bars at 175 mm in 175 mm thick slab if preferred (since ξ is then 1.33)

(l) Distribution steel:
$\rho_c \nless 0.12\%$, hence $A_{st} \nless 0.12 \times b \times 0.200/100 = 240 \times 10^{-6}$
Use 8 mm diameter high-yield bars at 200 mm spacing; area = 252×10^{-6} (Appendix G), placed above main bars.

(m) Shear and bond are considered later in chapter 5. However, shear reinforcement is usually neither necessary nor desirable in slabs.

General design considerations

Notice that in design the load due to self-weight must be estimated before the bending moment can be determined. Notice also that the depth of the slab was taken on the generous side: a safe initial ξ-multiplier of 1.2 was chosen initially which was appreciably less than the actual value of 1.55. Furthermore, d = 0.174 is considerably greater than 0.100, the ESR value for a singly reinforced section. This ensured that the values assumed initially were found to be adequate subsequently, with the bars finally found to be necessary, etc. This approach, which is necessary in design if numerous attempts are to be avoided, can lead to uneconomic solutions if carried to extremes. The student must appreciate that meticulous accuracy in only one aspect of design is usually unjustified, for all manual solutions, since several aspects must generally be considered. For example, concrete cover, deflections and shear strength as well as flexural strength must be satisfactory.

The maximum span/depth ratio given in CP 114:1957 for this simply supported slab would have been $0.9 \times 30 = 27$, from which the minimum overall depth is $4/27 = 0.148$: less than the value which has been chosen. Although greater economies are usually possible with limit state concepts, in this particular instance continuing experience has indicated the need for greater caution.

4.4 DOUBLY REINFORCED BEAM DESIGN

The moment of resistance of a singly reinforced beam at ESR can be increased slightly by increasing the tensile steel, as shown previously. It is much better, however, to add compressive steel as well. If the size of a beam is restricted, or a very high bending moment occurs along a relatively short length of the beam, it is possible to achieve a satisfactory design by using a doubly reinforced section.

Low compression-steel stress

Doubly reinforced beams are normally designed so that the tension steel and the concrete are stressed to their permissible values, as in the case of singly reinforced beams. The compression steel, however, is only stressed to α_e × (adjacent concrete stress). For example, if

$$d' = d_n/3,\ \alpha_e = 15 \text{ and } f_{dcb} = 10 \text{ MN/m}^2$$

then f_{sc} is given by

$$f_{sc} = \tfrac{2}{3}\,\alpha_e f_{dcb} = 100 \text{ MN/m}^2$$

This is less than the permissible stress in tension for mild steel and only about half that for deformed bars. It cannot be increased without overstressing the concrete. However, despite the generally lower stresses in the compression steel, the total quantity of steel required in a singly reinforced beam almost always exceeds that for a doubly reinforced beam of the same size and moment of resistance. The only exception is when d' becomes very large and nearly equals d_n.

Design procedure for doubly reinforced beams

The procedure for practical design of doubly reinforced beams is therefore as follows:

(i) Determine M_d for a singly reinforced beam at ESR having the same dimensions and permissible stresses.

(ii) Subtract M_d for the ESR beam from the applied moment.

(iii) Add sufficient compression steel and additional tension steel to take the excess moment.

Notice that since the assumed stresses in the compression and tension steel are not the same, greater areas of compression steel are required in order that the additional compressive force should equal the additional tensile force. Notice also, however, that the areas of steel are added in just the right proportions to keep the position of the neutral axis constant.

Maximum quantity of steel

Theoretically, although it is extravagant in steel, the moment of resistance of a beam of given dimensions can be increased almost indefinitely by increasing both the tension and compression steel ratios. In practice, however, the more steel that is crowded into the concrete member, the more difficult it is to ensure good concreting and the higher the shear and local bond stresses. Difficulties in compacting the concrete between the mass of steel result in air pockets which reduce the effective bond between steel and concrete, as well as reducing the concrete strength. The total steel in doubly reinforced beams should therefore be restricted to about 8%, i.e. 4% compression steel ratio and 4% tension steel ratio (Appendix CD).

Moments of resistance

Eqns (3.4.6) and (3.4.7) are not suitable for design since I_e is not known until the section is known. It is therefore more convenient to obtain expressions for M_{dc} and M_{dt} in terms of the lever arm, z. It is also convenient to assume that $d' \simeq d_n/3$ so that the compression steel force coincides with the concrete force. That is

$$z \simeq d - d' \simeq d - d_n/3$$

The simplified expressions for the moments of resistance are then given by

$$M_c = f_{dcb}\left[\tfrac{1}{2}bd_n + 2A_{sc}(\alpha_e - 1)/3\right] z \tag{4.4.1}$$

$$M_{dt} = f_{dst}\,A_{st}z \tag{4.4.2}$$

Example 4.4.1

Design a doubly reinforced concrete beam to resist a bending moment of 132 kNm if the breadth and effective depth are restricted to 300 mm and 450 mm respectively. Assume that

$$f_{dst} = 125\ \text{MN/m}^2 \quad f_{dcb} = 8.33\ \text{MN/m}^2 \quad f_{dsc} = 110\ \text{MN/m}^2$$

$$\alpha_e = 15.4 \text{ and } d' = \tfrac{1}{3}d_n$$

Solution By calculation or interpolation from Table 4.1.1(b), M_d for a singly reinforced ESR beam is

$$1.75 \times 10^6 bd^2 = 1.75 \times 0.30 \times 0.45^2 \times 10^6$$

$$= 106 \times 10^3$$

Lever arm for ESR beam is

$0.831 \times 0.45 = 0.374$

Hence A_{st} for ESR beam is

$106 \times 10^{3-6}/125 \times 0.374 = 2267 \times 10^{-6}$

The additional moment is

$(132 - 106) \times 10^3 = 26 \times 10^3$

$f_{sc} = (2/3) \times 15.4 \times 8.33 \times 10^6 = 85.5 \times 10^6$

$< f_{dsc} = 110 \times 10^6$

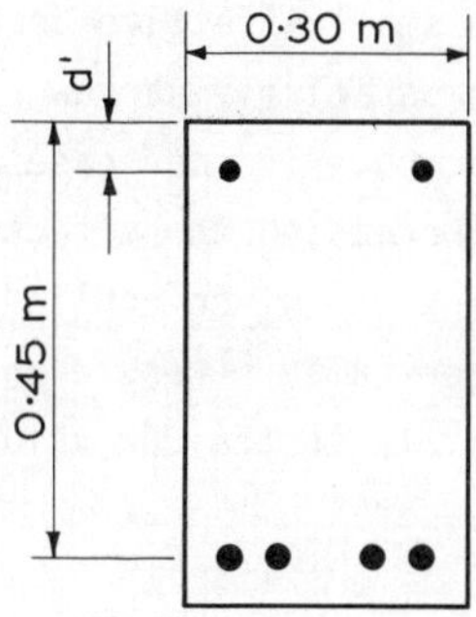

Fig.4.4.1 Example 4.4.1

Therefore

$A_{sc} = 26 \times 10^{3-6}/85.5 \times 0.374 = 813 \times 10^{-6} = 813\ \text{mm}^2$ (Answer)

since the compression steel force coincides with concrete force, i.e. $d' = \frac{1}{3}d_n$, and

$A_{sa} = 26 \times 10^{3-6}/125 \times 0.374 = 556 \times 10^{-6}$

Therefore

$A_{st}(\text{total}) = 2823\ \text{mm}^2$ (Answer)

Note: If $f_{dcb} = 7.5\ \text{MN/m}^2$ and $\alpha_e = 18.7$ then the answers would be:

$A_{sc}/A_{st} = 949/2850 = 0.333$ and $A_{st} + A_{sc} = 3849\ \text{mm}^2$.

Comparison with the answers to Example 3.4.3 shows that the design approach for doubly reinforced beams gives the optimum amount of compression reinforcement.

4.5 T-BEAM DESIGN

T-beams, unlike rectangular beams, are usually designed for steel ratios which are much lower than the ESR value. The large compressive flange area usually requires a relatively small effective depth in order to produce the required resistance moment in compression. A very substantial steel area, however, may be necessary to produce an equal tensile resistance moment for such a small effective depth. The steel may become overcrowded in the comparatively narrow rib, resulting in difficulties in concrete placing. In such cases it is preferable to increase the lever arm, by increasing the effective depth, so reducing the required area of steel for the given applied moment. This means that the concrete is no longer fully stressed and that in general the steel ratio does not approximate to the ESR value.

Assumptions

Since ESR conditions do not usually apply, it is necessary to reconsider the general analysis of T-beams and introduce simplifying assumptions as follows:

(i) The small compressive force due to the concrete area in the rib between the neutral axis and the soffit of the slab is negligible.

(ii) The additional lever arm due to the compressive force in the flange acting just above the mid-depth of the flange, instead of actually at the mid-depth, is also negligible.

Moment of resistance of the steel

The T-beam is usually considerably under-reinforced, hence the steel area is given simply by

$$M_{dt} = f_{dst}\, A_{st}(d - \tfrac{1}{2}h_f) \qquad (4.5.1)$$

Moment of resistance of the concrete

The strength of the concrete is usually more than adequate in the case of T-beams. However, if the moment is high then this may be checked from Eqn (4.5.3) which is derived as follows.

The average compressive stress in the flange is given by

$$f_{cb}(d_n - \tfrac{1}{2}h_f)/d_n$$

Equating longitudinal forces, for equilibrium of the section, gives

$$f_{st}A_{st} = f_{cb}(d_n - \tfrac{1}{2}h_f)bh_f/d_n$$

Substituting $f_{st}/f_{cb} = \alpha_e(d - d_n)/d_n$, putting $A_{st} = \rho bd$ and rearranging, gives

$$d_n = (\alpha_e\rho d^2 + \tfrac{1}{2}h_f^2)/(\alpha_e\rho d + h_f) \tag{4.5.2}$$

Once d_n is known, and since the lever arm is simply $(d - \tfrac{1}{2}h_f)$, then

$$M_{dc} = f_{dcb}\,\frac{d_n - \tfrac{1}{2}h_f}{d_n}\,bh_f(d - \tfrac{1}{2}h_f) \tag{4.5.3}$$

where $M_{dc} \gg M_{dt}$ generally.

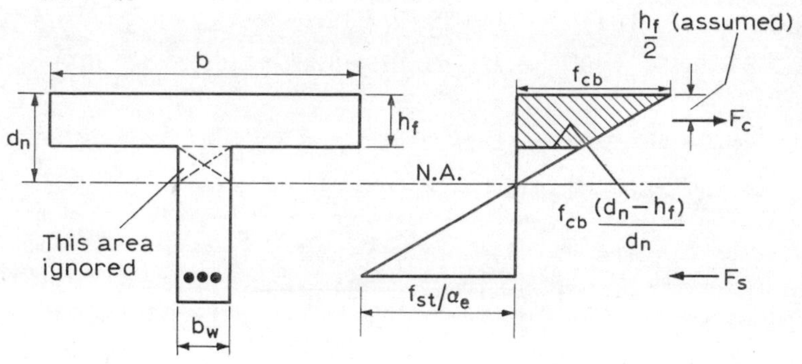

Fig.4.5.1 Cross-section and concrete stress profile for a typical T-beam

Example 4.5.1

Calculate the required reinforcement for a simply supported T-beam with a web (or rib) 275 mm wide and extending 450 mm below a 150 mm deep slab. The midspan moment = 200 kNm. The span is 5.4 m and the distance between centres of beams is 2.4 m. Assume an effective depth of 520 mm. Refer to Appendix CB for flange width and use the usual assumptions for elastic design. Let f_{dst} = 212 MN/m^2, f_{dcb} = 10.0 MN/m^2 and α_e = 16.2.

Solution From Appendix CB,

$$b \nmid 0.275 + 0.2 \times 5.4 = 1.355$$

Assuming that the section is under-reinforced, that the resultant concrete force acts at mid-depth of the slab and ignoring the concrete in the web:

$$A_{st} = 200 \times 10^{3-6}/212(0.520 - 0.075) = 2.12 \times 10^{-3}$$

$$= 2120 \text{ mm}^2 \qquad \text{(Answer)}$$

To check the concrete strength at ESR:

$$d_n/(d - d_n) = \alpha_e f_{dcb}/f_{dst} \text{ or } d_n = d(1 + f_{dst}/\alpha_e f_{dcb})^{-1}$$

$$f_{dst}/\alpha_c f_{dcb} = 212/162 = 1.31,$$

therefore

$$d_n = 0.520/2.31 = 0.225$$

Hence M_{dc} at ESR for the flange only is

$$1.355 \times 0.150 \times 10.0 [(0.225 - 0.075)/0.225] \times (0.520 - 0.075) \times 10^6$$

i.e.

$$M_d = 603 \times 10^3$$

This is more than adequate.

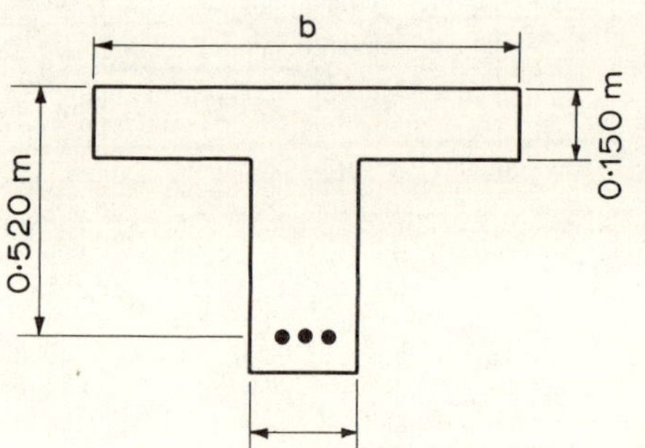

Fig.4.5.2 Example 4.5.1

4.6 CONTINUOUS BEAM AND SLAB DESIGN

Beams continuous over three or more supports frequently occur in floors where secondary beams frame into main beams or main beams frame into columns. Slabs spanning between T-beams are taken as continuous over the T-beam ribs. Similarly, bridge beams may span over several piers. When determining the distribution of moments it is often sufficiently accurate to assume that the beam or slab is free to rotate about the supports. That is, the stiffness of the column or any other member is ignored.

Theorem of three moments

A loaded continuous beam or slab takes a shape as shown in Fig. 4.6.1, the curvature changing in direction at points such as K, L, M and N. This change in curvature means that a negative (or hogging) bending moment occurs at the supports while a positive or sagging bending moment occurs near midspan.

The negative bending moments at the three supports A, B and C, can be obtained by using the theorem of three moments. This theorem can be expressed as follows for any two continuous spans on fixed supports:

$$\frac{M_A L_1}{I_1} + 2M_B\left(\frac{L_1}{I_1} + \frac{L_2}{I_2}\right) + \frac{M_C L_2}{I_2}$$

$$= -6\left[\frac{A_1\bar{x}_1}{I_1 L_1} + \frac{A_2(L_2 - \bar{x}_2)}{I_2 L_2}\right] \qquad (4.6.1)$$

I_1, I_2 are the moments of inertia of the beams in spans L_1, L_2; A_1, A_2 are the areas of the free bending moment diagrams for the loading on spans L_1, L_2 respectively; and $\bar{x}_1$, $\bar{x}_2$ are the distances of the centroids of A_1, A_2, respectively, from the left-hand supports.

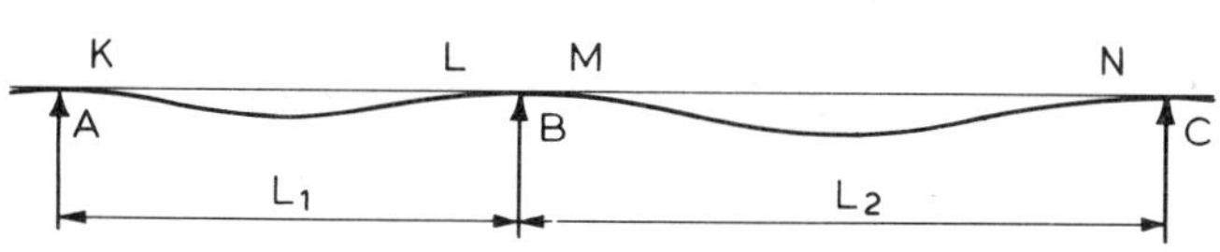

Fig.4.6.1 Deformed shape of loaded continuous beam

Uniform loading and sections

If the loading is uniformly distributed throughout and the moment of inertia of the beam is constant, then Eqn (4.6.1) reduces to

$$M_A L_1 + 2M_B(L_1 + L_2) + M_C L_2 = \tfrac{1}{4}w(L_1^3 + L_2^3) \qquad (4.6.2)$$

If the beam consists of two spans only, and both ends are free, then $M_A = M_C = 0$ and M_B can be calculated directly. If the beam spans across several supports, the theorem can be applied to groups of three adjacent supports in turn along the beam, and simultaneous equations obtained for all the support moments. Where all spans are equal then, by symmetry, pairs of support moments are equal. If span L_1 is loaded and L_2 unloaded then Eqn (4.6.2) becomes

$$M_A L_1 + 2M_B(L_1 + L_2) + M_C L_2 = \tfrac{1}{4}wL_1^3 \qquad (4.6.3)$$

Bending moment diagram

Consider a uniformly loaded beam continuous over two supports as shown in Fig. 4.6.2. Let all three spans be equal. The 'free' bending moment diagram is parabolic with a maximum value of $wL^2/8$. The resultant BMD is given by the difference between the negative BMD (given by the support moments) and the free BMD.

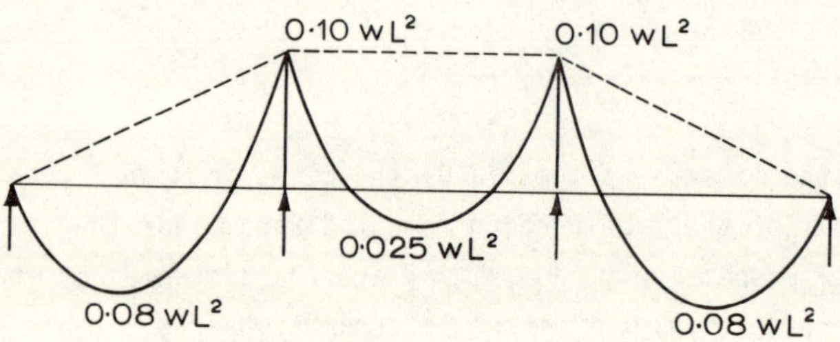

Fig.4.6.2 Bending moment diagram for a uniformly distributed load on beam with three equal spans

Maximum moment adjustment

It has long been recognized that the following factors tend, in practice, to modify the elastic support moments in an RC beam as follows:

(i) Concrete does not behave as a purely elastic material. This moderates the moments throughout a beam in such a way that high moments are reduced and low moments increased.

(ii) Supports have a finite width. This always reduces the support moment.

(iii) The beam is partially fixed at the supports, and is not completely free to rotate about them as is assumed in deriving the theorem of three moments. This effect, like (ii), always reduces the support moment.

Inelastic behaviour of concrete

Collapse load considerations for reinforced concrete indicate that very large plastic deformations, and therefore redistribution of moments, can occur before failure. A distribution of moments based

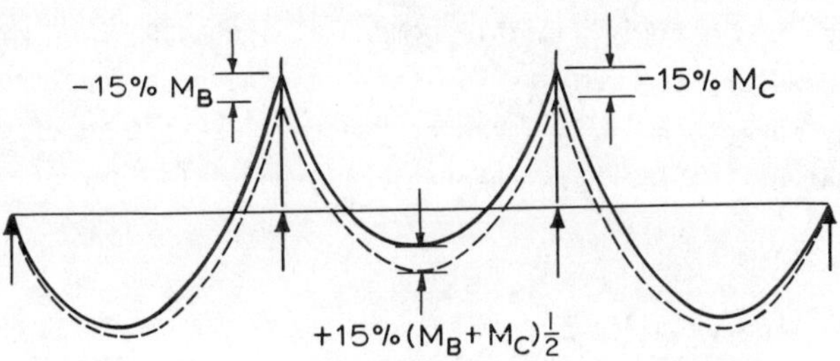

Fig.4.6.3 Bending moment diagram after 15% moment redistribution

solely on elastic theory can be misleading for conditions at failure. The old code, CP 114, recognized this fact and permitted a maximum adjustment allowance of 15% for moments based on elastic behaviour. However, if the maximum theoretical support moment is reduced by 15%, then the positive moments in the two adjacent spans must be increased accordingly, as shown in Fig. 4.6.3. Similarly, the support moments could be increased by 15% if the span moment is critical for a particular loading condition.

Loading conditions

The self-weight (or dead load) of a structure is such that all spans are always loaded. The superimposed (or live) load is such that it can vary in magnitude between zero and its maximum value at any location, and must therefore be considered separately. However, extreme bending moments are produced by entire spans being either fully loaded or fully unloaded. The most severe moments are assumed to be given by one of the following cases:

(i) Alternate spans loaded, all other spans unloaded

(ii) Any two adjacent spans loaded, all other spans unloaded.

Case (i) gives the highest midspan moments, while case (ii) gives the highest support moments. It should be noticed, however, that even higher support moments can be achieved for four or more spans, if, in addition to the two adjacent spans being loaded, alternate spans on both sides are also loaded (e.g. spans 1, 3, 4 and 6 for a 6-span beam produces the maximum possible moment at the support between spans 3 and 4). However, the moments for this case never exceed those for case (ii) by more than 4%, and can be ignored. CP 114 in effect allowed an adjustment of 19% for this theoretically possible form of loading.

Width of supports

If the continuous member frames into very massive supports—for example, a very thin slab resting on a very wide main beam—then it is preferable to assume that the slab is fixed-ended and not free to rotate about the support. However, it is far more usual to ignore any fixity at an intermediate support, whether it is a slab supported by beams or a beam framing into columns.

Bending moment coefficients

The different spans of a continuous beam or slab are frequently equal to within 15% of the longest span. If all spans are assumed equal, then a bending moment analysis assuming elastic distribution and free rotation about supports can be carried out for a given number of spans. Each support or midspan moment can therefore be represented by some coefficient multiplied by the total load on the appropriate span. Standard coefficients for a given number of approximately equal spans can therefore be tabulated for both dead and superimposed loads as shown in Appendix I, Tables 11 and 12.

The standard coefficients can be modified by making adjustments to the support moments. 10%, 20% and 30% (to CP 110 requirements) reductions have been included in Appendix I. The use of standard coefficients is very satisfactory for most purposes, except that it is usually advisable to allow for some fixity at the end supports (25% of the full fixed end moment is often assumed).

Elastic shear forces

The maximum shears for a continuous beam or slab differ slightly from the shear determined for a number of single separate spans. Standard coefficients can therefore be used as for bending moments, as shown in Table 6.3.2, which is sufficiently accurate for most purposes.

Distribution steel in slabs

If the ratio of the longer to the shorter side of a rectangular slab is greater than about 2½, the longitudinal bending is relatively small and the slab effectively spans in one direction only. Distribution steel, given by Appendix CD as $\rho_c \nless 0.12\%$, should always be provided for this nominal longitudinal bending. Also, since high concentrated loads can produce some bending in the longitudinal direction, it is wise to check that the distribution steel is also $\nless$ 10% of the main steel and that early thermal crack control is adequate.

End supports

Unlike the case for intermediate supports, fixity at end supports should be allowed for. The moments induced at the intermediate supports of a continuous beam of equal spans, when a unit moment is applied at the end support, are as given in Table 13. The moment induced over each support is found by multiplying the end moment by the appropriate coefficient. These values are then combined with negative moments obtained from Tables 11 and 12 to give the actual support moments. The maximum positive moments in the spans are not materially affected by partial fixing of the nominally free ends unless the end moment is large, as shown in Fig. 4.6.4, although the position is displaced slightly.

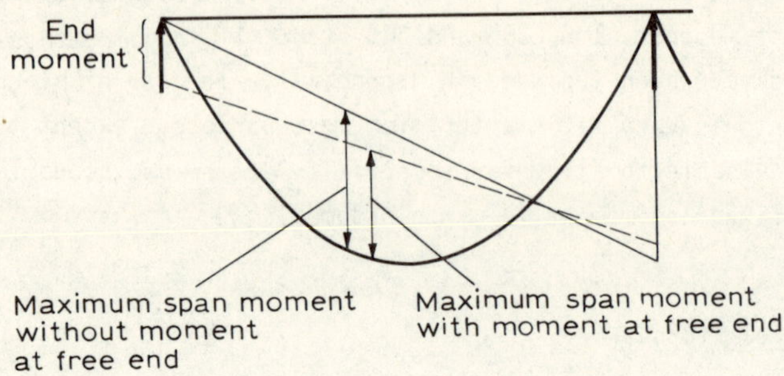

Fig.4.6.4 Effect of moment at free end

The end moment is calculated from the relative stiffnesses of the various members at the junction. This is considered in more detail in chapter 9 for beams framing into columns. If full end fixity can be assumed then the appropriate end moment for one end fixed can be obtained from Appendix L. Alternatively, an approximate probable value can be assumed (e.g. $wL^2/24$ is typical).

Continuous beam design

A beam spanning three or more columns or other supports is treated as a continuous beam. If it is a rectangular beam monolithic with a floor slab then the midspan and support sections are different. It is a T-beam section for positive bending moments near midspan and a rectangular section for negative bending moments at the supports. The top flange of the beam does not assist the resistance to bending if it is in tension (which it is over the supports of a continuous beam). Bond stresses should be checked for the main reinforcement and all beams must contain at least nominal shear reinforcement (see chapter 5).

Continuous slab design

A slab spanning three or more T-beam ribs or other supports is treated as a continuous slab and the bending moments and shears are calculated as for continuous beams. The negative moments over the supports can be provided for by either of the following methods:

(i) Tensile bars in the bottom of the slab, when no longer required for positive bending moment, are bent up and carried over the supports in the top of the slab as shown in Fig. 4.6.5(a). It is common practice to bend up half the number of tensile bars and carry the remaining half through to the support in the bottom half of the slab.

(ii) Two separate mats of reinforcement run continuously over all supports, one in the bottom and one in the top of the slab as shown in Fig. 4.6.5(b). The slab is, in fact, doubly reinforced. This method increases the steel cost but reduces bar bending. It is the most convenient method if the live load exceeds about twice the dead load, since negative moments can exist at midspan in this case. It is also better for early thermal crack control.

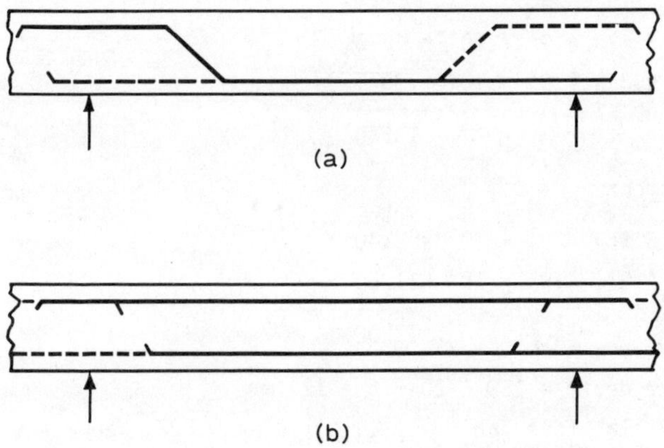

Note: For illustration purposes the bars in the alternate rows are shown as broken lines.

Fig.4.6.5 Alternative methods of reinforcing continuous slabs

Shear and bond stresses in continuous slabs

Shear reinforcement is usually neither necessary nor desirable in slabs, whether continuous or not. If a slab is not strong enough in shear (see chapter 5) the depth should be increased to obtain the required strength. Bond stresses (see chapter 5) should be checked as for a single span.

Example 4.6.1

A floor slab 150 mm deep overall carries a superimposed load of 4.8 kN/m^2 over continuous spans of 4.9 m.

If f_{dcb} = 10 MN/m^2 and f_{dst} = 125 MN/m^2, what areas of steel are required at midspan and supports for the central spans? Assume a maximum bar diameter of 12 mm and a nominal cover of 20 mm. Refer to Appendix I for the bending moment coefficients and assume 10% moment redistribution and long-term E_c to CP 110.

<u>Solution</u>

Self-weight of slab = 24 × 0.150 = 3.6 kN/m^2

Reference to Appendix I indicates coefficients of 0.054 and 0.071 for dead load, and 0.077 and 0.096 for superimposed load, at midspan and supports respectively.

Midspan moment (per metre width):

$$0.054 \times 3.6 \times 4.9^2 \times 10^3 = 4670$$
$$0.077 \times 4.8 \times 4.9^2 \times 10^3 = 8870$$
$$13540$$

Support moment:

$$0.071 \times 3.6 \times 4.9^2 \times 10^3 = 6140$$
$$0.096 \times 4.8 \times 4.9^2 \times 10^3 = 11060$$
$$17200$$

Effective depth = 150 − 20 − 6 = 124 mm

For an ESR slab (see Table 4.1.1.(b)), δ_z = 0.822, hence z = 102 mm

At midspan:

$$A_{st} = 13.54 \times 10^{3-6+3}/125 \times 102 = 1062 \times 10^{-6}$$
$$= 994 \text{ mm}^2$$ (Answer)

At support:

$$A_{st} = 17.20 \times 10^{3-6+3}/125 \times 102 = 1349 \times 10^{-6}$$
$$= 1349 \text{ mm}^2$$ (Answer)

Practical design †

Continuous floor beams in conventional beam and slab construction are usually designed as inverted doubly reinforced rectangular beams at the supports and as T beams in the spans. In practice, extensive use is often made of design charts; hence since those in Appendix E are for collapse the procedure is illustrated as for limit state design rather than elastic design. Thus for limit state design of continuous beams follow procedure (a) to (m) given in 4.3 except as amended below.

(b) (Continued) Apply partial safety factors for load as necessary. (e.g. 1.4 for dead load and 1.6 for superimposed load for maximum vertical loads at collapse.)

(d) (Continued) See Appendix I for continuous members

(e) Determine required value of M_u/bd^2 at support from (a) and (b).

(f) Select grades of concrete and steel and hence appropriate design chart (e.g. Fig.E2 for grade 20 concrete and grade 425 steel). Hence for M_u/bd^2 from (e), select suitable values of ρ' and ρ

(g) Calculate $A_{sc}(= \rho' bd^2)$ and A_{st} $(= \rho bd^2)$

(h) For T beam at mid-span determine M_u for span from (b). Calculate z $(= d - \frac{1}{2}h_f)$ and A_{st} $(= M_u/0.87f_{yt}z)$ Calculate flange breadth (Appendix CB) and check that compression reinforcement is unnecessary. Check ξ and ζ_2 for span/depth ratio in (a).

Exercises

1. What is meant by the equal strength steel ratio (ESR) for a reinforced concrete beam? Calculate the moment of resistance for a rectangular beam 300 mm wide and of 450 mm effective depth, at ESR, for f_{dst} = 205 MN/m², f_{dcb} = 12.5 MN/m² and α_e = 14.5. What is the ESR value for this beam?

2. One support section of the beam described in question 1 is subjected to a bending moment of 182 kNm. Determine the areas of main reinforcement required at this support section. Assume d' = 20 mm and f_{dsc} = 170 MN/m².

† Omit on first reading

5 Shear and bond

Shear forces must necessarily accompany any changes in bending moment which occur along the length of a beam. The effects of these shear forces include both diagonal tension and the formation of bond stresses between the reinforcement and the concrete. The behaviour of beams subjected to combined bending and shear is therefore much more complex than the behaviour of beams subjected to pure bending only. Elastic theory provides a useful introduction to the subject even though all the expressions for shear and bond given in CP 110 are based on the limit state of collapse.

5.1 ELASTIC SHEAR STRESS DISTRIBUTION

The vertical and horizontal shear stresses which exist in a reinforced concrete beam can be determined by using elastic theory assumptions. It is therefore convenient to reiterate some of the conventional results for homogeneous materials before applying elastic theory specifically to reinforced concrete beams.

Complementary shear stress

Consider a small cube of concrete in the web of a beam subjected to a vertical shear force. Let the cube sides be of length δx as shown in Fig. 5.1.1 and let the shear force, V, produce a shear stress, v,

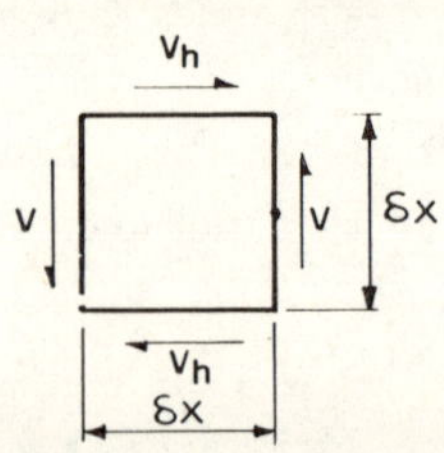

Fig.5.1.1 Complementary shear stresses

in the vertical plane. If there is no change in V over the length δx, the shear forces on opposite faces of the cube must be equal in magnitude and opposite in sign, and therefore form a couple. The shear force on each vertical face is $v(\delta x)^2$. To maintain equilibrium there must also be a pair of shear forces on the horizontal faces of the cube. If the shear stress on each horizontal face is v_h, the shear forces are given by $v_h(\delta x)^2$. Taking moments about one corner of the cube

$$v(\delta x)^2 \delta x = v_h(\delta x)^2 \delta x \quad \text{i.e. } v = v_h \tag{5.1.1}$$

That is, a vertical shear stress at any point produces an equal and opposite horizontal shear stress at that point.

Shear force and bending moment

Consider a length δx of a beam which is symmetrical about a vertical axis. Let the loading be w per unit length, and let the changes in shear and bending moment be δV and δM respectively, as shown in Fig. 5.1.2. Taking moments about the steel at the right-hand section (point Z), for the external forces acting on the short length of beam, gives

$$(M + \delta M) + \tfrac{1}{2}w(\delta x)^2 = M + V\delta x$$

Neglecting second-order terms in the limit gives

$$V = dM/dx \tag{5.1.2}$$

That is, shear force equals rate of change of bending moment: a well-known result.

General shear stress equation

The net horizontal force acting on the elemental strip δy (see Fig. 5.1.2(d)) due to the change in flexural strain along length δx of beam is given by

$$\delta F_{cy} = \delta f_{cy} b \delta y \tag{5.1.3}$$

Now the concrete stress due to bending at the level y in Fig. 5.1.2 is given by

$$f_{cy} = (M/I_e)y$$

where I_e is the second moment of area of the equivalent concrete section. Similarly, the change in stress at level y with change in

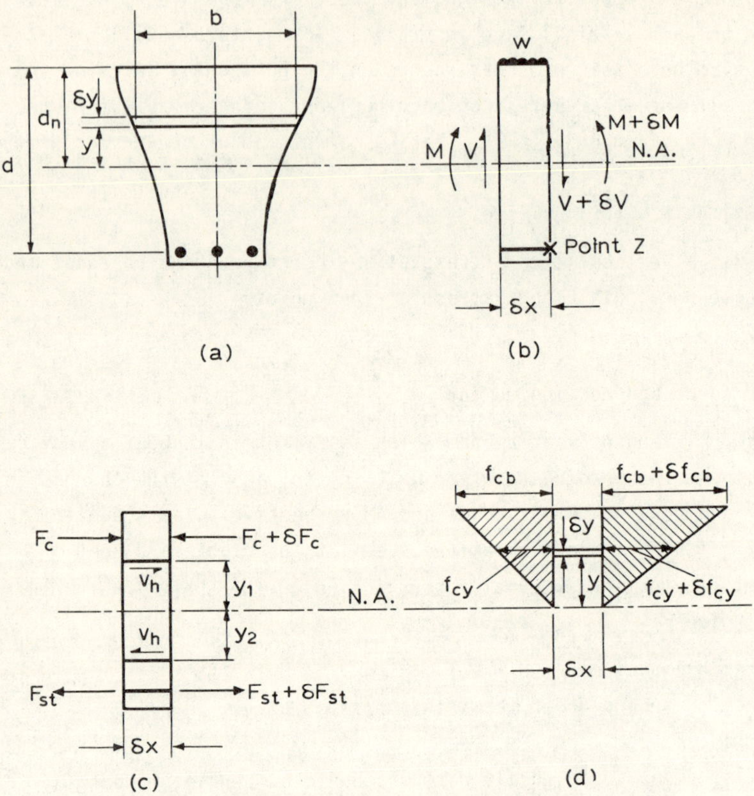

Fig.5.1.2 Shear stresses and normal stresses in singly reinforced concrete beam. (a) Cross-section. (b) External loads, bending moments and shear forces. (c) Internal forces and shear stresses. (d) Normal stresses in concrete.

moment is given by

$$\delta f_{cy} = (\delta M/I_e)y \tag{5.1.4}$$

Substituting for δf_{cy} from Eqn (5.1.4) and putting $\delta M = V\delta x$ in Eqn (5.1.3) enables δF_{cy} to be expressed in terms of V as follows:

$$\delta F_{cy} = (\delta M/I_e)by\delta y = (V\delta x/I_e)by\delta y \tag{5.1.5}$$

The equilibrium of the portion of the beam above level y_1 in Fig. 5.1.2(c) is given by summing the resultant horizontal forces on the strips above this level, and equating this force to the horizontal shear force at level y_1:

$$\sum_{y_1}^{d_n} \delta F_{cy} = v_h b_1 \delta x$$

where b_1 is the breadth at level y_1. Substituting for δF_{cy} from Eqn (5.1.5) gives

$$(V\delta x/I_e) \sum_{y_1}^{d_n} by\delta y = v_h b_1 \delta x$$

Putting $v_h = v$ and $\sum by\delta y = A\bar{y}$, where $A\bar{y}$ is the first moment of area of the portion above level y_1, gives

$$v = \frac{VA\bar{y}}{b_1 I_e} \qquad (5.1.6)$$

Rectangular section

Plotting Eqn (5.1.6) for a rectangular homogeneous section gives a parabolic shear stress distribution as shown in Fig. 5.1.3(a).

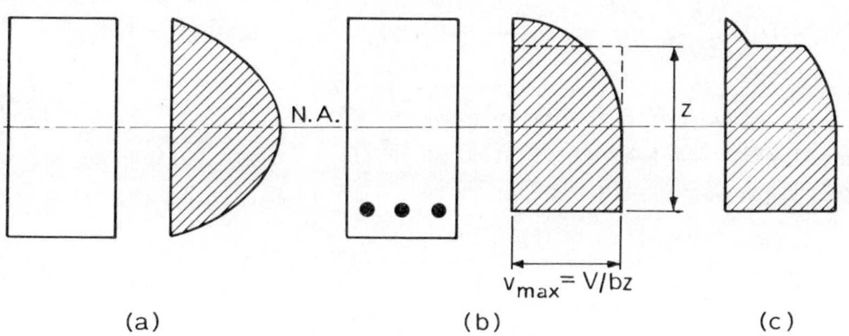

Fig.5.1.3 Elastic shear stress distributions in beams. (a) Homogeneous beam.(b) RC beam. (c) Doubly reinforced beam

In the case of reinforced concrete, the concrete is assumed to have cracked below the neutral axis, due to tensile bending stresses. Reference to Fig. 5.1.2(c) indicates that the resultant horizontal force in the steel is simply δF_{st}, so by considering the equilibrium of the portion below some level y_2:

$$F_{st} = v_h b_2 \delta x \quad (5.1.7)$$

Now $M = F_{st}z$, hence $\delta F_{st} = \delta M/z$. But $\delta M = V\delta x$ and $v = v_h$, hence

$$v = V/b_2 z \quad (5.1.8)$$

This equation is independent of y_2, so the shear stress below the neutral axis is constant for a constant beam breadth.

Maximum shear stress

The shear stress distribution above the neutral axis has been shown to be parabolic and given by Eqn (5.1.6). The maximum value is therefore at the neutral axis. For a rectangular beam

$$A\bar{y} = \tfrac{1}{2}bd_n^{\,2}$$

at the neutral axis, hence

$$v_{max} = Vd_n^{\,2}/2I_e \quad (5.1.9)$$

Now from Eqns(3.2.5), (3.2.7) and (3.3.3),

$$M = \tfrac{1}{2}f_{cb}bd_n z = f_{cb}I_e/d_n$$

Hence

$$bz = 2I_e/d_n^{\,2}$$

Substituting for d_n in Eqn (5.1.9) gives

$$v_{max} = V/bz \quad (5.1.10)$$

This value agrees with the value given by Eqn (5.1.8) for all positions at or below the neutral axis. The shear stress distribution for a singly reinforced beam is therefore as shown in Fig. 5.1.3(b). For a T- or L-beam, then, the equation becomes

$$v_{max} = V/b_w z \quad (5.1.11)$$

Difficulties of elastic theory

Equations (5.1.6) and (5.1.8) were derived by assuming that shear stresses act on the vertical faces considered in Fig. 5.1.2. However, it has already been shown that extensive vertical cracking of the concrete below the neutral axis must occur even at working loads. The shear must therefore be transferred to the reinforcement via the 'struts of the concrete between the cracks. The shear stresses, at least below the neutral axis, must be much less uniform and continuous than those indicated by elastic theory.

Nominal shear stress

In view of the above difficulties, Eqn (5.1.10) gives a misleading precision which is not justified. Furthermore, at or near to collapse, the distribution of stresses is affected by extensive cracking throughout the beam and other factors become significant; hence this value by itself can only give a very approximate indication of the strength. CP 110 has therefore discarded the apparently precise elastic value in favour of a 'nominal shear stress' given by

$$v = V/bd \qquad (5.1.12)$$

Example 5.1.1

A cantilever 150 mm wide has an effective depth of 250 mm and is reinforced with four 16-mm diameter bars. It carries a design shear force of 75 kN at the support. Calculate the nominal shear stress.

Solution The nominal shear stress is

$$V/bd = 75 \times 10^3/0.15 \times 0.25 = 2\ \text{MN/m}^2 \qquad \text{(Answer)}$$

Notice the simplicity of this solution. It would have been necessary to calculate the lever arm, z, in order to obtain the elastic shear stress value, V/bz.

5.2 DIAGONAL TENSION

One effect of shear forces is to produce tension across a diagonal plane. This diagonal tension, as it is called, is now considered in more detail and shown to be more significant than the shear stresses themselves.

Shear strength

Concrete is much stronger in compression and pure shear (sometimes called punching shear) than it is in tension. It follows, therefore, that the principal tensile stresses induced by the vertical and horizontal shear stresses are much more critical than the shear stresses themselves. Consider, for example, the beam distorted by the load W in Fig. 5.2.1. The effect of the load W is to shorten the diagonals BD and

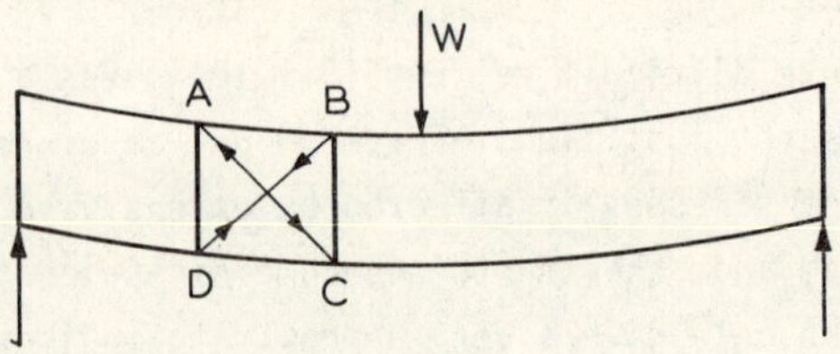

Fig.5.2.1 Diagonal tension and compression in deformed beam

to extend the diagonals AC: i.e. compressive and tensile stresses act normal to the 45^{0} planes indicated by AC and BD respectively.

Principal stresses

The maximum shear force in a simply supported beam occurs near the supports. The ends of such a beam are under tensile stress on one diagonal plane and under compressive stress on the other, which is at right angles to the former. The principal stress contours for a homogeneous beam are shown in Fig. 5.2.2. Notice that the bending

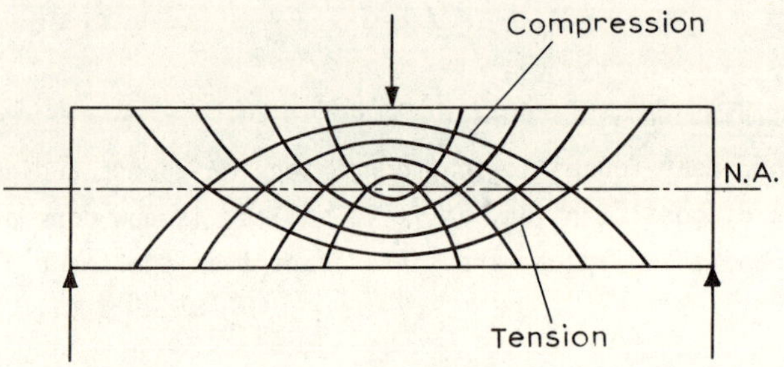

Fig.5.2.2 Principal stress contours for a homogeneous beam

stresses are a maximum at the extreme fibres and decrease to zero at the neutral axis, whereas the shear stress is a maximum at the neutral axis and decreases to zero at the extreme fibres. The maximum principal tensile stresses are given at the points of maximum shear and minimum compressive bending stresses, i.e. at or below the neutral axis.

Diagonal tension and compression

Consider small triangular portions of concrete at the neutral axis as shown in Fig.5.2.3. It is clear that the shear stresses on the faces

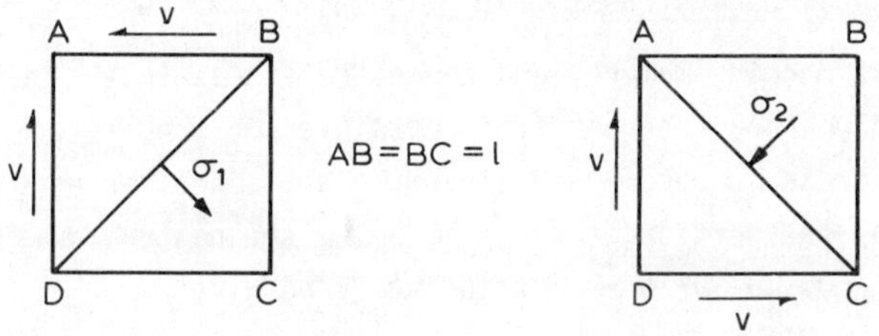

Fig.5.2.3 Principal stresses at neutral axis of beam

of the cube produce a principal tensile stress σ_1 on the diagonal plane BD given by

$$v l AB/\sqrt{2} + v l AD/\sqrt{2} = \sigma_1 l BD$$

Hence

$$\sigma_1 = v \qquad (5.2.1)$$

Similarly, the principal compressive stress σ_2 on the diagonal plane AC also equals v. However, since concrete is weak in tension it tends to fail across the plane BD. It is important to remember that the diagonal tension stress is equal in magnitude to the shear stress. Critical shear stresses really represent critical diagonal tension stresses of the same magnitude.

Notice also that the diagonal tension is the principal tensile stress at the level of the neutral axis. Above the neutral axis the principal tensile stress is reduced by the flexural compressive stress, and the plane on which it acts becomes flatter. Below the neutral axis the principal tensile stress is increased by the flexural tensile stress, and the plane on which it acts becomes more nearly vertical.

5.3 BOND BETWEEN CONCRETE AND STEEL

One of the basic assumptions made in the analysis of reinforced concrete is that good bond exists between the concrete and the reinforcement. Reinforced concrete members can act satisfactorily as composite units only if the bond between the concrete and the steel is adequate. Shear forces are present wherever the bending moment changes, and such changes imply a change in the tensile force in the main reinforcement. This in turn means that bond stresses must be present to transmit forces from the reinforcement to the concrete.

Local or flexural bond stress

If the applied bending moment, and hence the stress in the main steel, changes rapidly along the beam it is possible for the flexural (or local) bond stress to become undesirably high. Considering the equilibrium of the portion of the beam below the neutral axis in Fig.5.1.2(c) and *for collapse* putting $v = V/bd$ gives

$$\delta F_{st} = vb\delta x = V\delta x/d \qquad (5.3.1)$$

This horizontal shear stress produces the change in tension of the steel. It must therefore be transmitted into the steel by the action of the bond stress between the concrete and the steel. No bond stress problem arises in the upper compressive zone of the beam, although the horizontal shear stress acts between successive layers of the beam in a similar manner.

If f_{bs} is the local bond stress and $\sum u_s$ is the sum of the perimeters of the tensile reinforcement, then

$$\delta F_{st} = f_{bs} \sum u_s \delta x = V\delta x/d$$

Rearranging,

$$f_{bs} = V/\sum u_s d \qquad (5.3.2)$$

The maximum local bond stress values are shown in Table CD3 and are 20% to nearly 50% higher than the anchorage bond values for tension which are given in Table CD2.

Anchorage bond stress

The axial stress at the end of any bar must be zero. The stress at any intermediate section, distance L from the end, must be resisted by an anchorage (or average) bond stress f_b, between the concrete and steel, which acts over the length L. Equating the tensile force at the section to the bond force

$$F_{st} = f_b \pi \Phi L$$

where Φ is the bar diameter. The anchorage bond stress is therefore given by

$$f_b = F_{st}/\pi\Phi L = f_{st}\Phi/4L \qquad (5.3.3)$$

Development length

The minimum development length is given by rearranging Eqn (5.3.3) as follows

$$L_{min} = (f_{st}/4f_b)\Phi \qquad (5.3.4)$$

The development length is therefore given by the bar diameter multiplied by an appropriate coefficient. If

$$f_{st} = 250 \text{ MN/m}^2 \text{ and } f_b = 1.2 \text{ MN/m}^2 \text{ then}$$
$$L = 52\Phi$$

The maximum bond stress, and therefore the bond length, varies with the grade of concrete as shown in Table CD2. The bond can be increased by about 80% by indenting the surface (Type 2) or by about 40% by using a square twisted bar (Type 1) thanks to the additional mechanical bond so provided. Where the full bond distance cannot be provided, the assumed stress (tensile or compressive) in the bar must be reduced.

Example 5.3.1

Calculate the flexural bond stress for the main bars at the support section for the beam in Example 5.1.1. What bond length would be required at the support if the anchorage bond stress is not to exceed 1.4 MN/m² and the design stress in the bars is 217 MN/m² (i.e. 250/1.15)?

Solution

$$\sum u_s = 4 \times \pi \times 16 \times 10^{-3} = 0.201$$

$$f_{bs} = V/(d\sum u_s) = 75 \times 10^3/0.25 \times 0.201$$
$$= 1.49 \times 10^6$$
$$= 1.49 \text{ MN/m}^2 \qquad \text{(Answer)}$$

From Eqn (5.3.4)

$$L = 217 \times 16 \times 10^{6-6-3}/4 \times 1.4 = 0.620 = 620 \text{ mm} \qquad \text{(Answer)}$$

End anchorage

Consider a simply supported beam subjected to a single point load as shown in Fig. 5.3.1. The shear force between A and C is constant. If the perimeter of the tensile reinforcement is constant then the flexural bond stress is also constant. From the bending moment diagram the theoretical tensile stress in the steel is a maximum at midspan and zero at the supports A and B. It is clear that if the maximum flexural bond stress can exceed the maximum anchorage bond stress, the development length requirement is such that the bar must extend beyond the point of theoretical zero stress. This additional length of bar is known as the end length or end anchorage. The maximum required end length is given when the flexural bond stress is constant at the maximum value.

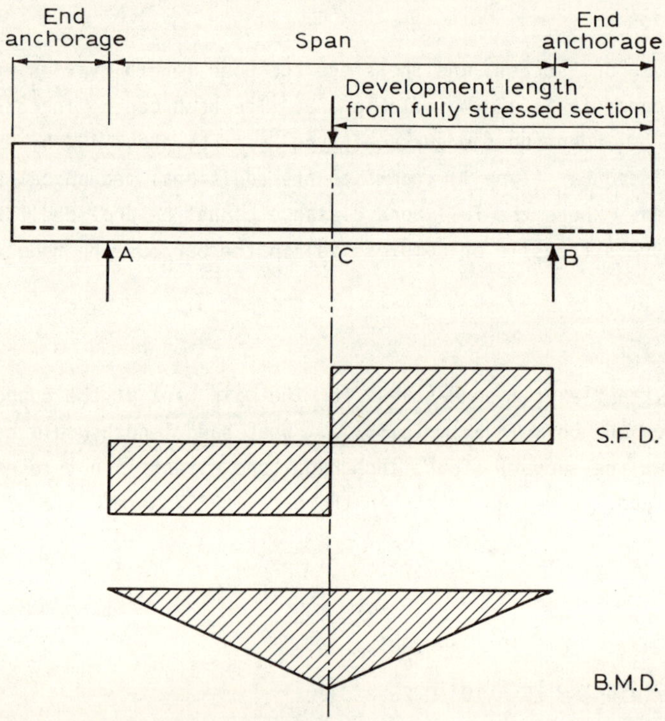

Fig.5.3.1 End anchorage for steel reinforcement

The minimum end anchorages for various grades of concrete and steel which are adequate for the worst possible cases are then given by Eqn (5.3.5) (below). Averaged values are shown in Table 5.3.1. It is therefore convenient (and advisable) to provide at least these end lengths in all cases and then simply check that the flexural bond stress is not exceeded at the most critical section (which is usually adjacent to a support, as explained in the next section).

$$L_a \nless {}_*f_{yt}(1/f_b - 1/f_{bs}) \times \tfrac{1}{4}\Phi \qquad (5.3.5)$$

CP 110 specifies that the minimum end anchorage should generally be not less than 12Φ, a value exceeded only once in Table 5.3.1. Anchorage lengths less than 12Φ should only be used if local bond stress does not exceed ½ maximum value (see Appendix CD).

Table 5.3.1 Minimum end lengths to ensure satisfactory development lengths in all cases

Characteristic cube strength:	20	25	30	≮40
Plain bars (250 MN/m^2)	13Φ	12Φ	12Φ	8Φ
Square twisted bars (425 MN/m^2)	11Φ	11Φ	9Φ	8Φ
Ribbed bars (425 MN/m)^2	6Φ	6Φ	5Φ	5Φ

It should be noted that the end anchorages for plain bars have been increased by up to 40% compared with the former CP 114 values. The values for square twisted and ribbed bars are generally slightly higher, and over 20% lower, respectively, than the CP 114 values.

Types of anchorage

The common forms of anchorage for both tension and compression bars are shown in Fig.5.3.2 for mild steel. The requirements for other angles or for high yield steel are included in Appendix CD.

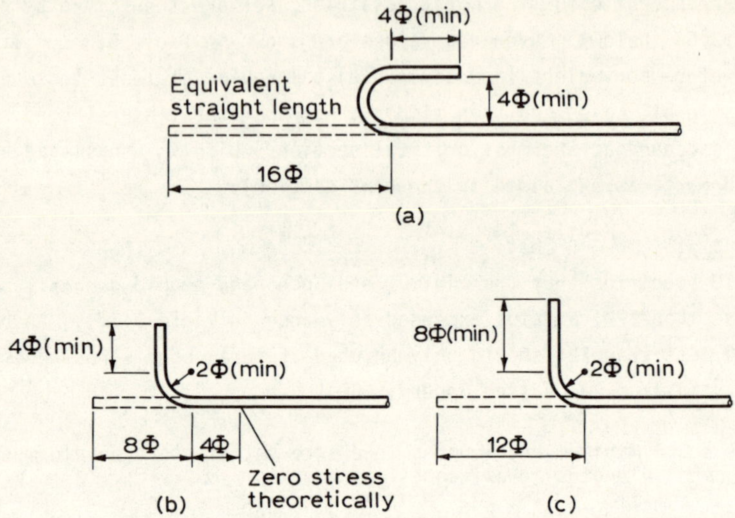

Fig.5.3.2 Typical anchorages for mild steel. (a) Standard U hook (tensile bars only). (b) Standard L hook. (c) L hook with equivalent straight length of 12Φ

Laps in bars

If the normal maximum length of a bar is exceeded, or if the ease of placing the steel so requires, one or more overlaps must be introduced. The stress in the bars at the lap must be transmitted from one bar to the other through the bond stress between the bars and the surrounding concrete. The lap length must therefore be at least equal to the required bond length for the smaller diameter bar. In addition it should be noted that the axial forces in the reinforcing bars at a lap are displaced slightly, which produces a lateral bursting force which could cause the concrete cover to break away. This must be prevented by the provision of stirrup reinforcement, which is considered later.

5.4 CURTAILMENT OF REINFORCEMENT

The section of a beam and the quantity of main reinforcement is determined from the maximum bending moment only. It is clear, however, that where the bending moment is considerably less than the maximum some of the bars are superfluous. Such bars can be curtailed if suitable end anchorage is provided beyond the point at which they can be cut off theoretically according to the bending moment diagram.

Use of bending moment diagram

The points at which main tensile bars are no longer required in the bottom of a beam can be obtained very conveniently from the bending moment diagram. At the points of curtailment of successive sets of bars the area of the remaining bars should provide sufficient moment of resistance to take the bending moment at that point. If the lever arm is assumed to remain constant as bars are curtailed, then the bending moment diagram can be divided into equal increments of the bending moment at midspan as shown in Fig. 5.4.1. This assumption is on the safe side since the lever arm increases as the steel ratio decreases for either elastic theory or ultimate load theory.

Maximum local or flexural bond stress

The flexural bond stress is a maximum at points of maximum shear (i.e. maximum slope of the BMD) combined with minimum tensile reinforcement. The maximum flexural bond stress for a simply supported beam therefore

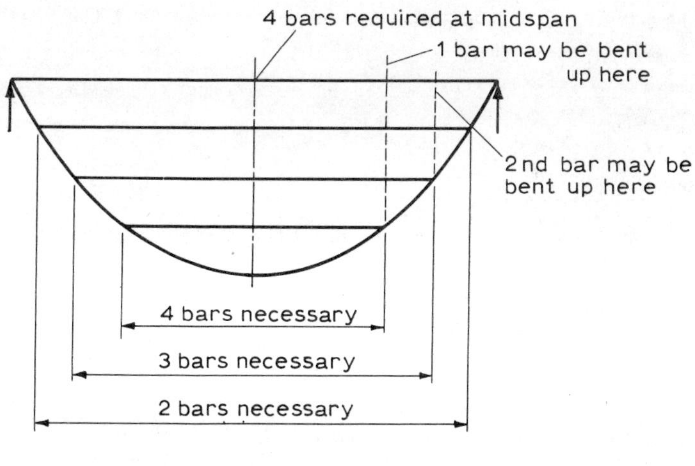

Fig.5.4.1 A simply supported beam with four bars for midspan moment

occurs at the supports. It is clear that if too many bars are curtailed, the value of $\sum u_s$ becomes too small and f_{bs} becomes excessive in Eqn (5.3.2). It is usually found that nearly 50% of the tensile bars need to be run through to the support in a simple beam and about 33% for a continuous beam. The simplified Code requirements (see next page) require not less than 50% and 30% respectively.

Standard parabolic BMD

The bending moment diagram for a uniformly loaded beam is parabolic. The points at which bars may be either curtailed or bent up for such a beam can therefore be obtained from a standard curve. Furthermore, if the negative moment at the support is assumed to be equal to the positive moment at midspan, this simple approach can be extended to the end and interior spans of a continuous beam, as shown in Appendix J.

Bent-up bars

In a continuous beam the maximum amounts of main steel are required in the bottom of the beam at midspan and in the top of the beam at the supports. It is therefore often convenient to bend up some of the bars from the bottom of the beam into the top. The number and spacing of bars available for bending up are determined from the bending moment diagram and the maximum flexural bond stress. The actual positions at which the bars are bent up usually differ slightly from the positions at which the bars first become available, in order to provide a more uniform spacing to suit shear strength requirements (see later).

Bond length—tension bars

The bond length must be considered for those bars running through the support in the bottom of a beam or slab, as well as those bars which are stopped short. Similarly, the curtailment of tensile bars running through the top of a continuous beam at the supports is also affected by the required bond length.

Bond length—compression bars

The considerations for compression bars in beams are similar to those for tension bars, except that the maximum anchorage bond stress in compression is about 25% greater than the tensile value (see Table CD2) since the lateral expansion, due to the Poisson's ratio effect, improves the bond for compression bars. Notice also that since the stresses in the reinforcement are lower in compression than in tension, the required anchorage lengths are appreciably lower for compression bars.

Stopped-off bars

Bars which are stopped off, rather than bent up, require an adequate end length beyond the points at which bars become available from the bending moment diagram.

Simplified code requirements for curtailment

The curtailment of reinforcement is governed by the bending moment diagrams and bond considerations as discussed above. CP 110 also permits, however, an alternative simplified approach which, for either beams or one-way slabs, merely requires compliance with one of the requirements given in Appendix CD.

Bond length—lightweight concrete

Bond strength varies according to the inclination of the reinforcement, for all concretes, due to internal water-gain effects [5.1]. At present, CP 110 only differentiates between horizontal and vertical bars for lightweight concrete (Appendix CL). BS 5337 applies a reduction factor of 0.7 for horizontal bars which are in sections in direct tension (Appendix CW).

5.5 BEAMS WITHOUT SHEAR REINFORCEMENT

It will have been noticed that the maximum bond stresses given by CP 110 are in terms of collapse values rather than permissible values. The same bond lengths could equally well have been obtained by considering permissible stresses in bond and tension, so the difference is largely immaterial in this instance. However, the shear strengths of beams according to elastic theory and collapse theory cannot be made directly comparable. This is largely because some factors which affect the collapse strength are ignored completely by elastic theory. All the expressions for shear strength are therefore in terms of the limit state of collapse.

Mutual effect of stresses

It is clear that a beam fails, in general, due to the combined effects of bending and shear. This should be allowed for if these effects are arbitrarily separated by considering shear stresses and bending stresses individually. There is no doubt that the presence of shear stresses can decrease the ultimate moment of resistance of a beam but the problem is very complex. However, it has been shown experimentally that within limits the effect of bending stresses on the shear strength of a section is sufficiently small to enable this effect to be virtually ignored in normal design. For example, Taylor has shown [5.2] that when the maximum applied moment varied between 50% and 100% of the nominal flexural strength of the member, the shear strength only decreased by about 10%.

Quantity of main steel

The nominal shear strength increases with the quantity of longitudinal reinforcement, irrespective of type. It follows, therefore, that beams with ordinary mild steel bars can sustain slightly higher shears than corresponding beams with an equivalent (that is, reduced) quantity of high-yield steel bars. This increase in strength is primarily due to dowel action of the bars which is considered later.

Concrete strength

The nominal shear strength increases with the cube strength, but at a reduced rate. Based on the January 1969 report of the Shear Study Group of the Institution of Structural Engineers, CP 110 gives values for the critical shear stress, v_c which allow for different quantities of main steel and different cube strengths of concrete as shown in Table CC1 in Appendix C. v_c is the maximum nominal shear stress (or diagonal tension stress) for unreinforced beams which can be taken by the concrete section without shear reinforcement when the member is subjected to the maximum design loads. That is

$$v_c = V/bd$$

where V is the design shear force for collapse, given by $\gamma_f V_k$

Partial safety factors for shear

Shear-compression failures of individual members cannot be predicted with the same precision as a flexural failure, hence the safety margin against this type of failure should err on the high side. Again, shear failures are brittle and tend to occur suddenly, whereas an under-reinforced beam gives ample warning of imminent flexural failure. Reference has been made previously to the Russian Code which has a partial safety factor for 'service' which can allow for these considerations. The American Code also has a factor which can allow specifically for deficiencies in knowledge or special considerations in service: that is, the particular location of the member or the particular stresses being considered. CP 110 reduces the possibility of a shear failure by increasing the partial safety factor for material strength of concrete when subjected to shear. This increased value of γ_m (i.e. $\gamma_{f_3}\gamma_{c_1}$) is already included in Table CC1 for critical shear stress in order to avoid confusion with the values of γ_m used for other calculations such as flexure.

Example 5.5.1

A beam 150 mm wide has an effective depth of 250 mm and a steel area of 924 mm². It carries a design shear force of 30 kN. Is shear reinforcement required if $f_{cu} = 25$ MN/m²? Refer to Table CC1.

Solution The steel ratio is

$$\rho = 924/150 \times 250 = 2.46\%$$

Hence from Table CC1, for $f_{cu} = 25$ MN/m², $v_c = 0.87 \times 10^6$,

$$v = 30 \times 10^3/0.15 \times 0.25 = 0.8 \times 10^6$$

No shear reinforcement is required, theoretically, since $v_c > v$. However, nominal stirrups should be used, as discussed later.

5.6 SHEAR REINFORCEMENT FOR DIAGONAL TENSION

It has been shown already that the diagonal tension which results from shear forces is much more critical for the concrete than the direct effect of the shear itself. If the critical shear stress values (which are really critical 'diagonal tension' stress values) are exceeded, shear reinforcement must be provided to take the excess tension.

Types of shear reinforcement

Ideally, the tensile reinforcement in a beam should follow the lines of principal tension shown in Fig. 5.2.2. However, since this is impractical for normal RC beams, the two main types of shear reinforcement are as follows:

(i) *Links or stirrups* Vertical stirrups are the most usual form of shear reinforcement. They are relatively light bars which are bent into a rectangular shape so as to pass round the tensile reinforcement in the bottom of the beam as well as the compression reinforcement in the top. Where no compression reinforcement is present, top bars of light section, termed hanger bars, should generally be provided. The diameter of such bars is usually governed by the distance they must 'hang' while the steel is being fixed. The terms 'links' and 'stirrups' are often interchangeable. However, in the present text links will imply closed ties (i.e. Fig. 5.6. 7(b) and (c) only) whereas stirrups may b either open or closed.

(ii) *Bent-up bars* It is often possible to bend up the main tensile bars, from the tensile flange into the compression flange, near the ends of the beam where the shear is large and the bending moment is reduced. The angle of bend is normally 45° to the line of the main bars. The bars must be bent up through a depth equal to, or greater than, the spacing. Bent-up bars can produce excessive compression and hence bursting of the concrete near the bends which could only be restrained adequately by providing ties at appropriate heights across the legs of the stirrups. This is usually impracticable. Hence bent-up bars should only be used in moderation and only in association with stirrups.

Vertical links or stirrups

Vertical stirrups are commonly used as shear reinforcement, either alone or in conjunction with bent-up bars. A vertical stirrup system of pitch (or spacing), s is shown in Fig. 5.6.1. If it is assumed that the

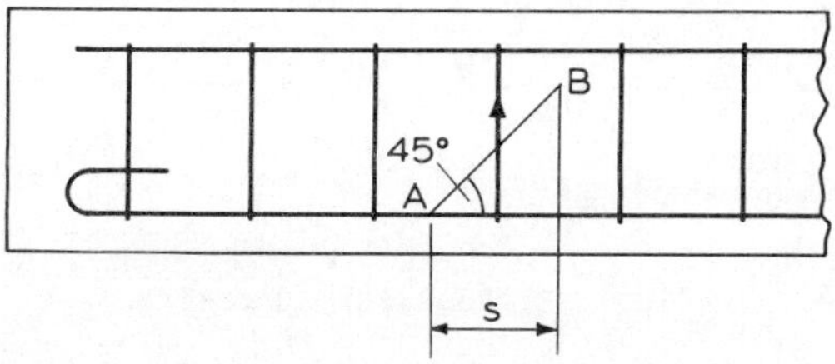

Fig.5.6.1 Vertical stirrups across an uncracked section

vertical stirrups act in conjunction with the main horizontal bars, then *the stirrups need only take the vertical component of the diagonal tension.* Hence the force in each stirrup, F_{sv}, is given by

$$F_{sv} = vbAB \cos 45^{o} = vsb.$$

since v equals the diagonal tension stress, and s is the spacing of the stirrups. Substituting for $v = v_{max} = V/bz$ from Eqn (5.1.10), putting $F_{sv} = f_{sv}A_{sv}$, and rearranging, gives

$$V = (z/s)f_{sv}A_{sv} \qquad (5.6.1)$$

This is the result to be expected from elastic theory assuming that $s \ngtr z$ and that the concrete is incapable of taking any shear at all. However, since concrete can normally resist some diagonal tension unless it is cracked, it is preferable to consider the actual cracked section.

Formation of diagonal crack

Instead of using elastic theory it is possible to consider the mechanism of the diagonal crack as the basis for the analysis. When the diagonal tension stress reaches the ultimate tensile strength of the concrete the latter fails and a diagonal crack forms. The portions of the beam on either side of the inclined crack tend to rotate in opposite directions about some point O in the compression zone above, as shown in Fig.5.6.2(a).

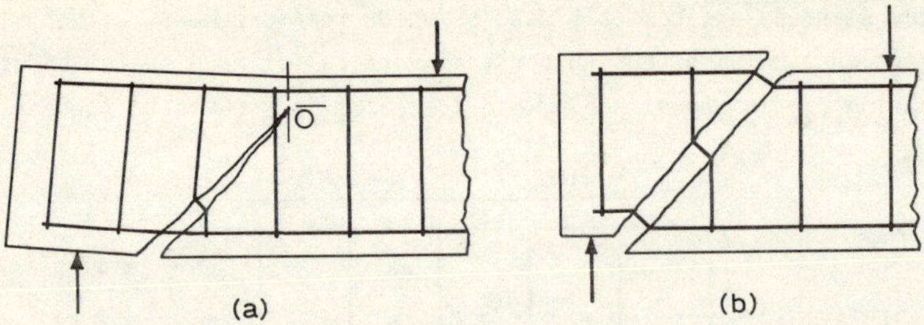

Fig.5.6.2 Vertical stirrups across cracked sections

Although shear-compression failures actually follow a complex curve, it is convenient to assume that the failure plane follows a straight line.

Vertical links or stirrups at collapse

The concrete in the compression zone can be assumed to shear or crush, finally, as shown in Fig.5.6.2(b). Just before this occurs the shear force V at the head of the section is resisted by the vertical forces in both the steel and the concrete. If it is assumed that the maximum average tensile stress which can be resisted by the compression zone of the concrete is given by an average value of v_c acting across the entire diagonal section (see Fig.5.6.3) then the vertical components

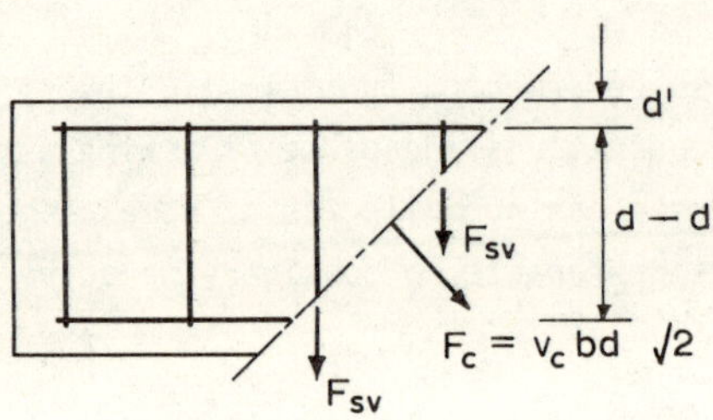

Fig.5.6.3 Combined resistance of stirrups and concrete across a diagonal section

of the forces in both the steel and the concrete can be equated to the total shear force as follows:

$$V = v_c b(d\sqrt{2})/\sqrt{2} + F_{sv}(d - d')/s$$

Dividing throughout by bd and transposing gives

$$v - v_c = 0.87 f_{yv} A_{sv}(d - d')/bds \qquad (5.6.2)$$

where A_{sv} = total cross-sectional area of stirrups (i.e. both legs for 2-leg stirrups)

f_{yv} = characteristic strength of stirrup reinforcement ($\not>$ 425 MN/m^2)

b = width of beam (or web)

s = spacing of shear steel along the longitudinal axis of member, which should not exceed 0.75d.

Notice that this derivation is based upon the vertical equilibrium of forces at the crack. It does not depend upon elastic theory assumptions in any way. It does assume, however, that the main steel is more than adequate to take the horizontal component of the diagonal tension force. CP 110 modifies Eqn (5.6.2) as indicated below and in Appendix CC.

Inclined links or stirrups at collapse

The approach is the same as for vertical stirrups. That is, the main steel takes the horizontal component of the diagonal tension, and the vertical components of the design strengths of the steel and concrete equal the vertical component (i.e. the design shear force).

Consider triangle ABC in Fig.5.6.4.

$$\frac{a}{\sin\theta} = \frac{s}{\sin[180 - (\theta + 45)]} = \frac{s}{\sin(\theta + 45)}$$

or

$$a = s \sin\theta/(\sin\theta \cos 45 + \cos\theta \sin 45).$$

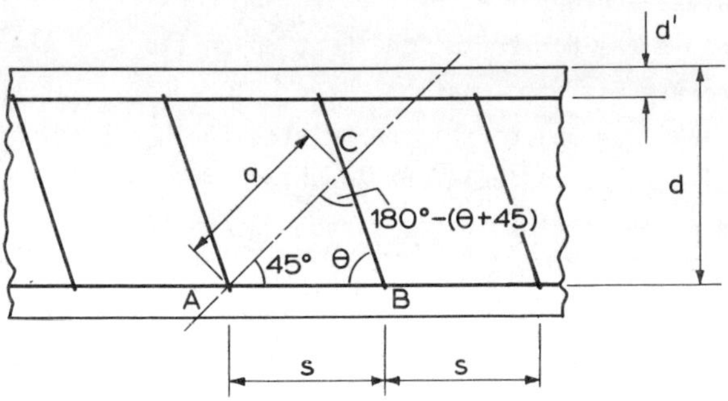

Fig.5.6.4 Inclined stirrups across a diagonal section

The number of links intersected by a plane at 45° is

$$(\surd 2)(d - d')/a = (d - d')(\sin\theta + \cos\theta)/s \sin\theta$$

Hence the vertical component of the design yield strength of the stirrups which intersects a plane at 45° is given by

$$(0.87 f_{yv} A_{sv} \sin\theta)(d - d')(\sin\theta + \cos\theta)(s \sin\theta)^{-1}$$

The shear stress to be resisted by the stirrups is the actual shear stress less the shear stress which can be resisted by the concrete. Since $v = V/bd$

$$v - v_c = (0.87 f_{yv} A_{sv})(d - d')(\sin\theta + \cos\theta)(sbd)^{-1}$$

The Unified Code assumes that $(d - d')/d$ can be taken as unity if the shear stress which is assumed for the concrete is reduced accordingly. Hence for safe design

$$v - v_c \not> (0.87 f_{yv} A_{sv})(\sin\theta + \cos\theta)(sb)^{-1} \qquad (5.6.3)$$

When $\theta = 90$ (i.e. vertical links) then

$$v - v_c \not> 0.87\, f_{yv} A_{sv}/s_v b \qquad (5.6.4)$$

Truss analogy

CP 114:1957 [5.3] suggested that the resistance to shear at any section of a beam reinforced with inclined bars may be calculated by assuming that the inclined bars form the tension members of one or more single systems of lattice girders in which the concrete forms the compression members. With this analogy the imaginary concrete 'struts' should bisect the obtuse angle formed by the bent-up bars in order to maintain equal tensions in the horizontal and bent-up portions of the bars. Since the concrete 'struts' must be taken between the centres of bends in the lower and upper bars, this restricted the bar spacings to particular values (i.e. 'single' or 'double' 'systems'). For closer spacings then either the shear resistance is arbitrarily considered to remain constant or the equilibrium requirement that the tensions in adjacent portions of the bent-up bars is ignored. Unfortunately, CP 110 has perpetuated the truss analogy theory and appears to favour violating the equilibrium requirement [5.4].

The classical theory [5.5] resolved the force in the shear reinforcement in the direction of the principal tensile stress (i.e. at 45° for RC beams). Unfortunately, by this theory, horizontal bars placed at the neutral axis should be just as effective as vertical bars. The theory given here effectively follows the classical theory except for considering nominal shear stresses—and hence no longer restricts the analysis to the theoretical maximum elastic shear stresses at the neutral axis. Thus the presence of the main reinforcement can be accounted for and the primary purpose of the shear reinforcement, to take the vertical component of the diagonal tension force, emphasized. For bars inclined at 45° to 90° it gives the same result as the classical elastic theory except for replacing the lever arm by the effective depth.

Example 5.6.1

The steel ratio for the web reinforcement, ρ_v, for stirrups is given by the total cross-sectional area of all the arms of a stirrup, A_{sv}, divided by the area of concrete appropriate to the stirrup and normal to it. Determine the optimum angle of inclination of the stirrups such that ρ_v is a minimum for a given shear.

Solution Using the usual notation and, referring to Fig.5.6.4 for convenience,

$$A_{sv} = \frac{V_e s}{0.87 f_{yv}(d - d')(\sin\theta + \cos\theta)}$$

where V_e is the excess shear, and

$$\rho_v = \frac{A_{sv}}{bs\ \sin\theta} = \frac{V_e}{0.87 f_{yv} b(d - d')\sin\theta\ (\sin\theta + \cos\theta)}$$

The optimum inclination of the stirrups is given by considering f_{yv}, b, d − d' and V_e as constants. Let

$$X = \sin\theta\ (\sin\theta + \cos\theta)$$

$$\frac{d\rho_v}{d\theta} = 0 \quad \text{when} \quad \frac{dX}{d\theta} = 0$$

i.e. when

$$\sin\theta\ (\cos\theta - \sin\theta) + \cos\theta(\sin\theta + \cos\theta)$$

$$\tan 2\theta = -1 \quad \text{i.e.} \quad \theta = 67\tfrac{1}{2}^{\circ} \qquad \text{(Answer)}$$

Notice that this solution again assumes that there is ample main steel to take the horizontal component of the diagonal tension. In practice it is far more convenient to use vertical stirrups for ease of positioning, and inclined stirrups are seldom used.

Minimum links or stirrups

The concrete alone may be strong enough to take the shear, especially in the midspan portion. However, nominal (or minimum) stirrups in accordance with CP 110 (see Appendix CC), should always be used in beams and similar members for the following reasons:

(i) Premature failure occurs if the concrete cover bursts away and the bond to the main bars is lost. This is prevented by nominal stirrups.

(ii) Shrinkage and thermal stresses, which are normally ignored, can produce appreciable tension. Shallow sections such as slabs and footings often have adequate longitudinal steel suitably positioned whereas beams may have little steel in the top (or bottom). The shear strength of a vertically cracked plain concrete section is nil! Stirrups should always take a proportion of the total shear and help to transfer the longitudinal stresses to the face containing the main steel.

(iii) Stirrups act as ties to hold the beam reinforcement in position while the concrete is being placed.

(iv) Shear failures in concrete beams without secondary reinforcement are essentially brittle and can occur very suddenly. Stirrups help to provide increased warning of failure.

Minimum shear reinforcement (see Appendix CC) need not be provided in members where the breadth is large (i.e. $b > d$) and $v < v_c$ since the shear strength is greater than that for deeper members and so provides some reserve of strength. Similarly, in walls and members of minor importance and where the shear stress is $< \frac{1}{2}v_c$, minimum stirrups can also be omitted.

Example 5.6.2

Calculate the maximum nominal shear stresses for collapse for the beam in Example 8.2.2 (see chapter 8) and hence check the strength in shear at the following positions using CP 110 recommendations:

(i) At midspan (ρ = 2.2%)

(ii) At a section 0.6 m from a support, without any stopping-off of bars

(iii) At a section 0.6 m from a support, where the longitudinal reinforcement has been stopped off such that $\rho = 1.1\%$

Hence determine the required spacing for 6-mm diameter mild steel stirrups at sections (i), (ii) and (iii) if $f_{yv} = 250$ MN/m².

Solution

$R_A = 1.6 \times 36(3.6 - x) \times 10^3/3.6 = 16(3.6 - x) \times 10^3$

Hence x = 1.8 and 0.6 for maximum shear at midspan and 0.6m from support, respectively.

(i) At midspan:

$V = 16 \times 1.8 \times 10^3 = 28.8 \times 10^3$

Therefore

$v = 28.8 \times 10^3/0.15 \times 0.242 = 0.793 \times 10^6$

and, for $\rho = 2.22\%$ and $f_{cu} = 40 \times 10^6$, from Appendix CC:

$v_c = 0.96 \times 10^6$

Therefore only nominal stirrups are necessary. From Appendix CC

$s \ngtr 0.87 \times 250 \times 2 \times \frac{1}{4}\pi \times 6^2 \times 10^{-6}/0.150 \times 0.44 = 0.186$

(or, more simply $s \ngtr A_{sv}/0.002b_t = 0.188$)

Also

$s \ngtr \frac{3}{4}d = 0.75 \times 0.242 = 0.182$

Hence let s = 180 mm (to nearest 5 mm) (Answer)

(ii) 0.6 m from support:

$V = 16 \times 3.0 \times 10^3 = 48 \times 10^3$

Therefore

$v = 48 \times 10^6/36.3 = 1.32 \times 10^6$

$v_c = 0.96 \times 10^6$ as before; hence stirrups are necessary.

$v - v_c = (1.32 - 0.96)10^6 = 0.36$

This is less than 0.44, hence use nominal stirrups as in (i). (Answer)

(iii) $v = 1.32 \times 10^6$ as in (ii).

$v_c = 0.77 \times 10^6$ for $\rho = 1.11\%$

Therefore

$v - v_c = 0.55 \times 10^6$

hence

$s \ngtr 0.87 \times 250 \times 2 \times \frac{1}{4}\pi \times 6^2 \times 10^{-6}/0.150 \times 0.55$

$= 0.149$

s = 145 mm (to nearest 5 mm) (Answer)

Notice that all CP 110 values for shear and bond are expressed in terms of the limit state of collapse. This limit state is considered in detail in the next chapter.

Bent-up bars for shear

It has been shown previously that when bars are no longer required as main reinforcement in the bottom of the beam they can sometimes be bent up to act as shear reinforcement, as well as provide additional main steel in the top of the beam at the support. The effective zone for bent-up bars is from the bend in the tension zone of the first bar to the bend in the compression zone of the last bar, as shown in Fig.5.6.5. To ensure that the failure plane cuts at least one bar, the spacing should not exceed 0.75d. The analysis for bent-up bars is exactly the same as for inclined stirrups. The provision of minimum stirrups in all cases avoids any heavy local stresses which would occur if there were a sudden transition from (all) stirrups to (all) bent-up bars. Independent shear bars, such as the end bent-up bar in Fig.5.6.5 may be used if necessary. However, the cost of bending and placing such bars

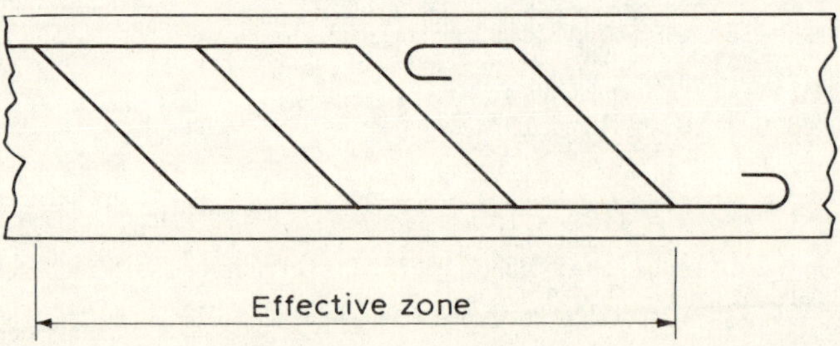

Fig.5.6.5 Effective zone for bent-up bars

is not usually justified nowadays and additional vertical stirrups should be used instead. Notice that if θ is not greater than 45°, the presence of the main steel does not increase the shear resistance by contributing a horizontal force, since the horizontal component of the force in the shear steel itself is more than adequate. However, CP 110 does not permit this possibility.

End anchorage—bent-up bars

Maximum tension occurs throughout the inclined portion of tensile bars bent-up for shear, so that the bond length is measured from the bend. Notice that this requirement may make it difficult adjacent to a support, as shown in Fig.5.6.6. Vertical stirrups must sometimes be used entirely on their own in such cases. The bent-up bar can be brought nearer to the support by bending down the end of the bar into the tensile flange, but this can cause congestion of steel since tensile bars are also running through to the support.

End anchorage—links or stirrups

Stirrups are also fully stressed at the neutral axis. Hence the arms of the stirrups must possess adequate bond length beyond the neutral

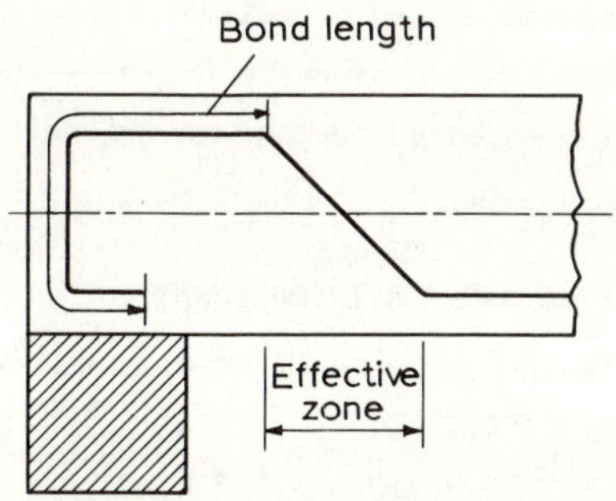

Fig.5.6.6 Development length for bent-up bar

axis and it is not always satisfactory simply to hook the stirrups over longitudinal bars as in Fig.5.6.7(a) and (b). If large diameter stirrups are used in a shallow beam then (c) may be necessary. Notice that (b) only provides additional bond length for one arm: it is the most commonly used stirrup, as compared with (a), because of its convenience in giving increased rigidity to the cage of reinforcement before concreting.

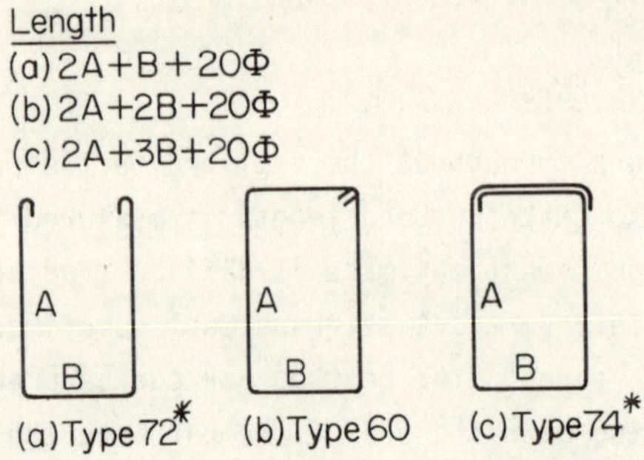

Fig.5.6.7 Various stirrups with differing bond lengths (See also Fig. H2)

Concentrated loads on corbels

If a concentrated load is applied near to a support the concentrated load itself can inhibit failure on an inclined plane due to the vertical compressive stress which is present. The increased shear stresses permitted by CP 110 are given in Appendix CC. The method of loading a member is considered further in the next section.

5.7 FAILURE OF BEAMS WITH SHEAR REINFORCEMENT

It has been assumed that the main function of shear reinforcement is to accommodate the excessive diagonal tension which can be developed in the concrete. The way in which the formulae have been derived was based upon this principle. It is now necessary to study the actual behaviour of beams subjected to combined bending and shear in order to re-assess the justification of this basic assumption.

Stress redistribution

Redistribution of stress must take place once cracking occurs. If a shear crack develops as shown in Fig.5.6.2(a) then the average shear stress across the remaining portion of the section, above point O, must be greater than that before the crack formed. The magnitude of this stress increase depends on the shape of the crack and the characteristics of the beam, with the bond stress adjacent to the cracked section reaching some maximum value.

Dowel action

Notice that the shearing force at the section gives rise to dowel action. The dowel force combined with the flexural tension force produces a resultant which may cause yielding of the main reinforcement.

Bond splitting

This dowel force can also cause 'bond splitting' of the concrete at the level of the reinforcement, causing the bond strength between concrete and steel to be greatly reduced over an appreciable length. Unless the original bond lengths are considerably in excess of the minimum values considered adequate, premature flexural failure can occur because of bond failure accelerated by shear.

Method of load application

The usual method of loading a beam in the laboratory can give shear strengths that are extremely high compared with those for an actual structure. This is because the loads applied to the beam in practice are usually either uniformly distributed, or transmitted by secondary beams which frame into the member. A laboratory test, however, usually

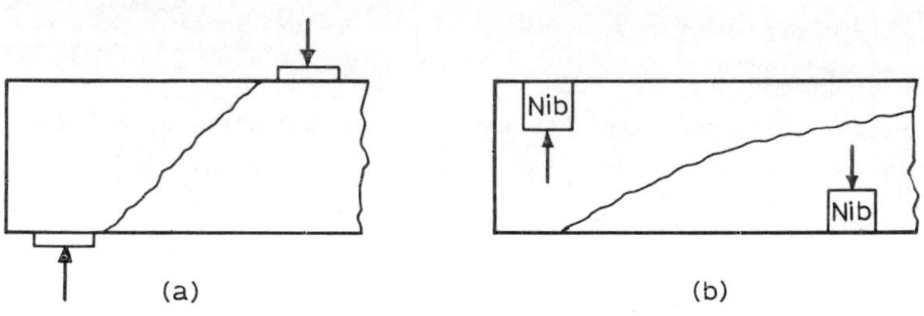

Fig.5.7.1 Different modes of failure due to different methods of load application

applies one or two point loads via plates placed on the top surface of the beam, with supports pressing against the bottom surface. The latter method induces vertical compressive stresses which can greatly enhance the shear resistance of the beam, especially at the critical sections of short spans (see Fig.5.7.1). The permissible shear stresses given in some of the earlier codes of practice were based on laboratory tests using point loads as in Fig.5.7.1(a). They were therefore too high and some catastrophic failures occurred. The lesson to be learnt from these failures is that the testing techniques used in the laboratory should be reasonably representative of the conditions in practice.

Modes of failure

The various types of failure due to the combined action of shear and flexure are as follows:

Diagonal tension	(i) Shear-compression (ii) Constrained shear
Bending at a diagonal plane	(i) Shear-bond (ii) Shear-tension
Diagonal compression	(i) Web failure (ii) Support failure

Different authorities sometimes use different terms or attach different meanings to particular terms. However, the terms given above together with the meanings given below will be used here since they appear to give the most satisfactory description of the various modes of failure.

Diagonal tension

Failure due to diagonal tension can occur in two ways as follows:

(i) *Shear compression* The shear crack is inclined and is frequently a continuation of a normal bending crack. Failure eventually occurs with the extension of the crack becoming almost horizontal before turning up again into the compression zone as the concrete crushes in a manner similar to that in a flexural failure. A shear crack develops as an extension of a vertical crack which turns into an inclined crack near the neutral axis (see point 1 in Fig.5.7.2), and propagation upwards into the zone of compression becomes flatter. The crack then tends to 'crack-back' with further increase in load, as shown by point 2. Propagation is usually gradual until some point (3) is reached, whereupon sudden failure occurs.

(ii) *Constrained shear* A diagonal crack is observed in the region of the neutral axis, and as the failing load is reached the crack suddenly extends up and through the compression zone with extreme violence. This type of failure occurs if concentrated loads are applied at the top (or bottom) of the beam on a short span as shown in Fig.5.7.1(a). This form of failure is associated with very high shear stresses, which rarely occur in practice except for corbels, etc.

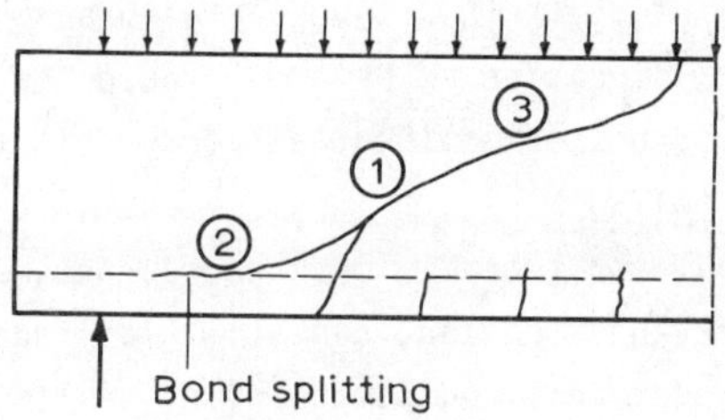

Fig.5.7.2 Typical shear-compression failure

Bending at a diagonal plane

The diagonal crack widens due to the rotation of each portion about the point O shown in Fig.5.6.2(a) until the steel which crosses the crack is no longer able to prevent a flexural failure at a diagonal plane. Flexural failure on a diagonal plane can occur in one of two ways:

(i) *Shear bond* Bond splitting occurs along the level of the reinforcement due to dowel action: the reduced effective bond length produces premature flexural failure due to slipping of the main reinforcement. Shear-bond failures can be avoided by providing nominal links and adequate development lengths for the main reinforcement, as required by CP 110. Even very light stirrups can greatly offset the effect of reductions in bond due to splitting, by their own frictional contact force.

(ii) *Shear-tension* The main steel yields due to the combined effects of bending and dowel action. Calculations for bending at normal sections ensure that the areas of longitudinal reinforcement are satisfactory for the maximum moments. The provision of adequate development lengths for curtailed bars ensures that the flexural resistance in regions of high shear is slightly more than is necessary for pure flexure. Hence a shear-tension failure is unlikely before a flexural failure across a normal plane occurs elsewhere.

Diagonal compression

Local failure of the concrete can occur due to inclined compression as well as inclined tension. It is difficult to recognize as such since the outward manifestation of this type of failure is often inclined cracking in the same direction as before because the inclined compression acts at right angles to the inclined tension. This type of failure is likely to occur as follows:

(i) *Web failure* A T- or I-beam with a narrow web, suitably reinforced to take the inclined tension, can still fail due to crushing of the concrete under inclined compression. A limited number of tests carried out by Taylor [5.2] indicated that, for beams of a particular depth but varying cube strengths and web breadth, the nominal shear stress at the first sign of web crushing was approximately one-fifth of the cube strength. CP 110 avoids such failures by stipulating that the shear stress should not exceed $0.75f_{cu}$ which $\simeq f_{cu}/5\gamma_m$ for higher concrete grades only.

(ii) *Support failure* Compression failure of the concrete can occur at the supports in short deep beams where considerable arching action can take place. This theory was first proposed by Faber

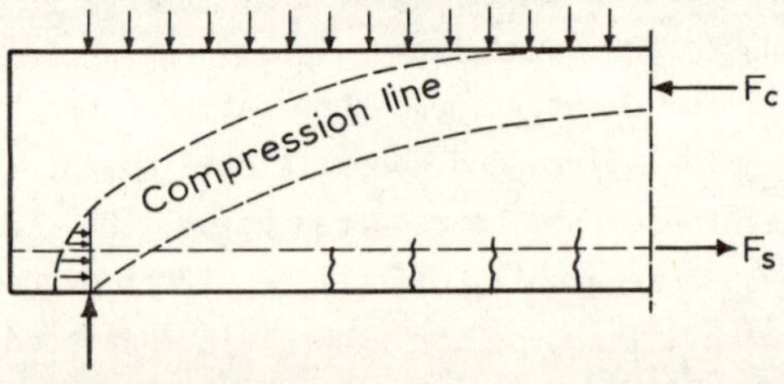

Fig.5.7.3 Possible compression failure due to arching action

who assumed that the longitudinal reinforcement slips over its entire length to induce stresses as shown in Fig.5.7.3. The compressive strength of the concrete should be checked in such cases, if necessary.

Discussion of shear failures

Shear-compression failures are the most important type to consider in design, so the earlier emphasis on diagonal tension is justified. Provided that the designer is aware of the other possibilities, suitably prudent precautions can be taken by following CP 110 requirements.

It is clear that although the actual behaviour of beams subjected to combined bending and shear is complex, these precautions are sufficient to eliminate failure modes other than shear-compression. The uncertainty in estimating shear compression failures is now not so very much greater than that for pure bending effects. However this larger uncertainty, together with the fact that shear failures can occur suddenly and with considerable violence, requires that the partial safety factor for material strength should still be greater for concrete in shear.

Comparison with CP 114:1957

Comparison with CP 114:1957 shows that the old code was relatively unsafe for beams without shear reinforcement, whereas for beams with shear reinforcement it was extravagant on steel, often requiring twice the amount that was really necessary. CP 114:1957, before amendment, implied ultimate nominal shear stresses at failure of approximately 1.18 and 1.93 MN/m^2 for cube strengths of 20.7 and 41.4 MN/m^2, respectively, regardless of how the main steel was curtailed. Assuming that suitable values for ρ and γ_m are, say, 1% and 1.8 respectively, CP 110 implies corresponding values of 1.08 and 1.35 MN/m^2 for f_{cu} = 20 and 40 MN/m^2 respectively. Although few failures occurred in actual structures, in the UK this was probably due to the liberal use of nominal stirrups in detailing (which CP 114 did not actually require) and a load factor which was intended to be at least 2. Users of the American Code were not generally so conservative in detailing and one or two catastrophic failures occurred before the ACI Code was amended.

References

5.1 Hughes, B.P. and Ash J.E., 'Water gain and its effects on concrete', Concrete vol. 3, No. 12, December 1969, pp 494-6

5.2 Taylor, R., 'Some shear tests on reinforced concrete beams without shear reinforcement', Magazine of Concrete Research vol. 12, No.36, November, 1960, pp 145-54.

5.3 Scott, W.L., et al. 'Explanatory handbook of the BS Code of practice for reinforced concrete,' Concrete publications, London, 1957, pp 154.

5.4 Bate, S.C.C., et al. 'Handbook on the unified Code for structural concrete', (CP 110:1972), Cement and Concrete Association, London, 1972, pp 153.

5.5 Morsch, E., 'Concrete-Steel construction', Translated by E.P. Goodrich, Eng. News Pub. Co., New York, 1910.

Exercises

1. A doubly reinforced cantilever is 3 m long and carries a live point load of 20 kN at its free end. The beam measures 150 mm wide by 300 mm deep overall, and provides a dead load of 1.1 kN/m. The tension and compression bars are positioned at a distance of 50 mm from the tension and compression faces of the concrete respectively. Assume that $\gamma = 1.6$ for live load and 1.4 for dead load.

(a) Calculate the maximum nominal shear stress developed under the design loads for the limit state of collapse. If v_c equals 0.95 MN/m^2, calculate the required spacing of two-leg vertical links of 6-mm diameter steel with a characteristic strength of 250 MN/m^2. Is the calculated spacing necessarily satisfactory?

(b) Explain the meaning of the term 'local bond strength'. If the beam requires five 20 mm diameter tensile bars to carry the maximum bending moment, what local bond stress is developed? γ_m for steel is 1.15

(UB2 1969 modified, 40 min)

2. A beam of width 250 mm and effective depth 450 mm has a steel area of 2500 mm^2. The section carries working load shear forces of 100 kN dead load and 150 kN live load. Determine the nominal shear stress if the partial safety factors are 1.4 for dead load and 1.6 for live load. If shear reinforcement is provided in the form of two-leg 12-mm diameter links inclined at 45° to the horizontal, calculate the required pitch.

If all the reinforcement consists of square twisted bars such that $f_{yt} = f_{yv} = 425$ MN/m^2, calculate the theoretical maximum 'equivalent diameter' of main bars which could be used if the maximum flexural bond stress is not to exceed 3.4 MN/m^2. Also calculate the anchorage length required for the links.

Assume γ_m for steel is 1.15, f_{cu} is 40 MN/m^2 and v_c (from Table CC1) is 0.96 MN/m^2 and maximum anchorage bond stress (from Table CD2) is 2.6 MN/m^2.

3. A uniform rectangular reinforced concrete beam AB on a simple span of 6 m carries characteristic concentrated imposed loads of 56.25 kN at each third span point (i.e. 2m from each support). The characteristic dead load (including self-weight) can be taken as 12 kN/m.

(a) Determine the maximum midspan moment for collapse and hence design a suitable singly reinforced rectangular beam of breadth 250 mm for the midspan section.

(b) Also determine whether nominal 8 mm diameter 2-leg vertical stirrups at 350 mm spacing are adequate at the support sections, if the critical shear stress for the concrete alone is 0.45 MN/m^2 for the appropriate amount of reinforcement.

Assume characteristic strengths of 20 MN/m^2 for the concrete and 425 MN/m^2 for both main and secondary reinforcement. The minimum cover to the main steel is 25 mm and the partial safety factor for material strength for the steel is 1.15. Assume a rectangular design stress block for concrete in flexure of maximum depth ½d and stress 0.4 × 20 MN/m^2. The safety factors for load are as follows

Dead load : 1.4

Imposed load : 1.6 (UB2 1975, 50 min)

(For part (a) see chapter 8).

6 Limit states of collapse

The ultimate limit states which are usually of most concern are the limit states of collapse where it is assumed that the force actions are still acting in essentially the same manner as for service conditions. e.g. a beam or a slab still acts as a beam or a slab (and not as a catenary or a membrane).

Plastic deformations occur in reinforced concrete members as they approach the limit state of collapse. Permanent deformations are produced and the basic assumption of elastic theory, that stress is proportional to strain, no longer applies. The distribution of moments throughout the structure, the moment of resistance of each section, and the margin of safety against a catastrophic failure due to overloading, can only be predicted accurately by considering the actual conditions at collapse.

6.1 INTRODUCTION TO COLLAPSE LOAD ANALYSIS

The basic principles of limit state theory, including the general assumptions for structural analysis which apply to all limit states, have been outlined in chapter 1. Reinforced concrete as a structural medium has been introduced with special reference to elastic theory, but it should now be considered with specific reference to collapse. Although elastic theory is useful for the serviceability limit states, it is not valid for sections at collapse. Even for collapse, however, it will be seen that the assumed distribution of forces and moments is based on the values obtained from an elastic analysis.

General requirements at collapse

The structure must be designed so as not to collapse. Collapse of the structure, or part of the structure, could arise from rupture of one or more critical sections, from overturning, or from buckling due to elastic or plastic instability (including the effects of sway where appropriate). The design resistance to bending, shear, torsion and axial loads at every section should not be less than the appropriate value at that section produced by the most severe arrangements of design loads on the structure. For example, the maximum overturning moments due to the design loads should not exceed the counter-balancing moments. Hence notice that the partial safety factor for dead load given in Appendix B for case II is as high as 1.4 where the load is assisting the overturning action, and only 0.9 where it is preventing the overturning action.

Example 6.1.1

A continuous reinforced concrete beam, ABCD, rests on two supports B and C. BC = 6m and cantilevers AB and CD = 2m. The characteristic dead load (including self-weight) is 35 kN/m throughout and the characteristic imposed load is 15 kN/m throughout. Calculate the maximum and minimum span and support design moments for the limit state of collapse. Refer to Appendix B.

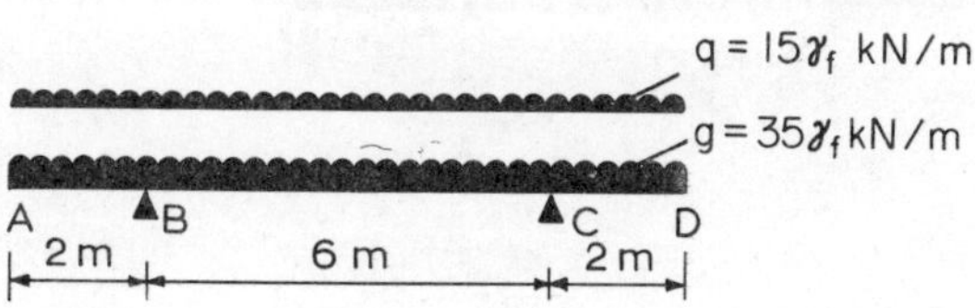

Fig.6.1.1 Example 6.1.1

Solution

Span or cantilever loading From Table B1, the partial safety factor γ_f for collapse under dead and imposed loading (case I) is 1.6 or 0 for imposed loading and 1.4 or 1.0 for dead loading. Hence the maximum value of w is

$(1.4 \times 35 + 1.6 \times 15) \times 10^3 = 73 \times 10^3$

and the minimum value is

$1.0 \times 35 \times 10^3 = 35 \times 10^3$

Moments The maximum midspan moment is

$(73 \times 6^2/8 - 35 \times 2^2/2) \times 10^3 = (328.5 - 70) \times 10^3$

$= 258.5$ kNm (Answer)

The minimum midspan moment is

$(35 \times 6^2/8 - 73 \times 2^2/2) \times 10^3 = (157.5 - 146) \times 10^3$

$= 11.5$ kNm (Answer)

The maximum cantilever moment is 146 kNm. (Answer)

The minimum support moment is 70 kNm. (Answer)

Example 6.1.2

Repeat Example 6.1.1, if the characteristic imposed load is now only 10 kN/m for AB and BC and zero for CD.

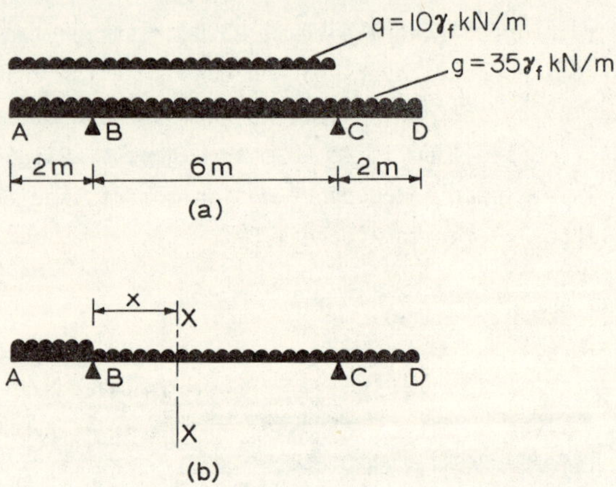

Fig.6.1.2 Example 6.1.2.(a) General loading conditions. (b) Loading for minimum span moments

<u>Solution</u> For cantilever AB, the maximum value of w is

$(1.4 \times 35 + 1.6 \times 10) \times 10^3 = 65 \times 10^3$

The maximum moment (in magnitude) is

$(65 \times 2^2/2) \times 10^3 = 130$ kNm (Answer)

and the minimum is

$(35 \times 2^2/2) \times 10^3 = 70$ kNm (Answer)

(Notice that the maximum and minimum moments algebraically are -70 and -130 kNm).

For cantilever CD, the moment is 98 kNm maximum and 70 kNm minimum (Answer)

For span BC the maximum moment is

$(65 \times 6^2/8 - 70) \times 10^3 = 222.5$ kNm (Answer)

The maximum span moment for BC for the loading shown in Fig.6.1.2(b) is no longer at midspan.

For span BC, the minimum (algebraic) moment is given by

$R_B \times 6 = 65 \times 2 \times 10^3 \times 7 + 35 \times 6 \times 10^3 \times 3 - 1.4 \times 35 \times 2 \times 10^3 \times 1$

$R_B = (910 + 630 - 98) \times 10^3/6 = 240 \times 10^3$

$M_{xx} = 240 \times 10^3 x - 65 \times 10^3 \times 2 \times (x + 1) - 35 \times 10^3 \times x^2/2$

$\quad = 10^3(110x - 130 - 17.5x^2)$

$dM_{xx}/dx = 0$ when $110 = 35x$ or $x = 3.14$

Therefore

$M_{xx} = 45 \times 10^3$

which is positive and therefore does not induce reversed bending.

Stress-strain curves

Typical stress-strain curves for short-term loading of concrete in uniaxial compression, and steel in tension, are shown in Figs. 6.1.3 and 4. Typical stresses under working load conditions are contrasted

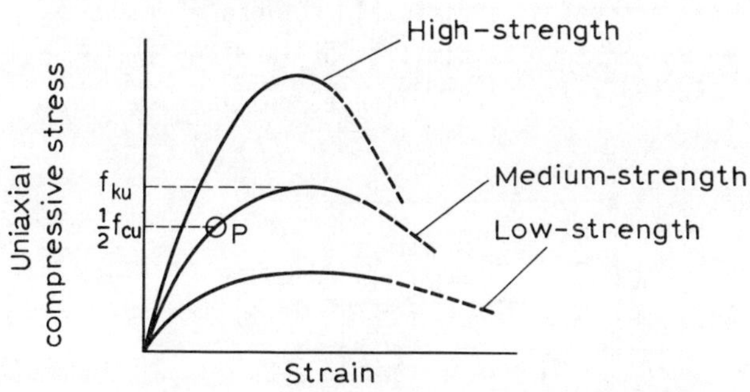

Fig.6.1.3 Typical stress-strain curves for concrete in compression

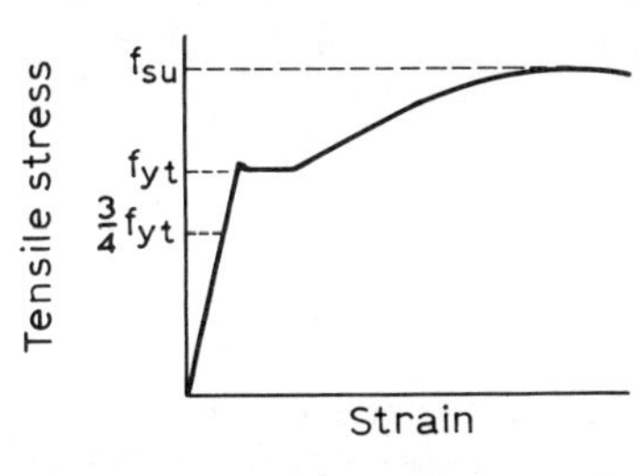

Fig.6.1.4 Typical stress-strain curve for steel in tension

against typical stresses at collapse. Design stress-strain curves as given in CP 110 are shown in Figs D1 to 3 (Appendix D). Notice that the design stress-strain curve for concrete relates to uniaxial tests. Hence it is assumed that the characteristic uniaxial strength, f_{ku}, is approximately 0.78 f_{cu}. The falling branch of the actual stress-strain curve in Fig. 6.1.3 is neglected, so the coefficient is reduced further to 0.67 to allow for this as shown in Fig. 6.1.5. Hence the design strength, $_{*}f$, equals $0.67f_{cu}/\gamma_m \simeq 0.45f_{cu}$ since γ_m = 1.5 for load cases I to III.

Plastic theory

It is necessary to distinguish between two distinct aspects of plastic theory for reinforced concrete as follows:

(i) The increased resistance of a cross-section after its extreme fibre commences to yield.

(ii) The increased resistance of a redundant structure due to moment redistribution.

The ultimate resistances of individual sections of members are considered later in chapter 7. Limit theory and the analysis of framed structures for the limit state of collapse are discussed in the next section.

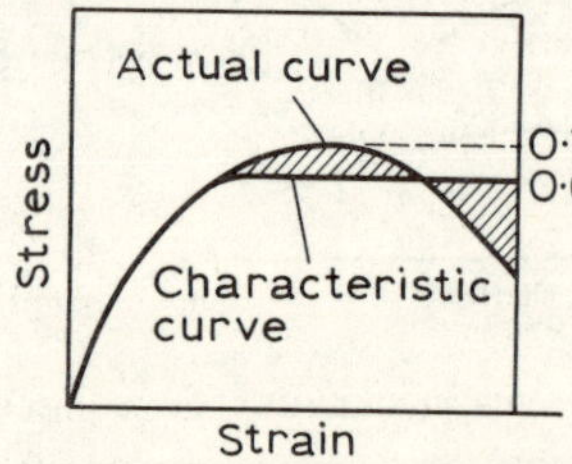

Fig. 6.1.5 Fitting characteristic curve to actual stress-strain curve for concrete

6.2 LIMIT THEORY

Limit theory refers to the changing distribution of moments which occurs throughout a structure as plastic hinges are formed. The ultimate strength of a structure can therefore be increased if redistribution of moments throughout the structure enables more sections to reach their ultimate capacity. Although the onset of yielding introduces large deflections, and should be avoided under working load conditions due to undesirable cracking and fatigue complications, the following implications of limit theory should be fully appreciated.

(i) A statistically indeterminate structure does not collapse when the first section yields.

(ii) A large reserve of strength usually exists between initial yielding and ultimate collapse of the structure.

Plastic hinge conception

The curvature of a member increases with the applied bending moment. Fig. 6.2.1(a) shows a typical M-r^{-1} curve for an underreinforced beam or slab, where r^{-1} is the curvature which, if constant along the member, equals the change in slope over unit length of the member, θ. The significant part of the M-r^{-1} curve for limit design is that portion

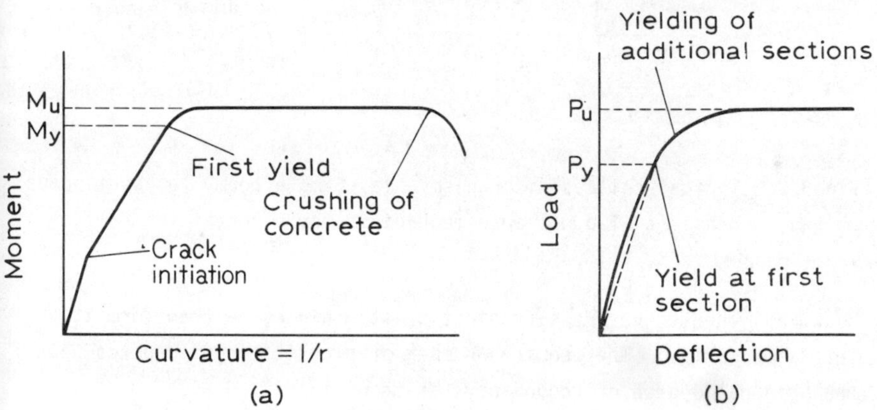

Fig.6.2.1 Typical moment-curvature and load-deflection diagrams. (a) Moment at a section. (b) Deflection of member

beyond the first yield moment, for which M_u remains substantially constant for a wide range of r^{-1} values. This means that if a beam section develops a moment M_u it then behaves as a plastic hinge. It continues to resist the moment M_u, and further loading merely produces an increased curvature rather than an increased moment, so that other parts of the structure must then take the additional loading. The changes in moment produced elsewhere on the member due to the subsequent loading are just the same as if a real hinge existed at the plastic hinge point.

Collapse mechanism

The ultimate strength considered in limit theory is that which brings the structure to the verge of collapse, assuming perfect plastic hinge action. A member collapses under moment loading (other than by

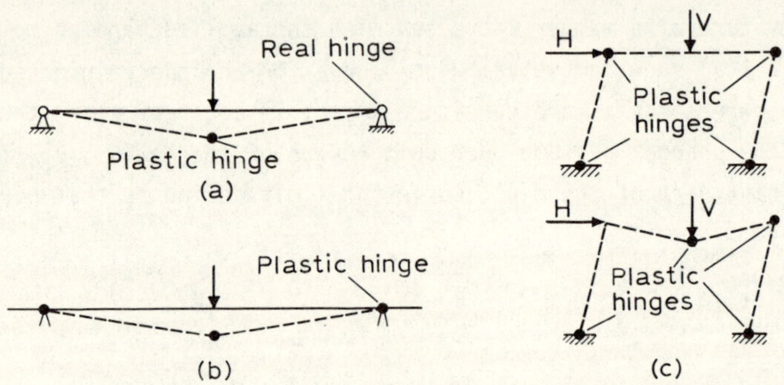

Fig.6.2.2 Typical collapse mechanisms. (a) Simple beam. (b) Continuous beam. (c) Two collapse mechanisms for a portal frame

buckling) when there are sufficient plastic hinges to transform it into a mechanism. The required number of plastic hinges depends upon the original degree of redundancy, as shown in Fig. 6.2.2.

Moment redistribution

A study of working load conditions should usually involve an elastic analysis. Successive sections will normally pass the M_y value and reach the M_u value as increasing overloads are applied. Only a very special frame would have its maximum moments so equalized that all the plastic hinges required to form a mechanism would develop at any one loading stage and so lead to an immediate collapse. For any normal frame the collapse pattern can be radically different from the original elastic moment pattern, and can be established quite independently of either the elastic or elasto-plastic stages.

Uniqueness theorem

The bending moment diagram which satisfies the following conditions is unique:

(i) Sufficient plastic hinges are formed to convert the structure to a mechanism.

(ii) The BMs are in equilibrium with the applied loads.

(iii) The ultimate moment is not exceeded at any point in the structure.

Upper bound (or limit) to load factor

Any arbitrary choice of hinge positions leads to an estimate of the collapse load which is greater than, or equal to, the actual. A structure which is on the verge of collapse must have developed a sufficient number of plastic hinges to change it into a mechanism. The true ultimate load for the structure can be no larger than the collapse load for this particular mechanism. It may be smaller for some other (more probable) mechanism. This approach therefore establishes an upper bound (or limit) on the value of the true load factor. If all possible mechanisms are investigated, the lowest collapse load is the true one. For this purpose it should be noted that the moments are all statically determinate when M_u values are used at each plastic hinge in a mechanism.

Lower bound (or limit) to load factor

Any arbitrary equilibrium state which nowhere causes a moment exceeding the ultimate leads to a collapse load which is less than, or equal to, the actual. Greenberg and Prager have shown that any moment diagram satisfying statics may be used as the basis of a lower-bound solution [6.1]. It is usually easier to develop an 'arbitrary' moment diagram than it is to establish the elastic moment values. Different moment diagrams indicate different ultimate loads. Higher values of the lower bound are obtained when more and more sections for the assumed moment diagrams are at the M_u value. The moment diagram can therefore be 'equalized' towards values establishing the true collapse load. The final solution can be checked by the upper-bound method above by assuming plastic hinges at the points of high moment.

Limitations

Limit theory assumes that

(i) Failure by shear or bond is prevented.

(ii) Sudden flexural compression failure is prevented.

The application of limit theory to normal reinforced concrete design has been retarded due to uncertainties concerning these assumptions. The M-θ curve for a concrete beam, unlike that for a mild steel I-beam, involves a gradual rather than a sudden change to plastic conditions. How is the diagonal tension and bond resistance affected by large deflections? What is the maximum permissible steel ratio consistent with an adequate plastic hinge formation? These complications have now been largely resolved (see, for example, reference [6.2]). CP 110 makes allowances for the rotational capacity of member sections and, in addition, has followed the lead of the Danish Code in permitting limit design for slabs for which the method is well suited. The application of limit theory to continuous structures is discussed in the next section.

Example 6.2.1

According to elastic theory a single point load W at midspan of a uniform propped cantilever produces a moment of 3WL/16 at the support and 5WL/32 at the centre, where L is the span. Determine the additional load which can be carried assuming that the beam section has adequate rotational capacity at maximum moment and that the negative moment of resistance at the support is equal in magnitude to the positive moment of resistance at midspan.

Solution Let the ultimate moment of resistance of the beam section be M_u throughout.

The maximum load according to elastic theory is given when the maximum moment at any one section reaches the moment of resistance as shown by stage I. Hence

$M_u = 3WL/16$ or $W = 16M_u/3L$

If the support section is capable of yielding while still resisting its maximum moment of M_u, then the propped cantilever can sustain additional loading by subsequently acting as a simple beam. Collapse

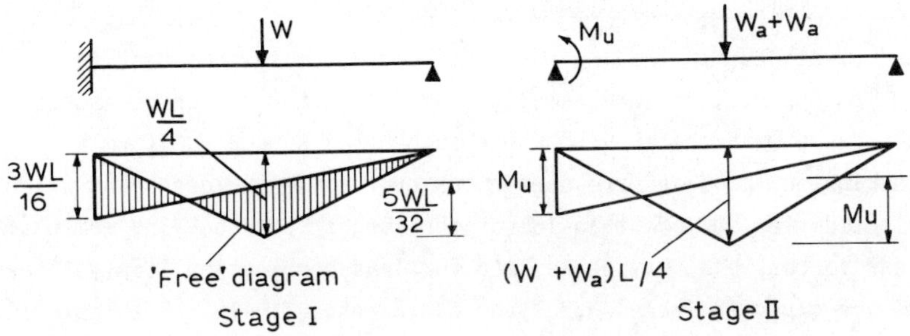

Fig.6.2.3 Example 6.2.1

finally occurs when the midspan section also reaches its maximum moment of resistance and the beam becomes a mechanism, i.e.

$M_u = 5WL/32 + W_aL/4$

where W_a is the additional load necessary to produce failure of the beam after the support section becomes a plastic hinge. But

M_u at midspan = M_u at support = $3WL/16$

Hence

$W_a = W/8$ (Answer)

6.3 ANALYSIS OF CONTINUOUS STRUCTURES AT COLLAPSE

The bending moments, axial loads and shear forces at sections in individual members of framed structures may be calculated by the application of elastic theory using the relative stiffnesses based on any of the sections described in section 1.8 and any of the frame, sub-frame or beam assumptions in Appendix CB. The moments determined from an elastic analysis for any particular loading may then be redistributed by an amount not exceeding 30% of the maximum moment provided that equilibrium between the external forces and moments and the design loads is maintained, and provided that the rotational capacities of the members are adequate for shedding the excess moment at the sections concerned. In addition, the moment capacity provided at any section must nowhere be less than 70% of the undistributed moment. The ability of individual sections to deform plastically under any excess moment is considered later.

Collapse moments for continuous beams

The moments due to loading for continuous beam systems carried on simple supports, or for beams in framed structures, can be assessed in accordance with the above. Where such beams support uniformly distributed loads and are continuous over three or more spans which do not differ by more than 15% of the longest, the bending moments and shear forces which may be assumed for load combination I only (Appendix B), are given in Tables 6.3.1 and 6.3.2 respectively. See Table I4 for simplified CP 110 moments (and shear forces) only if $Q_k \nmid G_k$

Table 6.3.1 Simplified moments for continuous beams

F_{var} = variable (design) load (= $0.4G_k + 1.6Q_k$)
G_k = characteristic dead load
Q_k = characteristic imposed load
ℓ = effective span

Type of load	Near middle of end span	Penultimate support	Middle of interior spans	Interior supports
Dead load	$+ G_k\ell/12$	$- G_k\ell/10$	$+ G_k\ell/24$	$- G_k\ell/12$
Variable load				
Maximum	$+ F_{var}\ell/10$	$- F_{var}\ell/24$	$+ F_{var}\ell/12$	$- F_{var}\ell/40$
Minimum	$- F_{var}\ell/50$	$- F_{var}\ell/9$	$- F_{var}\ell/24$	$- F_{var}\ell/9$

No redistribution of moments is permitted for the moments given in Tables 6.3.1 or I4. More accurate determinations of the moments can be obtained using the coefficients given in Tables I1-I3 and described in section 4.6. The curtailment of the reinforcement is carried out as indicated in section 5.4. The bending moment diagram given by Table 6.3.1 can be used for this purpose, if desired, for approximately equal spans. The principle of redistributing moments to a certain value (e.g. 10%, 20% or 30% according to the rotational capacity of the sections) is illustrated in Example 6.3.1.

Shear forces for continuous beams

The shear forces given in Table 6.3.2 are satisfactory for most purposes, even when the use of coefficients is felt desirable for the bending moments. See Table I4 only if $Q_k \ngtr G_k$.

Table 6.3.2 Simplified shear forces for continuous beams

Type of load	Outer support	Penultimate support: Outer side	Penultimate support: Inner side	Interior support
Dead load	$0.4\ G_k$	$0.6\ G_k$	$0.5\ G_k$	$0.5\ G_k$
Variable load	$0.45\ F_{var}$	$0.6\ F_{var}$	$0.6\ F_{var}$	$0.6\ F_{var}$

Redistribution of moments

Adjustment of the theoretical elastic moment diagram has been referred to in section 4.6. CP 110 requires that the moment capacity of any section must not be reduced to less than 70% of the maximum value for that section required by an elastic analysis. This requirement helps to ensure that no section is likely to suffer local damage at working loads, especially near points of contraflexure as indicated by any portion of the elastic maximum bending moments diagram. (Fig.4.6.3 clearly shows the significant displacement in points of contraflexure.)

CP 110 places no limit by which support moments may be increased other than the independent further criterion that the elastic moments at any section should not be reduced by more than 30% of the numerically largest moment given anywhere by the elastic moments diagram for that particular member covering all appropriate combinations of ultimate load. For normal situations this clause is less critical than the 70% moment capacity requirement.

Greater adjustment is possible with CP 110 than with CP 114 such that 10% redistribution to CP 110 can give much lower span moments than 15% to CP 114 (i.e. simple $\pm$ 15% adjustment of the support moments). Both CP 110 and CP 114 of course require that equilibrium between the internal forces and external loads is maintained: which simply means that the overall height of the bending moment diagram remains the same after adjustment for any particular loading. Even for CP 110 it is convenient to start by assuming that support moments in all cases are at 70% (or 90% etc.) of their unadjusted maximum values, even though some readjustment for particular cases may be necessary subsequently to avoid excessively high negative span moments. Astill has suggested [6.3] that it is only necessary to find the maximum support moments by elastic analysis, apply β_{red} (i.e. reduce by 10, 20 or 30% etc.) and then, span by span, superimpose the 'free' bending moment diagrams for maximum and minimum loads for each span in turn. That is, the total range of moments for any one span is simply the range of the free moments for the maximum (total) loading and the minimum (total) loading on that span. However, in order to illustrate the adjustment of individual loading cases Example 6.3.1 does not use this useful simplification.

The coefficients in Appendix I apply to the maximum hogging moments at the actual support sections, whereas the sagging moments are the maximum sagging moments in the span (and need not be at the actual midspan). When using these coefficients it will usually be adequate (and on the safe side) to add the maximum dead and variable load sagging moments (i.e. to assume that they coincide at the same section) to obtain the maximum combined span moment.

The actual moment redistribution at midspan can be less than the β_{red} assumed for the supports and this can be taken into account when considering the neutral axis depth for the midspan section (see chapter 8). This reduced β_{red} can also be obtained by using Appendix I: dividing the combined span moment coefficient (for nominal β_{red}) by the corresponding combined span moment coefficient for zero redistribution for the particular section gives $1 - \beta_{red}$.

More accurate maximum combined span moments can be obtained by actually drawing the bending moment diagram envelopes and superimposing. (Appendix J can conveniently be used for parabolic moment diagrams).

Example 6.3.1

Determine the maximum moments at the midspans and the supports of a four-span beam ABCDE continuous over supports B, C and D. The spans are all equal. Consider both fixed and variable loads. Assume 10% maximum redistribution of moments to CP 110 and consider either alternate spans loaded or adjacent spans loaded for the variable load. Express the results in terms of coefficients (× load × span).

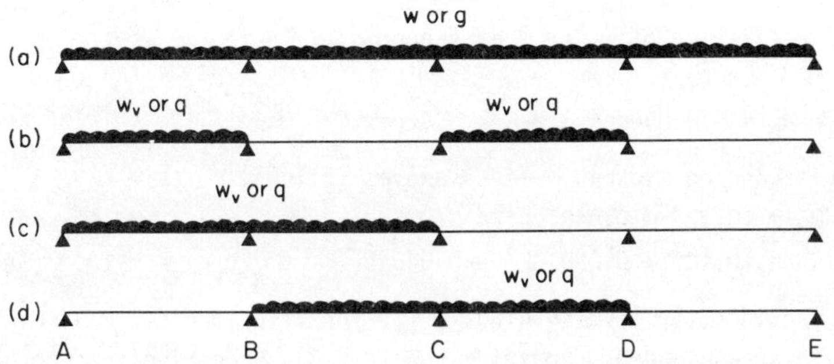

Fig.6.3.1 Example 6.3.1

<u>Solution</u> The area A under the curve for the free bending moment diagram for a uniformly distributed loading w is given by

$$(2/3)(wL^2/8)L = wL^3/12$$

If L and I are common for all spans and $\bar{x} = L/2$ then the theorem of three moments given by (Eqn (4.6.1) becomes

$$M_A + 4M_B + M_C = -3(A_1 + A_2)/L$$

This equation can be applied to adjacent spans in turn.

(a) Consider the fixed loading,g, with all spans loaded.

$$A_1 = A_2 = A_3 = A_4 = gL^3/12$$

Also

$$M_A = M_E = 0 \text{ (end supports)}$$

Hence

$$4M_B + M_C = -gL^2/2$$

$$M_B + 4M_C + M_D = -gL^2/2$$

$$M_C + 4M_D = -gL^2/2$$

Now by observation of the above

$$4(M_B + 4M_C + M_D) - (4M_B + M_C) - (M_C + 4M_D) = 14M_C$$

Hence

$$14M_C = -gL^2 \quad \text{or} \quad M_C = -gL^2/14 = -0.071gL^2$$

Also

$$4M_B = -gL^2(7-1)/14 \quad \text{or} \quad M_B = -3gL^2/28 = -0.107gL$$

and

$$M_D = -3gL^2/28 = -0.107gL^2$$

For 10% relaxation of moments at supports,

$$M_C = -0.064gL^2$$

$$M_B = M_D = -0.096gL^2 \qquad \text{(Answer)}$$

The corresponding midspan moments are as follows:

$$AB = DE = (0.125 - 0.048)gL^2 = 0.077gL^2 \qquad \text{(Answer)}$$

$$BC = CD = (0.125 - 0.080)gL^2 = 0.045gL^2 \qquad \text{(Answer)}$$

(b) Consider the live loading, q (or variable loading, w_v), with alternate spans loaded. Applying Eqn (4.6.1) successively as before except for putting $A_2 = A_4 = 0$:

$$4M_B + M_C = -qL^2/4 \quad \text{etc}$$

and

$$M_C = -qL^2/28 = -0.036qL^2$$

$$M_B = M_D = -3qL^2/56 = -0.054qL^2$$

Therefore the corresponding elastic midspan moments are as follows:

$$AB = (0.125 - 0.027)qL^2 = 0.098qL^2$$

$$BC = (0 - 0.045)qL^2 = -0.045qL^2$$

$$CD = (0.125 - 0.045)qL^2 = 0.080qL^2$$

$$DE = (0 - 0.027)qL^2 = -0.027qL^2$$

Therefore the elastic midspan moments (using symmetry of loading for spans BC and DE only loaded) are given by

$$AB = DE = (-0.027 \text{ to } +0.098)qL^2$$

$$BC = CD = (-0.045 \text{ to } +0.080)qL^2$$

The support moments are relatively small and need to be increased if the span moments are to be reduced. If the support moments are increased to 90% of their unadjusted maximum elastic values: i.e. $0.9M_B = -0.104qL^2$ from (c) and $0.9M_C = -0.096qL^2$ from (d) (see later), then the adjusted midspan moments are as follows.

$$AB = DE = (-0.052 \text{ to } +0.073)qL^2$$

$$BC = CD = (-0.100 \text{ to } +0.025)qL^2$$

However, the adjustments in this case are too great since the negative midspan moments are now relatively large and the positive moments are relatively small.

As an alternative consider the unadjusted maximum elastic span moments. The 90% values are as follows.

$AB = DE = 0.9 \times 0.098qL^2 = 0.088qL^2$

$BC = CD = 0.9 \times 0.080qL^2 = 0.072qL^2$

The adjusted midspan moment for AB becomes $0.088qL^2$ if M_B is increased to $-0.074qL^2$ only and, in addition, that for CD becomes $+0.070qL^2$ if M_C equals $-0.036qL^2$.

The negative moments at midspan for the unloaded spans BC and DE are then $-0.055qL^2$ and $-0.037qL^2$. Again by using symmetry of loading the adjusted midspan moments can now be given as follows:

$AB = DE = (-0.037 \text{ to } +0.088)qL^2$ (Answer)

$BC = CD = (-0.055 \text{ to } +0.070)qL^2$

The negative span moments are now much more reasonable and cannot be reduced further without increasing the sagging moments above the 90% maximum elastic moment envelope.

(c) Consider live loading q with adjacent spans AB and BC loaded. Applying Eqn (4.6.1) successively as before, except for putting $A_3 = A_4 = 0$, and solving gives:

$$M_C = -qL^2/28 = -0.036qL^2$$
$$M_B = -13qL^2/112 = -0.116qL^2$$
$$M_D = +qL^2/28 = +0.009qL^2$$

The above gives the maximum elastic moment for M_B for all cases.

For 10% relaxation of moments at support B

$M_B = -0.104qL^2$ (Answer)

The elastic midspan moments (using symmetry of loading for spans CD and DE only loaded) are given by:

$AB = DE = (+0.005 \text{ to } +0.067)qL^2$

$BC = CD = (-0.014 \text{ to } +0.049)qL^2$

If support moment M_B only is reduced to $-0.104qL^2$ then the adjusted midspan moments are given by:

$AB = DE = (+0.005 \text{ to } 0.073)qL^2$ (Answer)

$BC = CD = (-0.014 \text{ to } +0.055)qL^2$

All these span moments are within the ranges given by case (b).

(d) Consider live loading q with adjacent spans BC and CD loaded. Applying Eqn (4.6.1) successively as before, except for putting $A_1 = A_4 = 0$, and solving gives:

$M_C = -3qL^2/28 \quad = -0.107qL^2$

$M_B = M_D = -qL^2/28 = -0.036qL^2$

The above gives the maximum elastic moment for M_C for all cases.

For 10% relaxation of moments at support C:

$M_C = -0.096gL^2$

If support moment M_C only is reduced to $-0.096qL^2$ then the adjusted midspan moments are given by:

$AB = DE = -0.018qL^2$ (Answer)

$BC = CD = +0.059qL^2$ (Answer)

All these span moments are again within the ranges given by case (b).

The worst moments for fixed and variable loading are therefore those shown in Fig. 6.3.2 for the midspan and support sections.

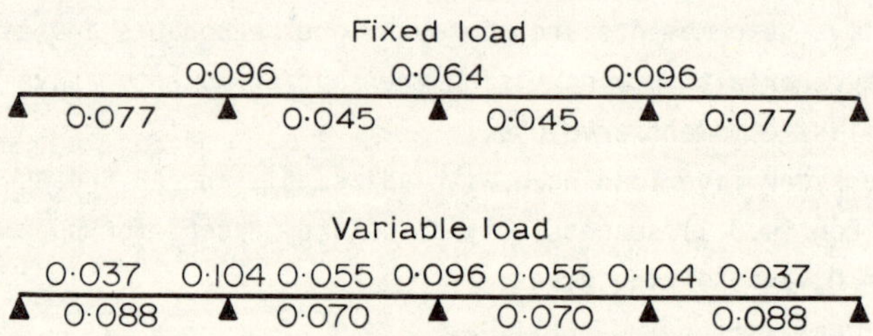

Fig.6.3.2 Example 6.3.1

Notice that care must be taken to ensure that the different loading cases are not confused. Notice also that midspan sections were considered for simplicity. The moments for these may not equal the maximum span moments, particularly for end spans. For example, for fixed loading, maximum span moment occurs 0.404 L from end support for 10% relaxation and equals $0.082wL^2$ which is appreciably greater than the midspan moment of $0.077wL^2$ obtained above.

Notice also that 10% of the numerically largest elastic maximum moment given anywhere by the elastic maximum moments diagram in Example 6.3.1 was $0.1 \times 0.116qL^2$, hence any moment could have been reduced by $0.012qL^2$ for any particular loading case, but the need did not arise.

When considering a 70% redistribution then, clearly, reductions of 30% in both the support and midspan variable load moment coefficients cannot be achieved whilst still retaining equilibrium. Various approaches can be adopted, but if all maximum support moments are reduced by 30% this has the advantages of minimizing both the hogging moments at midspan and the deviations from the 70% maximum elastic moments envelope near points of contraflexure. This approach has therefore been used for the coefficients given in Appendix I. Fig. 6.3.3 shows the bending moment diagrams which would be obtained if

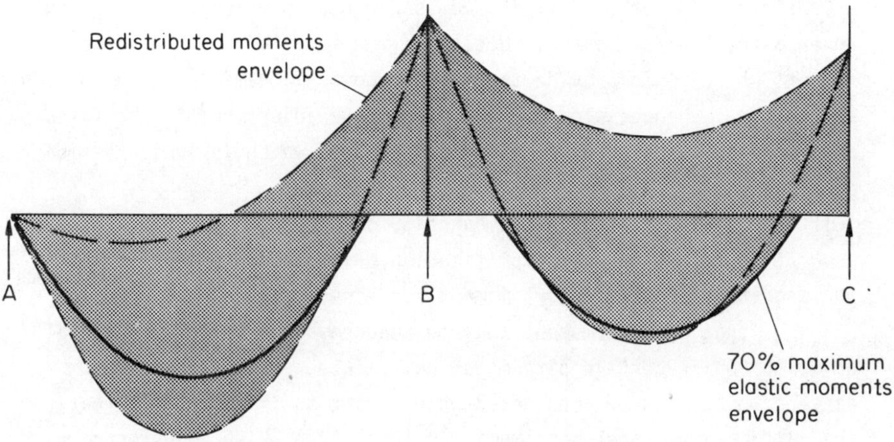

Fig.6.3.3 Typical 30% moment distribution

Example 6.3.1 was repeated for a 30% moment redistribution with $g_k = q_k = q$ (i.e. $g = q$ and $w_v = 0.4g_k + 1.6q_k = 2q$).

6.4 ECONOMIC ASPECTS OF MOMENT REDISTRIBUTION

CP 110 permits up to 30% redistribution of moments provided that the rotational capacity of individual sections is assured. However, assuming that the rotational capacity is satisfactory, how does one select the maximum value for the redistribution of moments, in a particular instance? Is it better to design for the maximum 30% or for only 10%? This is a difficult question to answer since the choice depends mainly on the economics for a particular situation. The following example, however, indicates that it would rarely be desirable, even on economic considerations, to exceed the maximum redistribution value of 30%.

Example 6.4.1

The 'most economical' design for a continuous reinforced concrete beam could be considered to be given by either of the following alternatives:

(a) Adjustment of the support moments so that the area enclosed by the bending moment diagram and the axis (all areas considered positive) is a minimum.

(b) Adjustment of the support moments so that they are equal in magnitude to the maximum span moment.

Consider a uniformly distributed load w per unit length, acting on an encastré beam of span L. Determine the magnitude of the support moments which would provide the most economical beam according to each alternative. Hence calculate the percentage adjustment to the fixed end moments, as given by an elastic analysis, which is indicated in each case.

If economy is related to the quantity of steel can you explain why assumption (a) is an oversimplification which tends to overestimate the magnitude of the support moment for economical design? CP 110 proposes that support moments may be reduced by up to 10% in general or up to 30% in certain circumstances. State, very briefly, the dangers which must be considered when trying to implement alternative (b). If maximum steel economy is in fact given by an adjustment which is intermediate between those represented by (a) and (b) do you think that a reduction in the elastic support moments of only 10% in the above encastré beam would be a very satisfactory compromise?

(UB3 1969, 45 min)

Solution

Free span moment = $wL^2/8$

End moments M_C and M_D for full fixity = $wL^2/12$

Hence let adjusted end moment = $\lambda wL^2/12$

The maximum span moment for adjusted end moments is

$$wL^2/8 - \lambda wL^2/12 = (1 - \tfrac{2}{3}\lambda)wL^2/8$$

(a) The area enclosed between BMD and axis in Fig.6.4.1 is $2A_1 + A_2$.

The bending moment at a section distance x from support, M_x, is

$$\tfrac{1}{2}wx(L - x) - \lambda wL^2/12$$

$M_x = 0$ when $x^2 - Lx + \lambda L^2/6 = 0$, i.e. when

$$x = \tfrac{1}{2}L \pm \tfrac{1}{2}L\sqrt{(1 - \tfrac{2}{3}\lambda)}$$

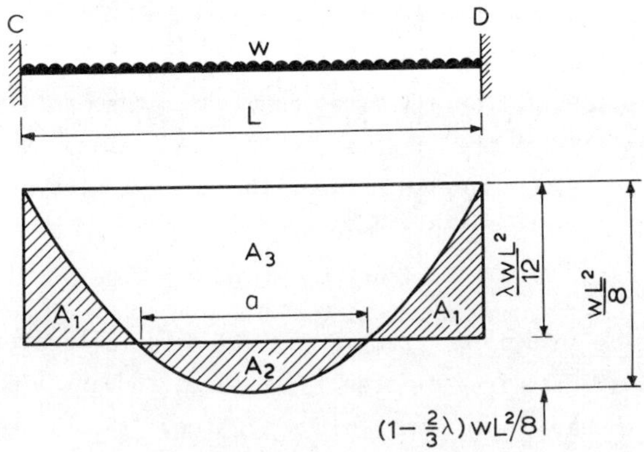

Fig.6.4.1 Example 6.4.1

Let a = length of beam under sagging moment, hence

$$a = L\sqrt{(1 - \tfrac{2}{3}\lambda)}$$

Now the area enclosed by a parabola is $\frac{2}{3}$ × (enclosing rectangle). Therefore

$$A_2 = \tfrac{2}{3}(1 - \tfrac{2}{3}\lambda)wL^2a/8 = (wL^3/12)(1 - \tfrac{2}{3}\lambda)^{3/2}$$

and

$$2A_1 = \lambda wL^3/12 - (\tfrac{2}{3}wL^3/8 - A_2) = (wL^3/12)(\lambda - 1) + A_2$$

i.e. the area A_3 forms a rectangle with $2A_1$ and a parabola with A_2. Therefore

$$2A_1 + A_2 = (wL^3/12)(\lambda - 1) + 2A_2$$
$$= (wL^3/12)[\lambda - 1 + 2(1 - \tfrac{2}{3}\lambda)^{3/2}]$$

$$\frac{d}{d\lambda}(2A_1 + A_2) = 0 \text{ when } 1 + 3(1 - \tfrac{2}{3}\lambda)^{\frac{1}{2}}(-\tfrac{2}{3}) = 0$$

i.e. when

$$(1 - \tfrac{2}{3}\lambda)^{\frac{1}{2}} = \tfrac{1}{2} \text{ or } \lambda = 9/8$$

Hence this criterion requires that the elastic end moment is increased by 12.5%. (Answer)

(b) For equal moments at supports and midspan

$$\lambda wL^2/12 = \tfrac{1}{2} \times wL^2/8 \qquad \text{or} \qquad \lambda = 3/4$$

i.e. this criterion requires that the elastic end moment is decreased by 25%.

Alternative (a) does not allow for the fact that the excess support moment requires compression steel as well as additional tension steel.

That is, the steel area per unit moment required for moments in excess of the balanced value is at least twice that for smaller moments.

Alternative (b) permits a 'balanced' section to be used which is just adequate at both midspan and supports. The adjustment of -10% allowed by CP 110 for all cases, and which does not introduce difficulties with the limit state of local damage, would seem to be a satisfactory compromise between the two extremes of + 12.5% and -25%. See also section 8.4 later.

Alternative solution A simpler solution for alternative (a) is obtained as follows. The area of steel is assumed to vary linearly with bending moment. Hence the minimum volume of steel is obtained by equating the lengths of beam which must resist positive and negative bending moments *irrespective* of the actual form of the BMD. In Fig.6.4.1, therefore, a = 1/2 since any vertical displacement of the axis from this position produces a net increase in $2A_1 + A_2$.

Therefore

$$a = L/2 = L\surd(1 - \tfrac{2}{3}\lambda) \quad \text{or} \quad \lambda = 9/8 \text{ as before.}$$

References

6.1 Greenberg, H.J., and Prager, W., 'Limit design of beams and frames', Transactions of the American Society of Civil Engineers, vol. 117, 1952, pp 447-84.

6.2 Cranston, W.B., and Cracknell, J.A., 'Tests on reinforced concrete frames', 2: 'Portal frames with fixed feet', Technical Report TRA 420, Cement and Concrete Association, London, 1969, pp.35.

6.3 Astill, A.W., 'Redistribution of moments', Concrete, vol.7, October, 1973, pp 46-8.

Exercise

1. A rectangular, reinforced concrete beam ABC is simply supported at A and C and is continuous over support B. Span AB = span BC = 9 m.

The uniformly distributed loadings and the partial safety factors for load respectively are as follows:

Dead load total: 8kN/m and 1.4

Live load: 12 kN/m and 1.6

Calculate the design moments at the central support B and at midspans AB and BC assuming a maximum redistribution of moments from an elastic analysis of 10%.

(UB3/H 1970, part-question only)

7 Analysis of beam sections at collapse

Analysis usually concerns determining certain unknowns when the section (or member) is fully defined: that is, when the concrete dimensions, and the areas and positions of the steel, are known. The present chapter considers the determination of the collapse load for a given section subjected to pure bending.

7.1 BASIC ASSUMPTIONS FOR INDIVIDUAL SECTIONS

The basic assumptions which can be made when determining the collapse load for a member in bending are summarized below. It is important to visualize the section as carrying the collapse (or ultimate) load and about to fail.

(i) Plane sections remain plane in bending.

(ii) Good bond exists between concrete and steel. Hence the strain in the reinforcement, whether compressive or tensile, is given by the theoretical strain in the adjacent concrete.

(iii) The stress distribution in the concrete in flexural compression is given by the stress-strain curve for concrete in uniaxial compression. The stress in the reinforcement is also given by the appropriate stress-strain curve.

(iv) The tensile strength of the concrete, and any direct tension stresses due to shrinkage, etc., are ignored.

(v) The ultimate flexural compressive strain in the concrete at the outermost fibre provides a satisfactory failure criterion.

Strain distribution

The basic assumption (already used in elastic theory) that plane sections remain plane after bending, is again made. Whilst the validity of this assumption has been verified by numerous investigators, the following limitations should be appreciated.

(i) *Imperfect bond* The linear strain distribution across a section refers to the *average* strain. Cracking occurs in the tension zone of a reinforced concrete beam at a series of localized points (since the bond strength is not infinite) so that rapid fluctuations of strain occur in both the concrete and the steel at the regions of discontinuity.

(ii) *'Deep' or 'wide' beams* If the depth of a beam is more than about one-third of the span, the shear strains become significant compared with the normal strains due to bending. This results in a non-linear distribution of strain. Similarly, distortion of the plane section can also occur in beams or slabs where the applied bending moment is not uniformly distributed across the breadth of the section. Local distortions of the plane section occur in the region of concentrated loads or reactions.

Stress distribution

The stress at any point can be determined from the strain at that point and the appropriate stress-strain relation. The 'appropriate' curves for concrete and steel are taken to be those in direct compression and direct tension respectively. The stress-strain curves for concrete in axial compression and pure bending have, in fact, been shown by Hognestad, Hanson and McHenry [7.1] to be similar.

The following additional assumptions affecting the stress distribution are also made.

(i) *Direct tensile stresses* Temperature and shrinkage stresses are negligible. If they are not, then separate calculations are necessary in order to take these effects into account.

(ii) *Tensile strength of concrete* All concrete in flexural tension is ignored. Calculations can be made taking the tensile strength of the concrete into account, but such refinements are rarely justified. Experimental evidence suggests that cracks can occur throughout the tensile zone and can even exist above the neutral axis if high temperature fluctuations or shrinkage effects occurred before the application of the load. Ignoring the tensile strength of the concrete partly compensates for (i) above.

Ultimate compressive strain

The compressive strain in the concrete at the extreme fibre, at collapse, is assumed to be given by the empirical value of Hognestad et al. or, more approximately, by a value of 3.5 millistrain.

7.2 FLEXURAL MOMENT AT COLLAPSE

The area under the stress-strain curve for the concrete can be referred to as a *stress block*. The shape of the stress block for the compression side of the neutral axis of a beam varies according to the level of loading as shown in Fig. 7.2.1. The present section first describes these different shapes and then shows how the ultimate flexural moment can be related to the ultimate stress block by the introduction of suitable stress block factors. This approach is due to Hognestad, Hanson & McHenry [7.1].

Triangular stress block

It is assumed that plane sections of the beam remain plane after bending, hence the distribution of strain due to bending is linear as shown in Fig. 7.2.1(a). Initially, stress is approximately proportional to strain, hence the stress block is as shown in Fig. 7.2.1(b) for fairly small loads.

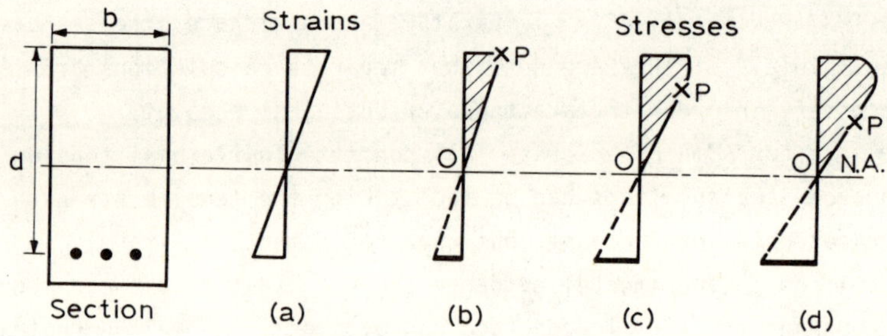

Fig.7.2.1 Strain and stress distributions for beam section subjected to an increasing moment

Curved stress block

Higher loads produce a stress block as shown in Fig.7.2.1(c). The strain is still proportional to the distance from the neutral axis, but since the stress-strain curve for concrete becomes curved at higher stresses, as shown in Fig.6.1.3, this same curve should be plotted (to a suitable scale) as in Fig.7.2.1(c) and (d).

Ultimate stress block

Just before failure the stress block becomes similar in shape to the entire stress-strain curve shown in Fig.6.1.3. The point P in Fig.6.1.3 is reproduced in Fig.7.2.1 for comparison.

Stress block factors

Whatever the actual shape of the stress block, its area can be related by some factor λ_1 to the area of the enclosing rectangle. Again, the position of the resultant concrete compressive force can be related by some factor λ_2 to the depth of the neutral axis, x, as shown in Fig.7.2.2. Similarly, let the maximum stress in the concrete be $\lambda_3 f'_c$, where f'_c is the crushing strength of a 150 mm × 300 mm cylinder.

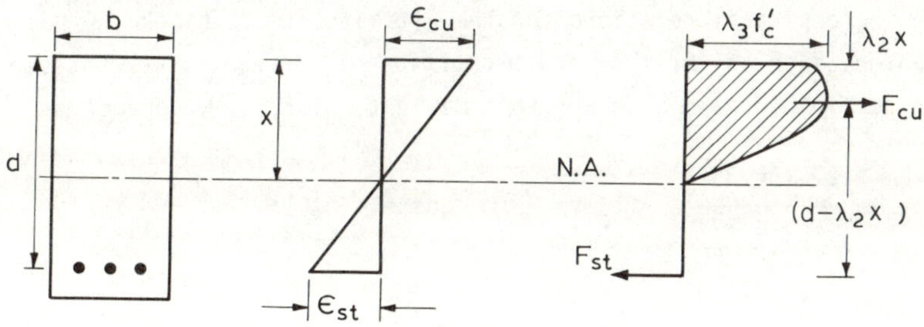

Fig.7.2.2 Strain distribution and stress block factors at collapse

Longitudinal forces

The maximum compressive force in the concrete is given by

$$F_{cu} = \lambda_1 \lambda_3 f'_c bx$$

The tensile force in the steel is given by

$$F_{st} = f_{st}A_{st}$$

where f_{st} can be less than, equal to, or greater than f_{yt}. Equating the horizontal forces, F_{cu} and F_{st}, gives

$$x = f_{st}A_{st}/\lambda_1\lambda_3 f'_c b \tag{7.2.1}$$

Ultimate bending moment

The ultimate bending moment of the section is given by taking moments about the line of action of either the steel force or the concrete force:

$$M_{uc} = \lambda_1\lambda_3 f'_c bx(d - \lambda_2 x) \tag{7.2.2}$$

and

$$M_{ut} = f_{st}A_{st}(d - \lambda_2 x) \tag{7.2.3}$$

Substituting for x from Eqn (7.2.1) in Eqn (7.2.3) and putting $M_{ut} = M_u$ gives

$$M_u = f_{st}A_{st}\left(d - \frac{\lambda_2 f_{st}A_{st}}{\lambda_1\lambda_3 f'_c b}\right) \tag{7.2.4}$$

Empirical values for the stress-block factors

Hognestad et al. found that age had very little effect upon the following empirical relations for the ultimate stress block. Other variables, such as the type and proportion of aggregate, were assumed to be negligible also although this may not be strictly correct.

$$\lambda_1\lambda_3 = \frac{27 + 0.35f'_c}{22 + f'_c} \quad (f'_c \text{ in MN/m}^2 \text{ units})$$

$$\lambda_2 = 0.5 - f'_c/550$$

$$\varepsilon_{cu} = (4 - f'_c/45) \times 10^{-3}$$

Also, $f'_c = 0.78f_{cu}$ is a typical relation between the standard 150-mm diameter cylinder strength and the 150-mm cube strength.

7.3 MODES OF FAILURE AND ULTIMATE CONCRETE STRAIN

Although the ultimate flexural moment has been determined in terms of the ultimate stress block factors in Eqn (7.2.4), it is evident that its actual value cannot be calculated unless the steel stress f_{st} (and hence x) is known. It is therefore necessary to examine the alternative modes of failure which can occur.

Balanced section

A balanced section is one for which the steel yields just as the concrete reaches its ultimate strain. In other words, the beam fails simultaneously in flexural compression and tension. The position of the neutral axis in Fig. 7.2.1 has been shown, for convenience, to be constant at all stages of loading. While this may be approximately correct for a balanced design, it is **not true** in all cases.

Under-reinforced section

The steel becomes over-stressed and therefore yields. The strain increases from ε_{s_1} to ε_{s_2} without any increase in stress, hence the neutral axis rises and the concrete finally fails in compression as shown in Fig.7.3.1(b). Although the final collapse occurs due to crushing of the compression zone, this is a secondary effect. The primary cause of failure is yielding of the tensile reinforcement, which allows the cracks in the beam to widen and progressively reduce the depth of the compression zone.

Over-reinforced section

The steel remains elastic while the concrete strains without increasing its moment (due to the descending portion of the concrete stress-strain curve). The steel strains elastically from ε_{s_1} to ε_{s_3} so that the

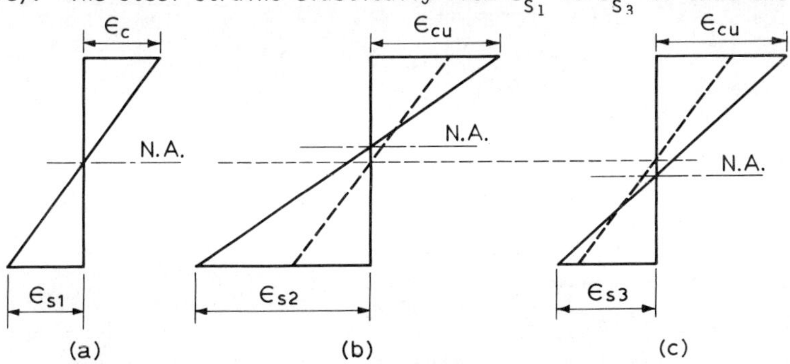

Fig.7.3.1 Variations in position of neutral axis for under- and over-reinforced sections

neutral axis falls while the concrete rapidly strains to failure.

Ultimate concrete strain

Regardless of the exact mode of failure, it is evident that the strain at the extreme concrete fibre reaches its ultimate value before flexural failure occurs. Hence in all cases

$$\varepsilon_{cu} = (4 - f'_c/45) \times 10^{-3}$$

as shown in Fig.7.3.2.

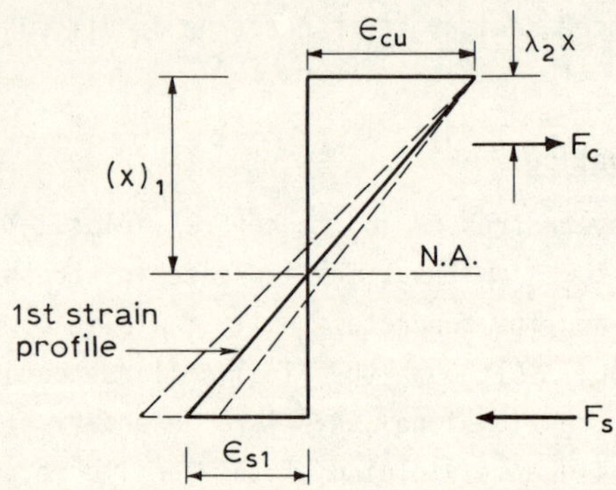

Fig.7.3.2 Tentative strain profiles for collapse

7.4 ULTIMATE FLEXURAL STRENGTH ANALYSIS

If the ultimate concrete strain is eventually reached in all cases then one point on the actual strain profile for the concrete section at collapse is known. This fact can be utilized to determine the complete strain profile at failure and hence the ultimate flexural strength.

Tentative strain profiles

If the correct strain profile at failure has a value at the extreme concrete fibre equal to the ultimate concrete strain, ε_{cu}, then the following trial and error method can be used. Draw any arbitrary strain profile using ε_{cu} as shown in Fig.7.3.2. Calculate F_s from the tentative strain value ε_{s_1} and the stress-strain curve for the steel. Calculate F_c from the expression $\lambda_1\lambda_3 f'_c bx$. If the assumed profile is correct then F_s balances F_c. If not, repeat for different profiles until a balance is in fact obtained. The ultimate moment at collapse can then be calculated. This approach is basically simple but can be laborious. Consideration is now given, therefore, to making a reasonably sensible first estimate for x.

Neutral axis depth for balanced section

A balanced section is achieved if the steel yields just as the concrete reaches its ultimate strain. From the design curves:

(i) Ultimate concrete strain = 0.0035

(ii) Minimum yield strain in steel exceeds 0.002, and can be taken as approximately equal to ε_{cu} for convenience.

It follows from Fig.7.4.1 and simple geometry that for a balanced section $x \simeq \frac{1}{2}d$. A more accurate, but less convenient, value would be

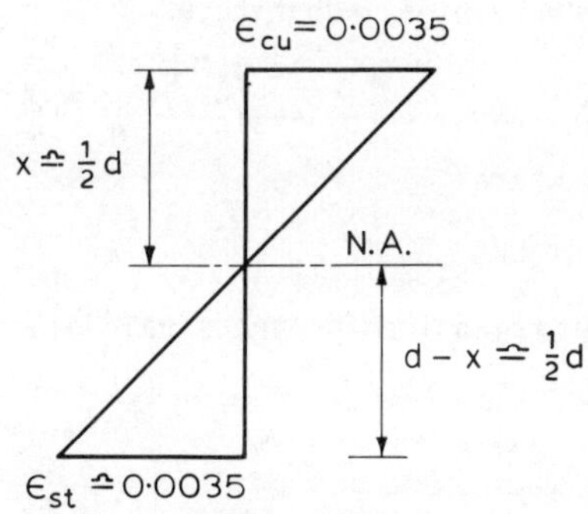

Fig.7.4.1 Strain profile at collapse for approximately balanced section

$x \simeq 0.6d$. Putting $f_{st} = f_{yt}$ in Eqn (7.2.1) provides the equation

$$x = f_{yt}A_{st}/\lambda_1\lambda_3 f'_c b$$

which is only true for a balanced section, i.e. $\frac{1}{2}d < x < 0.6d$ approximately. However, it can be used for any section in order to assess if it is over- or under-reinforced, as shown in the following procedure.

Recommended procedure for collapse-load analysis of beams

The collapse-load analysis of sections is a trial-and-error process. It is most conveniently carried out for sections in flexure as follows:

1. Determine a first approximation $(x)_1$ from
   $x = f_{yt}A_{st}/\lambda_1\lambda_3 f'_c b$
2. (i) If $(x)_1 < \frac{1}{2}d$ the beam is under-reinforced (i.e. A_{st} is less than balanced value).
   (ii) If $(x)_1 > \frac{1}{2}d$ the beam is over-reinforced (i.e. A_{st} exceeds balanced value).
3. (i) If under-reinforced, (x) is low, since actual $f_{st} > f_{yt}$. A grossly under-reinforced beam has a value given by
   $(f_{su}/f_{yt})(x)_1$
   Estimate a suitable value between these limits and check to see if forces balance for $(x)_2$.
   (ii) If over-reinforced, $(x)_1$ is high, since actual $f_{st} < f_{yt}$. If slightly high, choose a value intermediate between $(x)_1$ and $\frac{1}{2}d$. If very high, substitute
   $f_{st} = E_s\varepsilon_{st}$
   in the equation for x to obtain
   $\varepsilon_{st} = \lambda x$ where
   $\lambda = \lambda_1\lambda_3 f'_c b/A_{st}E_s$
   From similar triangles for the strain diagram (see Fig.7.4.1)
   $\varepsilon_{st}/\varepsilon_{cu} = (d - x)/x$
   or
   $(\lambda/\varepsilon_{cu})x^2 + x - d = 0$
   and hence calculate $(x)_2$. See if forces balance for $(x)_2$.
4. If forces do not balance within 2% calculate
   $(x)_3 = (F_{st}/F_c)(x)_2$
   (i) If under-reinforced, see if forces balance for $(x)_3$
   (ii) If over-reinforced, $(x)_3$ over-corrects, hence try mean of $(x)_2$ and $(x)_3$.
5. If forces still do not balance within 2% for $(x)_3$ estimate subsequent values for x by interpolation. This should be unnecessary, however, if reasonable estimates and slide-rule calculations have been made earlier.

6. Once the forces balance, and have therefore been determined correctly, the ultimate moment of resistance can be found by taking moments.

Compression steel

Consider the resultant concrete and steel forces in a balanced section of a doubly reinforced beam as shown in Fig.7.4.2. The total tensile force in all the steel is given by

$$F_s = f_{st}A_{st} - f_{sc}A_{sc}$$

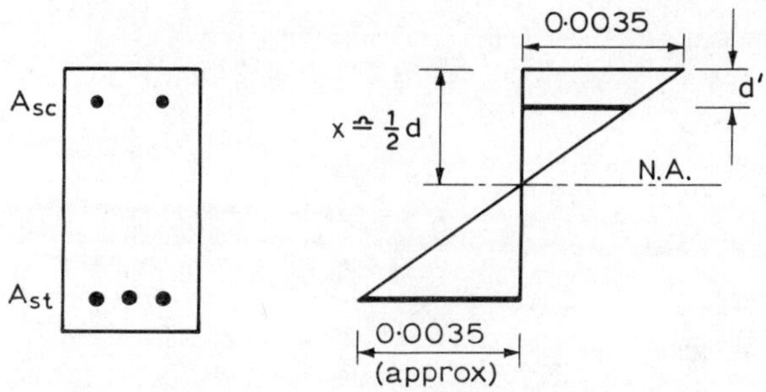

Fig.7.4.2 Strain profile for approximately balanced section of doubly reinforced beam

If $d' \ngtr \frac{1}{4}d$ the compressive steel has a strain of 2 millistrain and f_{sc} equals f_{yc} which is rather less than f_{yt}. The total compressive force in the concrete, F_{cc}, is given by

$$F_{cc} = \lambda_1\lambda_3 f'_c bx$$

Equating forces and rearranging gives

$$x = \frac{f_{st}A_{st} - f_{sc}A_{sc}}{\lambda_1\lambda_3 f'_c b} \qquad (7.4.1)$$

The same general procedure given previously for singly reinforced beams can therefore be applied to doubly reinforced beams, provided that

$$\sum F_s (= F_{st} - F_{sc})$$

is substituted for $f_{yt}A_{st}$ in 1 and for F_{st} in 4.

Notice that a compressive force has a positive sign when substituted into F_{sc} (since F_{sc} is specified as a compressive force) and a negative sign when substituted into F_s (in accordance with the general sign convention (1)—see Appendix Y).

Example 7.4.1

Calculate the collapse moment of resistance of the following beam by using the analytical approach of Hognestad, Hanson and McHenry:

Rectangular section 0.30 m deep × 0.15 m wide

Three 32-mm bars 0.20 m from compression face

Three 32-mm bars 0.25 m from compression face

The characteristic cube strength, f_{cu} is 37.5 MN/m². Assume that

$$f'_c = 0.78 f_{cu}$$

$$\lambda_1\lambda_3 = (27 + 0.35f'_c)/22 + f'_c)$$

$$\lambda_2 = 0.5 - f'_c/550$$

$$\varepsilon_{cu} = (4 - f'_c/45) \times 10^{-3}$$

where f'_c is in MN/m² units.

The load-strain curve for 20-mm diameter bars is as shown in Fig. 7.4.3(a). The load-strain curve for 32-mm diameter bars is similar with the different ordinate scale as shown. Assume similar curves for compression up to 0.23% strain.

<u>Solution</u>

$$f'_c = 0.78 \times 37.5 \times 10^6 = 29.2 \times 10^6$$

$$\lambda_1\lambda_3 = (27 + 0.35 \times 29.2)/(22 + 29.2) = 0.727$$

$$\lambda_2 = 0.5 - 29.2/550 = 0.447$$

$$\varepsilon_{cu} = (4 - 29.2/45) \times 10^{-3} = 3.35 \times 10^{-3}$$

Hence

$$\lambda_1\lambda_3 f'_c b = 0.727 \times 29.2 \times 0.15 \times 10^6 = 3.18 \times 10^6$$

The force per 32-mm bar at 'yield' (e.g. 2 millistrain) is 195×10^3. If the steel is just yielding as section fails then

$$x = f_{yt}A_{st}/\lambda_1\lambda_3 f'_c b = (6 \times 195/3.18) \times 10^{3-6} = 0.368$$

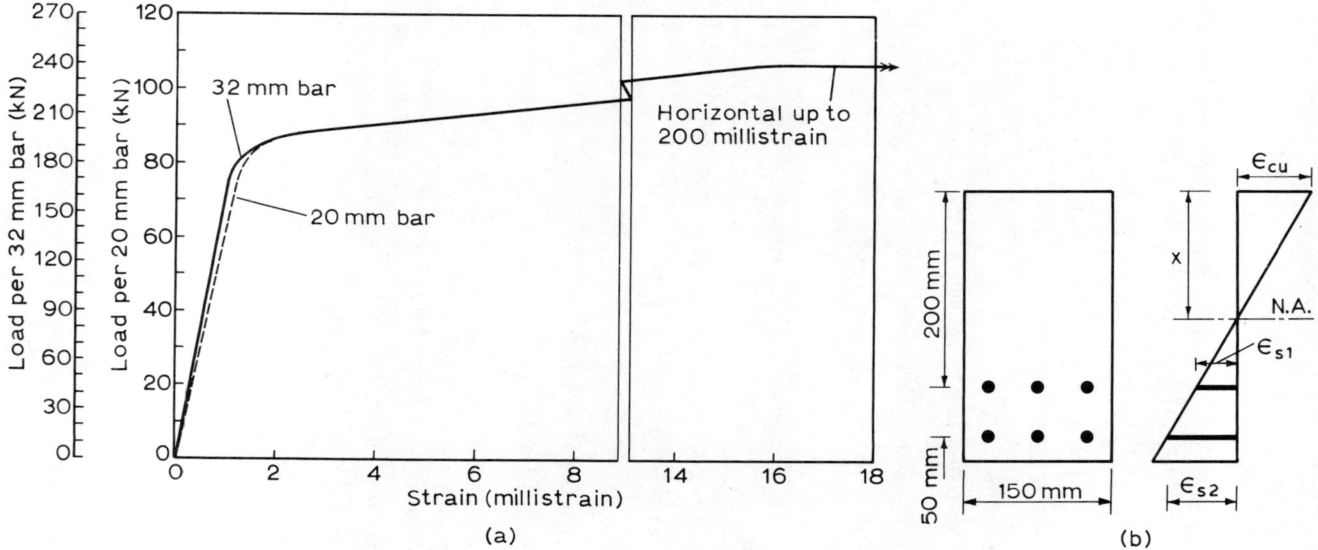

Fig.7.4.3 Example 7.4.1.(a) Load-strain curves for 32 mm and 20 mm bars. (b) Section and strain profile

Try $(x)_1 = 0.368$

For a balanced section

$x = \frac{1}{2}d = 0.1125$

Since $(x)_1 \gg 0.1125$, section is grossly over-reinforced and the steel is therefore behaving elastically. From the load-strain curve.

$$E_s = (160/\tfrac{1}{4}\pi \times 32^2 \times 1) \times 10^{3+6+3} = 200 \times 10^9$$

From similar triangles, the average steel strain is given by

$$\varepsilon_s = [(0.225 - x)/x] \times 3.35 \times 10^{-3}$$

For elastic steel strains:

$$F_{st} = \tfrac{1}{4}\pi \times 32^2 \times 6 \times 200 \times 10^{9-6}\varepsilon_s = 965 \times 10^6\varepsilon_s$$

But $F_{st} = F_{cu} = 3.18 \times 10^6 x$. Therefore

$$(3.18/965)x^2 + 3.35 \times 10^{-3}(x - 0.225) = 0$$

Hence

$x = 0.1895$

Try $(x)_2 = 189.5$ mm

$$\varepsilon_{s_1} = (10.5/189.5) \times 3.35 \times 10^{-3} = 0.185 \times 10^{-3}$$

$$\varepsilon_{s_2} = (60.5/189.5) \times 3.35 \times 10^{-3} = 1.07 \times 10^{-3}$$

Hence

$$\sum F_s = 3 \times (30 + 172) \times 10^3 = 606 \times 10^3$$

$$F_{cu} = 3.18 \times 10^6 x = 603 \times 10^3$$

Therefore

$\sum F_s = F_{cu}$ (within 2%)

$$\lambda_2 x = 0.447 \times 0.1895 = 85 \times 10^{-3}$$

Therefore, taking moments about the concrete force:

$$M_u = [90(200-85) + 516(250-85)] \times 10^{3-3} \qquad \text{(Answer)}$$

$$= 95.5 \text{ kNm}$$

T-beams

In the case of a T-beam, if x is less than the thickness of the compression flange, proceed as for a rectangular beam. This is the more usual case. If x exceeds the thickness of the compression flange an approximate solution only is possible. Divide up the compressive force as shown in Fig.7.4.4 such that

$F_c = F_{c_1} - F_{c_2}$

F_{c_1} = compressive force for an ordinary rectangular beam of breadth b.

F_{c_2} = compressive force under the flange projections.

The stress-strain relation for the area below the soffit of the flange is often approximately linear, hence a suitable estimate for F_{c_2} is given by

$$F_{c_2} \geq \tfrac{1}{2}E_c\varepsilon_{c_2}(b - b_w)(x - h_f)$$

where ε_{c_2} is the concrete strain at the soffit of the flange at failure. Alternatively

$$F_{c_2} < \lambda_1\lambda_3 f'_c(b - b_w)(x - h_f)$$

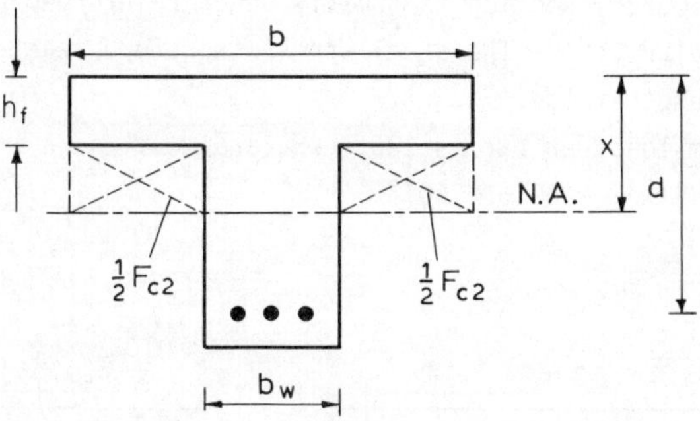

Fig.7.4.4 T-beam section

Example 7.4.2

Calculate the collapse moment of resistance of the following beam by using the empirical factors of Hognestad et al. and the load-strain curves given in Example 7.4.1:

T-beam with flange 650 mm wide and 150 mm deep, overall depth 550 mm and web 150 mm wide, reinforced as follows:

Three 20-mm bars 400 mm from compression face

Three 20-mm bars 450 mm from compression face

Three 20-mm bars 500 mm from compression face

There are also four 10-mm bars (equivalent in area and strength to one 20-mm bar) 25 mm from top edge of flange. The concrete cylinder strength, f'_c, is 30 MN/m^2.

Solution

$$f'_c = 30 \times 10^6$$
$$\lambda_1\lambda_3 = (27 + 10.5)/(22 + 30) = 0.721$$
$$\lambda_2 = 0.5 - 30/550 = 0.445$$
$$\varepsilon_{cu} = (4 - 30/45) \times 10^{-3} = 3.333 \times 10^{-3}$$
$$\lambda_1\lambda_3 f'_c = 0.721 \times 30 \times 0.65 \times 10^6 = 14.06 \times 10^6$$

If $x < 150$ mm, beam can be considered as rectangular. The load for 20-mm diameter bar at 'yield' (2 millistrain) is 86.7×10^3.

$$f_{yt}(A_{st} - A_{sc})/\lambda_1\lambda_3 f'_c b = [86.7(9 - 1)/14.06] \times 10^{3-6}$$
$$= 49.3 \times 10^{-3}$$

Try $(x)_1 = 49.3 \times 10^{-3}$

Since $\frac{1}{2}d \simeq 0.225$, section is grossly under-reinforced and is acting as a rectangular beam. The steel stress therefore reaches its ultimate strength value.

The maximum load per bar is 106.7 kN, and

$(106.7/86.7)(x)_1 = 60.6 \times 10^{-3}$

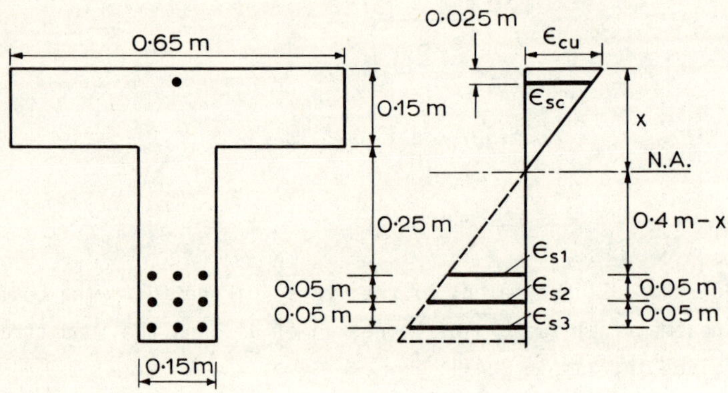

Fig.7.4.5 Example 7.4.2

Try $(x)_2 = 60.6 \times 10^{-3}$

From Fig.7.4.5

$$\varepsilon_{s_1} = (339.4/60.6) \times 3.333 \times 10^{-3} = 18.65 \times 10^{-3}$$
$$\varepsilon_{s_1} < \varepsilon_{s_2} < \varepsilon_{s_3} \quad \text{hence}$$
$$\sum F_{st} = 9 \times 106.7 \times 10^3 = 960 \times 10^3$$

$\varepsilon_{sc} = (35.6/60.6) \times 3.333 \times 10^{-3} = 1.955 \times 10^{-3}$ hence

$F_{sc} = 86.5 \times 10^{-3}$

$F_{cu} = 14.06 \times 10^6 x = 852 \times 10^3$

Therefore

$\sum F_s = 873.5 \times 10^3 \neq F_{cu}$

and

$$(\sum F_s/F_{cu})(x)_2 = (873.5/852) \times 60.6 \times 10^{-3}$$
$$= 62 \times 10^{-3}$$

Try $(x)_3 = 62 \times 10^{-3}$

$\varepsilon_{s_1} = (338/62) \times 3.333 \times 10^{-3} = 18.18 \times 10^{-3}$

$\varepsilon_{s_3} > \varepsilon_{s_2} > \varepsilon_{s_1}$

so $\sum F_{st}$ equals 960×10^3 as before.

$\varepsilon_{sc} = (37/62) \times 3.333 \times 10^{-3} = 1.99 \times 10^{-3}$

Hence from the curve

$F_{sc} = 86.7 \times 10^3$

$F_{cu} = 14.06 \times 10^6 x = 872 \times 10^3$

Therefore $\sum F_s = 873.3 \times 10^3$

Hence

$\sum F_s = F_{cu}$ (within 2%)

$\lambda_2 x = 0.445 \times 62 \times 10^{-3} = 27.6 \times 10^{-3}$

The depth of the resultant is

$(872 \times 27.6 + 86.7 \times 25)/958.7 \times 10^3 = 27.4 \times 10^{-3}$

Therefore $M_u = 959(450 - 27.4) = 405 \times 10^3$

$= 405$ kNm (Answer)

Reference

7.1 Hognestad, E., Hanson, N.W., and McHenry, D., 'Concrete stress distribution in ultimate strength design Journal of the American Concrete Institute, vol. 27, No.4, 1955, pp 455-79.

Exercises

1. Sketch moment rotation curves for under- and over-reinforced beams subject to bending moments.*

Using the analytical method of Hognestad, Hanson and McHenry, determir the ultimate flexural moment of the beam shown in Fig.A. Is this beam under- or over-reinforced? The following information is given:

$\lambda_1\lambda_3, \lambda_2$ and ε_{cu} as in Exercise 2 below.

The appropriate stress-strain curve for the reinforcement is also given in Fig.A. The cylinder compressive strength is

f'_c = 5,000 lbf/in^2 (= 34.5 MN/m^2)

1 kip/in^2 = 6.895 MN/m^2

1 in = 25.4 mm (UB3/0 1968, 36 min)

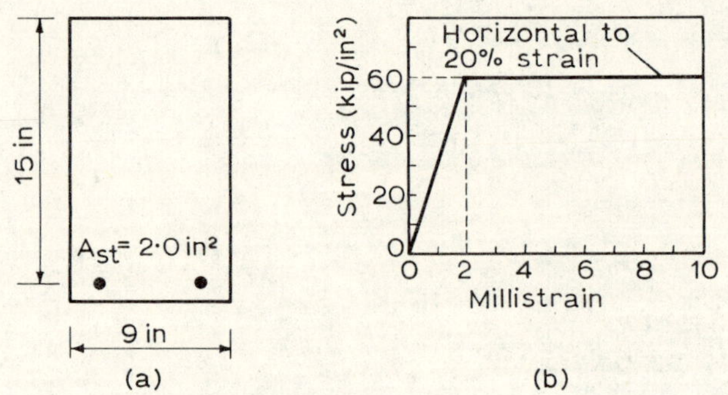

Ch.7 Fig.A Question 1.(a) Section (b) Stress-strain diagram

2. (a) Determine by the method of Hognestad, Hanson and McHenry the ultimate bending moment of the beam shown in Fig.B. The tensile stress-strain curve for the reinforcement reaches a maximum of 420 MN/m^2.

* Refer to chapter 8 for first part of question 1 and for part (b) of question 2.

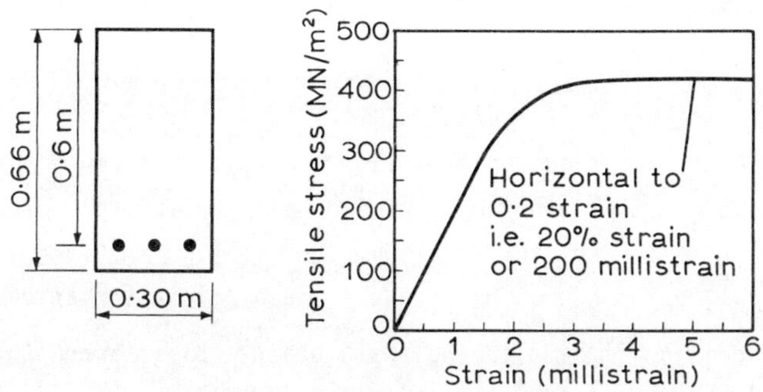

Ch.7 Fig.B Question 2

The empirical factors (MN/m^2 units) according to the usual notation are:

$$f_c' = 0.78f_{cu}$$
$$\lambda_1\lambda_3 = (27 + 0.35f_c')/(22 + f_c')$$
$$\lambda_2 = 0.5 - f_c'/550$$
$$\varepsilon_{cu} = (4 - f_c'/45) \times 10^{-3}$$

The concrete cube strength is 30 MN/m^2 and the reinforcement comprises three 32-mm diameter bars. Is this beam under- or over-reinforced?

(b)* Using the values of 420 MN/m^2 and 30 MN/m^2 as characteristic strengths of the steel and concrete respectively, determine the design strength of the section in bending by means of the design chart (Fig.E3). Why are the bending moment values different for (a) and (b)?

(UB3/0 1969, 45 min)

* Refer to chapter 8 for first part of question 1 and for part (b) of question 2.

8 Design of beams at collapse

The determination of a suitable section to resist a given moment is a design problem. It must not be confused with the analysis of a given section which is already fully defined.

A practical design approach can adopt one of two alternatives:

(i) Make additional (simplifying) assumptions. This approach was used in chapter 4 for elastic theory and will be considered first.

(ii) Analyse a large number of sections and use the information to draw up design charts. Once the underlying principles are fully understood and the appropriate charts are available, designs can be carried out very rapidly. This is the normal method for use by an experienced designer.

8.1 GENERAL DESIGN CONSIDERATIONS FOR COLLAPSE

Practical design must consider all the factors which affect the serviceability and safety of a structure. Limit state theory requires that each limit state is considered individually. This means that a section or member designed for one limit state is then checked by analysing the particular section for compliance with the requirements of other limit states. In theory it is possible to design in accordance with any limit state and then analyse in accordance with the remainder, making any adjustments that are necessary in the process. In practice it is desirable to try and reduce any subsequent adjustments for the other limit states to a minimum. The accumulated experience which is available for working load conditions greatly exceeds that which is available for structures which have failed. Hence it has been possible to evolve simple arbitrary rules which often ensure compliance with the limit states of local damage and deflection. The normal procedure will therefore be to design according to the limit states of collapse as well as complying with a number of arbitrary rules given in CP 110. The design will then be found, in general, to comply with the other limit states as appropriate, and subsequent adjustments or re-designs should usually be unnecessary. Design requirements for the limit state of collapse are now considered in detail.

Warning at failure

An under-reinforced beam undergoes extremely large deflections as the load approaches its ultimate value, since the steel yields and the neutral axis rises. Such a beam gives ample warning of approaching failure. An over-reinforced beam, on the other hand, fails at a relatively small deflection since the steel is still in the elastic range as the concrete crushes with explosive violence. The latter beam is therefore a potentially dangerous beam to use in any structure.

Moment-rotation curves

The moment-rotation characteristics of over- and under-reinforced concrete beams are as shown in Fig. 8.1.1. The over-reinforced beam has little plasticity and is typical of a brittle failure. This lack of plasticity means that if a redundant structure is constructed of such members, collapse occurs as soon as any single section reaches its

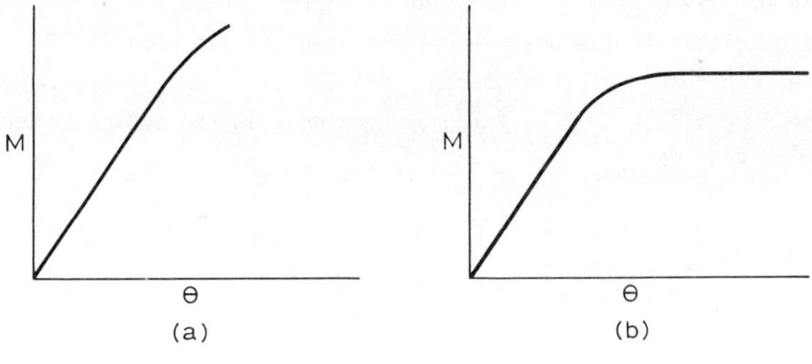

Fig.8.1.1 Moment rotation characteristics of (a) over- and (b) under-reinforced beams

ultimate moment. The under-reinforced beam, however, can undergo a very large deformation and still resist a moment very near the ultimate value. This therefore enables the full plastic strength of the structure to be realized.

Neutral axis position

The position of the neutral axis for a balanced section has been shown to be at approximately one-half of the depth of the beam. If it is arbitrarily stipulated that $x \ngtr \frac{1}{2}d$ then the steel strain at failure always exceeds the yield strain, and an under-reinforced (and therefore inherently safer) type of failure is assured. Furthermore, CP 110 permits a decrease in the maximum elastic moments of up to 10% in this case: i.e. 10% redistribution of the bending moments. If x is made even smaller, the rotational capacity of the section is increased further. Up to 30% redistribution of moments throughout a structure is allowed if x is reduced to 0.3d.

Steel stress

The stress-strain curve for the steel is of the form shown in Fig.6.1.4. If the steel strain at failure exceeds the yield strain then, for all but the greatly under-reinforced beams, the steel stress at failure equals the yield stress. The slightly higher stress value obtained for the relatively few beams where the steel strain becomes very much greater can reasonably be ignored, however, since such variations are on the safe side. This has already been done in the design curve shown in Fig. D2.

Concrete stress block

For design purposes it is convenient to assume a stress block shape which is as simple as possible, while still giving results which are near to the actual. Many different shapes have been suggested from time to time, such as an ellipse or a trapezium, but the one which formed the basis for earlier British and American Codes was the rectangular shape first proposed by C.S. Whitney in 1937 [8.1].

Comparison between theory and design

Reference to Fig. 8.1.2 indicates that the rectangular stress blocks have, in effect, given simple values to the constant factors in the

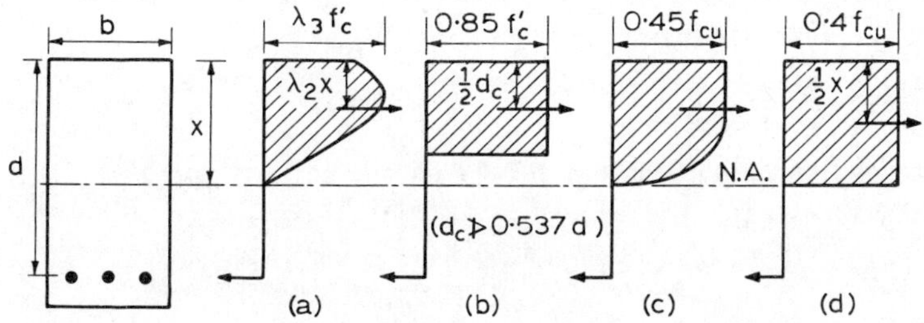

Fig.8.1.2 Stress blocks according to (a) Hognestad (b) Whitney (c) CP 110 and (d) CP 110 (simplified)

theory of Hognestad et al. For example, the simplified stress block assumed by CP 110 is rectangular with an average stress value of $0.4f_{cu}$ and the depth of the resultant is simply ½x. The ultimate moment values obtained are similar to Hognestad's more precise values provided that the beams are under-reinforced.

CP 110 requirements

The basic assumptions of CP 110 when using the simplified design approach are as follows:

(i) The concrete compressive stress block at failure is rectangular, of depth x and average stress $0.4f_{cu}$ (and 0.0035 maximum strain).

(ii) The steel stresses at failure for sections in pure flexure are the upper design yield stresses of $f_{yt}/1.15$ at adequate strain in tension and about $0.72f_y$ at 0.002 strain in compression.

(iii) In all cases $x \not> \frac{1}{2}d$ with moment redistributions up to 10%.

(iv) The tensile strength of the concrete is ignored.

(v) Ultimate axial stresses up to $0.1f_{cu}$ may be ignored.

For the design charts approach (i) and (ii) should refer to plane sections remaining plane and to the stress-strain curves in Figs. D1 to D3, and (iii) becomes as below.

If a redistribution of moments between 10 and 30% is assumed such that the ultimate resistance moment at a section is reduced then:

$$x \ngtr (0.6 - \beta_{red})d$$

where β_{red} is the ratio of the reduction in resistance moment, to the numerically largest moment given anywhere by the elastic maximum moments diagram for that particular member, covering all appropriate combinations of ultimate loads.

For beams reinforced only in tension $x \ngtr \frac{1}{2}d$. Also, in structures over four storeys in height and which rely on the columns for stability, the reduction in moment $\ngtr$ 10%.

8.2 DESIGN OF SINGLY REINFORCED CONCRETE BEAMS

Once the design assumptions are known, the moments of resistance can be determined by equating the longitudinal forces and taking moments about the lines of action of the steel and concrete forces.

Moment of resistance of steel

By using the simplified assumptions of CP 110 and equating longitudinal forces then, from Fig. 8.2.1,

$$0.4f_{cu}bx = 0.87f_{yt}A_{st}$$

$$x \simeq 2.2f_{yt}A_{st}/f_{cu}b \qquad (8.2.1)$$

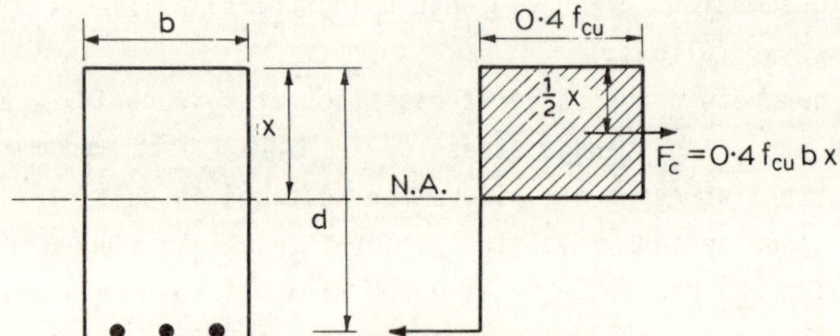

Fig.8.2.1 Simplified CP 110 assumptions for singly reinforced section

Taking moments about the concrete force gives

$$M_{ut} = 0.87f_{yt}A_{st}z \tag{8.2.2}$$

Since $z = d - 1.1f_{yt}A_{st}/f_{cu}b$ $(z \not> 0.95)$

and $A_{st} = \rho bd$ Eqn (8.2.2) can be expressed as

$$\frac{1.15M_{ut}}{bd^2f_{yt}} = \rho\left(1 - \frac{1.1\rho f_{yt}}{f_{cu}}\right) \tag{8.2.3}$$

Eqns (8.2.1) to (8.2.3) could have been obtained by substituting

$f_{st} = 0.87f_{yt}$ $\lambda_2 = \frac{1}{2}$ $\lambda_1\lambda_3f'_c = 0.4f_{cu}$ and $M_u = M_{ut}$

in Eqns (7.2.1) and (7.2.4)

Example 8.2.1

Calculate the collapse moment of resistance of the following beam by using the simplified method to CP 110 if $f_{cu} = 30$ MN/m² and $f_{yt} =$ 420 MN/m².

Rectangular section 300 mm deep and 150 mm wide
Four 10-mm bars 250 mm from compressive face
Four 10-mm bars 275 mm from compressive face.

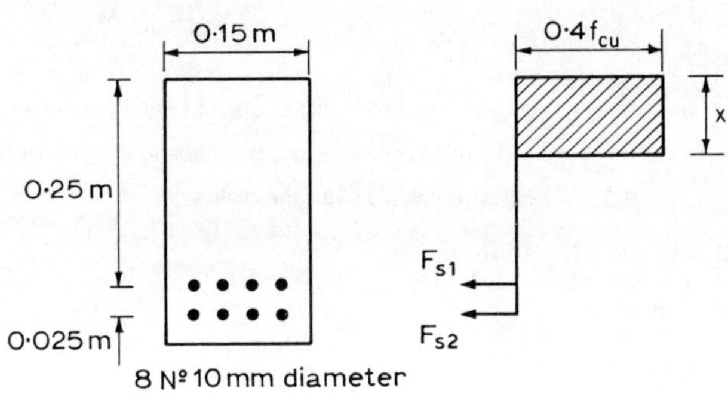

Fig.8.2.2 Example 8.2.1

Is the assumed stress block satisfactory (i.e. is it on the safe side) if the accurate analysis of Hognestad et al. indicates an ultimate moment of resistance of 57.6 kNm? What is the percentage difference?

Solution

$*f_{yt} = 420 \times 10^6/1.15 = 365 \times 10^6$

Therefore

$F_{s_1} = F_{s_2} = 4 \times \frac{1}{4}\pi \times 10^2 \times 365 \times 10^{-6+6} = 114.7 \times 10^3$

$f_{cu} = 30 \times 10^6$

hence by equating forces and transposing

$x = (2 \times 114.7/0.15 \times 0.4 \times 30) \times 10^{3-6} = 127.5 \times 10^{-3}$

Therefore

$M_u = 114.7(250 + 275 - 127.5) \times 10^{3-3} = 45.6 \times 10^3$

The percentage difference is

$100 \times (45.6 - 57.6)/57.6 = -20.8\%$ (Answer)

(Notice that the partial safety factor of 1.15 gives a reduction of 1 − 0.87, i.e. 13%, and therefore accounts for most of the difference.)

Maximum steel requirement

The requirement that $x \not> \frac{1}{2}d$ can be rewritten by substituting for x from Eqn (8.2.1) to give

$2.2f_{yt}A_{st}/f_{cu}b \not> \frac{1}{2}d.$

Since $A_{st} = \rho bd$:

$$\rho \not> f_{cu}/4.4f_{yt} \quad (8.2.4)$$

The limitation concerning the neutral axis position is therefore directly equivalent to limiting the amount of steel that can be used. If f'_c equals $0.78f_{cu}$ then the condition becomes

$\rho \not> 0.292f'_c/f_{yt}$

which compares with

$\rho \not> 0.40f'_c/f_{yt}$

in the American Code for f'_c values less than 5000 lbf/in² (34.5 MN/m²).

Collapse moment for balanced design section

The balanced condition is given when failure occurs simultaneously in flexural tension and compression. Since the basic assumptions vary for different design methods, Hognestad et al. theory etc., then the steel ratios for a balanced section also vary. For the simplified CP 110 design assumption of a rectangular stress block, the 'balanced' collapse moment is therefore given by putting x equal to ½d and considering the moment of resistance of the concrete, i.e.

$$M_{uc} = 0.4f_{cu}b(\tfrac{1}{2}d)(\tfrac{3}{4}d) = 0.15f_{cu}\ bd^2 \qquad (8.2.5)$$

Alternatively, this result could be obtained by substituting

$$\rho = f_{cu}/4.4f_{yt} \quad (\text{or } 0.23f_{cu}/f_{yt})$$

in Eqn (8.2.3).

Similarly, for cylinder strengths up to and including 5000 lbf/in² (34.5 MN/m²) the American Code gives:

$$M_{uc} = 0.306f_c'bd^2 \simeq 0.24f_{cu}bd^2$$

It is clear that the maximum tension-steel ratio permitted by CP 110 ensures that all sections are appreciably under-reinforced even if only 10% moment redistribution is assumed. This is a deliberate policy to ensure that failures in bending are always due to flexural tension, and are therefore gradual.

Collapse design and theory

Table 8.2.1 compares the values obtained by using the design assumptions of CP 110 with the analytical method of Hognestad et al. Notice that

Table 8.2.1 Comparison between design and analysis for reinforcing steel with characteristic yield strength of 420 MN/m²

Hognestad et al.		CP 110 (design chart)		CP 110 (simplified)	
Characteristic cube strength (MN/m²)					
20	30	20	30	20	30
Average stress factor (0.78 $\lambda_1\lambda_3$ for Hognestad et al.)					
0.673	0.605	0.412	0.403	0.4	0.4
Depth factor of resultant (λ_2 for Hognestad et al.)					
0.472	0.457	0.459	0.451	0.5	0.5
M_u/bd^2 ρ = 1.00%					
3.18	3.32	3.91	3.15	2.81	3.09
M_u/bd^2 x = 0.5d					
5.14	7.00	3.17	4.69	3.00	4.50
(ρ=1.84%)	(ρ=2.49%)	(ρ=1.13%)	(ρ=1.66%)	(ρ=1.10%)	(ρ=1.64%)

while both the line of action of the resultant and the average stress values of the Code differ appreciably from Hognestad's, the expressions for M_u/bd^2 are not nearly so variable since the failure is primarily due to the steel: the smaller the steel ratio, the smaller the difference. The maximum error occurs at the maximum permissible values of ρ. However the error is on the safe side and is therefore very satisfactory.

Example 8.2.2

Design a simply supported beam of 3.6 m span to carry a rolling point load of 36 kN (neglect dead weight). Assume that

f_{cu} = 40 MN/m² f_{yt} = 410 MN/m²

Use the partial safety factors given in Appendix B. Assume a breadth of 150 mm and consider only the midspan section in bending.

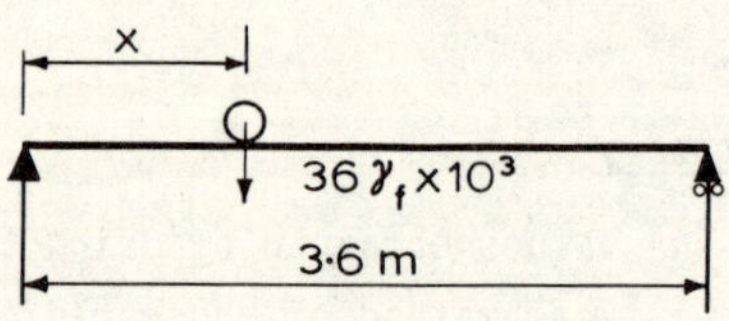

Fig.8.2.3 Example 8.2.2

Solution

M at midspan is

$$\tfrac{1}{4} \times 1.6 \times 36 \times 10^3 \times 3.6 = 51.8 \times 10^3$$

Therefore

$$0.15 f_{cu} bd^2 = 51.8 \times 10^3$$

$$d = \sqrt{\frac{51.8 \times 10^{3-6}}{0.15 \times 40 \times 0.15}} = 0.240 = 240 \text{ mm}$$

Let h = 270 mm and actual d = 0.242

$$A_{st} \times 0.87 f_{yt} \times \tfrac{3}{4} \times 0.242 = 51.8 \times 10^3$$

$A_{st} = 0.80 \times 10^{-3} = 800\ mm^2$

Say 4 No. 16 mm (804 mm^2) (Answer)

8.3 DESIGN OF DOUBLY REINFORCED RECTANGULAR BEAMS

Doubly reinforced rectangular beams may be used for two reasons:

(i) The applied moment exceeds the maximum permissible for a singly reinforced beam, because the size of the beam is restricted. The moment in excess of $0.15f_{cu}bd^2$ must therefore be taken by compression steel and additional tension steel.

(ii) The applied moment does not exceed the maximum permissible for a singly reinforced beam, but compression steel is already present for other reasons.

Additional longitudinal forces

Consider case (i) first. The additional steel for beams subjected to a high moment is given by considering the *additional* moment to be resisted and using, in effect, the steel beam theory. Let A_{sa} be the additional area of tension steel required, and A_{sc} be the area of compression steel. Equating the additional longitudinal forces gives

$$A_{sa} \cdot {*f_{yt}} = A_{sc} \cdot {*f_{yc}}$$

or

$$A_{sa}f_{yt}/1.15 = 2000A_{sc}f_{yc}/(2300 + f_{yc}) \simeq 0.72f_yA_{sc} \tag{8.3.1}$$

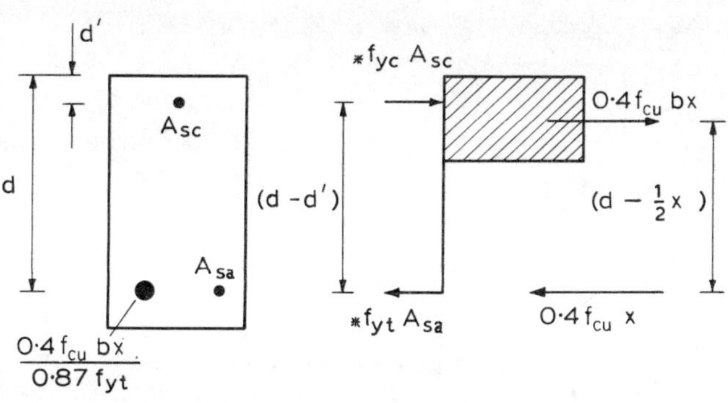

Fig.8.3.1 Simplified CP 110 assumptions for doubly reinforced section

If the same steel is used in tension and compression then

$$f_{yt} = f_{yc} = f_y$$

where f_y is the characteristic yield strength in tension.

Notice that f_{yc} must be in MN/m² in Eqn (8.3.1). Also, that the term $A_{sc} \cdot {}_*f_{yc}$ should really be $A_{sc}({}_*f_{yc} - 0.4f_{cu})$. However, this small difference is absorbed in ${}_*f_{yc}$ which is much less than ${}_*f_{yt}$.

Areas of steel

The area of compression steel is given by Eqn (8.3.1) and the total area of tension steel is given by

$$A_{st}f_{yt}/1.15 = 0.4f_{cu}bx + 0.72f_yA_{sc} \qquad (8.3.2)$$

Collapse moment of resistance

CP 110 assumes that the compression steel is matched by additional tension steel to maintain x/d at 0.5, hence M_u is given by

$$M_u = 0.15f_{cu}bd^2 + 0.72f_yA_{sc}\ (d - d') \qquad (8.3.3)$$

Minimum steel strains

The foregoing equations also assume that the steel strains reach upper yield: i.e. $0.002 + {}_*f_y/E_s$ in tension and 0.002 in compression. Thus for grade 460 steel the required tensile strain is 0.0040 and hence by similar triangles (e.g. from Fig. 7.4.2):

$$\frac{0.0035}{x} = \frac{0.0040}{d - x} \quad \text{or } x/d = 0.47 \simeq \tfrac{1}{2}$$

Hence if $x \not> \frac{1}{2}d$ the upper yield is nearly reached for all steels up to grade 460. For the compression steel, again from similar triangles:

$$\frac{0.0035}{x} = \frac{0.002}{x - d'}, \quad \text{or } d'/x = 0.43$$

Hence if $d'/d \not> 0.2$ (and $x = \frac{1}{2}$) the upper yield in compression is reached for all steels. However, if $d'/d > 0.2$ then the concrete can fail first and Eqn. (8.3.3) is not valid.

Beams under-reinforced in tension

CP 110 does not include design formulae based on case (ii) for the simplified assumptions, but if it did then from Fig.8.3.2 and since the compressive strength of the concrete is not fully utilized,

$$M_u = 0.4f_{cu}bx(d - \tfrac{1}{2}x) + 0.72f_yA_{sc} \qquad (8.3.4)$$

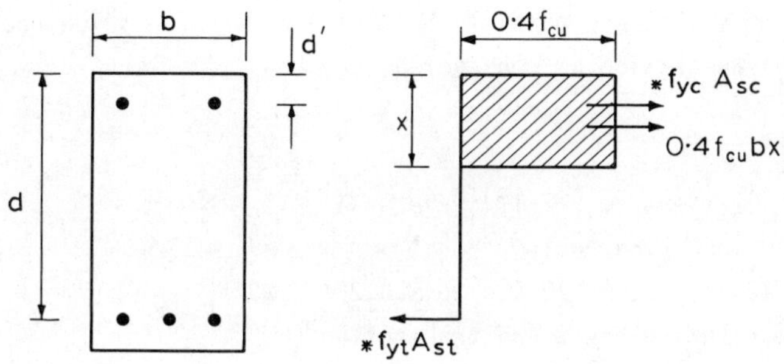

Fig. 8.3.2

If the ratio $d'/d \not> 0.2$ then x can be found from Eqn (8.3.4). CP 110 rejects this alternative because the steel strain is more likely to be critical now that x is less than the balanced value. Once x is known, A_{st} can be obtained from Eqn (8.3.2).

Example 8.3.1

Design a suitable section for the beam in Example 6.1.1 if it is to have the same cross-section throughout. Take

$f_{cu} = 30\ \text{MN/m}^2$, $f_y = 425\ \text{MN/m}^2$ and $*f_{yc} \simeq 0.72f_y$

and assume a breadth of 250 mm. Give the effective depth and the required areas of steel at both support and midspan sections. Assume a depth of 25 mm to any compression reinforcement.

Solution (This is a normal design problem)

$$M_u = 0.15f_{cu}bd^2 = 0.15 \times 30 \times 0.25 \times 10^6d^2$$
$$= 1.125 \times 10^6d^2$$

(a) For balanced support section,

$$d = \sqrt{\left(\frac{146}{1.125} \times 10^{3-6}\right)} = 0.36$$

(b) For balanced span section

$$d = \sqrt{\left(\frac{258.5}{1.125} \times 10^{3-6}\right)} = 0.48$$

A wide range of values are possible but let a compromise value nearer to the balanced span section be considered, say

$d = 0.45$

At support:

$146 \times 10^3 = A_{st}(425 \times 10^6/1.15)(0.45 - \tfrac{1}{2}x)$

Equating forces and transposing

$x = 0.87 \times 425A_{st} \times 10^{6-6}/0.4 \times 30 \times 0.25 = 123A_{st}$

Hence, by substituting for x and transposing

$61.5A_{st}^2 - 0.45A_{st} + 0.395 \times 10^{-3} = 0$

Therefore

$$A_{st} = \frac{0.45 \pm \sqrt{(0.45^2 - 4 \times 61.5 \times 0.395 \times 10^{-3})}}{123}$$

$= 0.00102$ (positive root is inapplicable)

$= 1020\ \text{mm}^2$ (Answer)

At midspan:

Balanced moment for section is

$0.15 \times 30 \times 0.25 \times 0.45^2 \times 10^6 = 0.228 \times 10^6$

A_{st} for balanced section is

$0.228 \times 10^{6-6}/0.87 \times 425 \times 0.75 \times 0.45 = 1827 \times 10^{-6}$

Additional moment is

$(258.5 - 228) \times 10^3 = 30.5 \times 10^3$

Therefore

$$A_{sa} = \frac{30.5 \times 10^{3-6}}{0.87 \times 425(0.45 - 0.025)} = 197 \times 10^{-6}$$

Therefore total tensile steel = 2047 mm^2 (Answer)

$*f_{yc} \simeq 0.72 \times 425 = 306 \times 10^6$

Therefore area of compressive steel is

$$A_{sc} = \frac{30.5 \times 10^{3-6}}{306(0.45 - 0.025)}$$

$= 235 \text{ mm}^2$ (Answer)

Example 8.3.2

Calculate the collapse moment of resistance of the beam in example 8.2.1 if it also has four 10-mm bars 25 mm from the compression face. Assume a rectangular concrete stress block of mean stress $0.4f_{cu}$ (i.e. as in CP 110) and that the design yield strength for the steel in compression is

$2000f_y/(2300 + f_y)$

where f_y is in MN/m².

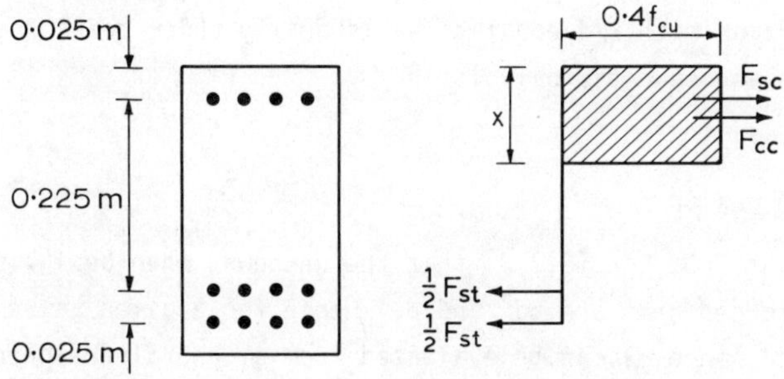

Fig.8.3.3 Example 8.3.2

<u>Solution</u> (This is an analysis problem.) Notice that the section need not conform to normal design (i.e. $x = \frac{1}{2}d$); hence the general approach as in case (ii) must be used.

$*f_{yc} = (2000/2720) \times 420 \times 10^6 = 309 \times 10^6$

Hence, since $d'/d \not> 0.2$,

$F_{sc} = \pi \times 10^2 \times 309 = 97.0 \times 10^3$

Equating forces and since $\Sigma F_{st} = 229.4 \times 10^3$:

$0.4f_{cu} \times 0.15x = (229.4 - 97.0) \times 10^3$

or

$x = 73.6 \times 10^{-3}$ (and $d'/x = 0.34 \not> 0.4$)

Taking moments about the compression steel:

$$M_u = 229.4 \times 237.5 \times 10^{3-3} - 0.4 \times 30 \times 0.15 \times 73.6 \times (36.8 - 25) \times 10^{6-3-3}$$

= 52.9 kNm (Answer)

8.4 DESIGN CURVES FOR REINFORCED CONCRETE BEAMS

It has been shown that the simplified assumption of a rectangular concrete stress block can give reasonably accurate values of the collapse moment of resistance provided that over-reinforced sections are avoided. For repetitive and rapid design, however, it is convenient to refer to design charts. These can be based on the design stress-strain curves given in Figs.D1 to D3. They can therefore be slightly more accurate for under-reinforced sections and can, in addition, cover over-reinforced sections as well should these be required in certain circumstances. (As noted in section 8.2 an over-reinforced section *by design* may not be actually over-reinforced *in theory*. A balanced section with x/d equal to ½d is really under-reinforced according to an accurate analysis.)

Variables in design

It is apparent from Eqn (8.2.3) that the unknowns when designing a singly reinforced beam are bd^2 and ρ. Hence for a given value of bd^2, ρf_y (and hence ρ) can be evaluated from a general design chart as shown in Fig.E1 (Appendix E) for given values of f_{cu} and f_y. This chart covers all the main grades of concrete and steel and also shows the appropriate cut-off positions for neutral axis depths of 0.3, 0.4, 0.5 and 0.6d, i.e. moment reductions of 30%, 20%, 10% and 0% respectively. Use x = 0.6d only with compression steel.

Design charts for doubly reinforced beams

A doubly reinforced beam can be designed by using Fig.E1 and the appropriate value of x/d (e.g. 0.5) to determine M_u for the given values of f_{cu} and f_y. A simple calculation for the additional moment to be resisted (see section 8.3) gives the required additional areas of tension and compression steel.

An alternative approach is to specify a value of d'/d and produce a separate design chart for each value of f_y and f_{cu}. Typical examples of such charts are given in CP 110. Although very convenient in use the required number of design charts is considerable. Figs.E2 and E3 are examples of such charts, except that they have been made a little more general by giving a limited range of steel grades on one chart, e.g. Fig.E2 is suitable for f_y = 410 or 425 MN/m².

Moment redistribution for optimum design

Notice from Figs. E2 and E3 that the greater the moment redistribution, the greater the compression steel required and the less the tension steel for a given moment. Figure E1 shows the onset of the flatter portions of the curves for different steel grades and indicates that for the optimum design to give minimum total steel the x/d ratios are slightly over 0.5 for mild steel and about 0.45 for f_y = 485 MN/m². This therefore supports the compromise optimum result from section 6.4 of redistributions between about 10 and 15%.

Example 8.4.1

Calculate the collapse moment of resistance of the beams in Examples 8.2.1 and 8.3.2 by using Appendix E. What are the approximate percentage increases compared with the simplified approach?

Solution

$f_{cu} = 30 \times 10^6 \quad d = 0.2625$

$f_y = 420 \times 10^6 \; A_{st} = 100\pi \times 10^{-6} \times 2$

$A_{st}/bd = 2 \times 100\pi \times 10^{-6+3} /0.150 \times 262.5 = 1.596\%$

(a) From Fig.E3 for $\rho = 1.6\%$:

$M_u = 4.55 \times 150 \times 0.2625^2 \times 10^{6-3} = 47.1 \times 10^3$ (Answer)

The increase is

$100 \times (47.1 - 45.6)/45.6 = 3.3\%$ (Answer)

(b) $A_{sc}/bd = 0.798\%$ i.e $\rho' \simeq 0.8\%$

$M_u = 5.08 \times 150 \times 0.2625^2 \times 10^{6-3} = 52.4 \times 10^3$ (Answer)

The increase is

$100 \times (52.4 - 52.3)/52.3 = 0.2\%$ (Answer)

The reader should now repeat Example 8.4.1 using Fig.E1.

Derivation of design charts

Design charts can be conveniently obtained by carrying out a series of calculations as follows:

1. Assume that the steel strains to some value ε_{st}. This gives a strain distribution as shown in Fig. 8.4.1(a). The values of both x/d and ε_{sc} can be determined since a value of d'/d is selected for the latter.
2. The steel stresses can be determined from the design stress-strain curve for the steel, and Fig. 8.4.1(b) shows the concrete stress block.
3. Calculate the compressive force in the concrete for the given value of f_{cu}. For example, if f_{cu} is 30 MN/m² then

   $F_c = 0.403 f_{cu} bx$
4. The concrete force equals the net tensile force in the steel. Hence for each percentage of compression reinforcement the value of A_{st}/bd can be determined.

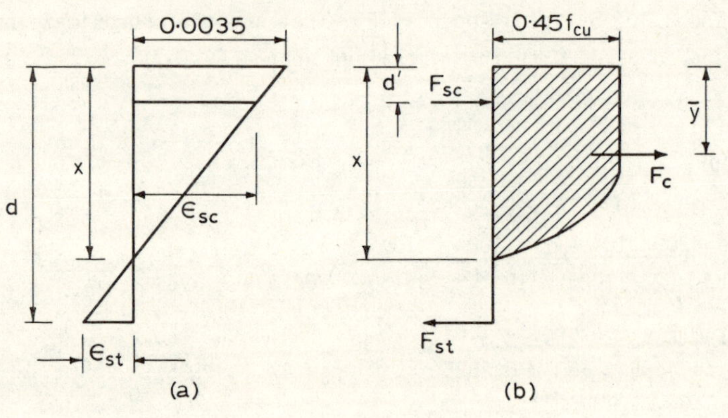

Fig.8.4.1 Stress block for deriving design charts

5. Determine M_u/bd^2 by taking moments. This gives a point on the design curve for the particular value of A_{sc}/bd.
6. Repeat for other values of ε_{st} and compression steel ratios. This is most conveniently carried out by using a digital computer programme.

Example 8.4.2

Compare the values of the magnitude and position of the concrete force as given by the design chart and simplified methods of CP 110 for a singly reinforced beam when

$f_{cu} = 30\ \mathrm{MN/m^2}$

<u>Solution</u> The centre of the concrete compression force is determined by reference to Fig. D1, Table D1 and Fig. 8.4.2 for which

$0.198\sqrt{(f_{cu})}/3.5 = 0.310$

Hence for a unit width of beam:

$F_{c_1} = 0.690x \times 0.45f_{cu} = 0.310f_{cu}x$

$F_{c_2} = \frac{2}{3} \times 0.310x \times 0.45f_{cu} = 0.093f_{cu}x$

$y_1 = 0.345x$

$y_2 = (0.690 + \frac{3}{8} \times 0.310)x = 0.806x$

$F_{c_1} + F_{c_2} = F_c = 0.403f_{cu}x$

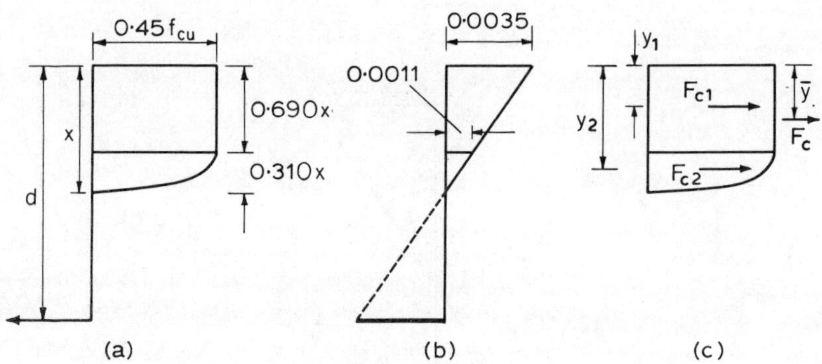

Fig.8.4.2 Example 8.4.2.(a) Concrete stress block. (b) Concrete strains. (c) Resultant concrete force.

This compares with $0.4f_{cu}x$ for simplfied rectangular stress block.

(Answer)

$$F_c\bar{y} = F_{c_1}y_1 + F_{c_2}y_2$$
$$= (0.310 \times 0.345 + 0.093 \times 0.806)f_{cu}x^2 = 0.182f_{cu}x^2$$

Hence

$\bar{y} = 0.451x$

This compares with $\bar{y} = 0.5x$ for simplified rectangular stress block.

(Answer)

Example 8.4.3

Determine the positions on the curve for $\rho' = 0$(i.e. a singly reinforced section) for the design chart given in Fig.E3, i.e.

$f_{cu} = 30\ MN/m^2 \qquad f_y = 420\ MN/m^2 \qquad d'/d = 0.10$

when the steel just reaches (a) lower yield and (b) upper yield.

Solution

(a) If steel has a stress of

$0.8f_y/1.15 = 292 \times 10^6$

then steel strain equals 1.46×10^{-3} from Fig.D2, and the strain distribution is given by

$x/d = 3.5/(3.5 + 1.46) = 0.706$

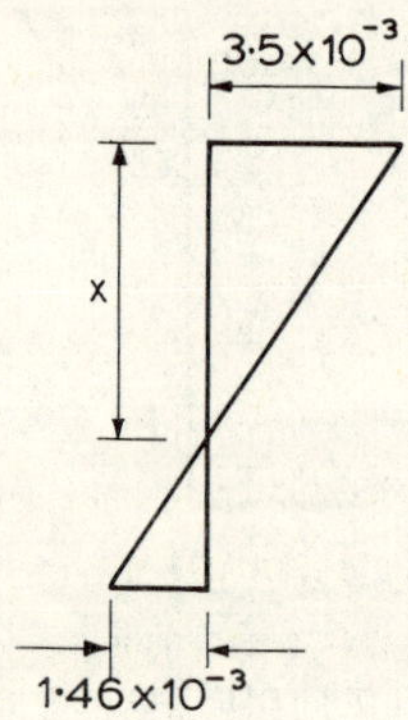

Fig.8.4.3 Example 8.4.3

This point has therefore been excluded from Fig. E3 since x/d exceeds 0.6. However the method of calculation is as follows:

The concrete force is (see Table 8.2.1)

$0.403f_{cu}bx = 8.54bd$

The steel force is

$$0.8 \times 420 \times A_{st}/1.15 = 292.2A_{st}$$

Equating forces gives

$$A_{st}/bd = 2.92\%$$

Taking moments about the concrete force ($\bar{y} = 0.451x$)

$$M_u = (0.8f_y/1.15)A_{st} \times (\text{lever arm}) = 292.2A_{st}(d - 0.318d)$$

We know A_{st}/bd, hence

$$M_u/bd^2 = 5.82 \qquad \text{(Answer)}$$

(b) At a steel stress of $f_y/1.15$ the steel strain is 0.00383. From the strain diagram we obtain

$$x/d = 0.477$$

Using a technique similar to the one used before:

$$A_{st}/bd = 1.58\% \text{ and } M_u/bd^2 = 4.53 \qquad \text{(Answer)}$$

8.5 DESIGN OF T- AND L-BEAMS

The strength of the concrete in compression in a T- or L-beam is usually more than adequate since the breadth of the flange (see section 3.5) is generally much greater than the breadth of the rib, b_w.CP 110 therefore ignores the rib when calculating the resistance moments of the section. Practical design details have been discussed in section 3.5.

Moments of resistance

Referring to Fig.8.5.1 and taking moments about the tension steel gives

$$M_{uc} = 0.4f_{cu}bh_f(d - \tfrac{1}{2}h_f) \qquad (8.5.1)$$

This expression is conservative provided that h_f is not taken to be greater than ½d.

Taking moments about the compression force gives

$$M_{ut} = 0.87f_{yt}A_{st}(d - \tfrac{1}{2}h_f) \qquad (8.5.2)$$

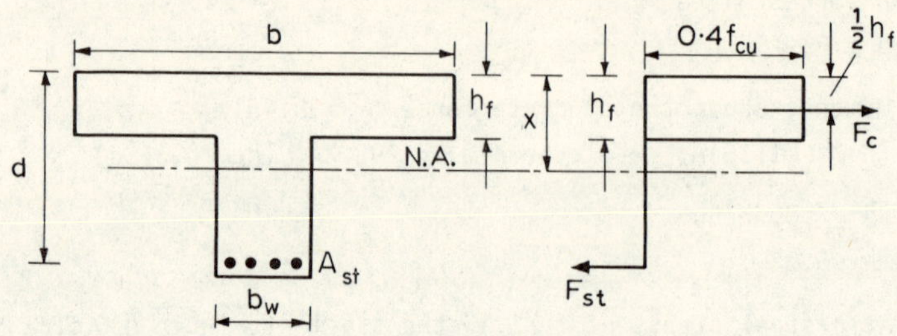

Fig.8.5.1 Stress block for T-beam design

Doubly reinforced T-beams

Where it is necessary for the resistance moment to exceed the value given in Eqn (8.5.1) it can be calculated on the basis of the assumptions given in 7.1 in conjunction with design stress-strain curves. If the resistance moment of the concrete is still inadequate even when the rib is included then the section can be doubly reinforced. However, if should be noted that the tension steel is usually very congested in such cases and it is far preferable to increase the beam size wherever possible.

Example 8.5.1

Determine the moment of resistance of a singly reinforced T-beam using the assumptions of Whitney's stress-block and including the resistance of the web.

Solution Equating longitudinal forces gives

$$A_{st}f_{yt} = 0.85f'_c(b - b_w)h_f + 0.85f'_c b_w d_c$$

Taking moments about the tensile steel gives

$$M_u = 0.85f'_c(b - b_w)h_f(d - \tfrac{1}{2}h_f) + 0.85f'_c b_w d_c(d - \tfrac{1}{2}d_c)$$

According to Whitney, maximum concrete strength is given by

$$d_c = 0.537d$$

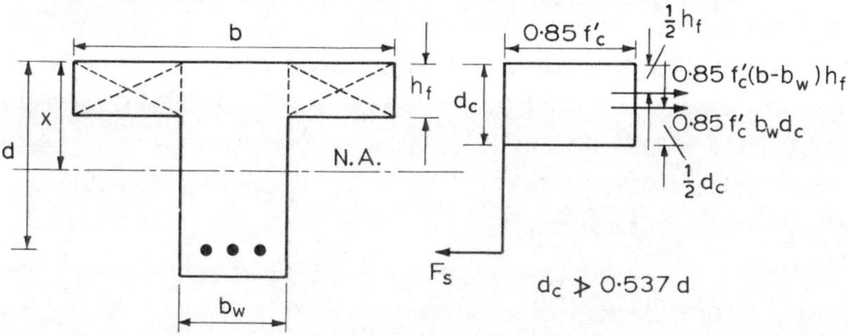

Fig.8.5.2 Example 8.5.1

Hence

$$\begin{aligned} M_u &= 0.85f'_c[(b - b_w)h_f(d - \tfrac{1}{2}h_f) \\ &\qquad + b_w d^2 \times 0.537(1 - \tfrac{1}{2} \times 0.537)] \\ &= 0.85f'_c bd^2[(1 - b_w/b)(h_f/d)(1 - h_f/2d) + 0.393b_w/b] \\ &= 0.85f'_c bd^2\lambda \end{aligned}$$

A design table can be drawn up to give λ for different values of the parameters b_w/b and d/h_f. CP 114 used this design approach of utilizing the parameters b_w/b and d/h_f in conjunction with the old CP 114 stress block.

Reference

8.1 Whitney, C. S., 'Design of reinforced concrete members under flexure or combined flexure and direct compression', Proc. A.C.I. vol. 33, 1937 pp 483-98.

Exercises

1. A simply supported beam spans 16 ft (4.88 m). It is subjected to a uniformly distributed dead load (including self-weight) of 300 lbf/ft (4.37 kN/m) and an imposed loading of 500 lbf/ft (7.29 kN/m). The midspan section is 8 in (203 mm) wide × 10 in (254 mm) in effective depth and is reinforced with three 7/8 in (22.2 mm) diameter mild

steel bars. Is it adequate for the above loading according to (a) limit state of collapse and (b) ultimate load theory as given in CP 114:1965?

(a) For the limit state of collapse assume a rectangular stress block with a mean stress of $0.4f_{cu}$, where f_{cu} is the characteristic cube strength = 4000 lbf/in² (27.6 MN/m²). The characteristic strength of the steel = 40000 lbf/in² (276 MN/m²). Assume the following partial safety factors:

(i) Loads

Dead load 1.4

Imposed load 1.6

(ii) Material strength

Steel 1.15

Concrete: included in factor of 0.4

(b) For ultimate load theory according to CP 114:1965 assume a rectangular stress block with a mean stress of $4f_{cu}/9$ where:

f_{cu} = works cube strength = 4000 lbf/in²(27.6 MN/m²).

The yield strength of the steel = 40000 lbf/in² (276 MN/m²).

Take the (overall) load factor to be 1.8

(UB3/H 1968, 36 min)

2. Refer to question 2, chapter 3 and complete part (b).

3. A reinforced concrete beam, JKLM, rests on simple supports at K and L where KL = 7 m and cantilever JK = cantilever LM = 2 m. The characteristic dead load (including self-weight) is 10 kN/m throughout. The characteristic imposed load consists of a single vertical load of 70 kN which can be applied anywhere along JKLM.

Determine, for the limit state of collapse, the maximum and minimum moments which can occur at the midspan of KL, and at either support. Design, for the support K (or L), a suitable singly reinforced section, assuming a beam breadth of 300 mm. Only the effective and overall depths and the area of tensile reinforcement need be determined, using the following data.

f_{cu} = 25 MN/m² and f_{yt} = 410 MN/m²

γ_m = 1.15 for steel

γ_f for dead load: 1.4 maximum and 1.0 minimum

γ_f for variable load: 1.6 maximum and 0 minimum

Rectangular design stress block for concrete in flexure has a maximum stress of $0.4f_{cu}$ and a maximum depth of ½d.

Assume a cover of 20 mm and a bar diameter of 20 mm.

(UB2 1974, 50 min)

9 Columns

Vertical members with heights much greater than their breadths are known as columns. They are used in buildings to support floor beams, in towers to support elevated tanks, and in trestle piers, etc. A strut need not be vertical but is still designed as a column. Piles and walls can also act as columns. RC columns have the advantage, in general, of acting as short columns: that is, columns which do not have a tendency to buckle. Special consideration is necessary for slender (or long) columns, for which additional bending effects become significant. Due consideration of slenderness limits avoids the use of columns which have excessive 'additional moments' and which may even fail by lateral torsional buckling under bending only.

9.1 GENERAL CONSIDERATIONS FOR COLUMNS AND WALLS

A column may be considered to carry either (i) a direct (i.e. axial) load, or (ii) an eccentric load (i.e. axial load combined with a bending moment). The stresses in an axially loaded column are purely compressive, whereas those for an eccentric load include bending stresses. Bending moments can be present due, for example, to beams framing into columns, or to wind loading.

Short columns

A short column as defined by CP 110 is one whose effective height does not exceed twelve times its lateral dimension (for either direction (see Appendix CD)). Short columns fail primarily by crushing of the concrete or tension of the steel, because buckling effects can be ignored.

Column types

There are two general types of column as follows:

(i) Rectangular

(ii) Round or octagonal

Each type has main reinforcement consisting of longitudinal bars. The secondary reinforcement usually consists of transverse links (or ties) although spiral or helical reinforcement can also be used in octagonal or round columns. Octagonal and round columns require more complicated formwork but the introduction of fibreglass shuttering, which is virtually indestructible for round columns and gives a 'glazed' finish to the concrete, greatly reduces the cost of repetitive work and can make round columns comparable in cost with rectangular columns. Columns resisting axial loads are normally made square or circular, while columns subjected to eccentric loading are preferably rectangular, the longer side being arranged to provide increased resistance to the bending moment.

Longitudinal reinforcement

The minimum bar diameter is 12 mm since the bar must be sufficiently rigid to stand up straight in the column boxes during fixing and concreting. The total cross-sectional area of the longitudinal steel must be not less than 1% and not more than 6% of the gross cross-sectional area of the column (see Appendix CD). If the column is lightly loaded the lower limit can be reduced to $0.15\ N/f_y$. If the column is cast horizontally the upper limit can be increased to not more than 8% since the steel congestion is not quite so critical, but the value at laps must not exceed 10% in all cases.

Transverse reinforcement

Transverse reinforcement is essential in order to resist the tendency of the longitudinal bars to buckle outwards and split the concrete. In rectangular columns it consists of links (also called ties or binders), which are similar to (closed) stirrups in beams. The Code (see Appendix CD) governs the diameter and spacing of the ties, and every longitudinal bar must be restrained in two directions by the links.

Continuous columns

Columns which continue up through a floor must be proportioned with due regard to the beams framing into the column. Column bars can either pass outside the beam bars, or bend inwards to miss the beam bars, to lap onto the bars for the next storey.

Braced columns

A column is assumed to be braced in a given direction if lateral stability of the structure is provided by walls or other suitable bracing designed to resist all lateral forces in that direction. A column is unbraced if the lateral stability of the structure in a given direction is provided by the columns. Hence a wall is braced if lateral stability perpendicular to the wall is provided by cross-walls.

In situ concrete walls

In situ walls may be necessary if moderate structural strength is required, as in walls providing lateral stability to a structure, retaining walls etc. They are generally more expensive, if used above ground level, than brickwork cavity walls, precast concrete block walls or precast concrete wall panels. If badly constructed they are also often inferior with respect to both durability in industrial atmospheres and moisture penetration. The limit state of local damage is important for in situ concrete walls (because of shrinkage and thermal movement (see section 11.3)) as well as the limit state of collapse. The limit state of deflection need not be checked for plain walls or, normally, for reinforced walls.

Precast concrete walls

Precast concrete walls can consist of either precast concrete blocks or precast concrete wall panels. Precast concrete wall panels are frequently used in conjunction with the many industrialized building systems which are available. It is essential to check the stability of a system of walls as a whole as well as the individual panels. Section 5 of CP 110 is entirely concerned with precast, composite and plain concrete construction (including large panel construction).

Reinforced concrete walls

A reinforced concrete wall is a wall which acts as a vertical load-bearing member, whose length between vertical edges is at least four times its thickness, and which has steel ratios (on gross area) of not less than 0.4% and 0.3% for vertical and horizontal (mild) steel respectively. RC walls can be designed as narrow columns in most respects (see Appendices CB and CD.) The shear strength and the limit state of deflection need not normally be checked for reinforced concrete walls.

'Plain' concrete load-bearing walls

A 'plain' concrete wall has less than 0.4% vertical steel (which is disregarded for strength) and transverse links are not required. It can either be in situ (dense or no-fines concrete) or precast and it can be braced, unbraced, short or slender. Wall panels are in general laterally supported or continuous along two or more edges and connections for lateral supports can either be hinged or provide resistance to rotation. If the lateral dimension is less than four times the thickness a 'plain wall' becomes a 'plain column' and the stresses should be reduced by a factor β_w (see Table CU1).

Analysis of plain walls

Floor and roof loads may be assumed to act at one-third the depth of the bearing from the loaded face on the basis that the beams and slabs are simply supported. Eccentricities of all vertical loads, together with any overturning moments due to horizontal loads, must be considered in determining the eccentricity of the resultant load from the centre line of the wall, at any given level, for an unbraced wall. For a braced wall, the eccentricity of the resultant load from the centre line of the bearing area immediately above a lateral support may be taken as zero.

The eccentricity of loading in the plane of the wall should either not exceed one-third of the wall length or vertical reinforcement should be provided ($\rho_c \nless 0.25\%$ or 0.3% for high yield or mild steel respectively) for length of wall which is in vertical tension under service loads. For details, including the analysis of plain sections, see Appendix CU.

Shear strength of plain walls

The shear strength of plain concrete walls should be considered at positions of maximum shear for the various combinations of design loads for the limit state of collapse. (See Appendix CU.)

9.2 ELASTIC ANALYSIS OF AXIALLY LOADED SHORT COLUMNS

An axial load is such that its resultant acts at the centroid of the section. The safe axial load for a column reinforced with longitudinal bars and transverse reinforcement depends upon the sum of the resistances of the concrete and steel. In the following, the axial load is first determined by means of an elastic analysis: this is then considered in the light of actual test results and shown to give misleading values.

Elastic analysis

Elastic theory assumes that the strains in the concrete and steel are equal. The permissible concrete stress in uniform compression, f_{dcc} needs to be lower than that for flexure f_{dcb}, since the latter occurs at the extreme fibre only. For conventional elastic theory f_{dcc} is approximately $\frac{1}{4}f_{cu}$. The concrete achieves its permissible value before the steel hence, referring to Fig. 9.2.1 and putting f_{dcc} equal to f_{cc},

$$F_d = f_{dcc}A_{cc} + \alpha_e f_{dcc}A_{sc}$$

$$= f_{dcc}A_c + (\alpha_e - 1)f_{dcc}A_{sc} \qquad (9.2.1)$$

where A_{cc} = actual (or net) concrete area

A_c = gross concrete area (=bh)

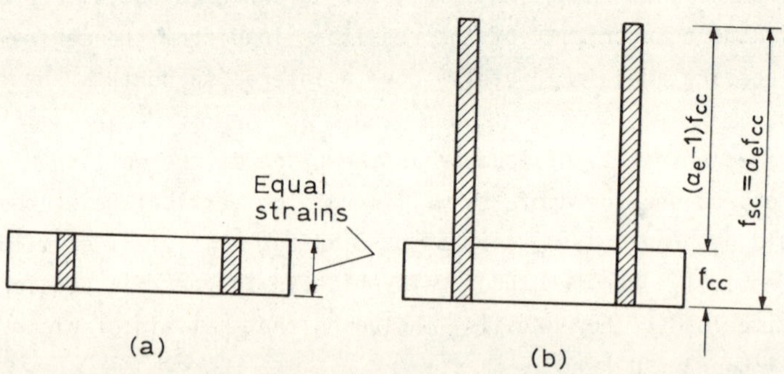

Fig.9.2.1 Elastic strains (a), and stresses (b), for column

Significance of the modular ratio

It has been shown already that concrete is not an elastic material and that the concrete modulus varies with rate, intensity and duration of loading. The concrete modulus is defined, for convenience, in terms of the total strain (elastic strain plus creep strain) so that its value is time-dependent. Hence the modular ratio is also time-dependent since only the steel is elastic up to the working load. The modular ratio for long-term loading has been assumed, for convenience, to be twice the short-term value although it can be larger than this. In the case of beams it was clear that variations in modular ratio only affect, directly, the lever arm. Such variations therefore have a secondary effect only upon the moment of resistance of the section. The resistance of a column section, however, is directly affected by changes in modular ratio, as shown by Eqn (9.2.1). Hence although large uncertainties could be tolerated in the modular ratio value for beams this is not possible for columns.

Stress redistribution in columns

An axially loaded column subjected to a constant load undergoes a gradual but significant redistribution of stress with time: the concrete stress decreases while the steel stress increases. Fig.9.2.2(a) shows a typical concrete strain-time curve where a load is applied suddenly at time $t = 0$, maintained constant, and then suddenly released at time $t = t_1$.

Effect of sudden sustained load

The instantaneous elastic concrete strain, ε_0, equals the initial steel strain. The concrete immediately commences to deform inelastically with time whereas the steel remains elastic. Part of the load initially carried by the concrete is therefore transferred to the steel as shown in Fig. 9.2.2(b). The increase in the load carried by the steel causes the latter to deform elastically, so that the column as a composite unit also strains with time. Notice that the apparent modular ratio at time t is very different to that at time t_0 since the stress, f_{cc}, decreases with time although the strain increases with time.

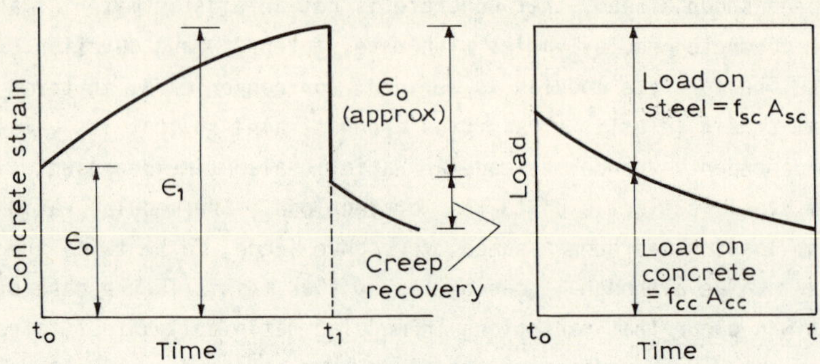

Fig.9.2.2 Stress redistribution in columns

Effect of sudden removal of load

If the load is removed quickly both the steel and the concrete undergo an instantaneous elastic recovery. However since the load carried by the steel has increased, while that for the concrete has decreased, the steel tries to recover much more than the concrete. The steel therefore retains some residual compression while the concrete becomes stressed in tension since the external load is now zero. This tension can cause cracking of the concrete.

Plastic theory for columns

The foregoing shows that even at moderate column loads the stresses in the concrete and steel are not necessarily given by the elastic theory. Considerable plastic flow or creep occurs in the constantly stressed concrete. Tests show that failure of a column occurs when both the concrete and the steel have reached their maximum load-carrying capacities. This limitation of conventional elastic theory was one of the first to be generally recognized. Eqn (9.2.1) was discarded by codes of practice in favour of an equation for ultimate load conditions even when elastic theory for beams was still retained as the main design method.

9.3 AXIALLY LOADED SHORT COLUMNS AT COLLAPSE

It has been shown that conventional elastic analysis is unsatisfactory for reinforced concrete columns. The limit state of collapse, on the other hand, can be designed for very satisfactorily. It is again important, however, to distinguish between the similarities and the differences which exist when considering columns as compared with beams.

Short braced axially loaded columns

The design stress for concrete in bending is $0.45f_{cu}$ (see Appendix D). For compression over the entire section however, rather than at the extreme fibre only, it is necessary to assume a lower value to allow for a nominal minimum eccentricity (of 0.05h). CP 110 uses the value of $0.40f_{cu}$ which conincides with the value for the rectangular stress block assumed for sections in flexure. The partial safety factor for the compression reinforcement is also increased from $(2300 + f_y)/2000$ to 1.5 (i.e. also about 10%) to give the following expression for the design load:

$$N = 0.4f_{cu}A_c + 0.67f_yA_{sc} \qquad (9.3.1)$$

where A_c = area of concrete
A_{sc} = area of compression steel
f_y = characteristic (tensile) strength of compression steel.

Load-contraction characteristics

Brittle compression failures in over-reinforced beams can be avoided by limiting the depth of the neutral axis, and hence the tension-steel ratio. For axially loaded columns, however, the compression is high throughout the entire cross-section and no tension is present. An alternative approach is therefore necessary. The load-contraction curve for an axially loaded column can become approximately 'flat-topped' if the load taken by the compression steel is capable of increasing beyond the design value after the concrete reaches its ultimate strain. That is, the compression steel can compensate for the decreasing proportion of the total load taken by the concrete at very high strains, if the amount of compression steel is adequate. CP 110 therefore specifies a minimum steel area of 0.15 N/fy or 1%, whichever is least, to ensure improved compression-strain characteristics (as well as providing a good vertical tie force for stability).

Short braced columns with low moments

If a short braced column supports an approximately symmetrical arrangement of beams and imposed load it may be designed without special calculations for the actual moments if the safety factors are effectively increased for the axial load as follows:

$$N = 0.35f_{cu}A_c + 0.60A_{sc}f_y \qquad (9.3.2)$$

Eqn (9.3.2) may be used provided that

(i) ℓ_{ex}/h and $\ell_{ey}/b \not> 12$

(ii) Beams carry a uniformly distributed imposed load

(iii) Beam spans differ by $\not> 15\%$

Wind loads do not affect braced columns, hence the column load is calculated from combination I for dead and imposed loads only (see Table B1).

9.4 COLUMN FOOTINGS

Suitable footings are necessary to distribute the loads from columns and walls onto the subsoil. CP 110 considers *pad footings*. which are supported directly by the subsoil, and *pile caps*. For very poor ground conditions it may be necessary to support the entire structure on a single mat or slab known as a *raft foundation*. At present the serviceability loads (i.e. $1.0G_k$ etc.) govern the size of bases, since bearing loads are usually considered in terms of permissible bearing stresses, whereas the collapse loads govern the structural design of the bases themselves.

Pad footings

The main types of pad footing are as follows:

(i) *Isolated footing*, usually square or rectangular, supporting an individual column.

(ii) *Combined footing*, supporting two or more columns.

(iii) *Strip footing*, supporting a continuous wall.

Critical stresses

The stresses in pad footings are conveniently considered in the order:

(i) Bearing pressure between column and top of footing

(ii) Bearing pressure between soil and base of footing

(iii) Diagonal tension stress for concrete

(iv) Bending stresses for concrete and steel

(v) Punching shear stress for concrete

(vi) Bond stress between concrete and steel

(*Note:* Pile caps are often ground beams (rather than slabs) due to the concentrated bearing between pile and pile cap and item (ii) especially would usually be inappropriate for pile caps).

Concrete bearing stress

The load from the column steel must be transferred into the footing by means of dowel (or 'starter') bars. If the grade of concrete in the column is much higher than that for the footing then additional dowel bars may be necessary. The bearing stress on the footing concrete may be increased from $0.4f_{cu}$ to $0.67f_{cu}$ if the column is fairly centrally placed on the footing (i.e. the central core of concrete in the base is effectively confined).

Soil bearing pressure

The permissible pressure on the ground determines the minimum overall area of the footing. The soil pressure is usually considered uniform for axial loading and trapezoidal or triangular for eccentric loadings (i.e. pressure varies linearly across the base). Notice that according to soil mechanics theory the pressure distribution is rarely uniform, so that the calculated shear and bending stresses in the concrete may not be those for the worst case. No redistribution of bending moments should be considered. It is therefore preferable to be conservative when designing bases, (especially for combined footings if reversed stresses are possible). The permissible soil pressure must exceed the combined effects of column load, self-weight of footing and any over-burden weight. It is usually convenient to consider the 'net soil pressure', given by the column load divided by the footing area, and to compare this with the permissible pressure less the self-weight and overburden pressure.

Diagonal tension

The critical sections for shear along a vertical section producing diagonal tension should be taken across the full width of the base at a distance from the column faces equal to 1½ times the effective depth of the base. In the case of rectangular footings, the vertical section extending across the full width of the base should be considered to resist a shearing force equal to the pressure on the area on the remote side of the section, as shown in Fig. 9.4.1. For a pile cap the provisions of §3.3.6.2 for concentrated loads on corbels and near beam supports (see Appendix CC) apply, where a_v is the distance from nearest face of column to edge of pile.

Bending stresses

The critical section for bending in the base should be taken at the face of the column or wall, the moment being given by the forces over the entire area on one side of the section. The reinforcement provided to resist these bending moments should, in general, be distributed

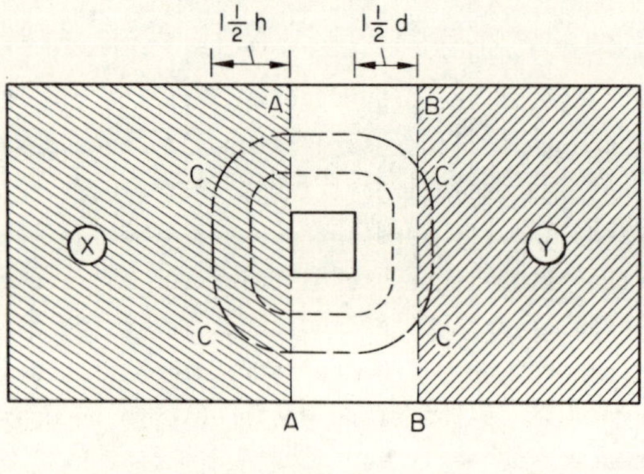

Fig.9.4.1 Critical sections for bending (AA) and shear (BB and CCCC) for rectangular pad footing

The moment due to shaded area X is used for bending and bond at face AA, and the shear force due to shaded area Y acts on section BB across the full width of the section. In rectangular bases, however, the reinforcement across the short span should be more closely spaced near the column. An amount

$[2/(\beta_1 + 1)] \times$ total area of steel

should be spread over a band centred on the column of width equal to the shorter dimension of the base. β_1 is the ratio of longer side to shorter side. The remainder of the steel should be spread evenly over the outer bands.

Punching shear

The critical perimeter around the loaded area for punching shear should be taken at a distance from the column faces equal to $1\frac{1}{2}h$. Hence the length of the perimeter CCCC in Fig. 9.4.1 resists a shearing force due to the base area on the remote sides of section CCCC. Stresses due to punching shear should be considered as for concentrated loads on slabs (see Appendix CC) and perimeter distance $\frac{3}{4}h$ from column faces should be provided with an exactly similar amount (if any) of shear reinforcement as required for the critical perimeter.

Bond stress

The critical section for local bond stress is usually the same section as the critical section for bending moment, i.e. at the face of the column or wall. A lower bond stress is sometimes used in two-way footings, since it is reasoned that cracks parallel to the bars may actually lie along these bars and lower their resistance to bond, due to bending in the other direction. Bond stresses should also be considered for sections at which the depth changes or at which reinforcement is curtailed.

Eccentric loads

Variable moments are frequently applied to footings, so that eccentric loading occurs and a trapezoidal soil pressure results. This modifies the design moments and shears, and, for punching shear, the half-perimeter CCCC with the more highly loaded remote area should be considered separately.

Strip footing

The design of wall footings is simpler than that of column footings since bending occurs in one direction only. Because of the uniform nature of a wall loading, it is sufficient to consider a typical one-metre length of wall. Plain concrete is often suitable for light loadings.

Footings on rock

The high bearing pressures possible on rock often necessitate purely nominal footings. It should be noted, however, that good rock is usually stiffer in compression than is reinforced concrete of normal proportions in bending, hence the latter cannot effectively distribute the loading. A rock footing requires an extra-large depth for shear stiffness, and little if any reinforcement for tension on its bottom face.

Combined footing

A single footing carrying two or more columns is known as a combined footing. They are particularly useful where there is:

(i) close proximity of two or more columns,

(ii) an exterior column which cannot be made symmetrical because of site boundary or other limitations.

The centroid of the combined footing is made to coincide with the resultant of the individual column loads. The cantilever projections on three sides of each column in a two column footing could therefore act very similarly to an isolated footing. It is usual, however, to consider strips of slab under the column as acting as transverse beams. There should, therefore, be a concentration of transverse steel for a band-width equal to the short-side dimension of the base under the columns, as shown in Fig. 9.4.2.

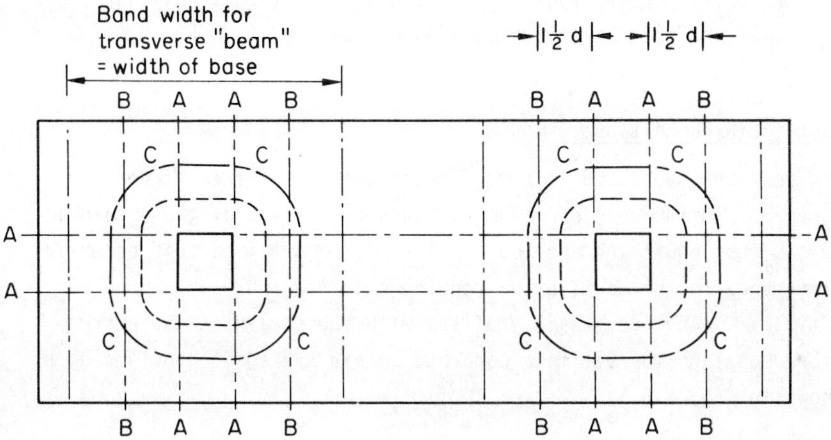

Fig.9.4.2 Critical sections for a combined footing

Piled foundations

Piles are often necessary on poor ground, and can be considered as concentrated reactions on the pile cap as indicated earlier. The designer must, however, allow for the fact that the location of piles can vary by as much as 300 mm or so from the intended position, due to difficulties in driving, when determining the forces and bending moments.

9.5 ANALYSIS OF COLUMN SECTIONS AT COLLAPSE

The two main types of flexural failure that occur in beams can also occur in eccentrically loaded short columns. They are:

(i) *Compression failure,* due to primary crushing of the concrete.

(ii) *Tension failure,* due to yielding of the steel and subsequent crushing of the concrete.

In either case, however, it is clear that the attainment of an ultimate concrete strain in the extreme fibre can be taken as the failure criterion for the column. The basic approach for the accurate analysis of column sections can therefore follow that of Hognestad, Hanson and McHenry [7.1] for sections in pure flexure. Alternatively, as shown later, any of the design assumptions can be used.

Equilibrium conditions

For sections subjected to pure flexure the sum of the internal forces is zero. In the case of columns, however, the sum of the internal forces must equal the applied axial load. Apart from this essential difference in the equilibrium requirements the approach can be very similar. Tentative strain profiles with the strain at one extreme fibre equal to the ultimate concrete strain are again considered, as shown in Fig. 9.5.1.

Known eccentricity of load

The basic problem for columns is to determine the collapse load when that load acts at a known eccentricity. For equilibrium of longitudinal forces,

$$N = F_c - \sum_1^n F_{sc}$$

$$= \lambda_1\lambda_3 f'_c bx + A'_{sc_1} f_{sc_1} + A_{sc_2} f_{sc_2} + A'_{sc_3} f_{sc_3} + A_{sc_4} f_{sc_4} \quad (9.5.1)$$

Notice that f_{sc_2} and f_{sc_4} are tensile, and therefore negative, for the neutral axis position shown in Fig. 9.5.1 when using the general sign convention (see Appendix Y).

Also, for equilibrium of moments,

$$Ne = F_c\,(\tfrac{1}{2}h - \lambda_2 x) + \sum_1^n F_{sc} y$$

$$= \lambda_1\lambda_3 f'_c bx(\tfrac{1}{2}h - \lambda_2 x) + A_{sc_1} f_{sc_1} y_1 + A_{sc_2} f_{sc_2} y_2 + A_{sc_3} f_{sc_3} y_3 + A_{sc_4} f_{sc_4} y_4 \quad (9.5.2)$$

Note: Eqn (9.5.2) has general applicability. For Fig. 9.5.1, y_2 and y_4 (also f_{sc_2} and f_{sc_4}) are negative and e, y_1 and y_3 (also f_{sc_1} and f_{sc_3}) are positive.

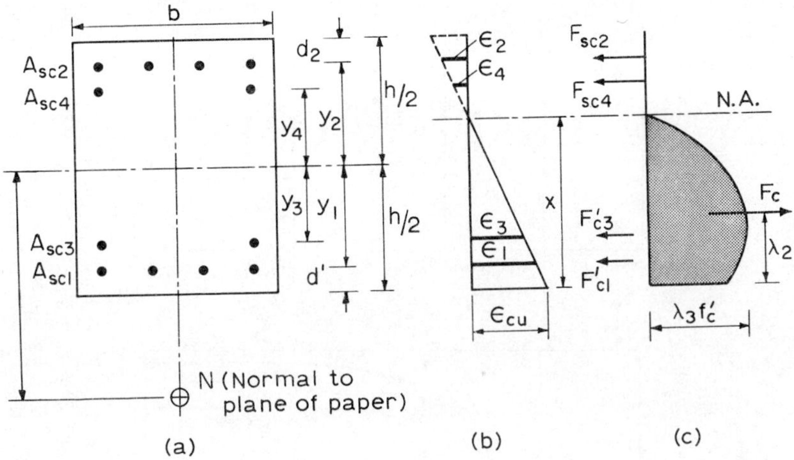

Fig.9.5.1 Column section at collapse.(a) Section.(b) Strains.(c) Forces.

<u>Strain profiles</u>

The stresses f_{sc_1}, f_{sc_2} etc, must be compatible with the strains ε_1, ε_2 etc., and due attention must be paid to the signs of both f_{sc} and y. Notice that when f_{sc} is negative y is negative in Eqn (9.5.2) so that all terms are additive. As for beams, a trial-and-adjustment procedure is used to find the strain distribution at failure. The essential requirement for columns is that the two values of N which must be made to coincide are obtained by considering forces and moments respectively: that is, the values of N obtained from Eqns (9.5.1) and (9.5.2) must be equal to within about 2%. A suitable solution is achieved most conveniently simply by inspection followed by trial-and-error adjustments to the value of x.

Example 9.5.1

Calculate the collapse load for a 0.5 m × 0.5 m column using the method of Hognestad et al. when the load acts 75 mm outside the midpoint of one edge as shown. The column is reinforced with six 32-mm diameter bars of load-strain characteristics in tension and compression as shown. The

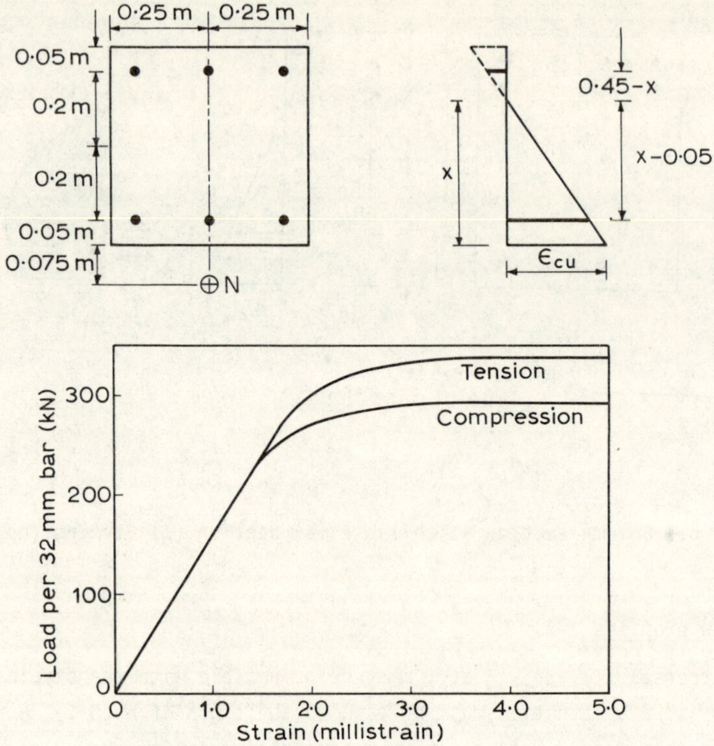

Fig.9.5.2 Example 9.5.1

cube crushing strength of the concrete is 37.5 MN/m². Assume that f'_c is equal to $0.78f_{cu}$ and refer to section 7.2 for the values of $\lambda_1\lambda_3$, λ_2 and ε_{cu}.

<u>Solution</u>

$$f'_c = 0.78 \times 37.5 \times 10^6 = 29.2 \times 10^6$$

$$\lambda_1\lambda_3 = (27 + 0.35 \times 29.2)/(22 + 29.2) = 0.731$$

$$\lambda_2 = 0.5 - 29.2/550 = 0.447$$

$$\varepsilon_{cu} = (4 - 29.2/45) \times 10^{-3} = 3.35 \times 10^{-3}$$

Eccentricity of load is fairly large, try

$$f_{st} = f_{sc} = f_{yc}$$

i.e. load per bar is 292×10^3. Hence

$$N = F_c + 0 \quad (1)$$

and

$$0.325N = F_c(\tfrac{1}{2}h - \lambda_2 x) + \sum F_{sc}y \quad (2)$$

$$F_c = \lambda_1\lambda_3 f'_c bx$$

$$= 0.731 \times 29.2 \times 0.5 \times 10^6 x = 10.67 \times 10^6 x$$

Therefore from (1) and (2)

$$F_c[0.325 - (0.25 - 0.447x)] = 6 \times 292 \times 0.2 \times 10^3$$

or

$$0.447x^2 + 0.075x - 32.8 \times 10^{-3} = 0$$

$$x = +0.20$$

If $(x)_1 = 0.20$ then $f_{st} > f_{sc}$; hence increase x.

Try $(x)_2 = 0.22$

$$\varepsilon_{st} = \frac{0.45 - 0.22}{0.22} \times 3.35 \times 10^{-3} = 3.50 \times 10^{-3}$$

$$\varepsilon_{sc} = \frac{0.22 - 0.05}{0.22} \times 3.35 \times 10^{-3} = 2.59 \times 10^{-3}$$

Hence

$$\sum_2^4 F_{sc} = -3 \times 337.5 \times 10^3 = -1013 \times 10^3$$

and

$$\sum_1^3 F'_{sc} = 3 \times 285 \times 10^3 = 855 \times 10^3$$

$$F_c = 10.67 \times 0.22 \times 10^6 = 2.35 \times 10^6$$

Resolving vertically

$$N = F_c + \sum(F'_{sc} + F_{sc})$$

$$= 2.19 \times 10^6$$

Taking moments about the centre line

$$0.325N = 1868 \times 10^3 \times 0.2 + 2350 \times 10^3(0.25 - 0.447 \times 0.22)$$

or

$$N = 2.25 \times 10^6 \neq 2.19 \times 10^6 \text{ obtained above.}$$

Now

$(2.25/2.19)(x)_2 = 0.225$

Increasing x will only affect the moments slightly in this case, so that over-correction appears unnecessary.

Try $(x)_3 = 0.225$

This gives $N = 2.25 \times 10^6$ (Answer)

9.6 COLUMNS IN CONTINUOUS FRAMES

Floor loadings produce bending and deflections in beams and slabs which, due to the stiffness of the connections, induce deflections in the columns also. These bending moments must be evaluated before the sections can be designed. Internal columns which support an arrangement of beams and loading which is approximately symmetrical can be designed as indicated in section 9.3. However, external columns and internal columns which are not symmetrical must be designed to take the appropriate bending moments in conjunction with the axial load. Transverse loads, such as wind pressure, can also produce bending moments in (unbraced) columns.

Structural stiffness

The stiffness of a member is given by dividing the second moment of area of the section by the length of the member, provided that the cross-section is constant throughout the length of the member. It is usual to consider the concrete section and to ignore the reinforcement, but any of the three sections listed in Appendix B may be used. The breadth of the flange for T-beams is that assumed for midspan calculations. For members which are assumed to be fully fixed at both ends the length of a member is taken, for convenience, as the distance between centres of supports.

Initial moments

The moments caused by the out-of-balance load on the members framing into the column are called initial moments. The moments in beams and columns in framed structures may be calculated using any of the frames, sub-frames or simplified sub-frames described in Appendix CB under structural frames. However, for columns only, it may be assumed that the far ends of members meeting at any joint are fixed and that the adjacent beam stiffnesses are halved. The moments in the columns are therefore as shown in Table 9.6.1. K_ℓ, K_u and K_b are the stiffnesses of the lower column, upper column and beam respectively. ΣK_b is the sum of the stiffnesses of the beams on each side of an internal column.

Table 9.6.1 Moments in external and internal columns

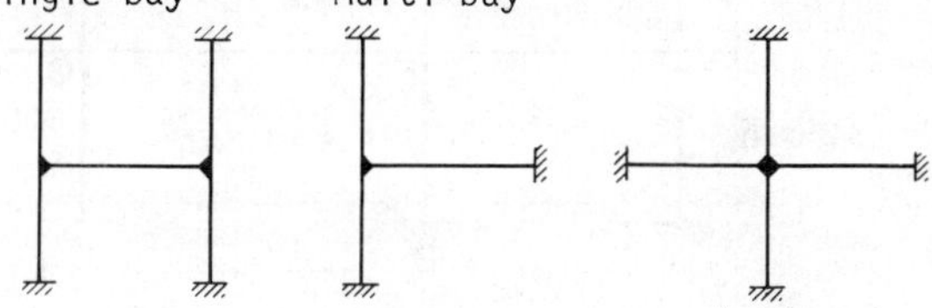

	External columns		Internal columns
	Single-bay	Multi-bay	
Moment at foot of upper column		$\dfrac{K_u M_e}{K_\ell + K_u + \frac{1}{2}K_b}$	$\dfrac{K_u \Delta M_e}{K_\ell + K_u + \frac{1}{2}\Sigma K_b}$
Moment at head of lower column		$\dfrac{K_\ell M_e}{K_\ell + K_u + \frac{1}{2}K_b}$	$\dfrac{K_\ell \Delta M_e}{K_\ell + K_u + \frac{1}{2}\Sigma K_b}$

M_e is the end moment for a beam framing into an external column, assuming full fixity at both ends. ΔM_e is the difference in end moments for beams on either side of an internal column.

Example 9.6.1

A three-bay three-storey framed building is shown in elevation in Fig.9.6.1. The characteristic dead and imposed loadings for the first floor beams are 14 kN/m and 20 kN/m respectively uniformly distributed. Calculate the maximum design end moments at the feet of columns KG and LH and at the heads of columns GC and HD if the columns are braced.

All external columns measure 300 mm × 250 mm and internal columns are 250 mm square. Beam EFGH measures 250 mm × 500 mm overall.

<u>Solution</u> Clear distance between vertical restraints is

$\ell_o = 3.3 - 0.5 = 2.8$

Hence

$\ell_o/b = 2.8/0.25 = 11.2$ i.e. < 12

$g = 14$ kN/m and $q = 20$ kN/m

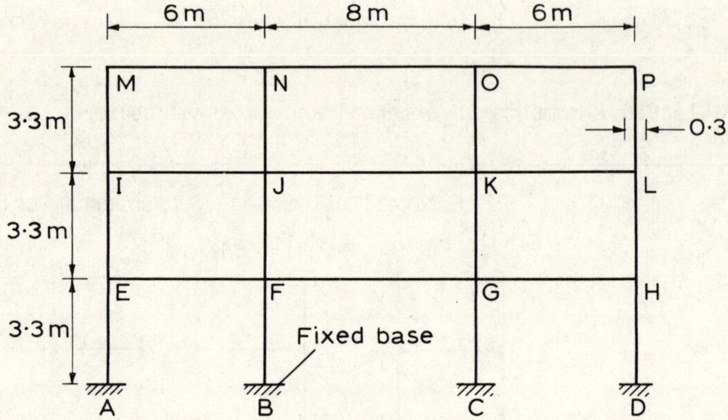

Fig.9.6.1 Example 9.6.1

External column:

The maximum design end moment is given by a loading of

$1.4g + 1.6q = (19.6 + 32)$kN/m

Therefore maximum M_e is

$wl^2/12 = 51.6 \times 6^2 \times 10^3/12 = 154.8 \times 10^3$

Internal column:

The minimum design end moment is given by a loading of

$1.0g = 14$ kN/m

The maximum out-of-balance moment is

$\Delta M_e = (51.6 \times 8^2 - 14 \times 6^2) \times 10^3/12 = 233 \times 10^3$

Second moments of area:

For external column

$I = bh^3/12 = 0.25 \times 0.3^3/12 = 562 \times 10^{-6}$

Similarly, I for internal column $= 326 \times 10^{-6}$

$\frac{1}{2}I$ for beam $= 1302 \times 10^{-6}$

Stiffnesses

The relative stiffnesses are based on distances between centres. For external columns (assumed fixed)

$K_u = E \times 562 \times 10^{-6}/3.3 = 170 \times 10^{-6}E = K_\ell$

For the end beam

$\frac{1}{2}K_b = E \times 1302 \times 10^{-6}/6 = 217 \times 10^{-6}E$

Therefore the maximum design moment at foot of column LH is

$$\frac{K_u M_e}{K_u + K_\ell + \frac{1}{2}K_b} = \frac{170 \times 154.8 \times 10^3}{340 + 217}$$

$$= 47.3 \text{ kNm} \qquad \text{(Answer)}$$

Similarly, at head of column HD

moment = 47.3 kNm (Answer)

For the internal beam

$\frac{1}{2}K_b = E \times 1302 \times 10^{-6}/8 = 163 \times 10^{-6}E$

$\frac{1}{2}\sum K_b = (163 + 217) \times 10^{-6}E = 380 \times 10^{-6}E$

$K_u = K_\ell = 326 \times 10^{-6}E/3.3 = 98.8 \times 10^{-6}E$

at node G. Therefore the maximum design moment at foot of column KG is

$$\frac{K_u \Delta M_e}{K_u + K_\ell + \frac{1}{2}\sum K_b} = \frac{98.8}{958.1} \times 233 \times 10^3$$

$$= 25.1 \text{ kNm} \qquad \text{(Answer)}$$

Similarly, at head of column GC,

moment = 25.1 kNm (Answer)

Short braced columns with applied moments

Where a column is subjected to significant bending moments about one or both axes, it is often sufficient to design the cross-section for single axis bending only, considering the axial load N together with the maximum moment about the critical axis, either M_x or M_y as appropriate. If biaxial bending needs to be considered, then the condition given in Appendix CB, with the values of α_n given in Table CB1, should be satisfied.

Short unbraced columns

If the columns are unbraced then lateral loads must be resisted and all three load combinations given in Appendix B should be considered as indicated in Appendix CB—structural frames—'frame supporting vertical and lateral loads'. When considering the design wind load only for combination (III), i.e. $1.2W_k$, it may be assumed that there are points of contraflexure at the centres of all beams and columns and that the total horizontal shear is divided between the columns in proportion *either* to their stiffnesses (e.g. external column resists about half the shear force of an internal column) *or* to the span of the bays assuming each bay acts as a simple portal [9.1]. Otherwise the design of the cross-section of an unbraced column is the same as that above for a braced column. Wind loadings are considered in Appendix V.

9.7 DESIGN OF COLUMN SECTIONS AT COLLAPSE

The assumptions made by CP 110 for sections subjected to flexure combined with axial load are exactly the same as for sections subjected to pure flexure. The stresses in the concrete in compression are therefore given by either of the methods explained in chapter 8; that is, either from the design stress-strain curves or from the simplified approach assuming a rectangular stress block of $0.4f_{cu}$ mean stress and depth d_c of concrete in compression.

Simplified approach for columns

When using the simplified approach for sections in pure flexure it could be assumed that in general all the tension steel was at the upper yield stress and that the compression steel also reached its yield value. In the case of columns this is not the case. The steel can have a stress which can vary, for different eccentricities, from yield in compression to yield in tension. When determining the depth to the neutral axis for the concrete, therefore, it is necessary to assume values for the stresses in the steel near each face. Having obtained a value for x it is then necessary to check that the assumed steel stresses are reasonable. If not, then it is necessary to recalculate x assuming stresses in the steel which are more compatible with the position of the neutral axis and the design stress-strain curves in Figs. D2 and D3.

A rigorous application of tentative strain profiles, even when used as above in association with simplified design stress-strain curves, is still essentially a method of analysis rather than a convenient method of design. However, if advantage is taken of the fact that portions of the design stress-strain curves are horizontal for large changes in strain then a more reasonable design approach can be evolved as indicated below. The symbol d_c will be used for the assumed depth of the rectangular concrete stress block since for small moments the neutral axis lies outside the section and $d_c = h$ and not x.

Column with nominal reinforcement

If the presence of the reinforcement is neglected, the ultimate axial load which can be carried by the concrete is given by the depth of concrete which is axially compressed by the load, i.e. if the load is at eccentricity e then $d_c = 2(\frac{1}{2}h - e)$ and:

$$N = 0.4f_{cu}b(h - 2e) \tag{9.7.1}$$

CP 110 further specifies that $e \nogtr (\frac{1}{2}h - d')$.

If N exceeds the value in Eqn (9.7.1) then the strength of the reinforcement can be included and Eqns(9.5.1) and (9.5.2) can be rewritten substituting:

$0.4f_{cu}bd_c$, $0.72f_y$, $\frac{1}{2}h - d'$, $-(\frac{1}{2}h - d_2)$ and $\frac{1}{2}(h - d_c)$ for

$\lambda_1\lambda_3f'_cbx$, f'_{sc_1}, y_1, y_2 and e respectively

as follows:

$$N = 0.4f_{cu}bd_c + 0.72f_yA'_{sc_1} + f_{sc_2}A_{sc_2} \tag{9.7.2}$$

$$M = 0.2f_{cu}bd_c(h - d_c) + 0.72f_yA'_{sc_1}(\tfrac{1}{2}h - d') - f_{sc_2}A_{sc_2}(\tfrac{1}{2}h - d_2) \tag{9.7.3}$$

If these equations are to be readily applied then it is convenient to disregard the strict strain compatibility requirements used in analysis and obtain simple step sequences as shown in Table 9.6.1.

Table 9.6.1 Simple step sequences for simplified design assumptions

e	f'_{sc_1}	f_c	d_c	f_{sc_2}
0	$0.72f_y$	$0.4f_{cu}$	h	$0.72f_y$
e_1	"	"	h	$0.72f_y$ to 0
e_2	"	"	h to $(h - 2d_2)$	0
e_3	"	"	$(h - 2d_2)$	0 to $-0.87f_y$
e_4	"	"	$(h - 2d_2)$ to 0	$0.87f_y$

Notice that only one parameter varies at a time and although it is still necessary to estimate the appropriate case (i.e. 0, e_1, e_2, e_3 or e_4) the trial and error sequence is greatly simplified.

Example 9.7.1

A column 500 mm × 500 mm square of grade 40 concrete carries a load of 2250 kN and a moment of 500 kN m. Determine the required area of symmetrically placed reinforcement of grade 425 if $d' = d_2 = 50$ mm. Take the design yield strengths in compression and tension as $0.72f_y$ and $0.87f_y$ respectively and assume the simplified rectangular concrete stress block of mean stress $0.4f_{cu}$.

Solution

$e = M/N = 500/2250 = 0.222$

From Eqn (9.7.1)

$N = 0.4 \times 40 \times 0.5(0.5 - 2 \times 0.222)10^6 = 448$ kN

This is much less than the required N of 2250 kN and hence the moment requirement (i.e. Eqn (9.7.3)) will be more critical in determining the required areas of reinforcement. Also, the eccentricity e of 0.222 is relatively large; hence cases (e_3) or (e_4) seem likely.

If $d_c = h - 2d_2 = 0.4$; $f_{sc_2} = -0.87f_y$ and $A'_{sc_1} = A_{sc_2} = \frac{1}{2}A_{sc}$

then from Eqn (9.7.3)

$$M = 0.2 \times 40 \times 0.5 \times 0.4(0.1)\ 10^6 + (0.72 + 0.87)f_y\ \tfrac{1}{2}A_{sc}(0.2)10^6$$
$$= 160 + (30.6 + 37.0)10^3A_{sc} \quad \text{kN m}$$

Hence required A_{sc}

$$= (500 - 160)\ 10^{-3}/67.6 = 5029\ \text{mm}^2$$

Try case (e_4)

From Eqn (9.7.2)

$$N = 0.4 \times 40 \times 0.5 \times 10^6 d_c + (0.72 - 0.87)425 \times \tfrac{1}{2}A_{sc} \times 10^6$$
$$= 8 \times 10^6 d_c - 31.9 \times 10^6 A_{sc}$$

or

$$d_c = (2.250 + 31.9 \times 5029 \times 10^{-3})/8 = 0.301$$

—which confirms that it is case (e_4). Reconsidering Eqn (9.7.3) gives:

$$M = 4d_c(h - d_c)10^6 + 67.6\ A_{sc} \times 10^6$$
$$= 240 + 340 \qquad \text{or } 580 \text{ kN m}$$

—which is more than adequate. Hence try reducing A_{sc}, say:

4 No. 25 mm diameter bars (= 1960 mm^2) in each face.

If $d_c = 0.3$ then:

$$M = 240 + 265 \quad \text{or} \quad 505 \text{ kN m}$$

and $N = 8 \times 0.3 \times 10^6 - 31.9 \times 3920 = 2275$ kN

These values are satisfactory. (Answer)

Design curves

Parts 2 and 3 of CP 110:1972 include design charts for rectangular and circular columns and are clearly the most convenient for normal design use. Fig. E4 in Appendix E is similar to the charts in CP 110 except that it covers a range of steel strengths (instead of a single strength) for a given concrete grade and given d/h ratio.

If the curve for concrete alone (i.e. zero reinforcement) is considered in Fig. E4, it can be seen that this represents an envelope—any combination of N/bh and M/bh^2 which falls within this envelope is satisfactory, whereas any combination which falls outside is not. When reinforcement is added to the section the envelope increases in size, and further combinations of N/bh and M/bh^2 are enclosed—the greater the reinforcement ratio, the larger the envelope.

The minimum accidental moment of 0.05 Nh should always be taken into account—hence the maximum axial load which a column is permitted to carry is reduced, hence the horizontal cut-off portion for all column design curves; for example, see portion AB in Fig.E4.

For an eccentric load less than:

$$N_{axial} = 0.4f_{cu}A_c + 0.67f_yA_{sc} \quad (9.7.4)$$

then the design curves decrease linearly with increasing eccentricity (e.g. BC). As the eccentricity becomes even greater some tension occurs in the section and the curve becomes non-linear (e.g. CD). At high eccentricities the tension steel reaches the lower yield stress and is given by a distinct change in slope of the curves (e.g. point D). At very high eccentricities the tension steel reaches the upper yield stress (e.g. point E) and failure occurs simultaneously in compression and tension. This represents the maximum moment for a column section. Further decreases in axial load result in a decrease in the moment that can be resisted once tension failures occur (e.g. beyond E).

Example 9.7.2

A rectangular column, 400 mm × 200 mm of grade 30 concrete, carries a load of 1000 kN and a moment of 200 kN about the major axis. Determine the required area of symmetrical 410 grade reinforcement if d_2 = d' = 40 mm

Solution

f_{cu} = 30×10^6; $f_y = 410 \times 10^6$; $d/h = 360/400 = 0.9$

e = $200/1000 = 0.2$;

N/bh = $1000 \times 10^3/0.2 \times 0.4 = 12.5 \times 10^6$

M/bh^2 = $200 \times 10^3/0.2 \times 0.4^2 = 6.25 \times 10^6$

Hence from Fig. E4

$\rho_c f_y = 21.0 \times 10^6$ and $\rho_c = 5.12\%$

Alternatively, from Fig. E6

$\rho_c f_y/f_{cu} = 0.697$ and $\rho_c = 5.10\%$

or from Chart 64, CP 110:Part 2

$\rho_c = 5.11\%$

Hence $A_{sc} = 4096$ mm²; say 3 No 25Φ and 2 No 20Φ (2098 mm²) per face.

If interpolation between the curves is based on radial lines through the origin (i.e. along constant e/h values) then errors in determining the required ρ_c values are minimized and well within design requirements (i.e. the maximum difference of 0.02% in 5% represents less than ½% error).

Example 9.7.3

Calculate the design collapse load at an eccentricity of 220 mm for a square column, 500 mm × 500 mm, of grade 40 concrete symmetrically reinforced with 4 No. 25 mm diameter grade 410 bars in each face parallel to the axis of bending. $d_2 = d' = 50$ mm.

Solution

$e = 0.22; \quad d/h = 450/500 = 0.9$

$f_{cu} = 40 \times 10^6$ and $f_y = 410 \times 10^6$ hence use Fig.E6.

$\rho_c = 2 \times 4 \times \frac{\pi}{4} 25^2 \times 10^{-6}/0.5^2 = 1.57\%$

and $\rho_c f_y/f_{cu} = 0.0157 \times 410/40 = 0.161$

e/h is constant at $0.22/0.5 = 0.44$

Hence for Fig. E6 for the radial line $e/h = 0.44$ and $\rho_c f_y/f_{cu} = 0.161$

$N/f_{cu}bh = 0.2375$

or $N = 2375$ kN (Answer)

Note: This compares with 2275 kN for the simplified result using grade 425 bars. Repeating Example 9.7.1 for $d_c = 0.31$ m gives $N = 2355$ kN and $M = 501$ kN m; thus with perseverance fairly similar results can be expected.

Columns with non-symmetrical reinforcement

The simplified step sequence method can be used, or possibly even tentative strain profiles with the simplified assumptions. For rectangular or circular columns a first estimate can be obtained from design charts, which can then be modified as necessary with the simplified step sequence method.

Redistribution of moments for columns

Notice that columns should be designed for the full moment arising from an elastic frame analysis. This is the case regardless of what is done with the beams. This is not incompatible with the general requirements for sections since in columns the neutral axis depth generally exceeds 0.6d.

9.8 DESIGN OF SLENDER COLUMNS AT COLLAPSE

The moments induced by the lateral deflection of a loaded column are termed *additional moments* and are significant only for slender columns: that is, where the effective height exceeds 12 times its lateral dimension for either direction. These additional moments affect mainly the critical sections (which are near mid height for braced columns and at the ends for unbraced columns) and should be added to the initial moments to give the *total moments*. Accidental eccentricities arising from constructional tolerances are significant only if the initial moment is small: hence the latter should again be taken as not less than 0.05 times the minimum depth of cross-section.

Instability or buckling

The classical Euler critical load, N_{crit} for a strut under axial load is given by

$$N_{crit} = \pi^2 EI/\ell_e^{\,2}$$

where the effective length ℓ_e is the actual length for a pin ended strut. If the material has a strength f_k, the maximum axial load, N_{uz} which can be resisted by a short column of area A is simply Af_k.

For a typical reinforced concrete square column if ℓ_e/h is under 12 the value of N_{crit}/N_{uz} is over 5 [9.2]. If ℓ_e/h is about 27 or 60 then N_{crit}/N_{uz} is about 1 or 0.2 respectively. Thus an instability failure is possible at the higher ℓ_e/h values. However, even when ℓ_e/h is about 27 the theoretical buckling failure will be preceded by a material failure due to the combination of axial load and bending moment. Any real column which is axially loaded will deflect laterally and result in an additional moment which, when acting in combination with the axial load itself reduces the value of N at ultimate (e.g. see Fig. E4 or any column interaction diagram).

The CEB-FIP recommendations [1.9] included the concept of additional moments and this has been developed by Cranston [9.2] to provide the equations given in CP 110. Thus the total column moment M_t is assumed to be given by:

$$M_t = M_i + Ne_{add} \qquad (9.8.1)$$

where M_i is the maximum initial moment in the column due to ultimate loads (but $\nless 0.05Nh$) calculated by using simple elastic analysis and e_{add} is the additional eccentricity given by the transverse deflection of the loaded column (the initial eccentricity e being given by M_i/N).

Thus by determining suitable values for e_{add} it is possible to ensure that the ultimate load for a material failure is adequately predicted. Furthermore, if a buckling failure should be possible at the upper slenderness limit of 60 permitted by the Code, then all that is necessary is to increase the allowance for e_{add}, since for all compressive failures the interaction diagrams show that an increase in applied moment reduces the maximum load which can be carried.

Marshall [9.8] has suggested that lateral torsional buckling of slender beams will not influence collapse provided that

$\ell d/b^2 \ngtr 500$

CP 110 sets the limit of $\ell_o h/b^2 \ngtr 100$ for cantilever columns and $\ell_o d/b_c{}^2 \ngtr 250$ beams, hence it can be assumed that these forms of instability are precluded and that complying with the usual consideration for collapse — material failure — in accordance with the CP 110 requirements is sufficient.

Additional eccentricity

The central deflection depends upon the maximum curvature $1/r_m$ and upon the distribution of the curvature. For example, from Appendix K, for a triangular distribution (i.e. maximum curvature and maximum moment at mid height only) $k = +\frac{1}{12}$ whereas for a rectangular distribution (i.e. maximum curvature throughout) $k = +\frac{1}{8}$.

Cranston [9.2] suggested a value for r_m^{-1} and an intermediate value of $\frac{1}{10}$ for k, to determine e_{add} as follows

$$r_m^{-1} = (1 - 0.0035\, \ell_e/h)/175h \qquad (9.8.2)$$

and $$e_{add} = \frac{1}{10}\, \ell_e^2\, r_m^{-1} \qquad (9.8.3)$$

Thus at ℓ_e/h values of 12, 30 and 60 additional eccentricities of 0.08h, 0.46h and 1.63h respectively are obtained (the latter being some 20 times greater than the former).

Additional moments

Substituting the values for r_m^{-1} and e_{add} from Eqns(9.8.2) and (9.8.3) into Eqn (9.8.1) gives the expression

$$M_t = M_i + \frac{Nh}{1750}\left(\frac{\ell_e}{h}\right)^2\left(1 - 0.0035\ \frac{\ell_e}{h}\right) \qquad (9.8.4)$$

This is the expression given in CP 110 for slender columns bent about a minor axis and the following points should be noted.

(a) The minor axis is the longer axis yy providing the smaller second moment of area I_{yy}.

(b) 'h' in Eqn (9.8.4) is defined as the overall depth of the cross-section in the plane of bending and is therefore 'b' for bending about a minor axis.
(*Note:* to avoid confusion in Appendix CB for the different cases, the breadth 'b' is consistently referred to as 'b' and is redefined as the depth normal to the minor axis).

(c) ℓ_e is the effective height either in the plane of bending or in the plane at right angles whichever is the greater.

Slender columns bent about a major axis can be considered as biaxially loaded, using the expressions given in Appendix C for M_{tx} and M_{ty} for slender columns bent about both axes, and simply putting $M_{iy} = 0$. However, if h is less than 3b, Cranston [9.2] has shown that a reasonable alternative is to substitute ℓ_e/b for ℓ_e/h in (9.8.4) to obtain

$$M_t = M_i + \frac{Nh}{1750}\left(\frac{e}{b}\right)^2\left(1 - 0.0035\right)\frac{\ell_e}{b} \qquad (9.8.5)$$

where 'h' is again defined in CP 110 as the overall depth in the plane of bending, and is therefore also 'h' for bending about a major axis.

Braced slender columns

Consider a braced column in single curvature subjected to initial end moments M_1 and M_2 as shown in Fig. 9.8.1(a). It is clear that the critical section for the maximum additional moment does not coincide with the maximum end moment and hence CP 110 allows the maximum initial moment M_2 to be reduced to

$$M_2 = 0.4\ M_1 + 0.6\ M_2 \qquad (9.8.6)$$

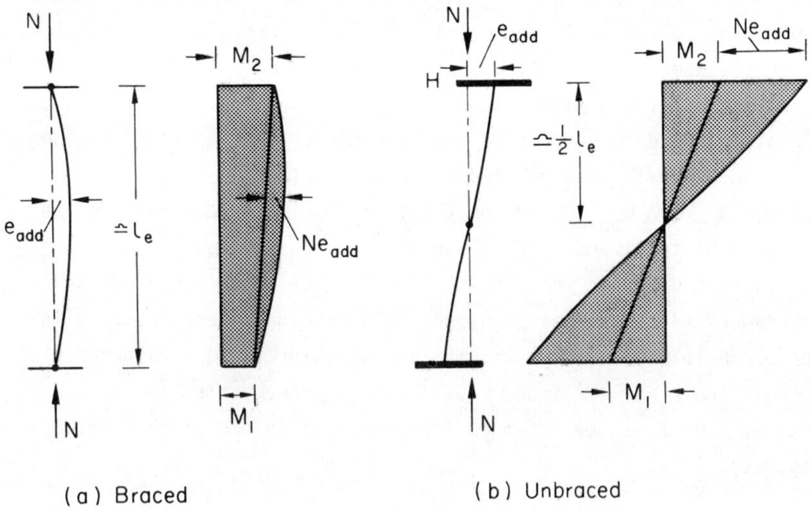

Fig.9.8.1 Loading and moments in slender columns

where M_2 is the larger initial end moment due to ultimate loads and is always assumed positive. If the column is bent in double curvature then M_1 is negative and M_i at the critical section near midspan is relatively small. In all cases M_i is taken as $\nless 0.4\, M_2$ and $M_t \nless M_2$ (and $\nless 0.05Nh$).

Unbraced slender columns

Consider an unbraced column in double curvature as shown in Fig.9.8.1(b). In this case the critical section for the maximum additional moment coincides with the maximum end moment, and the two are additive. Thus for unbraced slender columns it may be necessary to check that the beams (or bases if appropriate) into which the columns frame can resist these total moments.

Adjustments to additional moments

If N approximately equals N_{bal} (the axial load corresponding to the balanced condition of maximum compressive strain in the concrete of 0.0035 and tensile strain in the outermost layer of tension steel of 0.002) the curvature in the critically loaded cross-section is close to that assumed. However, as N increases upwards from N_{bal} the tension steel stress at ultimate reduces and then changes sign — so that the curvature and deflection also reduce. Hence the simple expression for K in Appendix CB under 'adjustments to slender columns'.

Since N_{uz} and N_{bal} depend on the steel ratio the section must be designed by trial and error. Furthermore, the maximum moment in the interaction diagrams (e.g. point E in Fig. E4) does not directly correspond since the design upper yield strain in tension steel is not normally 0.002. However, the design lower yield strain for all normal steels is less than 0.002; hence a conservative (higher) estimate for K can be obtained from the lower yield point (e.g. D in Fig. E4).

References

9.1 Gray, C.S., et al., 'Steel designer's manual', Crosby Lockwood, London, 1960, p.590.

9.2 Cranston, W.B., 'Analysis and design of reinforced concrete columns', Res. Rep. 20, Cement and Concrete Association, London, 1972 pp 54.

9.3 Marshall, W.T., 'A survey of the problem of lateral instability in reinforced concrete beams', Proc. Instn. Civil Engrs, vol. 43, July, 1969, pp 397-406.

Exercises

1. (a) Draw a typical stress-strain curve for concrete under uniaxial compression. Explain why elastic theory is usually justified for short-term working loads, and show how the non-linear behaviour of concrete is taken into account in collapse theory.

(b) A short concrete column has overall dimensions of 250 mm by 250 mm. It is reinforced by four 25-mm diameter bars in the corners of the column fixed so as to allow 25 mm cover on each face. Calculate the maximum axial live load that the column could carry under working load conditions (i) by elastic theory (ii) by collapse theory. Assume:

$f_{cu} = 45.0\ MN/m^2 \quad f_y = 420\ MN/m^2$

For elastic analysis:

$f_{dcc} = \frac{1}{4}f_{cu} \quad f_{dsc} = \frac{1}{2}f_y \quad \alpha_e = 13.2$

For collapse analysis:

$\gamma_f = 1.6$

The maximum steel stress is $0.67f_y$ and the mean concrete stress in the rectangular stress block is $0.40f_{cu}$. (UB2 1969, 40 min)

2. A mass concrete wall of weight w per unit volume is of height a and width a/2. It is subjected to a uniform horizontal pressure of p per unit area. Find p if the criterion of no tension is just satisfied at the base section. It is decided to increase the height of the wall by placing blocks centrally on top as shown in Fig. A. If the blocks weigh 4w per unit volume and the same pressure p acts

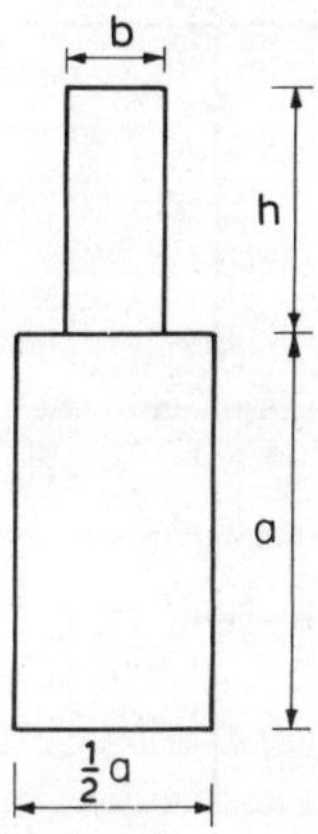

Ch.9 Fig.A Question 2

over the total height, find the dimensions b and h of the blocks to satisfy the two criteria:

(a) a factor of safety of 1.6 against the block sliding on the wall (coefficient of friction between block and wall is 0.1)

(b) no tension on the base section. (UL2 1969, 36 min)

3. The characteristic cube strength of the concrete for the column

analysed in Example 9.5.1 was found to be only 30 MN/m^2 when tested at 28 days due to the use of cement which had deteriorated due to storage in damp conditions. Determine the collapse load for the column by an accurate analysis according to the method of Hognestad et al. What is the percentage reduction in collapse load for this 20% fall in concrete cube strength?

4. Repeat question 3 using the simplified assumptions of CP 110 if the characteristic yield strength of the bars in tension is 420 MN/m^2. (The collapse load of the column when f_{cu} equals 37.5 MN/m^2 using simplified assumptions is 1.71 MN)

5. Fig.B shows part of a multi-storey framed building. The columns BE and ED are 0.25 × 0.25 metres square, and the beams AE and EC are 0.2 m wide by 0.4 m deep. Determine the reinforcement required for the column ED by considering the direct loads down column BE and the

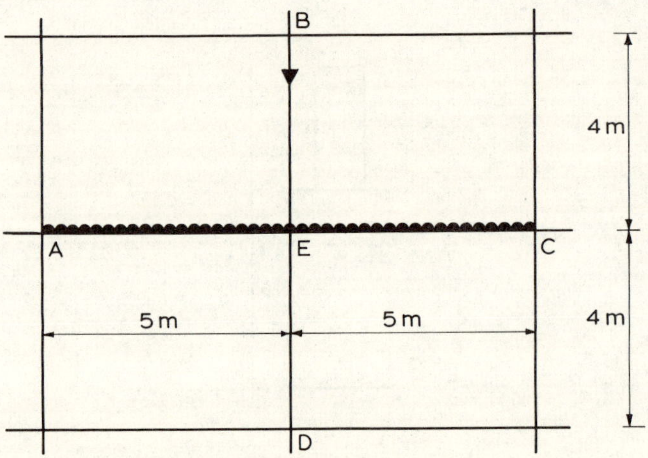

Ch.9 Fig.B Question 5

The characteristic loads acting on column BE from the floors above equal 300 kN (live) and 200 kN (dead). The characteristic loads acting on beam AEC equal 30 kN/m (dead) and 120 kN/m (live).

loads on the floor beams AE and EC. Neglect the self-weight of the columns and assume that the frame is braced. You may use the design chart for

f_{cu} = 40 MN/m^2, f_y = 410 MN/m^2 and d/h = 0.90

(UB3/O 1969 modified, 45 min)

6. Calculate the ultimate load of an 18-in (457-mm) square column, using the method of Hognestad et al., when the load acts 2 in (50.8 mm) outside the midpoint of one edge as shown in Fig.C. (1 kip is 1000 lbf, or 4448 N). The column is reinforced with eight 1-in (25.4-mm) diameter

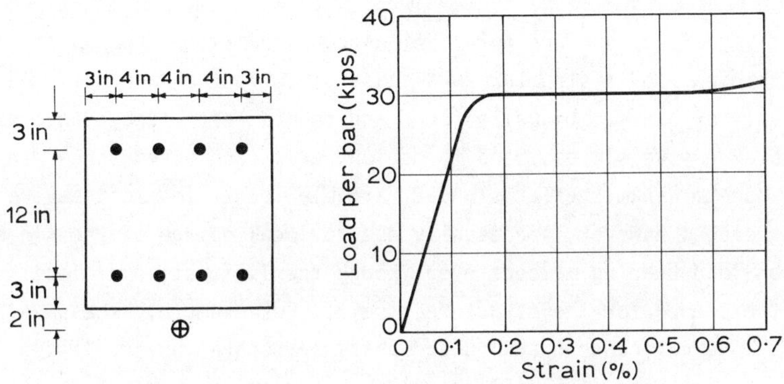

Ch.9 Fig.C Question 6

mild steel bars, each bar having a load-strain curve as shown. The cube crushing strength of the concrete, f_{cu} is 6000 lbf/in^2 (41.3 MN/m^2). Assume the following values for the standard factors $\lambda_1\lambda_3$ and λ_2, and the ultimate compressive concrete strain, ε_{cu} (f_c is in lbf/in^2 units and equals $0.78f_{cu}$):

$$\lambda_1\lambda_3 = (3900 + 0.35f'_c)/3200 + f'_c)$$

$$\lambda_2 \times 10^3 = 500 - f'_c/80$$

$$\varepsilon_{cu} \times 10^3 = 4 - f'_c/6500$$

(UB3/H 1968, 36 min)

10 Limit states of deflection

Excessive deflections and deformations can impair the appearance and the efficiency of a structure and cause discomfort or alarm to the occupants or users. Visual offence and reduced efficiency can occur due to excessive slopes, pond formation, etc., and disturbance can occur due to dynamic effects under variable load. The most severe consequences, however, are usually due to local damage of the finishes and non-load-bearing members even though the limit state of local damage is not exceeded for the structure itself. For example, excessive deflections can cause cracking and possible separation of plaster finishes, crushing of partition walls, or cracking of glazing units. The increased slenderness which is now possible for members designed for the limit state of collapse has made the consideration of deflections more important than in earlier codes.

The most important limit states of deflection and deformation are therefore as follows:

(i) Excessive deformability of the structure causing structural loads to be applied to non-load-bearing members.

(ii) Excessive deformation of the structure itself causing excessive slopes, sagging floors, etc.

(iii) Incompatible deformation of the structure, under the influence of shrinkage, creep and variable loads, with neighbouring load-bearing members (e.g. horizontal cracks in brickwork at the corners of slabs which are built in).

(iv) Excessive vibration under the effects of moving loads, e.g. fork-lift trucks operating on a suspended floor.

10.1 DEFLECTIONS AND DESIGN VALUES

The maximum deflections which are permitted by CP 110 under normal working loads are given, usually in terms of span or height, and are summarized in Appendix B. Experience has indicated that deflections are likely to be satisfactory if certain limiting span/effective-depth ratios are not exceeded. A suitable deemed-to-satisfy design procedure has therefore been evolved which is summarized in Appendix CE and is discussed later in 10.5. However, if it is desired to use higher span/effective depth ratios, or if the structure is unusual in any way, or if deflections are particularly important, then CP 110 also contains a procedure for calculating deflections (see Appendix CE(§A.2)) as an alternative to the usual design approach. The calculation of deflections is considered in 10.2 to 10.4 and may be omitted on first reading if desired.

Overall sag

The overall sag of floors, roofs and other horizontal members, given by the long-term deflection as measured below the level of the supports, is restricted in general to span/250. This ensures both satisfactory appearance and efficiency of the structure (e.g. satisfactory drainage from roofs, verticality of equipment placed on floors, etc.). In special cases deflection can be increased slightly if it can be shown that the increased deflection is unlikely to cause difficulties.

Pre-camber

The harmful effects of large deflections can be reduced by the provision of an initial camber (especially for prestressed concrete members). Such provision must be restricted, however, otherwise a satisfactory minimum stiffness may not be provided and dynamic effects under variable loads could become objectionable. The pre-camber of horizontal members should not exceed span/300 unless uniformity of camber between adjacent units can be ensured or no finishes are to be applied.

Damage to finishes

The acceptable deflection for partitions and finishes depends on their ability to absorb or resist strain. The Code specifies that the additional deflection after the application of finishes must not exceed either span/350 or 20 mm, unless provision is made for the movement. The limit of 20 mm is clearly necessary and becomes operative for spans over 7 m. These limits could be considered high compared with those of earlier codes and suitable adjustments may be necessary in certain cases. For example, deflections must be reduced if very brittle finishes or partitions are used. Designers must therefore use their discretion; either in selecting more restrictive limits, or in exceeding the normal limits if they can justify such action.

Lateral deflections

Lateral deflections of height/500 or less should ensure virtually complete freedom from difficulties with glazing panels and opening lights, etc. The lateral deflection of a tall slender structure, and possible vibration effects due to wind or machinery, should be considered.

Design values for deflection

The limit states of deflection are concerned with the probable maximum loading under normal working conditions throughout the life of the structure. Hence the characteristic loads can often be used directly since, in general, the partial safety factor for load is unity as shown in Table B1. The partial safety factor for material strength for both concrete and steel are also unity and the properties of the materials are assumed to be those associated with the design strengths.

10.2 ELASTIC DEFLECTIONS

The lateral deformation of an elastic member depends upon its flexural rigidity, EI. Several methods are available for determining elastic deflections and are adequately described in standard text-books on structures. Simple methods only are discussed here, as well as approximations which can be made for members with non-uniform sections.

Bending of elastic beams

Simple elastic theory for the bending of beams assumes small deflections and obtains the following relation:

$$-EI\frac{d^2y}{dx^2} = M_x \qquad (10.2.1)$$

where M_x is the bending moment at distance x from the origin, and y is measured downwards in the normal way as shown in Fig. 10.2.1.

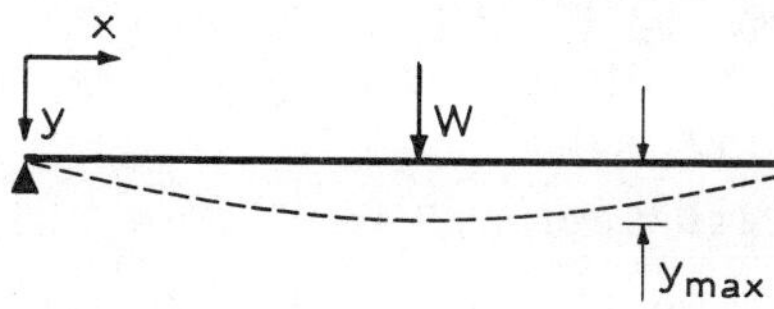

Fig.10.2.1 Elastic deflection of simple beam

Shape of deflected beam

Integrating Eqn (10.2.1) once with respect to the horizontal distance x gives the slope, dy/dx, at any position. Integrating twice gives an expression for the vertical deflection y in terms of x. Two constants of integration are introduced which are determined by reference to points with known conditions. For example, the slope at midspan for a symmetrical beam is zero.

Example 10.2.1

A uniform beam AB of span L and constant flexural rigidity EI is subjected throughout its length to a uniformly distributed load of w per unit length. Determine the maximum central deflection if it is (a) simply supported and (b) fixed ended.

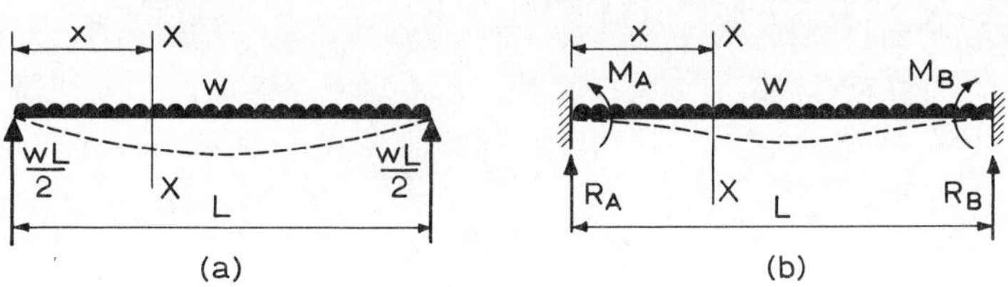

Fig.10.2.2 Example 10.2.1

Solution

(a) $M_x = \frac{1}{2}w(Lx - x^2)$

Substituting in Eqn (10.2.1) and integrating gives

$$-EI\frac{dy}{dx} = \frac{w}{2}\left(\frac{Lx^2}{2} - \frac{x^3}{3}\right) + A$$

The slope is 0 at x = L/2, hence

$A = -wL^3/24$. Integrating again gives

$$-EIy = \frac{w}{2}\left(\frac{Lx^3}{6} - \frac{x^4}{12}\right) - \frac{wL^3}{24}x + B$$

y = 0 when x = 0 or L, hence B = 0. Thus the maximum deflection (at x = L/2) is given by

$EIy = 5wL^4/384$

$y = 5wL^4/384EI$ (Answer)

(b) By symmetry

$M_A = M_B$ and $R_A = R_B = \frac{1}{2}wL$

$M_x = \frac{1}{2}w(Lx - x^2) - M_A$

Hence

$$-EI\frac{dy}{dx} = \frac{w}{2}\left(\frac{Lx^2}{2} - \frac{x^3}{3}\right) - M_Ax + A$$

Slope = 0 at x = 0 or L (fixed ends). Hence A = 0 and

$M_AL = \frac{1}{2}wL^3(\frac{1}{2} - \frac{1}{3})$ or $M_A = wL^2/12$

Integrating again gives

$$-EIy = \frac{w}{2}\left(\frac{Lx^3}{6} - \frac{x^4}{12}\right) - \frac{wL^2x^2}{24} + B$$

y = 0 when x = 0 or L, hence B = 0. Thus the maximum deflection (at x = L/2) is given by

$EIy = wL^4/384$

$y = wL^4/384EI$ (Answer)

Curvature

The curvature is the reciprocal of the radius, i.e. 1/r, and is given approximately by $-d^2y/dx^2$. The simple elastic beam formula, Eqn (10.2.1 is given in terms of r by

$M/I = E/r$

Hence

$$r^{-1} = \frac{d^2y}{dx^2} = \frac{M}{EI} \qquad (10.2.2)$$

Integrating M/EI twice w.r.t. x gives an expression for which the power of x is raised by two. It is therefore possible to substitute a particular value of r^{-1} in the final expression for maximum deflection, y_{max}, for a uniform rectangular beam, in order to obtain an expression of the form

$$y_{max} = KL^2r^{-1} \tag{10.2.3}$$

where K = a constant given by the laws of mechanics applied to a beam of constant cross-section

L = effective span of beam

r^{-1} = curvature of beam due to loading at a suitable section along the member (usually point of maximum bending moment).

This expression is given in CP 110.

Moment-Area Method

The *conjugate beam method* used in Example 3.5.3 is effectively another way of considering the *moment-area method*, which states that the displacement of any point on a member, from the tangent at another point on the member, equals the moment of the M/EI diagram between the two points taken about the first point.

Example 10.2.2

Repeat Example 10.2.1 using the moment-area method.

Fig. 10.2.3 Example 10.2.2.

Solution

(a) Moment of area of M/EI diagram between midspan and support, about support = $\frac{2}{3}.\frac{1}{2}L(wL^2/8EI)(\frac{5}{8}.\frac{1}{2}L)$ = $5wL^4/384EI$

Deflection at midspan numerically equals the displacement of the support from the (horizontal) tangent at midspan

i.e. $y = 5wL^4/384EI$ (Answer)

(b) Moment of area of M/EI diagram between midspan and support about midspan (since tangent at support is now horizontal)

$= \frac{2}{3}.\frac{1}{2}L(wL^2/8EI)(\frac{3}{8}.\frac{1}{2}L) - \frac{1}{2}L(wL^2/12EI)(\frac{1}{2}.\frac{1}{2}L)$

$= 3wL^4/384EI - wL^4/96EI$

Deflection at midspan numerically equals the displacement of midspan from the (horizontal) tangent at support.

i.e. $y = wL^4/384EI$ (Answer)

Example 10.2.3

Determine the value of K in Eqn (10.2.3) for the beams in example 10.2.1 for r^{-1}_{max} equal to the value of r^{-1} at midspan.

<u>Solution</u>

(a) $r^{-1}_{max} = M_{max}/EI = wL^2/8EI$

Therefore

$y_{max} = 5wL^4/384EI = (5/48)L^2r^{-1}_{max}$

$K = 5/48$ (Answer)

(b) $r^{-1}_{max} = M_{max}/EI = wL^2/24EI$

Therefore

$y_{max} = wL^4/384EI = (1/16)L^2r^{-1}_{max}$

$K = 1/16$ (Answer)

(Notice that the value of K depends on the position at which r^{-1} is calculated).

<u>Central deflection</u>

The values of K which are appropriate for the central deflection of a fully elastic beam due to a variety of loading conditions have been tabulated by Beeby & Miles [10.1] and are included in Appendix K. If it is assumed that the maximum deflection of a beam is never significantly different from the central deflection then it is possible to compute the deflection under more complex loading arrangements by obtaining values of K by superposition. The actual values of the maximum moments concerned must be carefully considered when using this approach, however, as shown below.

Example 10.2.4

Determine the central deflection for the fixed-ended beam subjected to a uniformly distributed load considered in Example 10.2.1 by suitable superposition of the data in Appendix K.

Solution The fixed end moments are $wL^2/12$ and the maximum moment for the free loading diagram is $wL^2/8$. Hence for fixed end moments of $-M$ at each end,

$$-M = wL^2/12 = \tfrac{2}{3}M_{free}$$

From Appendix K,

$$K = (5/48 - \tfrac{2}{3} \times \tfrac{1}{8}) \times M_{free}/M_{midspan}$$

$$M_{midspan} = wL^2/8 - wL^2/12 = wL^2/24$$

Hence

$$K = (1/48) \times (24/8) = 1/16$$

(which agrees with value previously obtained in Example 10.2.3) and

$$y_{midspan} = KL^2 r^{-1}_{midspan} = (L^2/16) \times (wL^2/24EI)$$

$$= wL^4/384EI \qquad \text{(Answer)}$$

Significance of midspan section

Eqn (10.2.3) for elastic deflections takes no account of variation in section properties along the span. The support and midspan sections, for example, usually contain varying amounts of reinforcement. A rigorous analysis is discussed elsewhere [10.1]. However, the use of values for these terms based on the section at midspan has been found to be satisfactory for both simple and continuous beams. It should be noted that for continuous beams the appropriate loading condition for the limit state of deflection is when the full live load is applied to the span under consideration and both adjacent spans are unloaded. This reduces the support moments to a minimum so that the effect of the support section is in general small and little accuracy is lost if the deflection is calculated on the basis of the midspan section only. The use of a numerical mean of the second moment of area of the support and midspan sections should therefore be avoided.

Support sections for cantilevers

The deflections of cantilevers should of course be based on the support section. The effective span, ℓ, of a cantilever should be taken as given in Appendix CB: which allows for partial rotation at the face of the support (since the bars at the support must dissipate their maximum stress into the supporting structure). In addition, the possibility of overall rotation at the support should also be considered.

10.3 SHORT-TERM DEFLECTIONS

Elastic theory assumes linear behaviour between stress and strain and normally avoids difficulties concerning the tensile strength of the concrete in flexure by assuming that it is negligible. However, experience has shown that elastic theory based on such an assumption tends to seriously over-estimate deflections, even though it is reasonably satisfactory for determining stresses at working loads for the concrete in compression. Because the concrete in reinforced concrete members cracks at relatively low tensile stresses the moment curvature curve, and hence the load deflection curve, tends to be bilinear, and it is important to distinguish between cracked and uncracked members. The concept of a *pseudo-cracked* section(which enables the tension in the concrete in the segments of a beam *between* the cracked sections to be taken into account)can give an approximation to the average deflection behaviour of an actual member with cracked sections separated by uncracked segments of significantly greater stiffness. Thus the short-term deflection is therefore determined according to whether or not the member is cracked or uncracked, from the short-term moduli associated with the design strengths of the materials.

Moment-curvature curves

An idealized bilinear moment-curvature relation can be represented as shown in Fig. 10.3.1 [10.1]. There is a sudden change in slope when the moment exceeds the concrete cracking moment. Since the moment varies linearly as the load, and from Eqn (10.2.3) the curvature varies linearly as deflection, it could be expected that the load deflection relation would be of similar form. However, since a beam is subjected to multiple and intermittent cracking, the load deflection curve is less clearly defined. Even for the load deflection relation, a stage can be reached where the crack pattern for the length of beam subjected to the maximum bending moment is effectively fully developed, and further increase in load does not introduce further cracks or reduce the crack spacing. The short lengths of concrete between the cracked sections are still long enough to prevent some bond slip with the reinforcement and hence tensile stresses up to the tensile strength of the concrete can be present between the cracks. Since a cracked section normally implies that the tensile strength of the concrete is ignored completely, sections for which this condition applies (i.e. sections in between cracked sections) will be termed *pseudo-cracked* sections for clarity.

(*Note:* If reference is made to Fig.6.2.1 the reader should visualize greatly increased horizontal scales for both curvature and deflection).

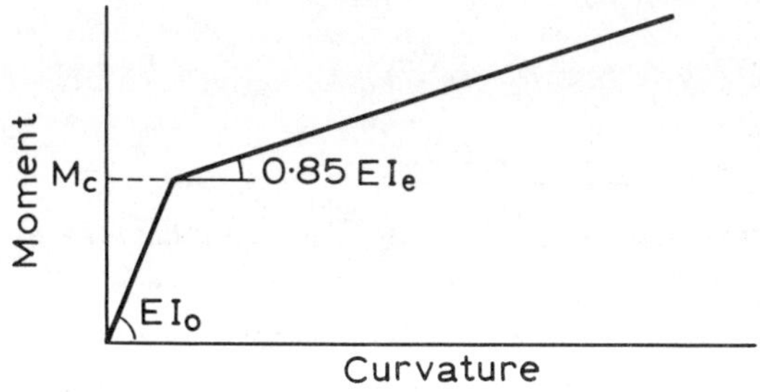

Fig.10.3.1 Bilinear moment-curvature relation

Curvature at uncracked section

The curvature at an uncracked section for short-term loading is given by

$$r^{-1} = M/E_c I_o \tag{10.3.1}$$

where E_c = concrete modulus for short-term loading

I_o = second moment of area of equivalent uncracked section (i.e. the gross section) at the maximum sagging moment (e.g. midspan).

Notice that the effective cracking moment, M_c, is given by the equation

$$M_c = f_{ct} I_o/(h - d_n) \tag{10.3.2}$$

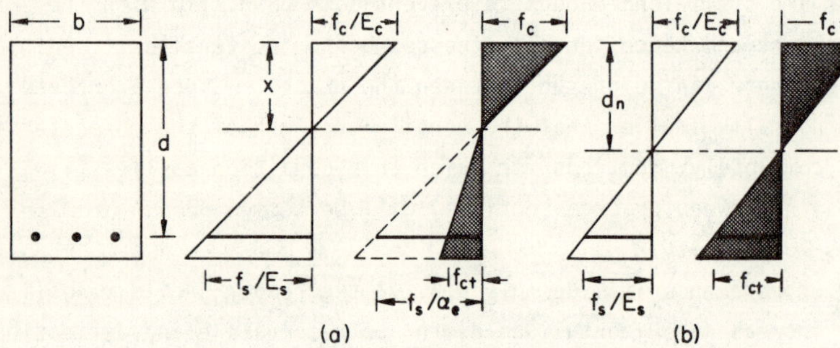

Fig.10.3.2 Strain and stress distributions for (a) pseudo-cracked and (b) uncracked sections

Curvature at pseudo-cracked section

The CP 110 assumptions for pseudo-cracked sections are given in Appendix CE(§A.2) and the assumed strain and stress diagrams for the concrete are shown in Fig. 10.3.2(a). Notice in particular that the concrete tensile stress, f_{ct}, is not given by the strain distribution and that the determination of the stresses using these assumptions requires a trial-and-error approach because the assumed value of f_{ct} is independent of M, f_c and f_s.

The concrete tensile stress f_{ct} at the level of the steel can be expressed in terms of an effective concrete tensile modular ratio α_{ec} $(= E_{ct}/E_c)$ as follows:

$$f_{ct}/\alpha_{ec} = f_s/\alpha_e = f_c(d-x)/x$$

Hence

$$\alpha_{ec} = \frac{f_{ct}}{f_c} \cdot \frac{x}{d-x} \tag{10.3.3}$$

Hence by reference to Fig. 10.3.2(a) and by taking first moments of the equivalent areas about the neutral axis:

$$\tfrac{1}{2} bx^2 + (\alpha_e - 1)\rho' bd(x - d') = \alpha_e \rho bd(d - x) + \tfrac{1}{2}\alpha_{ec} b(h - x)^2$$

Rearranging and dividing throughout by bd^2 gives

$$\tfrac{1}{2}(1-\alpha_{ec})\left(\frac{x}{d}\right)^2 - \left\{\alpha_e\rho+(\alpha_e-1)\rho'+\alpha_{ec}\left(\frac{h}{d}\right)\right\}\left(\frac{x}{d}\right)$$

$$- \alpha_e\rho+(\alpha_e-1)\rho'\frac{d'}{d} + \tfrac{1}{2}\alpha_{ec}\left(\frac{h}{d}\right)^2 = 0 \tag{10.3.4}$$

The effective second moment of area of pseudo-cracked section is given by:

$$I_c/bd^3 = \frac{1}{3}\left(\frac{x}{d}\right)^3 + \alpha_e\rho\left(1 - \frac{x}{d}\right)^2 + (\alpha_e-1)\rho'\left(\frac{x}{d} - \frac{d'}{d}\right)^2 + \frac{1}{3}\alpha_{ec}\left(\frac{h}{d} - \frac{x}{d}\right)^3 \tag{10.3.5}$$

The stresses are given by:

$$f_c = Mx/I_c \quad \text{and} \quad f_s = \alpha_e M(d-x)/I_c \tag{10.3.6}$$

and the curvature is given by:

$$r_b^{-1} = M/E_cI_c \tag{10.3.7}$$

A convenient trial and error approach is to let $\alpha_{ec} = 0$, solve Eqn (10.3.4) for x and then determine the corresponding values of I_c and f_c from Eqns(10.3.5) and (10.3.6). α_{ec} can then be obtained from Eqn (10.3.3) and new values obtained for x, E_c and f_c. If the second value of α_{ec} approximately equals the first value of α_{ec} then the curvature can be obtained from Eqn (10.3.7).

Example 10.3.1

A rectangular RC beam, 200 mm wide and 550 mm overall depth has 2 No. 20 mm diameter bars and 4 No. 25 mm diameter bars at depths of 50 mm and 50 mm and 500 mm below the compression face of the beam. The applied bending moment under service loads is 140 kN m, short-term E_c

is 28 GN/m² and E_s is 200 GN/m². Determine the instantaneous curvature if the short-term concrete tensile stress at the level of the tensile steel, f_{ct}, for a pseudo-cracked section is assumed to be 1 MN/m².

Solution

ρ = 1960/(500 × 200) = 1.96%; h/d = 550/500 = 1.1

ρ' = 628/(500 × 200) = 0.628%; d'/d = 50/500 = 0.1

α_e = 200/28 = 7.14

If α_{ec} = 0 (i.e. the tensile concrete is ignored) then from Eqn (10.3.4)

$$\tfrac{1}{2}\left(\frac{x}{d}\right)^2 + \{0.140 + 0.0386\}\left(\frac{x}{d}\right) - \{0.140 + 0.0386 \times 0.1\} = 0$$

or $\tfrac{1}{2}\left(\frac{x}{d}\right)^2 + 0.179\left(\frac{x}{d}\right) - 0.144 = 0$

i.e. $\frac{x}{d} = -0.179 \pm \sqrt{0.179^2 + 2 \times 0.144} = 0.387$, say 0.4

Hence from Eqn (10.3.5) for x/d = 0.4 and α_{ec} = 0

$$I_c/bd^3 = \tfrac{1}{3}(0.4)^3 + 0.140\,(1 - 0.4)^2 + 0.0386(0.4 - 0.1)^2$$

or $I_c = 0.075bd^3 = 1.88 \times 10^{-3}$

Hence $f_c = (140 \times 0.4 \times 0.5/1.88)10^{3+3} = 14.9 \times 10^6$

and resubstituting in Eqn (10.3.3), $\alpha_{ec} = \frac{1}{14.9} \times \frac{0.4}{1-0.4} = 0.045$

Resubstituting in Eqn (10.3.4)

$$\tfrac{1}{2}(0.955)\left(\frac{x}{d}\right)^2 + \{0.179 + 0.45 \times 1.1\}\left(\frac{x}{d}\right) - \{0.144 + \tfrac{1}{2} \times 0.045(1.1)^2\} = 0$$

or

$$\tfrac{1}{2}\left(\frac{x}{d}\right)^2 + 0.239\left(\frac{x}{d}\right) - 0.179 = 0$$

i.e. $\frac{x}{d} = -0.239 \pm \sqrt{0.239^2 + 2 \times 0.179} = 0.405$

Hence from Eqn (10.3.5) for x/d = 0.405 and α_{ec} = 0.045

$$I_c/bd^3 = \tfrac{1}{3}(0.405)^3 + 0.140(0.595)^2 + 0.0386(0.405-0.1)^2 + \tfrac{1}{3} \times 0.07(1.1-0.405)^3$$

or $I_c = 0.083bd^3 = 2.08 \times 10^{-3}$

Hence $f_c = (140 \times 0.405 \times 0.5/2.08)10^{3+3} = 13.5 \times 10^6$

$$(\alpha_{ec})_2 = \frac{1}{13.5} \times \frac{0.405}{0.595} = 0.05 \simeq 0.045$$

Hence take x as 0.405 × 0.5 and substitute in (10.3.7)

$$r_b^{-1} = \frac{13.5}{0.2025 \times 28} \times 10^{6-9} = 2.38 \times 10^{-3}$$

Instantaneous curvature for pseudo-cracked section = 2.38×10^{-3}

(Answer)

Notice that x/d only varied from 0.387 to 0.405 (i.e. under 5%) as α_{ec} varied from 0 to 0.045. Thus in design it is possible to assume that x/d is approximately constant at the $\alpha_{ec} = 0$ value and merely add to the resistance moment of the cracked section (i.e. for $\alpha_{ec} = 0$) the moment of the tensile concrete force about the neutral axis position (also for $\alpha_{ec} = 0$).

The tension force in the concrete is:

$$\tfrac{1}{2}f_{ct}b(h-x)^2/(d-x)$$

which acts at a distance of $\frac{2}{3}(h-x)$ from the assumed neutral axis position. Hence to a first approximation the actual moment M, for a pseudo-cracked section can be reduced by:

$$\tfrac{1}{3}f_{ct}b(h-x)^3/(d-x) \qquad (10.3.8)$$

when calculating the stresses and curvature from (10.3.6) and (10.3.7)

Notice also that design charts such as Figs F3 and F4 (after reference [5.4]) can further facilitate normal design if suitable computer programmes are not available.

If the above approximation is applied to Example 10.3.1 then the reduction in the applied moment

$$= \tfrac{1}{3} \times 0.2(0.55 - 0.387/2)^3/(0.5 - 0.387/2) \times 10^6$$

$$= 9.9 \times 10^3$$

Hence $f_c = (130.1 \times 0.4 \times 0.5/1.88)10^6 = 13.8 \times 10^6$

which compares with 13.5 MN/m^2 for the previous approach.

Calculation of short-term deflections

The above considerations enable the curvatures at midspan (or support) to be calculated assuming that the section is either pseudo-cracked or uncracked. Reference to Fig. 10.3.1 shows that the uncracked value corresponds to the initial slope (or its extension) whereas the pseudo-cracked value corresponds to the second linear portion (or its extension to the moment axis). The correct curvature for a given moment is therefore the larger of the two values (and which are coincident only when $M = M_c$). The moment and corresponding curvature diagrams for a simple beam subjected to two-point loading are shown in Fig. 10.3.3. The value of K in Eqn (10.2.3) is governed by the shape of the curvature diagram. Hence for a pseudo-cracked section it varies slightly from the elastic value(which is given by the broken lines in Fig. 10.3.3). However, by calculating the maximum curvature corresponding to the bilinear relation in Fig. 10.3.2 and then in effect assuming a linear relation between zero and this value, the values of K for a fully elastic beam can then be used.

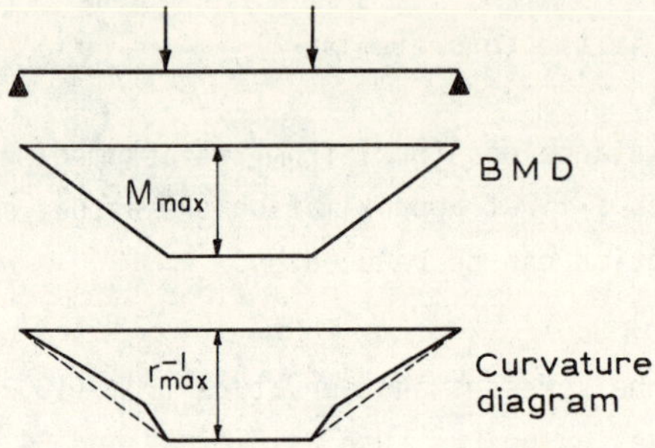

Fig. 10.3.3 Bending moment and curvature diagrams for simple beam under two-point loading and pseudo-cracked midspan section

Curvature diagrams

The approximation to a linear moment-curvature relation means that elastic theory and, in particular, the principle of superposition of curvatures can still be used. For example consider a fixed ended uniform beam subjected to a single concentrated load W at a third point as shown in Fig. 10.3.4(a). The bending moment diagram is as shown in (b) and, since EI is constant, the curvature diagram is of the same shape with curvatures of r_A^{-1}, r^{-1} and r_B^{-1}. Thus the curvature diagram can be obtained by the superposition of (c) and (d).

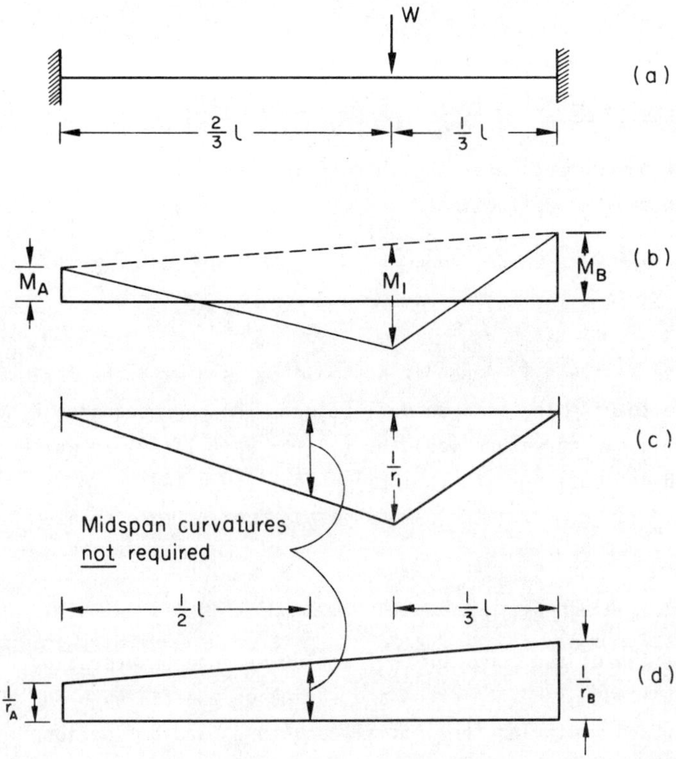

Fig.10.3.4 Curvature diagrams

Example 10.3.3

A fixed ended beam of constant section EI is subjected to a single third point load W. Determine the deflection at midspan using Appendix K.

Solution

From Appendix L (or by calculation)

$$M_A = -\frac{1}{3}\left(\frac{2}{3}\right)^2 W\ell; \qquad M_B = -\left(\frac{1}{3}\right)^2 \left(\frac{2}{3}\right) W\ell$$

$$\text{Therefore} \quad \frac{1}{r_A} = -\frac{4W\ell}{27EI}; \qquad \frac{1}{r_B} = -\frac{2W\ell}{27EI}$$

$$M_1 = \frac{2W\ell}{9} \quad \text{and} \quad \frac{1}{r_1} = \frac{2W\ell}{9EI}$$

From Appendix K

$$K = -\frac{1}{16} \text{ for an end moment } -M$$

$$\text{and } K = \frac{-4a^2 + 8a - 1}{48a} = \frac{23}{288}$$

for a maximum moment under the third point load of $W\ell(1-a)a$

Hence the midspan deflection

$$= \frac{\ell^2}{EI} \quad - \frac{1}{16}(4 + 2)\frac{W\ell}{27} + \frac{23}{288}\,\frac{2W\ell}{9}$$

$$= \frac{5W\ell^3}{1296}$$

(Answer)

Note: The K values in Appendix K should be used with the maximum curvature to give the midspan deflection. The corresponding K' value for the midspan curvature would be $(-4a^2 + 8a + 1)/24$, in which case:

midspan deflection = $K'\ell^2(Wl/6EI) = 23W\ell^3/(9 \times 144)$

—the same as above for a simple span.

10.4 LONG-TERM DEFLECTIONS

Determination of the long-term deflection depends on assessing the permanent loading and the creep and shrinkage coefficients for the concrete, and modifying the initial curvatures and deflections accordingly. Reference to Appendix CE (§A.2) indicates that the estimated loading can be seriously in error due to inaccurate assessment of the actual restraints at supports, neglecting the effects of finishes and partitions, etc. The loading effects are therefore considered first.

General load redistribution effects

The load due to partitions is usually included as part of the distributed live loading. However, if a partition of blockwork or brickwork is constructed over the whole span of a member, with no major openings near its centre, then the long-term deflection can cause the member to creep away, leaving the partition as a self-supporting deep beam. Again, if there is no initial gap at the top of a partition then it can take an increasing proportion of the permanent loading as the member above deflects eventually stopping the deflection altogether if the partition is sufficiently stiff. Thus it is reasonable to neglect the weight of such partitions when determining the permanent loading.

An in situ screed which is monolithic with the structure will also tend to act structurally. However, since screeds are usually laid after the removal of any propping the initial deflection will have already taken place before the screed is sufficiently strong to make a useful structural contribution. Bate et al. suggest that not more than 50% of the long-term deflection should be considered as reducible by the action of the screed [5.4].

Creep effects

When concrete is subjected to a constant load it undergoes an initial elastic strain and then a subsequent time-dependent creep strain which increases at a diminishing rate with time. The creep coefficient, ϕ, is the ratio of the creep strain to the initial (instantaneous) elastic strain. Thus if E_c is the short-term concrete modulus then:

$$\text{Long-term } E_c = E_c/(1+\phi)$$

The creep coefficient (which has been taken previously as 1 for simplicity) varies with age at first loading and duration of loading, and guidance has been given by Bate et al. [5.4] as indicated in Table 10.4.1. The coefficient is also affected by other factors such as aggregate properties, mix proportions, humidity and member size.

Table 10.4.1 Creep coefficients and stress in concrete at tensile steel level

Duration of load (days)	Age at loading (days) 7	14	28	56	$\nless 100$	pseudo-cracked tensile stress f_{ct} (MN/m^2)
0	0	0	0	0	0	1.00
10	0.39	0.33	0.27	0.23	0.19	0.95
100	1.92	1.65	1.37	1.17	0.96	0.77
1000	3.08	2.64	2.20	1.87	1.54	0.64
∞	3.85	3.30	2.75	2.34	1.93	0.55

As the modular ratio α_e increases with time then, as shown previously in Fig. 4.1.1, the neutral axis depth increases. This can be illustrated in terms of r_b^{-1} as shown in Fig. 10.4.1. Thus the effect

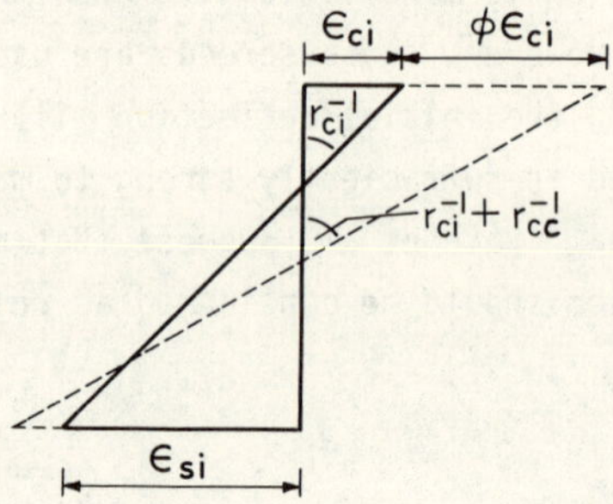

Fig. 10.4.1

of creep is to increase the compressive strain in the concrete by an amount $\phi\varepsilon_{ci}$ where ε_{ci} is the initial concrete strain. The strain in the tension steel also increases by an amount which is governed by the reduction in the lever arm. The curvature of the section increases by an amount r_{cc}^{-1} and hence the deflection increases by the factor

$$\left(r_{ci}^{-1} + r_{cc}^{-1}\right)/r_{ci}^{-1}$$

where r_{ci}^{-1} is the initial concrete curvature.

Shrinkage effects

The amount of shrinkage depends upon the curing conditions, size and shape of member, mix proportions and type of aggregate used. If concrete is unrestrained and not reinforced it can shrink uniformly. However, the movement of the concrete is restrained if reinforcement is present, in such a way that the steel becomes compressed and the concrete is subjected to tension, as shown in Fig.11.3.1 (chapter 11). Equation (11.3.3) shows that the presence of concentric reinforcement reduces the shrinkage strain ε_{cs} of concrete in the ratio

$$\varepsilon_{sc}/\varepsilon_{cs} = 1/(1 + \alpha_e\rho_c) \qquad (10.4.1)$$

where ε_{sc} is the elastic compression strain in the steel reinforcement.

In fact the reduction in shrinkage can be greater than this equation indicates because the bond developed between concrete and steel restrains the free shrinkage by creep effects.

In the case of a reinforced concrete beam it is the contraction strains between the top and bottom which cause the additional deflection. Thus for the doubly reinforced beam shown in Fig. 10.4.2. let the

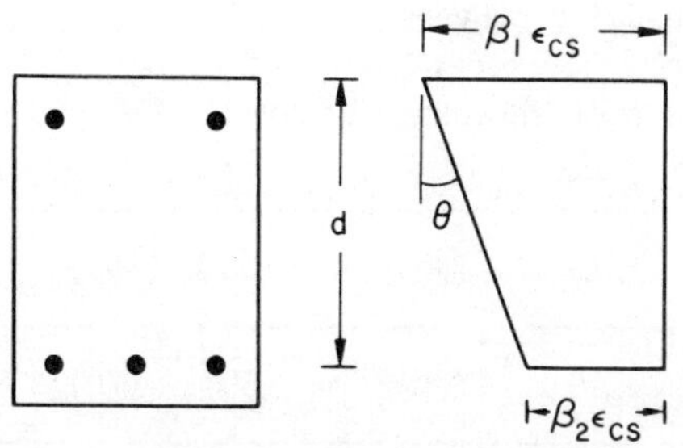

Fig.10.4.2 Rotation due to shrinkage

shrinkage strains be $\beta_1\varepsilon_{cs}$ and $\beta_2\varepsilon_{cs}$ at the levels of the compression face and tensile reinforcement respectively. The rotation θ gives the curvature r^{-1}_{cs} in terms of β_1,β_2 and ρ_o as follows

$$r^{-1}_{cs} = \theta = (\beta_1-\beta_2)\varepsilon_{cs}/d = \rho_o\varepsilon_{cs}/d \qquad (10.4.2)$$

Values of ρ_o are given in Table CE4 in Appendix CE.
The value of ρ_o for $\rho = \rho' = 1.0\%$ is 0.20 whereas for concentric shrinkage according to Eqn (10.4.1) with $\alpha_e = 15$, an actual value of 0.77 and a differential value of 0 would be expected. When looking at Table CE4 it is important to remember that the tensile reinforcement is assisted by the normal bending of the beam in restraining the shrinkage, thus with no compression steel and 3% or more of tension steel $\beta_1 - \beta_2 = 1.0$. With 2% compression steel (equivalent to 4% steel in the top half of the beam) and 4% tension steel (sufficient to reduce shrinkage contraction around tension steel to zero) then $\rho_o = 0.62$: the value obtained from Eqn (10.4.1) for β_1 assuming concentric reinforcement and $\alpha_e = 15$. Bearing in mind the complications due to eccentric contraction and the exaggerated restraint of the tension steel, the general form of Table CE4 can therefore be explained.

In the absence of information in CP 110, Bate et al. [5.4] have again given values based on the CEB recommendations [1.9]. They suggest an ε_{cs} value of 300 microstrain for sections less than 250 mm thick and a value of 250 microstrain for thicker sections. Concretes containing aggregates known to have high shrinkage can have values twice as large or even more but, conversely, many concretes can have less shrinkage—especially if external. The developments of shrinkage with time can be obtained very approximately from Table 10.4.2 [5.4], which has been extended for convenience.

Table 10.4.2 Percentage of shrinkage occurring up to a given age

Age (days)	10	(28)	100	(365)	1000	∞
%	10	(30)	50	(70)	80	100

If it is assumed that the tensile reinforcement reduces as the bending moment diagram, e.g. parabolically, then very approximately the appropriate K value to use for shrinkage will be the same value as for the loading.

Example 10.4.1

The beam in Example 10.3.1 carries a uniformly distributed dead load (including self-weight) of 6.8 kN/m on a simple span of 7 m and a uniformly distributed live load of 16 kN/m, of which 40% can be taken as permanent. Determine the long-term deflection. Assume that $\varepsilon_{cs} = 250 \times 10^{-6}$, that the creep coefficient is 2.75 and that the tensile stress f_{ct} for a pseudo-cracked section under long term loading is 0.55 MN/m².

Solution

The procedure is as outlined in Appendix CE(§A.2) i.e.

(1a) Calculate instantaneous curvature for total load

(1b) Calculate instantaneous curvature for permanent load

(2) Calculate long-term curvature under permanent load

(3) Calculate the shrinkage curvature

(4) Determine the total long-term curvature from

(1a) - (1b) + (2) + (3)

The midspan moments for design loads are

$(1.0 \times 6.85 + 1.0 \times 16.0)10^3 \times 7^2/8 = 140.0$ kN/m (total load)

$(1.0 \times 6.85 + 0.4 \times 1.0 \times 16.0)10^3 \times 7^2/8 = 81.2$ kN/m (permanent load)

(1a) As before, $r_{bi}^{-1} = 2.38 \times 10^{-3}$

(1b) The initial curvature for permanent load, r_{big}^{-1} is conveniently given by assuming that x/d, α_{ec} and hence I_c remains constant for the reduced loading.

Thus from (10.3.7)

$$r_{big}^{-1} = \frac{81.2}{140.0} \times r_{bi}^{-1} = 1.38 \times 10^{-3}$$

(2) Long-term $E_c = 28/(1 + 2.75) = 7.47$ GN/m^2

and $\alpha_e = 200/7.47 = 26.8$;

Therefore $\alpha_e \rho = 0.525$; $(\alpha_e - 1)\rho' = 0.162$

From Figs.F3 and F4 (or by calculation for $\alpha_{ec} = 0$)

$x/d = 0.560$ and $I_e/bd^3 = 0.194$

The approximate effective applied moment from (10.3.8) is given by

$$81.2 - \frac{10^3}{3} \times 0.55 \times 0.2(0.55 - 0.280)^3/(0.50 - 0.280)\text{kN/m}$$

$$= 77.9 \text{ kNm and } f_c \simeq \frac{77.9 \times 0.560 \times 0.5}{0.194\ bd^3}(10^3) = 4.50 \text{ MN/m}^2$$

For a closer estimate then from (10.3.3)

$$\alpha_{ec} = \frac{0.55}{4.50} \times \frac{0.560}{0.440} = 0.156$$

From Eqn (10.3.4)

$$\tfrac{1}{2}\left(\frac{x}{d}\right)^2 + 1.017\left(\frac{x}{d}\right) - 0.753 = 0 \quad \text{or} \quad \frac{x}{d} = 0.577$$

From Eqn (10.3.5)

$I_c/bd^3 = 0.202$ or $I_c = 5.06 \times 10^{-3}$

$$\text{Therefore } f_c = \frac{81.2 \times 0.577 \times 0.5}{5.06}(10^6) = 4.63 \text{ MN/m}^2$$

$$\text{and } r_{big}^{-1} = \frac{81.2}{7.47 \times 5.057}(10^{3-9+3}) = 2.15 \times 10^{-3}$$

(3) From Eqn (10.4.2)

$$r_{cs}^{-1} = \frac{\rho_o\ \varepsilon_{cs}}{d} = \frac{2.75 \times 250}{0.5}(10^{-6}) = 1.38 \times 10^{-3}$$

(4) Hence the total long-term curvature

$= (2.38 - 1.38 + 2.15 + 1.38)10^{-3} = 4.53 \times 10^{-3}$

Hence the midspan deflection

$$= \frac{5}{48} \times 7^2 \times 4.53 \times 10^{-3} = 23 \text{ mm} \qquad \text{(Answer)}$$

10.5 LIMITING SPAN/EFFECTIVE-DEPTH RATIOS

The methods of calculation given in CP 110 are reasonably satisfactory, but they can be too time-consuming for routine design and they can still overestimate the deflection of an actual structure due to the inaccuracies indicated in Appendix CE(§A.2). The Code therefore states that the requirements of the limit state of deflection may be deemed to be satisfied if the structure conforms to the limiting span/effective-depth ratios, based on practical experience, given in Appendix CE. The scope of the clauses giving limiting span-depth ratios are sufficiently extensive to cover almost all structures, thus enabling the calculation procedures to be restricted to use in abnormal circumstances or to justify the use of members of increased slenderness.

CP 110 Deemed-to-satisfy procedure

The normal CP 110 approach allows the limit state of deflection to be satisfied if the ratio of span to effective depth, l/d, of a member does not exceed the product of four appropriate factors, suitably defined. This approach is outlined in Appendix CE and the factors are discussed below.

Span/depth ratios for rectangular sections

The basic span/effective depth (l/d) ratios given in Table CE1 apply to all spans if the only requirement is that the deflection should be $\not>$ span/250. However, if finishes are to be applied or partitions erected then the lower l/d ratios should be used for spans exceeding 10 m. For a 10 m span then l/250 equals 40 mm and l/350 equals 29 mm. Since the additional deflection after finishes and partitions should be $\not>$ l/350 and $\not>$ 20 mm, the latter controls. Hence for 10 m span there can be 20 mm additional deflection as well as 20 (i.e. 40–20) mm deflection before the erection of partitions etc. Although the basic and special l/d ratios should generally satisfy the limits of l/350 and 20 mm, this does not necessarily ensure that particular finishes or partitions will not be damaged by deflections.

Support conditions

A member should be considered as 'simply supported' in Table CE1 if the reinforcement at the supports is insufficient to develop the bending moments implied by 'partial' or 'full' fixity. Although degree of fixity also depends on the relative loading of different spans, full fixity may be assumed in centre spans of continuous members and also in end spans when the combined stiffness of all other members connected to the end of the beam or slab is not less than the stiffness of the beam or slab being considered. The end span of a continuous beam or slab should otherwise be assumed to be partially fixed. A member should only be considered as continuous if its span does not differ by more than 15% from the spans of adjacent members.

ξ-multiplier for tension reinforcement

The basic and special span/depth ratios are based on 1% of 410 MN/m^2 steel stressed to $0.58f_y$ = 238 MN/m^2 under service conditions. However, the deflection increases with increasing tension steel ratios and increasing service stress as shown by Table CE2. That is, the greater the moment to be resisted (i.e. greater steel area and service stress) the lower the required ξ-multiplier.

Thus if M_u = design ultimate moment = $\beta_b M_e$

M_d = design serviceability moment

and M_e = ultimate elastic moment before moment redistribution

then $M_u = 0.87 f_y (A_{st})_{req} z_1$ and $M_d = f_s (A_{st})_{prov} z_2$

To a first approximation $z_1 = z_2$ and $M_e = 1.5M_d$ hence

$$1.5\beta_b = 0.87(f_y/f_s)\beta_A$$

or $f_s = 0.58 f_y \beta_A/\beta_b$

which is the expression given in CP 110. β_A and β_b are as defined in Appendix CE.

ψ-multiplier for compression reinforcement

It has been shown that the tension reinforcement restrains the concrete as it shrinks and dries out, causing a curvature which is additional to the loading curvature. This is reduced if the concrete shrinkage in the top of the beam is also restrained by compression reinforcement. This is accounted for in Table CE3.

Additional ζ-multipliers

The draft CP 110 code formerly allowed for the ratio of permanent to total load. However, by taking a datum of 0.75 the variations in practical cases gave multipliers which did not differ greatly from 1.0 and could be neglected in the interests of simplicity.

The code does not define the datum for normal shrinkage and creep, so Table 10.5.1, based on the draft code, can be used as a guide if desired.

Table 10.5.1 ζ_1-multiplier for high shrinkage or creep

	Shrinkage coefficient ε_{cs}		Creep coefficient ϕ	ζ_1
	$\ngtr 0.6 \times 10^{-3}$		$\ngtr 3.5$	1.0
Either	$> 0.75 \times 10^{-3}$	or	> 4.0	0.85
Both	$> 0.75 \times 10^{-3}$	and	> 4.0	0.70

The stiffness of a flanged beam when compared with a rectangular beam of the same width as the flange must be less as indicated by considering the pseudo-cracked section (i.e. the cracked tension zone contributes to the stiffness). This can be allowed for by a multiplier ζ_2 which varies linearly with web width from 0.80 for $b_w = 0.30b$ to 1.00 for $b_w = b$.

Factors involving the characteristic strength required of the concrete and the distinction between slabs spanning in one or both directions were considered in drawing up the code proposals and were shown to have little influence [10.1]. The differences in stiffness which occur between beams and slabs is reflected in the lower percentages of tension steel provided in slabs, so stiffness need not be considered separately. However, the increased flexibility associated with flat slabs without drops must be allowed for by an additional multiplier, $\zeta_3 = 0.90$.

Creep and shrinkage coefficients are increased by such factors as high cement content, high water content, high temperature, low humidity and early loading. The properties of the aggregate can also have a pronounced effect. In addition to its influence on creep and shrinkage coefficients, a temperature crossfall through a member can provide an additional increase in deflection.

The normal increases in creep coefficient associated with lightweight concrete are accounted for by an additional multiplier ζ_4 of 0.85 (see Appendix CL).

Example 10.5.1

Is the limit state of deflection satisfactory according to the span-depth ratio requirements in CP 110 for the beam described in Example 10.4.1, if f_y = 410 MN/m^2 and f_s under service loading is 238 MN/m^2? Would it also be satisfactory if high shrinkage aggregate were used?

Solution

Basic (l/d) ratio from Table CE1 is 20

ξ-multiplier from Table CE2 for f_y = 410 MN/m^2, f_s = 238 MN/m^2 and ρ = 1.96% is 0.84

ψ-multiplier from Table CE3 for ρ' = 0.63% is 1.17

ζ-multiplier is unity hence

span/effective-depth ratio $\ngtr 0.84 \times 1.17 \times 1.0(20) = 19.7$

The actual span/effective-depth is 7/0.5 = 14

Hence an effective depth of 500 mm is very satisfactory for normal aggregates. (Answer)

If high shrinkage aggregates are to be used then from Table 10.5.1

ζ_1 = 0.70 approximately

hence span/effective-depth $\ngtr 0.70 \times 19.7 = 13.8$

which is not quite satisfactory. Suitable adjustments could be made either to the effective depth or, more conveniently, by a slight increase in the compression steel ratio. (Answer)

Reference

10.1 Beeby, A.W., and Miles, J.R., 'Proposals for the control of deflection in the new Unified Code for structural concrete', Concrete vol. 3 No. 3, March, 1969, pp 101-10.

Exercise

1. A reinforced concrete beam of 10 in (254 mm) width and 20 in (508 mm) effective depth is singly reinforced with 4 in^2 (2581 mm^2) of steel throughout its length. Its effective span is 20 ft (6.1 m). Find the central deflection due to the self-weight alone according to elastic theory, assuming that the concrete in the tension zone is fully cracked over the whole span.

The beam carries a live load which is uniformly distributed over the central 10 ft (3.04 m) of the beam, thus extending 5 ft (1.52 m) either side of the centre line. If the deflection due to the live load alone is restricted to 3 times that calculated for the self-weight, find the load per foot run that the beam can carry.

Examine the basis of the deflection calculations and explain how they are likely to be in error.

Elastic moduli of steel and concrete are 30×10^6 and 5×10^6 lbf/in^2 (207 and 34.5 GN/m^2). Self-weight of concrete is 144 lbf/ft^3 (22.6 kN/m^3). (UL2 1969, 36 min)

11 Limit states of cracking and local damage

The limit states of local damage are concerned with damage to the structure which could occur due to the application of normal working loads, that is, loads which are not expected to exceed the characteristic values. The partial safety factors for load are therefore, in general, unity, as for the limit states of deflection. The partial safety factors for material strength are also unity, at least in principle. However, it is necessary to use a value of $\gamma_m = 1.3$ for concrete cube strength in order to obtain, approximately, the uniaxial compression strength. Local damage can occur theoretically due to either excessive tension or compression, in either the concrete or the steel. However, tension in the concrete and the limit states of cracking are usually of most practical interest. These include the limit state of crack formation and the various limit states of crack width. In most structures it is sufficient to ensure that the crack widths do not become excessive, but in very aggressive environments it may sometimes be desirable to ensure also that the limit state of crack formation is not reached.

11.1 SPALLING OF CONCRETE AND TENSION IN STEEL

The stresses under working loads in both concrete and steel must be restricted if compression spalling or microcracking of the concrete and tension in the steel are to be avoided. This was the approach of conventional elastic theory and it can provide a suitable basis for analysis for local damage. Suitable allowance can be made for creep by again assuming that the long-term modulus for concrete equals half the short-term modulus. However, for reasons which will become apparent, this particular limit state of local damage has limited practical significance.

General assumptions for serviceability

The analysis of sections for the serviceability limit states assumes that plane sections remain plane and that stress varies as strain. The short-term concrete modulus can be taken from Appendix A and the modulus of elasticity of the reinforcement as 200 GN/m^2. Due allowance should be made for creep in the concrete (e.g. take long-term modulus as half the short-term modulus for local damage) and the tensile strength of the concrete can be ignored except for uncracked (or pseudo-cracked—see deflection) sections.

Compression spalling of concrete

Spalling of the concrete must clearly be avoided under working load conditions. The difficulty is that spalling of the concrete rarely occurs much before final collapse. CP 110 does not give a definite requirement for reinforced concrete but states that the concrete stress should not exceed $\frac{1}{2}f_{cu}$ at transfer for prestressed concrete members in flexure (§4.3.3.3). BS 5337 includes a limit of 0.45 f_{cu} but this concerns the validity of the cracking formula (see Appendix CW). Although the value of $\frac{1}{2}f_{cu}$ is also intended to prevent excessive prestressing losses, it will be assumed for convenience to be suitable for preventing spalling in reinforced concrete.

Excessive tension in steel

Both CP 110 and BS 5337 stipulate that the steel strain in reinforced concrete beams should not exceed $0.8f_y/E_s$ (see Appendix B). Thus the steel stress must not exceed $0.8f_y$. The codes consider steel strain rather than stress because it is really concerned with the tensile strain (and hence crack widths) in the concrete. A short-term tensile stress of $0.8f_y$ does not produce visible damage in the steel any more than a short-term compressive stress of $0.5f_{cu}$ produces spalling of the concrete. Even if the full characteristic loads were applied throughout the working life of the structure (which is highly improbable) it is also very unlikely that the concrete would show any spalling at the end of this period.

Design stresses and design strengths

The design stresses for local damage for both concrete and steel have been assumed to be $\frac{1}{2}f_{cu}$ and $\frac{3}{4}$ respectively. These are the maximum stresse normally considered desirable for working load conditions. They should be regarded as quite distinct from the values for material strength. Hence the coefficients of $\frac{1}{2}$ and $\frac{3}{4}$ should not be confused with the reciprocals of the partial safety factors for material strength. For example, the partial safety factor of 1.3 given in Table B1 refers to material strengths only and is irrelevant in this instance. It is therefore not incompatible with the above coefficient of $\frac{1}{2}$ for which the reciprocal is 2 (and not 1.3).

Comparison with elastic theory

The permissible values used in conventional elastic theory were given by

$$f_{dst} = \tfrac{1}{2}f_y \text{ and } f_{dcb} = f_{cu}/3$$

which are only 2/3 of the values given above. Hence all stresses at working load are increased by 50% and the moments of resistance given in Table 4.1.1 are also increased by 50%. It is clear that this limit state for local damage permits much higher loads than conventional elastic theory. Hence although the limit state of collapse permits higher working loads than conventional elastic theory, they are usually less than those permitted by the limit state of local damage for spalling of concrete and tension in steel for beams (see Example 11.1.1). Thus for practical design, collapse should be considered first and, since it is generally more critical than local damage for spalling, the latter need not be considered.

Example 11.1.1

A simply supported beam carries a total uniformly distributed load w per unit length. The midspan section had been designed by conventional elastic theory to be at ESR. If f_{cu} = 30 MN/m², f_{yt} = 250 MN/m² and α_e = 15 (i.e. long-term E_c) compare the loadings which can be carried for the limit states of collapse and local damage. Assume: (a) all dead loading and (b) all live loading.

Solution Calculations according to elastic theory provide the values given in Table 4.1.1(c), i.e.

$\rho_e = 2.18$ and $M_d = 2.23bd^2$ (MN/m² units)

Hence for limit state of local damage, these values can be increased by 50% and the loading can equal 1.5w.

For collapse,

$_*w$ = either (a) $1.4g_k$ or (b) $1.6q_k$

where g_k or q_k is the working load permitted by limit state of collapse. Resistance moment at collapse is given by the steel, hence from Eqn(8.2.3)

$$M_{ut} = \frac{\rho f_{yt}}{1.15}\left(1 - \frac{1.1\rho f_{yt}}{f_{cu}}\right) bd^2 \quad = 3.79bd^2$$

after substituting the appropriate values into the equation. Hence in terms of the working load g_k or q_k, the moment of resistance is $2.71bd^2$ or $2.37bd^2$. But $M_d = 2.23bd^2$ corresponded to a working load of w, hence the working loads corresponding to the limit state of collapse are 1.22w and 1.06w respectively. The limit state of collapse therefore permits greater working loads in both of these two cases than conventional elastic theory but neither is as great as that permitted by the limit state of local damage. Thus

$w < 1.06w$ and $1.22w < 1.5w$ (Answer)

11.2 LIMIT STATES OF CRACKING FOR BEAMS

Excessive tension in the reinforcement can cause excessive cracking in the adjacent concrete. Beams subjected to flexural tension due to the action of transverse loads exhibit a series of distributed flexural cracks even at working loads. These cracks are in general harmless unless the width becomes excessive. CP 110 permits maximum surface crack widths of 0.3 mm generally, or 0.04 × nominal cover adjacent to main bars for severe conditions. BS 5337 only permits 0.3 mm surface cracks for *Class C* exposure—i.e. faces of members over 225 mm thick which are remote from either *Class A* or *Class B* exposure. BS 5337 permits 0.2 mm maximum crack widths for Class B exposure—i.e. continuous or almost continuous contact with liquids, and only 0.1 mm maximum crack widths for Class A exposure—i.e. moist or corrosive atmosphere or alternate wetting and drying (see Appendix CW).

Cracking in concrete beams under load

The distribution of flexural cracks in a beam is controlled more by the stress gradient than by the bond stress between the concrete and the steel. It follows therefore that little, if any, initial bond slip is likely adjacent to flexural tension cracks in beams. Base et al. [11.1] carried out a very extensive series of crack measurements on 133 beams and showed that although some slipping occurs it is convenient to regard the concrete at the surface of the reinforcement as a 'point of zero strain'. That is, the steel strain and the actual bond slip are relatively small, so the crack width at the surface of the reinforcement can be regarded as zero and the cracks may be assumed to increase in width towards the face of the beams. They also found that for stresses within the elastic range of the steel the crack widths were closely proportional to the steel stress. At very low levels of stress there was a transition from the apparently uncracked beam to the cracked beam, but the number of cracks, and hence the crack spacing, rapidly reached an almost constant value which varied little with further increase in load. Within the ranges examined the effect of bar size, steel ratio, concrete strength and method of curing could all be neglected, thus supporting the 'no-slip' theory. The width of a crack was therefore determined by the elastic recovery of the concrete due to relief of tension and by the restraining effect of the nearest reinforcing bar.

Work by Illston and Stevens [11.2] has shown that for singly reinforced beams under sustained working load, bond slip extends under load with a consequent deduction in tensile stiffening; the mean crack spacing, s_m, remaining constant and being given by:

$$s_m = 2.3\, c_e \qquad (11.2.1)$$

where c_e was the 'effective' cover, from concrete surface to the centre of the bar.

The mean crack width, w_m, approximately doubled during two years of sustained loading, and was given by:

$$w_m = 2.3\, c_e(\varepsilon_{st} + \varepsilon_{cs}) \qquad (11.2.2)$$

where the mean steel strain, ε_{st}, under service loads was given by an elastic analysis of the fully cracked section and an effective modulus taking account of the creep of the concrete.

Crack widths for rectangular beams

The general form of the expressions given in CP 110 and BS 5337 for the design surface crack width at a point on the surface of a mature concrete member is as follows

$$w_{max} = \lambda a_{cr}\varepsilon_m/K_1 \tag{11.2.3}$$

where λ and K_1 have different values according to CP 110 (see Appendix CE) and BS 5337 (see Appendix CW). The reason for the differences is that the probability of exceeding w_{max} is about 1 in 20 for BS 5337 (i.e. about 5% failures) whereas the design surface crack width for CP 110 permits considerably more failures.

Stiffening effect of concrete in tension zone

The position of the neutral axis and the magnitude of the calculated longitudinal strain, ε_c, are obtained using the assumptions given in section 11.1. With reference to Fig. 11.2.1, if P is distance y from the neutral axis and ε_s is the strain in the tension steel, then the strain in the concrete, ε_1, at P is

$$\varepsilon_1 = \varepsilon_s y/(d - d_n)$$

ε_s is calculated in the normal way, neglecting the stiffening effect of the concrete in the tension zone. However this stiffening effect should

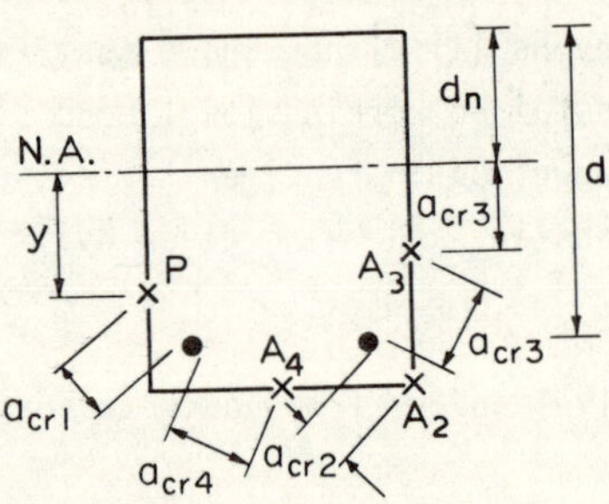

Fig.11.2.1 Determination of crack widths for rectangular beams

be allowed for when determining the average strain in the concrete, ε_m as follows:

$$\varepsilon_m = \varepsilon_1 - K_2 = \varepsilon_1 - \frac{0.7b_t hy}{A_s(h-d_n)f_s} \times 10^{-3} \tag{11.2.4}$$

where the symbols are as defined in Appendix CW and f_s is in MN/m² units. For CP 110 substitute $1.2/f_y$ for $0.7/f_s$ (see Appendix CE)

i.e. $0.7/f_s \simeq 0.7 \times 1.5 \times 1.15/f_y \simeq 1.2/f_y$

The coefficient of 1.2 was found experimentally [11.1].

Type of bar

Base et al. found that the crack widths for plain round bars were approximately 13% greater than for deformed bars. This was no doubt due to the fact that some slight bond slip did occur. However, it is clear that even large differences in bond strength have a relatively small effect. The codes ignore this small difference in assuming that Eqn (11.2.3) is valid for both types of bar.

Crack widths for T-beams

The 'no-slip' theory was verified for beams with steel ratios of 0.93% and over. For T-beams the values of a_{cr} can become very large and the following rule should be noted [11.3]: 'The primary crack pattern which forms is a function of the member geometry. The addition of reinforcement locally modifies this according to Eqn (11.2.3). If the bars are close enough together this modified pattern dominates.'

Protection against corrosion

Stirrup reinforcement, nominal or otherwise, tends to act as a crack inducer (since it is at right angles to the main reinforcement) and such steel is therefore vulnerable to attack. This possibility is covered by the general CP 110 requirement that crack widths should not exceed 0.3 mm. For members exposed to a particularly aggressive environment there is the additional requirement that the maximum crack widths at points nearest to the main steel in regions of maximum tensile strain shall not exceed 0.004 times nominal cover to the main reinforcement.

It is usually assumed that protection against corrosion is proportional to the depth of the concrete cover. Resistance to corrosion also depends on crack width which, according to Eqn (11.2.3), increases with depth of cover. However, reference to Fig. 11.2.1 shows that the most critical positions on the concrete surface are such points as A_2, A_3 or A_4 which depend more on the distribution of the bars than on the concrete cover. Furthermore it should be remembered that once the nominal cover has been selected, as for example when referring to Table A3 the cover to the main steel is then determined. The maximum crack width requirement is therefore considered in addition to the concrete cover.

Significance of CP 110 requirements for cracking

Notice that the requirement for the crack width directly over a bar reduces to (see Eqn (11.2.3) and Appendix CE)

$$w_{max} = 3c_{min}\varepsilon_m \not> 0.004c_{min}$$

since $a_{cr} = c_{min}$ in this case

Thus $\varepsilon_m \not> 1.3 \times 10^{-3}$

i.e. CP 110's requirement is effectively a limitation on the steel strain, and hence stress, under service conditions (of 260 MN/m^2)

When the distance a_{cr} is large compared with c_{min}, then for a point remote from a bar

$$w_{max} \simeq \frac{3}{2}(h - d_n)\varepsilon_m$$

If $h - d_n$ is sufficiently small then even when a_{cr} is large it is possible for w_{max} not to exceed 0.3 mm. This explains why excessive crack widths rarely occur in slabs under service loads, providing that the thickness is only 200 or 250 mm (see §3.11.8.2, Appendix CE).

Example 11.2.1

A rectangular beam, 250 mm wide × 440 mm overall depth has 2 No. 30 mm diameter bars 400 mm from the compression face with cover as shown in Fig. 11.2.2. Assume that E_s = 200 GN/m^2, E_c = 15 GN/m^2, f_y = 425 MN/m^2 and f_s = 250 MN/m^2. Are the crack widths satisfactory for protection against corrosion in an aggressive environment for the limit states of cracking (a) to CP 110 and (b) to BS 5337?
Refer to Appendix CE and Appendix F.

Solution

$$\rho = 2 \times \frac{\pi}{4}30^2/250 \times 400 = 1.41\%$$

$$\alpha_e\rho = (200/15) \times 1.41 = 0.188$$

Hence from Fig. F3, $d_n = 0.45d = 180$ mm
For the soffit of beam (P_1, P_2 or P_3)

$$y = 440 - 180 = 260 \text{ mm}$$

At P_1

$a_{cr_1} = \sqrt{\{(30 + 15)^2 + (40 + 15)^2\}} - 15 = 56$

Similarly for P_2

$a_{cr_2} = \sqrt{(70^2 + 45^2)} - 15 = 68$

For P_3

$a_{cr_3} = 30$ mm

Now ε_s is

$f_s/E_s = (250/200) \times 10^{6-9} = 1.25 \times 10^{-3}$
(which is satisfactory since $1.25 \times 10^{-3} < 0.8f_y/E_s$)
Therefore $\varepsilon_1 = \varepsilon_2 = \varepsilon_3 = (440 - 180)\varepsilon_s/(400 - 180) = 1.18\varepsilon_s$
Hence for P_1, P_2 and P_3 from Eqn (11.2.2)

$$\varepsilon_m = 1.18 \times 1.25 \times 10^{-3} - \frac{0.7\ (0.44 - 0.18)}{0.0141(0.44 - 0.18) \times 250} \times 10^{-3}$$

$$= (1.48 - 0.20) \times 10^{-3}$$

Hence from Eqn (11.2.1)

$w_{max} = \lambda a_{cr}\varepsilon_m/K_1$

(1) For CP 110 (from Appendix CE)

$\lambda = 3$ and $K_1 = 1 + 2(a_{cr} - c_{min})/(h - d_n)$

Thus for a_{cr_2}

$$w_{max} = \frac{3 \times 68 \times 1.28}{1 + 2(38)/(260)} \times 10^{-3-3}$$

$$= 0.202 \text{ mm}$$

Checking for P_1 and P_3 respectively

$w_{max} = 0.179$ mm or 0.114 mm

For an aggressive environment then P_3 gives

$0.114\text{mm} \not> 0.004 \times 30 \text{ mm} = 0.12$ mm

which is just satisfied.

Since maximum crack width of 0.20 mm $\not>$ 0.30 mm the crack widths are satisfactory to CP 110 (Answer)

Note 1: P_4, of equal distance from bar surface and from neutral axis (i.e. equidistant from points of zero strain), need not normally be considered. However, it could be calculated as follows.

$y = \sqrt{\{(400 - 180 - y)^2 + 55^2\}} - 15$

or $a_{cr} = y = 109$ mm

Therefore $\varepsilon_4 = 109\varepsilon_s/(400 - 180) = 0.495\varepsilon_s$

Hence for P_4

$$\varepsilon_m = 0.495 \times 1.25 \times 10^{-3} - \frac{0.7 \times 0.109}{0.0141 \times 0.26 \times 250} \times 10^{-3}$$

$$= (0.62 - 0.08)10^{-3}$$

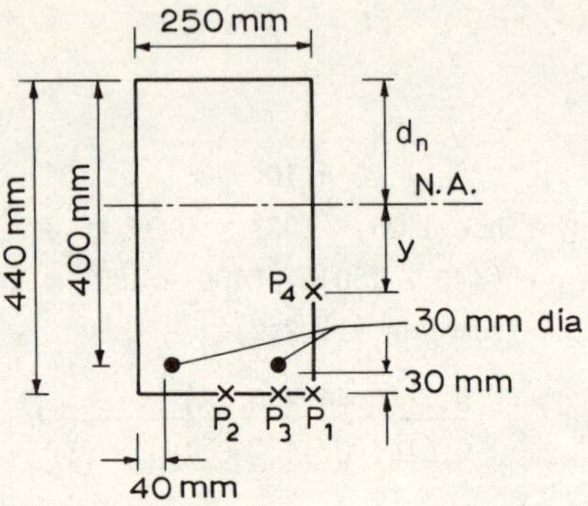

Fig.11.2.2 Example 11.2.1

Hence from Eqn (11.2.3)

$$w_{max} = \frac{3 \times 68 \times 0.54}{1 + 2\,(109 - 40)/(260)} \times 10^{-3-3}$$

$$= 0.072 \text{ mm}$$

which is much less than the crack width at P_2.

(ii) For BS 5337 (from Appendix CW)

$\lambda = 4.5$ and $K_1 = 1 + 2.5(a_{cr} - c_{min})/(h - d_n)$

Thus for a_{cr_2}

$$w_{max} = \frac{4.5 \times 68 \times 1.28}{1 + 2.5\,(38)/(260)} \times 10^{-3-3}$$

$$= 0.29 \text{ mm}$$

which is too wide even for Class B exposure. (Answer)

Note 2: Reference to the alternative design (§9.3) in Appendix CW shows a permissible stress of 100 MN/m^2 for deformed high yield bars which confirms that a stress of 250 MN/m^2 at working loads is likely to be much too high for BS 5337. If the crack width is recalculated for the above permissible stress then

$$f_s/E_s = 100/200 \times 10^{6-9} = 0.50 \times 10^{-3}$$

Therefore $\varepsilon_m = 1.18 \times 0.50 \times 10^{-3} - \dfrac{0.7}{0.0141 \times 100} \times 10^{-3}$

$$= (0.59 - 0.50)\ 10^{-3}$$

Hence from Eqn (11.2.3) for a_{cr}

$$w_{max} = \frac{4.5 \times 68 \times 0.09}{1 + 2.5(38)/(260)} \times 10^{-6}$$

$$= 0.02 \text{ mm}$$

which is more than satisfactory for Class A exposure.

For illustrative purposes only Example 11.2.1 considered a concrete cover of only 30 mm to the soffit. However, for a concrete grade of 30 then CP 110 would require a nominal cover of 40 mm (see Table A3). Similarly, BS 5337 would also require 40 mm. In both CP 110 and BS 5337 the required cover applies to all reinforcement, hence if 10 mm links are used the cover to the main bars should be increased to 50 mm in all cases.

Bar spacing guides for crack control

It is clear from the above that the closer the bar spacing, the finer will be the crack widths. Thus for normal design where crack width calculations may not be justified CP 110 gives an alternative procedure in which the maximum clear horizontal distances between adjacent bars near the tension face are specified for different steel strengths and amounts of redistribution (see Table CE5). The spacing rules for beams are illustrated in Fig. 11.2.3, for which $a_b \nsucc (a_b)_{max}$ given in Table CE5 and $a_{cr} \nsucc \frac{1}{2}(a_b)_{max}$. Bars less than 0.45 × maximum bar diameter should be ignored: thus, for example, 12 mm bars can count with 25 mm bars. For deep beams then the minimum bar diameter for the sides is related to the area of the beam for which cracking must be controlled without the bars yielding locally at the crack i.e. bar diameter $\nless \sqrt{(s_b b_w/f_y)}$. The requirements for solid slabs in flexure are given separately in Appendix CE from which it can be noted that spacings up to $2(a_b)_{max}$ and up to 5d are permitted for low steel ratios.

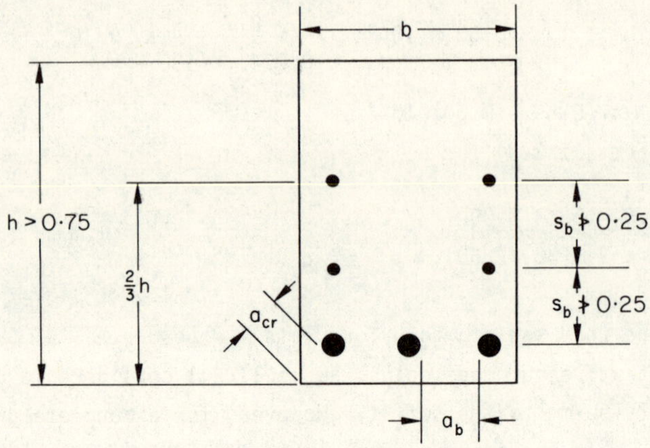

Fig.11.2.3 Bar spacings for a deep beam

Reference to Eqn (11.2.2) shows that at low steel ratios ($\rho_c = A_{st}/b_t h$) the stiffening effect of the concrete increases and ε_m reduces. Hence the larger clear distances between bars allowed by (c) and (d) in §3.11.8.2 (see Appendix CE).

11.3 EARLY THERMAL CRACKING IN CONTINUOUS WALLS

All structures are subject to temperature, creep and shrinkage effects but these are frequently inconsequential compared with the stresses due to gravitational and similar external loads. This is not so, however, when the external loading is small and only nominal reinforcement is considered necessary. The contraction which occurs due to shrinkage, heat of hydration of the cement, and seasonal variations in temperature, can produce excessive tension in lightly reinforced members due to the external restraint which invariably exists. Sections such as thin walls are particularly susceptible to drying shrinkage, for example, and must therefore contain a certain minimum quantity of steel, in both directions, if they are not to be rendered unserviceable by wide cracks. There should therefore be sufficient reinforcement to ensure that a number of distributed cracks can occur such that the limit states of crack width, appropriate to the particular conditions of exposure (as for beams), are satisfied. The reinforcement must ensure that

(i) All cracking is fully controlled, and
(ii) The width of a fully developed crack is not excessive.

Early thermal movement

All concrete expands after placing due to the heat of hydration of the cement. This early thermal movement takes place within a very few days of placing the concrete, hence any attempts to control the cracking must take this fact into account. The initial expansion is soon followed by the early thermal contraction as the concrete cools to the temperature of its surroundings. It is the control of this early thermal contraction which is of primary importance, since the mechanisms controlling the behaviour in the immature concrete are quite different to those considered in the previous section for the mature concrete. Once the crack pattern has formed, then allowances can be made for the subsequent shrinkage and seasonal temperature variations as indicated later.

Distribution of cracks

The early thermal movement in plain concrete can produce tension which exceeds the tensile strength of the concrete and which produces a wide, single crack. This cracking can be controlled by the introduction of a relatively small amount of reinforcement in the form of small-diameter bars or fabric. The function of the reinforcement is to distribute the cracks. Notice that if one wide crack is allowed to develop the tension is relieved at other sections along the wall and cracks are not introduced elsewhere. This situation for direct tension is therefore again different to that for flexural tension considered previously. A further difference is that steel ratios for beams are generally higher than those for wall and floor slabs. However, the essential difference is that extensive bond creep can be assumed and the ratio of bond strength to tensile strength for the immature concrete (when early thermal cracking occurs) is much lower than that for mature concrete.

Unrestrained shrinkage

Consider a reinforced concrete member with perfect bond between concrete and steel as shown in Fig. 11.3.1. If the free shrinkage strain for the

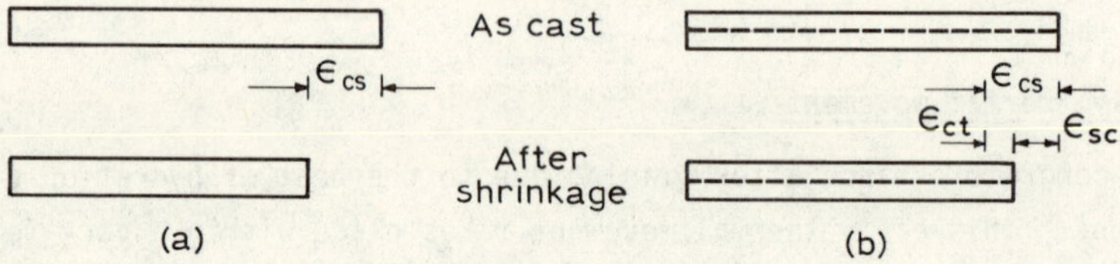

Fig.11.3.1 Strains before and after shrinkage for both plain and reinforced concrete·(a)Plain concrete·(b)Reinforced concrete·

plain concrete is ε_{cs} then the inclusion of reinforcement produces a tension strain, ε_{ct}, in the concrete and a compression strain, ε_{sc}, in the steel such that

$$\varepsilon_{cs} = \varepsilon_{ct} + \varepsilon_{sc} = (f_{ct}/E_c) + (f_{sc}/E_c) \qquad (11.3.1)$$

Equating internal forces and assuming that there is no external restraint:

$$A_c f_{ct} = \rho_c A_c f_{sc} \qquad (11.3.2)$$

Substituting for f_{ct} in Eqn (11.3.1), putting $\alpha_e = E_s/E_c$ and rearranging gives

$$f_{sc} = \varepsilon_{cs} E_s/(1 + \alpha_e \rho_c) \qquad (11.3.3)$$

Example 11.3.1

Calculate the theoretical steel and concrete stresses in a reinforced concrete column if E_s = 200 GN/m², α_e = 15, ρ_c = 1% and ε_{cs} = 250 microstrain. Assume (a) no restraint and (b) full restraint.

Solution

(a) From Eqn (11.3.3)

f_{sc} = 250 × 200 × 10³/(1 + 0.15) = 43.5 MN/m² (Answer)

From Eqn (11.3.2)

$f_{ct} = 0.01 \times 43.5 \times 10^3 = 435\ \text{kN/m}^2$ (Answer)

Notice that these stresses are relatively small even for an extremely high shrinkage strain. Although ε_{cs} and α_e can vary within wide limits, the stresses in unrestrained members, in practice, can be readily sustained by both the concrete and the steel.

(b) $f_{sc} = 0$ (Answer)

and

$f_{ct} = (200/15) \times 250 \times 10^3 = 3.33\ \text{MN/m}^2$ (Answer)

The calculated stress for the concrete is far too high and means that the concrete must crack in at least one position. It is now necessary to consider what happens adjacent to this first crack.

Fully restrained contraction cracks

As the heat of hydration is dissipated the first crack is formed and the local slipping between the concrete and steel must occur. The amount of slipping and hence the probable spacing between cracks depends upon the properties of the materials and the amount of steel present. If the ends are fully restrained then the steel cannot be in compression until the concrete cracks, since the overall strain is zero. When the concrete fractures, the steel at the crack is in tension and strains locally, resulting in compression strains elsewhere in the steel, since the uncracked lengths of concrete try to contract as shown in Fig.11.3.2.

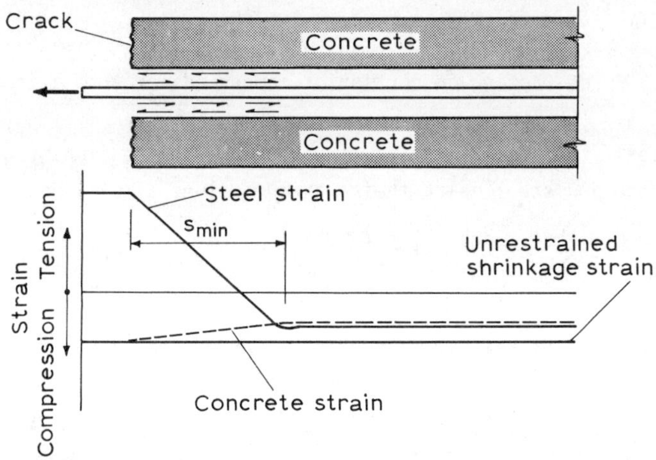

Fig.11.3.2 Probable strain distribution in concrete and steel adjacent to a crack

Even if the concrete achieved the full shrinkage strain of 250×10^{-6} quoted in Example 11.3.1, the steel would attain a stress of about 50 MN/m^2 only. Since the restrained concrete would have cracked at this strain, either a very large number of fine hair cracks form in the concrete or extensive bond slipping (i.e. 'bond creep') occurs along the bar.

Critical steel ratio

After the first crack has occurred the equilibrium of the portion of the member adjacent to the crack (e.g. between sections 1 and 2 in Fig. 11.3.3) is given by

$$A_s f_{st} = A_c f_{ct} - A_s f_{sc}$$

or

$$f_{ct} = \rho_c (f_{st} + f_{sc})$$

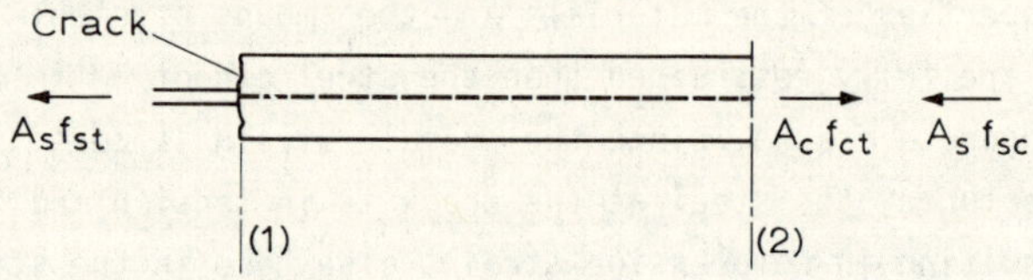

Fig.11.3.3 Forces acting on portion of member adjacent to a crack

The critical steel ratio, ρ_{crit}, is given when the tension stresses in the concrete and steel reach their maximum values simultaneously.

Hence when $f_{ct} = {}_*f_{ct}$ and $f_{st} = f_{yt}$

$$\rho_{crit} = {}_*f_{ct}/(f_{yt} + f_{sc}) \qquad (11.3.4(a))$$

${}_*f_{ct}$ and f_{yt} can be readily estimated. If required, determine f_{sc} as indicated below, or take as zero, in which case

$$\rho_{crit} = {}_*f_{ct}/f_{yt} \qquad (11.3.4(b))$$

Thus ρ_{crit} can be calculated or, alternatively, read off Fig. F1, for the appropriate grades of concrete and steel.

Effective period for full control

The heat of hydration is largely dissipated after about three days and base restraints can subsequently assist the effective control of cracking very significantly. The period which is most critical and which should be fully controlled by the steel alone may be taken as the first three days only. Hence $_*f_{ct}$ may be taken as the tensile strength of the concrete at three days and not some larger value, a suitable value being approximately $0.12(f_{cu})^{0.7}$

Compressive stress in steel for no slip*

If no bond creep occurs between the concrete and steel in regions of the wall which are remote from a crack, an equation such as (11.3.1) can apply even though the equilibrium condition is given by Eqn (11.3.4) instead of Eqn (11.3.2). Substituting $f_{ct} = {}_*f_{ct}$ in Eqn (11.3.1) for the limiting case and transposing gives

$$f_{sc} = E_s\varepsilon_{cs} - \alpha_e\,{}_*f_{ct}$$

Hence Eqn (11.3.4) becomes

$$\rho_{crit} = {}_*f_{ct}/(f_{yt} + E_s\varepsilon_{cs} - \alpha_{e*}f_{ct}) \qquad (11.3.5)$$

Although the final compression strain in the steel can exceed 200 microstrain, the value at about three days appears to be appreciably less. A reasonable simplification which includes a small margin of safety is therefore to assume that $f_{sc} = 0$.

Low steel ratios

If $\rho_c < \rho_{crit}$ the strength of the concrete exceeds the strength of the steel and the latter continues to yield in tension. It follows that the cracks tend to be widely spaced and unevenly distributed along the wall. A few very wide cracks occurring at quite random positions are clearly undesirable.

Moderate steel ratios

If $\rho_c > \rho_{crit}$ the strength of the steel exceeds the strength of the

* Omit on first reading

concrete so that cracks occur at intermediate sections and the maximum stress in the steel is less than the yield stress. The subsequent cracks in the concrete form when the bond force between steel and concrete equals the tensile strength of the concrete. That is,

$$f_b s \Sigma u_s \geq {}_{*}f_{ct} A_c$$

where Σu_s = total perimeter of steel reinforcement

u_s = perimeter of one bar (= $\pi\Phi$ for round bar)

f_b = mean bond strength between concrete and reinforcement adjacent to a crack

u_s/A $(=\pi\Phi/\frac{1}{4}\pi\Phi^2) = 4/\Phi$ for round bars so that

$$\Sigma u_s = A_s(u_s/A) = 4\rho_c A_c/\Phi$$

Substituting for Σu_s in Eqn (11.3.6) and rearranging gives

$$s \geq {}_{*}f_{ct}\Phi/f_b.4\rho_c$$

The minimum crack spacing, s, is given by the equality sign and the maximum crack spacing is twice this value (since a further crack can form once the development length on both sides of the potential crack position equals the minimum value of s), hence

$$\frac{{}_{*}f_{ct}\Phi}{f_b.2\rho_c} \geq s \geq \frac{{}_{*}f_{ct}\Phi}{f_b.4\rho_c} \qquad (11.3.7)$$

where ${}_{*}f_{ct}/f_b \simeq 1$ for plain round bars at early ages

$\simeq \frac{4}{5}$ for Type 1 deformed bars at early ages

$\simeq \frac{2}{3}$ for Type 2 deformed bars at early ages

<u>High steel ratios</u>

If $\rho_c \gg \rho_{crit}$ then s becomes very small. However, high steel ratios are only necessary for structural reasons. If such tensile loads are applied suddenly little 'bond creep' can occur and an even larger number of cracks tend to form. Eqn (11.3.7) is valid for gradually applied stresses in immature concrete.

Example 11.3.2

The horizontal reinforcement in a reinforced concrete wall, 240 mm thick, consists of 10-mm diameter mild steel bars at 100 mm centres near each face. Calculate the probable mean crack spacing for the wall if it is fully restrained at each end. Assume the following values:

$_*f_{ct} = f_b = 1.5$ MN/m² $f_{yt} = 250$ MN/m² $E_s = 200$ MN/m² $E_s = 200$ GN/m²

$\alpha_e = 20$ $\varepsilon_{cs} = 100$ microstrain.

Solution

$\rho_c = 2 \times \frac{1}{4}\pi \times 10^2/(240 \times 100) = 0.655\%$

$\rho_{crit} = 1.5/250 = 0.6\%$

Hence cracking is controlled by the steel since $\rho_c > \rho_{crit}$.
Substituting in Eqn (11.3.7) and simplifying gives

$0.764 \geq s \geq 0.382$

The probable mean spacing for intermediate cracks is

$0.382 \times 3/2 = 0.573$ m (Answer)

*Note**: If compressive stress in the steel is not ignored in this example then from Eqn (11.3.5)

$\rho_{crit} = 1.5/(250 + 0.2 \times 100 - 20 \times 1.5) = 0.625\%$

This value is higher in this case since $E_s\varepsilon_{cs} - \alpha_{e*}f_{ct}$ is negative and therefore invalid (i.e. $f_{ct} < {}_*f_{ct}$ and is *not* equal to $_*f_{ct}$)

Differential thermal movement

If there is no external restraint it is only the differential movement between concrete and steel which produces stress. However, since the coefficient of thermal expansion for the steel is very approximately equal to that for the concrete, any differential movement should be fairly small and cracking is highly unlikely. The combined shrinkage and differential thermal strain is given by

$\varepsilon_{cs} + (\alpha_c - \alpha_s)T$

where T is the fall in temperature, and this expression should be substituted for ε_{cs} in Eqns (11.3.3) and (11.3.5) if differential thermal strains are significant. Notice however that the actual thermal contraction of the concrete is fully effective in producing cracking in restrained members.

* Omit on first reading

Example 11.3.3

The coefficients of thermal expansion of mature concrete and steel in a reinforced concrete wall both equal 10 microstrain per ^{o}C. If $_*f_{ct} = 2$ MN/m^2 and $E_c = 16$ GN/m^2, calculate the minimum fall in temperature which would produce the first crack in the concrete, unaided by any shrinkage contraction, if the wall is fully restrained at each end.

Solution The ultimate tensile strain in the concrete is

$(2/16) \times 10^{-3} = 125$ microstrain

Minimum fall in temperature required to crack concrete is

$125/10 = 12.5^{o}$C (Answer)

Notice that this temperature drop is relatively small and can frequently occur in practice. Thermal effects can therefore produce very pronounced 'first' cracks. If $\rho_c < \rho_{crit}$, the 'first' crack tends to become excessively wide. If $\rho_c > \rho_{crit}$ the excess overall thermal contraction produces further (intermediate) cracks once the 'first' crack becomes fully developed.

Actual contraction strain

The crack spacing formula, Eqn (11.3.7), assumes that the actual contraction strain due to thermal and shrinkage movement is sufficiently large to produce a crack at all the positions which are theoretically possible. In practice the actual contraction strain is usually very much less, so that many of the cracks predicted by the theory do not in fact occur. Nevertheless it is the width of a fully developed crack which is of interest since it is the cracks of maximum width which cause most concern.

Width of fully developed crack

The total contraction is given by the sum of the thermal strain and the shrinkage strain. The tensile stress in the concrete builds up from zero at a cracked section to a maximum at a section which is a distance of s_{min} (or more) from the crack. Hence if the steel ratio exceeds ρ_{crit} and the total contraction strain exceeds the ultimate tension strain, the mean tension strain in uncracked concrete of length s_{min} adjacent to crack is $\frac{1}{2}\varepsilon_{ult}$. This is shown in Fig. 11.3.2. The concrete strain adjacent to the crack on the left hand side, relative to the unrestrained shrinkage strain, is zero. The strain increases linearly to a maximum value of ε_{ult}, hence the mean tension strain over s_{min} is $\frac{1}{2}\varepsilon_{ult}$. If a second crack occurs at the maximum distance of $2s_{min}$ from the first crack (i.e. $s = s_{max}$) then the average strain in the uncracked concrete of length $\frac{1}{2}s = s_{min}$ adjacent to the crack will reduce to $\frac{1}{2}\varepsilon_{ult}$ (since the strain again varies from zero up to ε_{ult} over the length s_{min}). Hence the maximum final crack width is given by:

$$w_{max} = s_{max}(\varepsilon_{cs} \pm \varepsilon_{te} - \tfrac{1}{2}\varepsilon_{ult}) \qquad (11.3.8)$$

Notice that $\varepsilon_{cs} \pm \varepsilon_{te} = \varepsilon_{cs} + T\alpha_c$ in all cases since it is the overall contraction of the concrete which determines the actual width of the cracks. T is the fall in temperature as before. This is in contrast to Eqns (11.3.3) and (11.3.5) where only the differential thermal strain can affect f_{sc} and therefore ρ_{crit}.

Construction joints

The first crack occurs in a given bay of concrete due to both shrinkage and thermal contraction. Consider, for example, concrete which is cast between two existing bays. The new bay undergoes a rise in temperature (due to heat of hydration), causing compression and therefore considerable creep in the immature concrete. When the temperature subsequently falls the concrete has had time to harden so that its rate of creep is less: hence tension stresses are induced. The bond strength between new and mature concrete is generally less than the tensile strength of the concrete. Hence the first crack is most likely to occur at the construction joint. Such cracks therefore tend to be wide and fully developed. They can become excessive if $\rho_c < \rho_{crit}$.

Intermediate cracks

The overall thermal contraction of a bay can be satisfied by relatively large cracks at one (or possibly both) construction joints at the end of the bay if extensive bond slipping can occur. However, provided that $\rho_c > \rho_{crit}$ these first cracks will be restricted in width to the fully developed value and the surplus contraction strain transferred to adjacent intermediate cracks. The possible development of intermediate cracks before the first crack is fully developed is unlikely. That is, intermediate cracks do not usually appear (if at all) until after the formation of a fully developed crack at the construction joint, since the shrinkage plus differential thermal strain only is usually less than the ultimate tension strain of the concrete. Although this conclusion can be modified by local restraints as discussed below it does explain why the worst cracks are usually observed either at the construction joints or at points of stress concentration (e.g. large changes in section).

Example 11.3.4

Determine the probable maximum crack spacing for the wall described in Example 11.3.2. Is the probable maximum crack width likely to exceed the value of 0.3 mm permitted by CP 110 if the coefficients of expansion of the steel and the mature concrete are given by

$\alpha_c = 12$ and $\alpha_s = 10$ microstrain per ^oC?

Assume that the concrete temperature falls by 40^oC and that creep effects effectively halve the thermal contraction of the immature concrete.

<u>Solution</u>

The probable maximum crack spacing = 0.764 m as before (Answer)

The ends are assumed to be fully restrained. Hence the overall contraction strain of the concrete is given by

$$\varepsilon_{cs} + \varepsilon_{te} = 100 + \tfrac{1}{2} \times 12 \times 40 = 340 \text{ microstrain}$$

This is more than sufficient to produce the first crack.* Intermediate cracks can also be formed once the first crack is fully developed since any further contraction can be transferred to adjacent cracks because $\rho_c > \rho_{crit}$. Hence the probable maximum width of each fully developed crack (from Eqn (11.3.8(b))) is given by

$$w_{max} = 0.764(340 - 150 \times \tfrac{1}{2}) \times 10^{-6} = 202 \ \mu m \simeq 0.2 \text{ mm}$$

* The ultimate tension strain in the concrete for Example 11.3.2 is $^*f_{ct}/E_c = 150$ microstrain

The probable maximum crack width is therefore satisfactory for both CP 110 (0.3 mm) and the CEB-FIP [1.7] code (0.2 mm). (Answer)

Local restraint

An upper lift of a wall is usually restrained along its base by the lift below. The shear force along the base is an additional term which could be included in Eqn (11.3.4) as follows:

$$\rho_{min} = ({}_{*}f_{ct} - R/A_c)(f_{yt} + f_{sc}) \tag{11.3.9}$$

where ρ_{min} is the minimum steel ratio required to give controlled cracking in a particular location and R the force due to local restraint. The local restraints can become smaller as the steel ratio approaches ρ_{crit}. The minimum steel ratio required to give controlled cracking in accordance with condition (11.3.7) therefore varies as R varies along the length of the member. The cracking can only be completely controlled by the steel, however, if ρ_c is not less than ρ_{crit}.

Design charts for cracking

The compressive stress which occurs in the steel in the regions remote from a crack has been shown to be small and can usually be neglected. Furthermore, if it is assumed that the direct tensile strength of concrete at 3 days is 50% of the estimated flexural strength values at 7 days formerly given in the draft CP 110, then the grade designation of the concrete in terms of the 28-day characteristic cube strength can be included as shown in Fig. F1.

In practice, base restraints will usually offer some slight assistance in most cases, hence a constant value of 100×10^{-6} can be assumed for $\frac{1}{2}\varepsilon_{ult}$ in Eqn 11.3.8. This has been done in the design chart given in Fig. F2.

Example 11.3.5

Repeat Examples 11.3.2 and 4 using the design charts in Appendix F. Determine the maximum crack width :-

(a) assuming average contraction strain equals

$$340 - \tfrac{1}{2} \times 150 = 265 \times 10^{-6}$$

(b) considering only the combined shrinkage and thermal strain of 340×10^{-6} and allowing a constant value of 100×10^{-6} for base restraint and $\frac{1}{2}\varepsilon_{ult}$.

Solution

$*f_{ct} = 1.5\ \text{MN/m}^2 \quad f_{yt} = 250\ \text{MN/m}^2$

Hence from Fig. F1, $\rho_{crit} = 0.6\%$ (Answer)

$\Phi = 10\ \text{mm} \quad \rho_c = 0.655\%$ using plain round bars

Hence from Fig. F2

(a) if $\varepsilon_{cs} \pm \varepsilon_{te} - \frac{1}{2}\varepsilon_{ult} = 265 \times 10^{-6}$ then

$w_{max} = 0.202$ mm (Answer)

(b) if $\varepsilon_{cs} \pm \varepsilon_{te} = 340 \times 10^{-6}$ then

$w_{max} = 0.185$ mm (Answer)

Minimum distribution reinforcement to CP 2007/BS 5337

The grades of concrete and the corresponding minimum steel ratios required for distribution reinforcement are shown in Table 11.3.1. Notice that the new code uses higher grades of concrete than CP 2007: 1960. The cement contents will tend to be somewhat higher, and hence impermeability of the well-compacted concrete should be improved. However, the main problem with water retaining structures is not the loss of water through well-compacted concrete but rather the much larger loss which can occur through honeycombed concrete and through the joints and cracks in the concrete. Assuming that honeycombed areas and poor detailing of joints are avoided then the major source of leakage is cracking in the concrete. The detrimental effect of increasing the cement content will be to increase the early thermal contraction following the hydration peak temperature. The need to provide adequate reinforcement to distribute the cracking due to the contraction therefore becomes even more important with BS 5337.

Notice that the minimum steel ratios in Table 11.3.1 for alternative design are virtually adequate for the worst cases for limit state design, and are therefore conservative for the other cases.

Table 11.3.1 Minimum distribution reinforcement to CP 2007/BS 5337 designs

Concrete grade (MN/m^2)	CP 2007:1960		BS 5337: 1976				
	Mild steel	High yield	Alternative design		Limit state design		
			M.S.	H.Y.	M.S.	410 MN/m^2	485 MN/m^2
21	0.30%	0.25%	-	-	-	-	-
25	0.30%	0.25%	0.50%	0.30%	0.31%* 0.46%	0.20%* 0.28%	0.20%* 0.24%
30	-	-	0.50%	0.30%	0.35%* 0.52%	0.21%* 0.32%	0.20%* 0.27%

* Values applicable if movement joints at appropriate spacings are used (see section 11.4)

Note: Values less than the above can be used for thick sections exceeding 500 mm in thickness and for ground slabs (see section 11.4).

Distribution reinforcement for Class B exposure

In addition to the requirement for the critical steel ratio (Eqn (11.3.4)) there is the requirement for the actual width of a fully developed crack (Eqn (11.3.8)).

The required steel ratios for given amounts and type of reinforcement in a continuous wall to give a crack width of 0.2 mm (i.e. Class B exposure) have been determined in Table 11.3.2 for a total contraction strain of 180 microstrain. The steel ratio required to satisfy Class B exposure rapidly increases with bar diameter—as expected from Eqns (11.3.7) and (11.3.8). Notice especially that even when a substantial amount of large diameter structural reinforcement is provided, this does not necessarily ensure that early thermal cracking is adequately distributed. For example, consider a retaining wall with counterforts which has 1% and 32 mm diameter plain round horizontal bars as structural reinforcement. Table 11.3.2 shows that the maximum crack width would be nearly 0.3 mm (i.e. 1.44 × 0.2 mm) and therefore too

Table 11.3.2 Distribution steel required for 180 microstrain and crack widths not exceeding 0.2 mm

Bar Size (mm)	6	8	10	12	25	32
Plain round	0.27%†	0.36%†	0.45%†	0.52%	1.13%	1.44%
Deformed (Type 1)	0.22%†	0.29%	0.36%	0.44%	0.90%	1.15%
Deformed (Type 2)	0.18%†	0.24%*	0.30%**	0.36%	0.75%	0.96%

† Too low even for Grade 25 (see Table 11.3.1)

* Too low for Grade 30 (and Grade 25 if f_y = 410 MN/m²)

** Too low for Grade 30 if f_y = 410 MN/m²

wide. For small diameter bars then ρ_{crit} can become more critical than the crack width requirement.

Thermal movement and shrinkage

It is common for concrete to set at temperatures of 30°C to 40°C, at the end of a hot summer's day, and subsequently fall to 0°C during the following winter. The anticipated thermal strain for such cases could approach 300 microstrain. The autogenous shrinkage of the concrete would be of the order of 75 microstrain only. Results indicate that the apparent shrinkage is small for normal structures exposed to normal climatic conditions and drying from one face only, even for one or two years [11.4,5]. This is probably due to a combination of only partial drying out and to tensile creep greatly reducing the resultant shrinkage strain,

It is evident that thermal movement can be considerably greater than the shrinkage movement. Hence it follows that, while elaborate procedures for curing the concrete for as long as possible can delay the eventual drying out (or partial drying out) of the concrete, relatively simple procedures can be adopted to obtain an actual reduction in the probable thermal strain. The emphasis in curing should therefore be on reducing the temperature of the freshly placed concrete, rather than on wet curing the concrete for extended periods after it has reached its peak temperature due to the heat of hydration.

Temperature rise

The rate of hydration of cement increases with temperature. It follows that the heat of hydration is liberated more quickly, and that the subsequent temperature rise is greater, in summer than in winter. The lowering of the temperature of the freshly placed concrete, to below that of the surrounding mature concrete, can prevent the early onset of cracking both by reducing the actual temperature rise and by offsetting some of the rise against the lower initial temperature relative to the surroundings.

Practical precautions

When placing concrete in summer it is desirable to reduce the temperature of the concrete as much as possible, and preferably down to 20°C, if ordinary Portland cement is being used. Rapid-hardening Portland cement should be avoided. Recommended precautions are:

(i) Shield the concreting materials and the concrete from direct sunlight wherever possible by erecting canvas canopies, with open sides, over the aggregate piles etc. The point of concrete placing should also be shielded if possible. If necessary, also consider adding ice to the mixing water or using evaporative cooling of the aggregates.

(ii) Fully effective water curing of the concrete should be undertaken for about 30 hours (i.e. till the end of the day after placing). This minimizes both shrinkage and compressive creep strain. The heat of hydration must be dissipated quickly if the maximum temperature rise (which can occur after 14 hours or even earlier) is to be reduced. This means that shutters need to be sprayed with cold water before they can be removed and that steel shutters are far preferable to timber shutters in this instance due to their superior thermal conduction. Once the shutters have been removed it is important to avoid thermal shock at the surface of the concrete: i.e. intermittent torrents of ice-cold water (rather than continuous spraying of cool water) must be avoided.

(iii) Adopt 'successive bay construction' rather than 'alternate bay construction', since the former enables one end to contract before the next bay is cast against it. The infill bays for alternate bay construction receive no relief in restraint at all. Successive bay construction tends, however, to produce a wider crack at the base of the lift since the shear restraint along the base tends to oppose the relief given by the free end.

Variable cracking behaviour

It is clear that variations in local restraints can produce large variations in the cracking behaviour of essentially similar members. If such members contain appreciably less steel than the critical value, as has been common in the past, then the behaviour could vary from completely uncontrolled to apparently controlled cracking, depending on the local restraint and the temperature at the time of casting.

Many retaining walls designed to CP 2007:1960 [11.6] have shown severe cracking while others have been satisfactory [11.7]. The principal reasons for this are that if the concrete was placed in cold weather and the overall restraint to movement was minimal there would be reduced tendency to crack anyway. Also, the distribution steel may or may not have been adequate. Table 11.3.1 shows, for example, that 0.3% mild steel in a thin continuous wall with full end restraint was far less than the 0.5% to BS 5337 [11.8] and wide cracks could occur. However, using the same 0.3% in a wall with relatively close movement joints compares with 0.31% for limit state design and would therefore have been virtually adequate. Design for movement joints is considered in section 11.4.

11.4 DESIGN FOR EARLY THERMAL MOVEMENT TO BS 5337

The previous section considered early thermal cracking and shrinkage in long continuous thin members fully restrained at the ends. However, there are other forms of restraint and degrees of restraint: the degree of restraint can be reduced by introducing movement joints, and the actual form of the external restraint has a fundamental influence on the cracking behaviour. Movement joints include expansion joints as well as complete and partial contraction joints. Other forms of cracking in immature concrete include warping at the ends of walls with significant base restraint, surface effects in thick sections and base restraint in ground slabs.

Movement joints

If movement joints are included in a long continuous wall then the restraint adjacent to the joints will be virtually zero for complete movement joints and partly reduced even for a partial joint. The main types of movement joints are described in Appendix CX.

Crack spacing and movement joint spacing

The probable maximum spacing of cracks, s_{max}, can be calculated and the probable minimum spacing of cracks, s_{min}, is simply $\frac{1}{2}s_{max}$. Hence if joints are provided at a spacing which does not exceed s_{max}, as shown in Fig. 11.4.1, the possibility of a crack occurring midway between two joints is eliminated. With such a close spacing of joints then the cracking need not be entirely controlled by the reinforcement and hence the steel ratio in this case can be reduced to $\frac{2}{3}\rho_{crit}$. By taking advantage of the reduced steel ratio of $\frac{2}{3}\rho_{crit}$, and using round bars of as large a diameter as possible (consistent with not exceeding about 300 mm bar spacing) the crack spacing, and hence the joint spacing, can be made as large as possible.

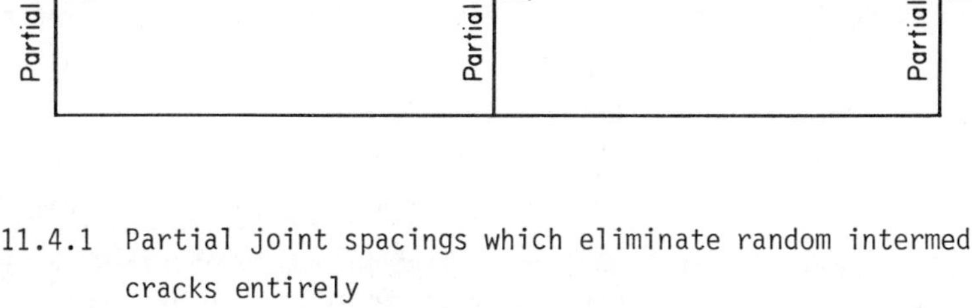

Fig.11.4.1 Partial joint spacings which eliminate random intermediate cracks entirely

Design for no cracking

Table 11.4.1 shows the maximum spacing of 100% partial contraction joints (i.e. 100% reinforcement passing through) for typical cases, including those for no cracking between joints. Notice how very close these spacings are, especially for a narrow wall with a double layer of reinforcement (i.e. a 250 mm wall). Notice also that the widest spacings can be given by a hot-rolled plain round bar of 410 MN/m^2 characteristic strength to BS 4449. These bars are called 'special' in Table 11.4.1 since they are not normally produced. However, if manufacturers could be persuaded to make two or perhaps three sizes available-10 mm, 12 mm and 16 mm - they could permit up to a 25% increase in joint spacing due to their reduced bond strength. The problem of avoiding confusion with mild steel bars could be overcome by providing a simple distinguishing feature such as a single continuous straight ridge, or flattening the bar slightly to make it oval.

In an attempt to achieve practical bay lengths for construction it was considered that if alternate partial joints were made complete contraction joints, the intervening partial joint need not have complete discontinuity in the concrete, i.e. the intermediate partial joint could simply have a reduced concrete cross-section (e.g. formed by battens fixed to the shutters) and need not be a construction joint. Thus the construction bay lengths could become twice the joint spacings given in Table 11.4.1. This is the present basis for option 5 in BS 5337 (see table CW2) with the additional proviso that in no cases should $\frac{2}{3}\rho_{crit}$ be taken as less than 0.2%.

Subsequent investigation has indicated that the effect of a free joint is generally given, for a wide range of wall thicknesses and vertical steel ratios, and horizontal steel ratios not less than $\frac{2}{3}\rho_{crit}$, by assuming that the end 2.40m adjacent to the complete contraction joint (or a free standing end) is free to contract inwards. Thus for a succession of complete contraction joints the joint spacing can be up to 4.80m and no cracks should occur. For alternate 100% partial/complete contraction joints the joint spacing could be increased to $\frac{1}{2}s_{max}$ + 2.4m e.g. for a 250mm Grade 30 concrete wall from 1.73m to 3.26m. For all 100% partial contraction joints the joint spacing would of course remain at s_{max} as shown in Fig. 11.4.1. A possible new option incorporating the new evidence and which could replace existing options 3, 4 and 5 (see Table CW2) is given in Table 11.4.2.

Table 11.4.1 Maximum spacing of partial contraction joints with distribution bars at not more than about 300 mm centres for no cracking, 0.1 mm cracks and 0.2 mm cracks for Grade 30 concrete.

	Reinforcement			Partial joint spacing		
Wall thickness	layers	type	size	no cracks	0.1 mm cracks	0.2 mm cracks
200 mm	Single	Round MS	16 mm	2.30 m	2.86 m	3.41 m
		Special	12 mm	2.84 m	3.40 m	3.95 m
		Type 1	12 mm	2.35 m	2.91 m	3.46 m
250 mm	Double	Round MS	12 mm	1.73 m	2.29 m	3.84 m
		Special	10 mm	2.36 m	2.92 m	3.47 m
		Type 1	8/10 mm	1.78 m	2.34 m	2.89 m
500 mm	Double	Round MS	16/20 mm	2.62 m	3.04 m*	3.73 m
		Special	12/16 mm	3.38 m	3.80 m*	4.49 m
		Type 1	12/16 mm	2.80 m	3.22 m*	3.91 m
Over 500 mm	Double	Round MS	20 mm	1.92 m	2.34 m*	3.03 m
		Special	16 mm	2.52 m	2.94 m*	3.63 m
		Type 1	16 mm	2.09 m	2.51 m*	3.20 m

180 microstrain contraction assumed except where shown* (240 microstrain)

Table 11.4.2 Possible new option replacing options 3, 4 and 5 if new criterion for complete movement joints was introduced.

Option	Method of control	Movement joint spacing	Minimum steel ratio	Comments
New 3	Close joint spacing	(a) Complete joints $\not> 4.8\text{ m} + \frac{w}{\varepsilon}$ (b) Alternate partial & complete joints $\not> \frac{1}{2}s_{max} + 2.4\text{ m} + \frac{w}{\varepsilon}$ (c) Partial joints $\not> s_{max} + \frac{w}{\varepsilon}$ where w = maximum crack width and $\varepsilon = \varepsilon_{cs} + \varepsilon_{te} - 100 \times 10^{-6}$	$\frac{2}{3}\rho_{crit}$	Avoid small size bar which restricts the joint spacing for 3(b) and 3(c).

Design for Class A exposure

A wall designed for no intermediate cracks between joints certainly satisfies Class A exposure requirements. However, since Class A permits crack widths up to 0.1 mm, a slightly wider spacing can be obtained by assuming that all the movement for a length as given for no cracking is accommodated at the adjacent movement joints, and that any increase in joint spacings is accommodated at the crack. Thus for crack widths up to 0.1 mm all the joint spacings shown in Table 11.4.1 can be increased by 0.1/0.180 = 0.556 m for 180 microstrain contraction. For thick walls 500 mm and over Table 11.4.1 assumes an increased contraction of 240 microstrain hence the joint spacings can only be increased by 0.42 m in these cases. BS 5337 in option 4 (see Table CW2) at present restricts the joint spacing for partial/complete joints to $1.25s_{max}$ maximum in all cases.

Design for Class B exposure

At present the BS 5337 restriction that the joint spacing for option 4 should not exceed $1.25s_{max}$ effectively precludes option 4 for consideration for Class B exposure. Option 3 (see Table CW2) can however be considered where it is assumed that if the joint spacing is increased to not more than 2 s_{max} and each alternate joint is a complete contraction joint (i.e. complete contraction joints at 4 s_{max}) then only 50% of the total contraction strain need be considered as being accommodated at the crack because of the conditions of partial restraint (see Appendix CW §15.1), provided that the minimum horizontal distribution steel is not less than ρ_{crit} as for a continuous wall with full end restraint. In other words the crack width can be calculated by using the actual contraction strain based on a length of only ½ s_{max} (and not s_{max}) because of the reasonably close proximity of the complete movement joints and the presence of adequate distribution reinforcement. This approach is useful for Class B exposure and Table 11.4.3 shows examples for option 3 for Grade 410 MN/m^2 steel at the critical steel ratio based on the dual criteria of 2 s_{max} maximum joint spacing (i.e. 4 s_{max} complete movement joint spacing) and 0.2 mm maximum crack width for an assumed length of ½ s_{max}.

If the new criterion for complete movement joints is introduced then the same new option 3(b) would be applicable to Class B as well as Class A exposure, since the restriction of a maximum joint spacing of 1.25 s_{max} for a minimum steel ratio of $\frac{2}{3}\rho_{crit}$ to be used would no longer be necessary. Thus the new option 3(b) in Table 11.4.2 could replace existing BS 5337 options 3, 4 and 5, as well as provide for new possibilities with option 3(a) as shown in Table 11.4.4.

Table 11.4.3 Maximum spacing of complete contraction joints for crack widths not exceeding 0.2 mm, Grade 30 concrete, Grade 410 steel, critical steel ratio and 180 microstrain contraction

Bar size (mm)	10	12	16	20
Plain Round (special)	6.31 m	7.57 m	8.89 m	8.89 m
Deformed (Type 1)	5.05 m	6.06 m	8.07 m	8.89 m
Deformed (Type 2)	4.21 m	5.05 m	6.73 m	8.41 m

Table 11.4.4 Maximum spacing of successive complete contraction joints if wall length of 2.4 m adjacent to each joint is free to move. (180 microstrain contraction)

Wall thickness	No cracks	0.1 mm crack	0.2 mm crack
< 400 mm	4.80 m	5.36 m	5.92 m
> 500 m	4.80 m	5.22 m	5.64 m

Semi-continuous construction

If complete movement joints are provided at 15 m spacing, or partial movement joints at 7.5 m spacing, then it can still be assumed that all thermal movement due to seasonal variations is accommodated at the movement joints. That is, only the contraction due to the fall from the hydration peak temperature need be considered. Thus for concrete with a coefficient of thermal expansion of 12 microstrain per ^{o}C the contraction strains (see Appendix CW) are:

For summer concreting: $\frac{1}{2} \times 12 \times 30 \times 10^{-6} = 180 \times 10^{-6}$

For winter concreting: $\frac{1}{2} \times 12 \times 20 \times 10^{-6} = 120 \times 10^{-6}$

In general it will be necessary to assume summer concreting since the time of construction cannot usually be predicted at the design stage.

If few or no movement joints are provided then the full seasonal movement must be allowed for and continuous construction assumed. Since the ratio of bond strength to concrete tensile strength is much higher, the crack spacing in the mature concrete is at least halved. Thus the crack width for a seasonal temperature fall T_2 is only $\alpha T_2(\frac{1}{2}s_{max})$ even for continous construction.

Example 11.4.1

Determine the maximum distance between complete movement joints for the 400 mm thick retaining walls of a water-retaining structure for which $\rho \nless \frac{2}{3}\rho_{crit}$ and the bar spacing must not exceed 300 mm, if:

(a) no cracks are permitted

(b) cracks must not exceed 0.1 mm for an effective total contraction of 180×10^{-6}

(c) as (b) but for $\rho = \rho_{crit}$

Assume special round bars, $f_{yt} = 410$ MN/m^2 and grade 25 concrete for which $_*f_{ct} = 1.15$ MN/m^2 and $_*f_{ct}/f_b = 1.0$

<u>Solution</u>

(a) $\frac{2}{3}\rho_{crit} = \frac{2}{3} \times \frac{1.15}{410} = 0.187\%$ (or use Fig.F1)

Therefore assuming bars in both faces at 300 mm spacing, required diameter is given by

$$\frac{\pi}{4}\Phi^2 = 0.2 \times 0.3 \times 0.00187 = 112 \times 10^{-6}$$

or $\Phi = 11.95$ mm—use 12 mm bars at 300 mm spacing

Therefore actual $\rho = 0.188\%$

and $s_{max} = 1.0 \times 12/2 \times 1.88 = 3.19$ m

Therefore required joint spacing = 3.19 m and required complete movement joint spacing = 6.38 m (Answer)

Note: BS 5337 states that generally spacing should not exceed 300 mm. Thus if $\rho = 0.187\%$ i.e. spacing = 302 mm and $s_{max} = 3.21$ m. This compares with $s_{max} = 2.84$ m for grade 30 concrete in Table 11.4.1.

(b) If 0.1 mm cracks are permitted then additional length between joints, (assuming that the contraction up to length s_{max} is taken entirely by the joints) is given by

$$\frac{0.1}{180} \times 10^{-3+6} = 0.556 \text{ m}$$

Therefore required complete movement joint spacing

= 6.38 + 2 × 0.556 = 7.49 m (Answer)

(c) If joint spacing does not exceed 2 s_{max} then width of intermediate cracks is effectively halved by the expansion joints.

Hence $w_{max} = \frac{1}{2} \times 180 \times 10^{-6}s$

or $s = \frac{0.1}{90} \times 10^{-3+6} = 1.111$ m

i.e. complete movement joint spacing = 4s = 4.444 m (Answer)

Notice that the spacing is closer than that in (a), showing that this approach is only useful for Class B exposure (i.e. 0.2 mm crack width) for which the complete movement joint spacing would be 8.89 m. However,

$s = 2.222$ m exceeds the value of s_{max} for $\rho_{crit} = \frac{1.15}{410} = 0.280\%$ which is

$$s_{max} = \frac{1.0 \times 12}{2 \times 2.80} = 2.14$$

Thus for Class B exposure complete movement joint spacing

$= 4\ s_{max} = 8.56$ m

(compare with Table 11.4.3 for Grade 30 concrete)

Note: The maximum spacing for the approach in (b) is $1\frac{1}{4}\ s_{max}$ i.e. $1\frac{1}{4} \times 6.38 = 7.98$ m (which is less than $6.38 + 2 \times 2 \times 0.556 = 8.60$ m)

Induced contraction joints

Partial contraction joints where some of the reinforcement is discontinued are sometimes referred to as induced contraction joints. Some engineers favour having all induced joints with 50% steel curtailment. It is suggested therefore that the spacing for a 50% partial joint/50% partial joint system (i.e. with $\frac{1}{3}\rho_{crit}$ continuous throughout the joints) could be assumed to be approximately given by the spacing for an alternate 100% partial joint/complete joint system with the same $\frac{2}{3}\rho_{crit}$ horizontal reinforcement. The reinforcement at each induced point should yield if necessary and so still ensure only a partially restrained condition.

Forms of restraint

The form of restraint considered so far has been end restraint due to direct tension in a thin wall or slab, as shown in Fig. 11.4.2(a). However two other forms of restraint which require consideration are direct tension in a thick wall or slab with end restraint, Fig.11.4.2(b), and warping in a wall with significant base restraint as in Fig.11.4.2(c).

Thick sections

The appearance of the cracking in a thick wall in direct tension will be similar to a thin wall. That is, cracks will occur at the construction joints at either end and, possibly, one, two or more cracks within the bay depending upon the bay length, etc. The development of cracking

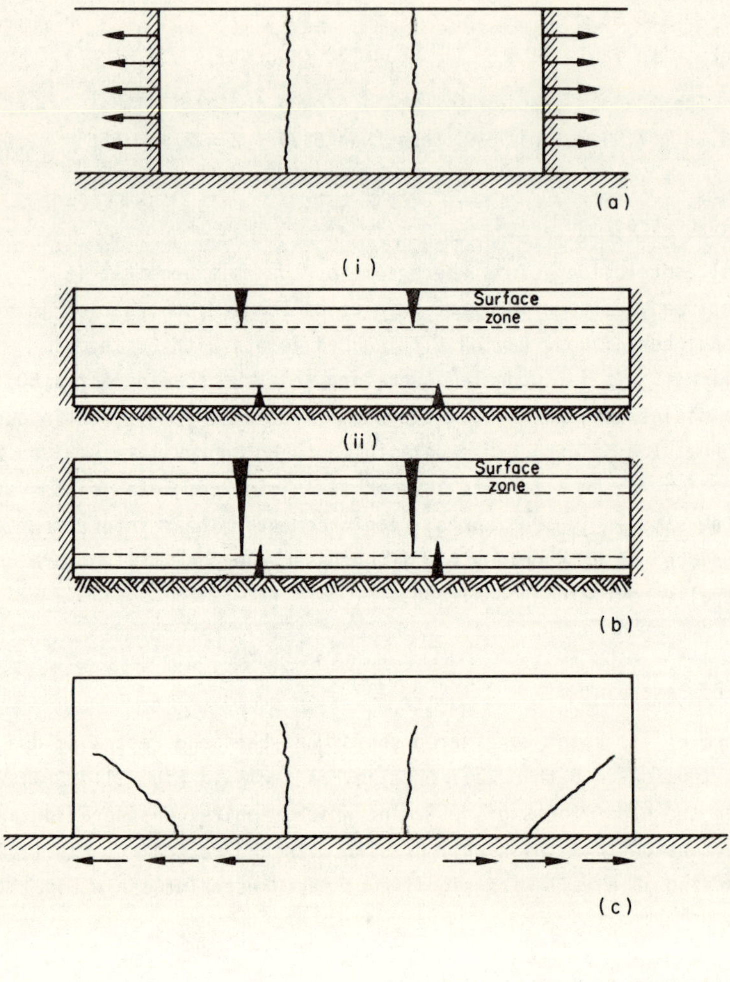

Fig. 11.4.2(a) Tension cracking due to end restraint in a wall
(b) Controlled cracking in a thick base slab
(c) (i) Initially (ii) finally

in thick walls, however, is progressive from the surface and this fact can be utilized to effect economies in reinforcement. Fig.11.4.2(b) shows a thick base slab where the conditions for the top face can be similar to either face of a wall. Although the temperature gradients at the surface depend upon the air temperature and surface insulation, if any, BS 5337 states that a 'surface zone' thickness of 250 mm can be considered in all cases and reinforced accordingly. Thus a wall (which has faces on both sides which can be exposed to significant thermal gradients) which exceeds 500 mm thickness need therefore only contain an amount of distribution reinforcement given by ρ_{crit} based on a thickness of 500 mm. The assumptions are that the reinforcement is at a depth of 125 mm or less so that as the surface cools the surface zone is subjected to tension and subsequently behaves as a continuous thin (i.e. 250 mm thick) wall. Hence once the cracks have developed in the surface they can extend into the core, as the latter cools, without the need for additional distribution reinforcement [11.9]. The temperature rise (and hence the subsequent contraction) for the core is greater than for the surface zones: hence the temperature rise to be considered should be at least 10°C higher than the average value assumed for the entire section.

Example 11.4.2

The raft to a North Sea oil platform is to be constructed in two stages. The first stage consists of forming a concrete cellular structure. The second stage consists of placing concrete in a single 3.6 m high lift to fill the bottom of each cell which on plan measures approximately 11 m square internally. Assume that the early temperature rise for the concrete is a maximum of 60°C and the effective coefficient of thermal expansion on cooling for the immature concrete is 5 microstrain per ^{o}C. Assume also that the ultimate tensile concrete strain offsets the shrinkage strain.

Calculate the predicted maximum crack width in the second stage if the top structural reinforcement consists of high bond Y40 bars at 330 mm both ways with a minimum cover of 65 mm. In addition, B503 mesh will be placed directly on top of the main bars such that the 8 mm diameter wires at 100 mm centres lie parallel to the lower Y40 bars and 8 mm diameter cross wires at 200 mm centres align with the upper Y40 bars as shown in Fig. 11.4.3. The Y40 bars can be assumed

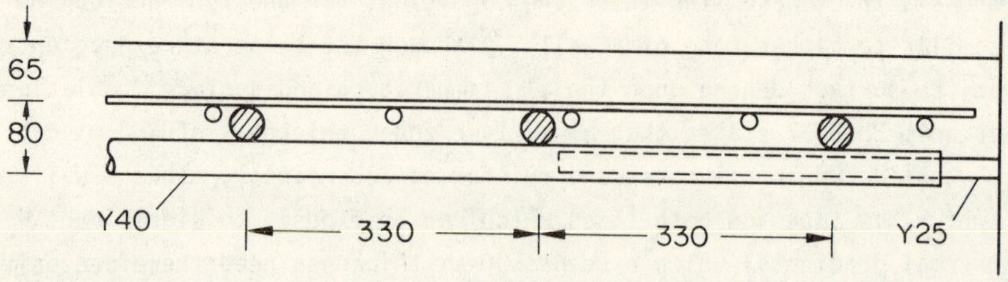

Fig. 11.4.3 (Example 11.4.2)

to lie at the mid-depth of their respective surface zones.

If high bond Y25 starter bars at 200 mm centres project 500 mm from the 1st stage perimeter wall at the level of the lower Y40 bars, determine also the predicted maximum crack width at the perimeter. Check that the steel area, and the bar length, are satisfactory.

Assume the following values.

Early bond strength between concrete and all bars = 2.85 MN/m^2

Early tensile strength of concrete = 1.9 MN/m^2

Design yield strength of Y25 and Y40 bars = 425 MN/m^2

Design yield strength of 8 mm wires = 485 MN/m^2

(UB3/H, 1975, 50 min)

Solution

The net effective contraction strain

$= \frac{1}{2} \quad 10 \times 60 \times 10^{-6} = 300 \times 10^{-6}$

Consider a one metre width of the surface zone in each case.

Surface zone depth for lower layer of Y40 = 2(65 + 40) + 40 = 250 mm

Equating tensile strength of concrete in surface zone to the bond strength gives

$$1 \times 0.25 \times 1.9 \times 10^{6} = \pi(40 \times \frac{1000}{330} + 8 \times \frac{1000}{100})\ s_{min} \times 2.85 \times 10^{6-3}$$

$$\text{or } s_{min} = \frac{0.25 \times 1.9 \times 10^{3}}{40\pi(3.03 + 2) \times 2.85} = 0.264 \text{ m}$$

Alternatively the fabric could yield, in which case the tensile strength of the concrete less the yield strength of the fabric equals the bond strength for the main bars, i.e.

$$1 \times 0.25 \times 1.9 \times 10^{6} - \frac{\pi}{4}(8)^{2} \times 485 \times 10^{6-6} \times \frac{1000}{100}$$

$$= \pi(40 \times \frac{1000}{330})\ s_{min} \times 2.85 \times 10^{6-3}$$

$$\text{or } s_{min} = \frac{0.25 \times 1.9 - 0.244}{1.085} \times 10^{6-6} = 0.213 \text{ m}$$

Hence the fabric does not yield and the worst case is given by $s_{min} = 0.264$ and the crack width is given by

$$\begin{aligned} w_{max} &= s_{max} \times 300 \times 10^{-6} \\ &= 2 \times 0.264 \times 300 \times 10^{-6} \\ &= 0.158 \times 10^{-3} \quad \text{i.e. } 0.16 \text{ mm} \end{aligned}$$

(Answer)

For the Y25 starter bars then

$$1 \times 0.25 \times 1.9 \times 10^{6} = \pi(25 \times \frac{1000}{200})s_{min} \times 2.85 \times 10^{6-3}$$

$$\text{or } s_{min} = \frac{0.25 \times 1.9 \times 10^{6-3}}{125\pi \times 2.85} = 0.424 \text{ m}$$

Crack width at perimeter is given by s_{min} (not s_{max}) since contractions occur on one side of the crack only.

Hence $w_{max} = s_{min} \times 300 \times 10^{-6} = 0.127 \times 10^{-3}$

Check strength of bars:

$$\rho_{crit} = \frac{1.9}{485} = 0.39\%$$

$$\text{Actual } \rho_c = \frac{\frac{\pi}{4}(25)^2}{200 \times 250} = 0.98\%, \text{ which exceeds } 0.39\%$$

Also starter bar length = 500 mm, which exceeds 424 mm

Hence maximum crack width at perimeter = 0.13 mm (Answer)

Warping Cracks

Figure 11.4.2(c) shows the form of cracking which can occur in a bay which is not restrained at the ends, either because the adjacent bays are to be placed subsequently, or because the construction joints coincide with complete movement joints (i.e. the distribution steel is not continuous through the joints). The shear forces which are exerted by the hardened base along the bottom of the immature concrete induce extensive creep movements to occur which give rise to two effects. First, the movements gravitate towards localized planes of weakness: hence a relatively few wide cracks occur rather than a large number of fine cracks. (Fine cracks only would form if the immature concrete was attached to the mature base in such a way that little or no bond creep was possible.) Second, the upper region of the wall can be virtually unrestrained, so that the base restraint reduces from a maximum at the base to a negligible value at some height l_w above the base.

A simplified analysis is possible for this case as follows. If the restraint is assumed to reduce linearly with height l_w then the contractions in length of the wall increase with height. Further, if a vertical section, distance l_w from one end of the wall, is assumed to remain vertical and cleavage along the base only is considered then the wall tends to warp as shown in Fig. 11.4.4.

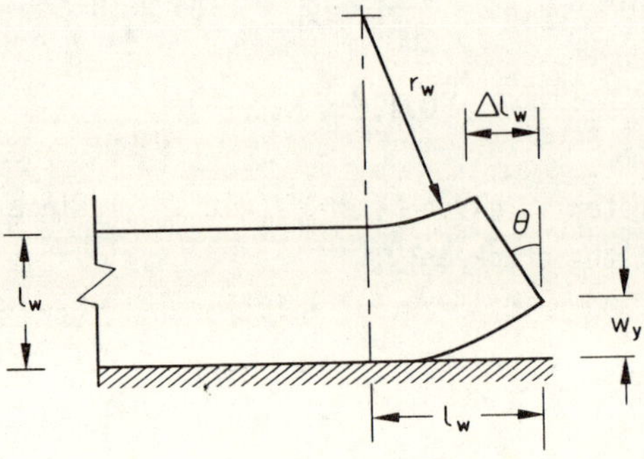

Fig.11.4.4 Warping movement at end of a bay with base restraint only

If Δl_w is the contraction in the wall at height l_w the angle which the end of the wall makes with the vertical is given by $\Delta l_w / l_w$. Therefore assuming small angles, the vertical movement at the end of the bay = $l_w \theta/2 = \frac{1}{2}\Delta l_w$, i.e. the vertical contraction is only half the horizontal contraction.

If horizontal reinforcement is provided as for walls with end restraint the horizontal component of the contraction is suitably controlled. The vertical component is controlled if the vertical reinforcement also satisfies the requirements for ρ_{crit}. The height l_w in Fig. 11.4.3 is given by the smaller of s_{max} for the vertical reinforcement and, provided that the steel is suitably anchored at the top of the wall (e.g. inner and outer face bars are looped together at the top), by the actual wall height, h_w. The warping cracks occur at approximately 45^0 with the horizontal hence the maximum value of $\frac{1}{2}\Delta l_w$ is given by considering a length l_w which equals the height l_w as obtained above (provided that bay length exceeds $2l_w$).

Example 11.4.3

A counterfort retaining wall, 350 mm thick and 8.5 m high, contains the equivalent of 25 mm diameter horizontal bars at 225 mm spacing. The vertical reinforcement consists of 16 mm bars at 300 mm in each face. What are the probable maximum widths of horizontal and vertical cracks if the thermal contraction for summer concreting is approximately 300 microstrain? Assume mild steel reinforcement (f_{yt} = 250 MN/m²) , a concrete cube strength of 30 MN/m², and an early tensile strength of 1.3 MN/m². If the predicted crack width should not exceed 0.3 mm, suggest suitable modifications.

<u>Solution</u>

$$\rho_{crit} = {}^*f_{ct}/f_{yt} = 1.3/250 = 0.52\%$$

(High because mild steel and high quality concrete are proposed)
Actual steel ratios = 0.62% (horizontal) and 0.382% (vertical)

The vertical steel is seen to be inadequate and hence uncontrolled (i.e. wide) horizontal and inclined cracks could occur. However, if high bond, high yield steel is used instead of mild steel for the vertical bars ρ_{crit} is then less than 0.32% and the vertical reinforcement can now control the cracking.

Therefore for horizontal mild steel bars,

$$s_{max} = \frac{25 \times 100}{2 \times 0.624} = 2.0 \text{ m}$$

and for vertical high bond bars,

$$s_{max} = \frac{2}{3} \times 16 \times \frac{100}{2 \times 0.382} = 1.4 \text{ m.}$$

Therefore w_{max} = 0.60 mm for vertical cracks (i.e. horizontal bars).
and w_{max} = 0.21 mm for horizontal cracks (i.e. vertical bars).

The vertical cracks are excessive and can be reduced by using high bond horizontal bars of smaller diameter. For example, if 16 mm high bond bars with the same horizontal steel ratio of 0.624% are used then w_{max} becomes less than 0.26 mm for the vertical cracks and would be satisfactory.

Ground slabs

Useful experience has been obtained in road and other construction concerning the amounts of reinforcement required in ground slabs with long lengths between movement joints [11.10,11.11]. However, for free joints up to 15 m apart, and for transverse steel for widths up to 15 m, it will be found that the virtually trivial amounts which have been used in road construction fall short of the amounts required to control cracking in reservoir floors advocated here, although the differences are not large if the relaxations permitted by BS 5337 and discussed below are taken into account. The 'surface zone' of most significance for a 'thick' slab occurs at the top surface. The rate of heat loss from the bottom surface is much slower, so that even for a truly 'thick' slab the bottom surface zone is much less critical as shown in Fig. 11.4.2(b). If fabric reinforcement is used for the top surface zone and provided with minimum cover of 40 mm, such that it may be considered as centrally placed in a surface zone which is only 100 mm thick, then slabs over 100 mm thick need only contain fabric reinforcement with ρ_{crit} based on 100 mm thickness. Furthermore, since the heat loss downwards from these relatively thin 'thick' slabs can be ignored, no additional reinforcement is required in the bottom of the slabs. The critical steel ratio based on 100 mm thickness can therefore be used for all slabs from 125 mm to 200 mm thick, with 45 mm maximum concrete cover in the top of the slab and with complete movement joints up to 15 m apart. One layer of A252 square mesh fabric (8 mm Φ at 200 mm each way) is thus adequate for Class B slabs up to 200 mm thick. For ground slabs over 500 mm thick assume top surface zone is 250 mm thick and bottom surface zone is only 100 mm thick (see Appendix CW).

If longer lengths between joints are considered, then the minimum requirements for road construction practice should be satisfied in addition to the minimum reinforcement indicated above, i.e. although the mature concrete conditions for a road are more severe than for reservoir floors, since the former is subjected to temperature fluctuations and stress reversals throughout its life, it is prudent for waterproof construction to satisfy normal road construction practice for long bay lengths if the latter are appropriate.

From more recent experience with unreinforced road slabs of 250 to 300 mm thickness it could be argued that no reinforcement (other than substantial dowel bars) is required for ground slabs on a sliding layer. Too close a comparison with road construction practice, however, can be misleading for two reasons. First, air entrained concrete of fairly low workability is invariably used for road construction and is therefore more cohesive than other fresh concretes. Second, the site control is generally such that any plastic shrinkage for the low workability concrete is negligible due to the rapidity with which curing membranes and protective covers can be applied. In a reservoir floor, air entrained concrete need not be used, and even if it is, it is not uncommon for some plastic shrinkage to occur for other than the very best quality control. If the air entrained concrete is pumped the workability needs to be much higher than that necessary for normal road construction, and under these circumstances it is extremely difficult to prevent plastic shrinkage occurring for summer placing in exposed positions. One example of a 150 mm thick Grade 30 non-air-entrained concrete reservoir ground slab, with minimum reinforcement to road construction standards and a sliding layer with partial contraction joints 4.3 m apart, was found to contain cracks up to 0.7 mm wide-due no doubt to the superposition of plastic shrinkage on the normal early thermal contraction. With minimum reinforcement to BS 5337, the maximum crack widths would have been little more than 0.2 mm despite the deficiencies in quality control. This was confirmed by the lack of serious cracks transverse to the longitudinal direction for those slabs which were laid in long bays and had longitudinal reinforcement at least equal to the minimum suggested.

11.5 STRUCTURAL DESIGN OF WATER-RETAINING STRUCTURES TO BS 5337

Structures which are to contain liquids should be designed to avoid wide cracks not only from thermal and shrinkage movements but also under the action of the normal working loads. The crack widths which are appropriate for limit state design are the same as before: 0.1 mm for Class A, 0.2 mm for Class B and 0.3 mm for Class C exposure. The conventional 'alternative design' requires, however, that cracking under load is (theoretically) avoided—i.e. in limit state terms that the limit state of crack formation is taken as the criterion rather than the limit states of 0.1 mm, 0.2 mm or 0.3 mm crack width. Thus for the alternative design the uncracked section is considered and the tensile stress limited to a suitable permissible value. The alternative design, which will be considered first, also requires that the 'strength' of a section is checked by conventional elastic theory for the cracked section.

It is clear that for all water-retaining construction the concrete should be well compacted and impermeable and that any joints intended to accommodate relative movement between adjoining parts of a structure require special provision for maintaining the watertightness of the joint.

Tensile strength of concrete

Consider an axially loaded reinforced concrete tension member. Provided that the concrete does not crack, the same approach that has been used for axially-loaded columns can also be applied here. If F is the applied load and f_{ct}, f_{st} are tensile stresses in the concrete and steel respectively, then

$$F = f_{ct}A_c + f_{st}A_s$$

Putting $f_{st} = \alpha_e f_{ct}$, $A_s = \rho_c A_c$ and rearranging gives

$$f_{ct} = \frac{F}{(1 + \alpha_e \rho_c)A_c} \qquad (11.5.1)$$

Permissible tensile stresses for no cracking

The maximum permissible stress in the steel is severely restricted if the concrete is to remain uncracked. For example, when $\alpha_e = 15$ and $f_{ct} = 2.02$ MN/m^2 then f_{st} cannot exceed 30.3 MN/m^2: an extremely low proportion of the steel strength. Nevertheless, this is what is required by the alternative design for 'resistance to cracking' calculations. The permissible concrete stresses for resistance to cracking are given in Table CW3.

If the concrete cracks, the steel should readily take the additional load, and the size of these cracks must be controlled in order to prevent rapid corrosion of the reinforcement and to contain the liquid effectively. Cracking of the concrete in liquid-retaining structures can allow constant leaching out of lime, rendering the steel in moderate cracks liable to corrosion. If cracks should occur they should be sufficiently fine so as not to endanger, seriously, the serviceability of the structure. Autogenous healing of fine cracks in water-retaining structures can occur if the pressure difference across the section can be made relatively small for a sufficiently long period—for example, if a reservoir can be filled gradually over a period of weeks or even months.

Uncracked concrete beams

Consider the equivalent concrete section for an uncracked beam section and express all stresses in terms of the extreme concrete fibre stress in tension, f_{ct}, as shown in Fig. 11.5.1. Let the depth of the neutral

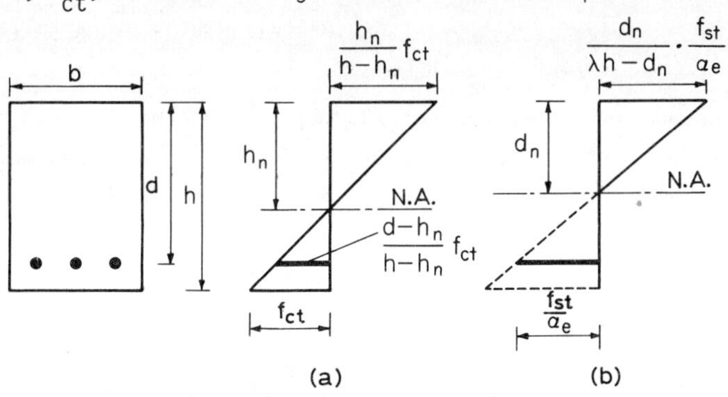

Fig.11.5.1 Concrete stress distribution in uncracked and cracked beam sections (a) Uncracked. (b) Cracked.

axis of the uncracked section be h_n. Taking first moments of area about the neutral axis and assuming that $\alpha_e - 1$ approximately equals α_e gives

$$bh(h_n - \tfrac{1}{2}h) = \alpha_e \rho_c bh(d - h_n)$$

Let $\lambda = d/h$, hence

$$h_n/h = (\tfrac{1}{2} + \alpha_e \rho_c \lambda)/(1 + \alpha_e \rho_c) \qquad (11.5.2)$$

Now

$$I_e = bh^3\{1/12 + (\xi - \tfrac{1}{2})^2 + \alpha_e \rho_c (\lambda - \xi)^2\}$$

where $\xi = h_n/h$

Hence the moment of resistance of concrete in tension is given by

$$M_{dct} = \frac{f_{ct} I_e}{(1-\xi)h} = \frac{f_{ct} bh^2}{1-\xi} \{\tfrac{1}{3} - \xi + \xi^2 + \alpha_e \rho_c (\lambda - \xi)^2\} \qquad (11.5.3)$$

<u>Cracked concrete beams for 'strength'</u>

The probability of a crack occurring in an 'uncracked' beam to the alternative design for 'resistance to cracking' is relatively high, and hence it is desirable to ensure that the consequence of a crack, should one occur, should not be serious. The alternative design therefore requires that the section is also designed so that the stresses in the steel and concrete do not exceed the permissible values in Tables CW5 and CW4 when the concrete is assumed to have no tensile strength (see Appendix CW §9.1—basis of design).

Elastic theory can be used to obtain the depth to the neutral axis of a cracked section in terms of the overall depth, h, as follows:

$$d_n/h = \{\sqrt{(1 + 2\lambda/\alpha_e \rho_c)} - 1\}\alpha_e \rho_c \qquad (11.5.4)$$

This equation corresponds to Eqn (3.2.4), except that there d_n was in terms of the effective depth, d. If stresses are expressed in terms of the maximum tensile steel stress, f_{st}, then the moment of resistance of the steel for the cracked section, from Eqn (3.2.6) is:

$$M_{dt} = f_{st} \rho_c bh(d - d_n/3) \qquad (11.5.5)$$

Example 11.5.1

If the permissible concrete stress to prevent cracking, f_{dct}, for flexural tension is 2.02 in MN/m^2 units, determine the moment for the limit state of local damage for a water-retaining structure for the uncracked midspan section considered in Example 11.1.1(a). Assume that $\lambda = d/h = 0.8$. What load can be carried by the beam before the section cracks, expressed as a proportion of the elastic theory working load w referred to in Example 11.1.1?

<u>Solution</u>

$$\rho = 2.18\% \text{ therefore } \rho_c = 2.18 \times 0.8\% = 1.74\%$$
$$\alpha_e\rho_c = 15 \times 1.74\% = 0.262 \text{ and}$$
$$\alpha_e\rho_c\lambda = 0.8 \times 0.262 = 0.209$$

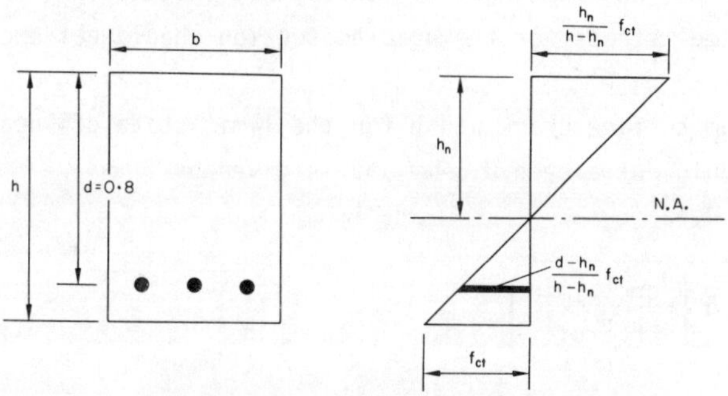

Fig.11.5.2 Example 11.5.1

Hence from Eqn (11.5.2)

$$h_n/h = (0.5 + 0.209)/1.262 = 0.562 = \xi$$

Now

$$I_e = bh^3\{1/12 + (\xi - \tfrac{1}{2})^2 + \alpha_e\rho_c(\lambda - \xi)^2\} = 0.102\ bh^3$$

Therefore

$$M_{dct} = f_{dct}I_e/(1 - \xi)h = 2.02 \times 0.102\ bh^3/0.438\ h$$
$$= 0.470\ bh^2 \text{ (MN/m}^2 \text{ units)}$$

Proportion of elastic working load w is given by

$$(0.470bh^2/2.23bd^2)w = (0.470/2.23 \times 0.8^2)w = 0.329w \qquad \text{(Answer)}$$

Example 11.5.2

The retaining wall of a rectangular reinforced concrete reservoir is 400 mm thick at the base and contains vertical reinforcement of 32 mm diameter high-yield deformed bars at 250 mm centres with an effective depth of 330 mm. Analyse this section according to the following alternative criteria for BS 5337:1976 to see if the head of fresh water (weighing 9.81 kN/m^3) can be increased to a maximum depth of 4 m. Assume a characteristic cube strength f_{cu} of 30 MN/m^2 and a characterist yield strength of the steel of 425 MN/m^2. Also assume a long-term concrete modulus of 12.5 GN/m^2 and a steel modulus of 200 GN/m^2.

(i) Under working loads the permissible concrete flexural tensile stress for the uncracked section should not exceed 2.02 MN/m^2 and the steel stress for the cracked section should not exceed 100 MN/m^2.

(ii) The maximum surface crack width for the limit state of local damage should not exceed 0.1 mm and is given by

$$w = \frac{4.5 a_{cr} \varepsilon_m}{1+2.5\left\{\dfrac{a_{cr} - c_{min}}{h - d_n}\right\}}$$

where $\varepsilon_m = \varepsilon_1 - \dfrac{0.0007}{\rho_c \; f_s}$

(See also list of symbols as in Appendix CW §8.2)

(UB3/H, 1975, 50 min)

Solution

Maximum moment at base (working load) for 1 m length of wall

$= \frac{1}{2} \times 9.81 \times 4^2 \times \frac{4}{3} \times 10^3 = 104.64 \times 10^3$

Steel for 1 m length $= 4 \times \frac{\pi}{4} (32)^2 \times 10^{-6} = 3217$ mm^2

$\alpha_e = 200/12.5 = 16$

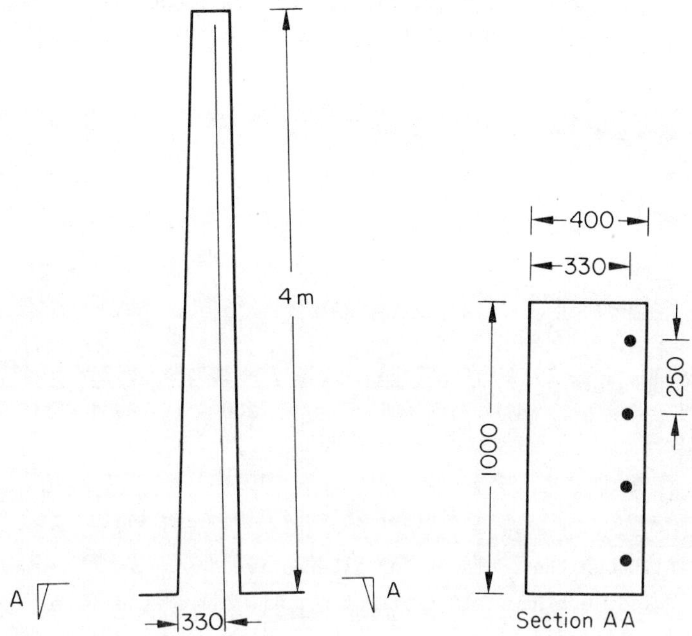

Fig.11.5.3 Example 11.5.2

(i) For the uncracked section let h_n be the depth to the neutral axis

Therefore taking first moments of area about the neutral axis:

$$1 \times h \left(h_n - \frac{h}{2}\right) = (16 - 1) \times 3217 \times 10^{-6} (0.33 - h_n)$$

hence $h_n = 0.2139$

Therefore $I_e = \frac{1}{12}(0.4)^3 + 0.4(0.0139)^2 + 0.0483 (0.1161)^2 = 0.00606$

$$f_{st} = \frac{104.64 \times 0.1161 \times 10^3}{0.00606} \times 16 = 32.1 \times 10^6$$

$$f_{ct} = \frac{104.64 \times 0.1861 \times 10^3}{0.00606} = 3.21 \times 10^6$$

For the cracked section let d_n be the depth to the neutral axis. Therefore taking first moments of area about the neutral axis,

$$1 \times d_n^2/2 = 16 \times 3217 \times 10^{-6}\ (0.33 - d_n)$$

$$\text{or } \tfrac{1}{2}d_n^2 + 0.0515d_n - 0.017 = 0$$

hence $d_n = 0.140$

Therefore $I_e = \frac{1}{3}(0.140)^3 + 0.0515(0.33 - 0.14)^2 = 0.00277$

Therefore $f_{st} = \dfrac{104.64 \times 10^3 \times 0.19}{0.00277} \times 16 = 114.7\ \text{MN/m}^2$ (Answer)

(ii) From (i) $f_{st} = 114.7 \times 10^6$ and $\rho_c = 3217 \times 10^{-6}/0.40 = 0.804\%$

Therefore $\varepsilon_m = \dfrac{0.4 - 0.14}{0.33 - 0.14} \times \dfrac{114.7 \times 10^{6-9}}{200} - \dfrac{0.7 \times 10^{-3+3}}{8.04 \times 114.7} = 26 \times 10$

Therefore $w = \dfrac{4.5 \times 127.26 \times 26 \times 10^{-6-3}}{1 + 2.5(127.26 - 54)/(0.4 - 0.14)10^3} = 0.008\ \text{mm}$

Although the criteria for alternative design to (i) are not satisfied the limit state of crack width of 0.1 mm for (ii) is satisfied. (Answer)

Notice that although the permissible steel stresses in 'strength' calculations for deformed bars have been increased from 85 to 100 MN/m^2 in BS 5337, limit state design permits higher stresses still and therefore even greater economies.

Direct tension in mature concrete

The requirements of BS 5337 are deemed to be satisfied if the steel stress for the cracked section under service conditions does not exceed the appropriate values in Table CW1 (limit state design) or Table CW5 (alternative design) and the clear distance between bars does not generally exceed 300 mm. Although CP 110 implies a crack width formula for members in direct tension [5.4], BS 5337 does not. For a member loaded in direct tension there is a greater tendency for bond slipping to occur, since additional cracks at adjacent sections are not induced by a strain gradient as in flexural members. Even for flexural members some slipping occurs with time as indicated in Eqn (11.2.2). Hence the above deemed-to-satisfy clauses.

Where structural loading calculations are necessary the required steel ratios are usually significantly greater than ρ_{crit}, hence the only further check for early thermal and shrinkage contraction is to ensure that the bar diameter is not too large for the required maximum crack width.

The alternative design method also gives permissible concrete stresses in Table CW3 for 'resistance to cracking' calculations for direct tension.

Limit state design procedure for flexural members

Example 11.5.2 has shown that limit state design can effect considerable economies in water-retaining structures compared with the traditional alternative design. Equally, however, it is clear that the direct use of the crack-width formulae is not convenient for routine design. Reference to design charts or tables such as those given in the explanatory handbook to BS 5337 [11.12] is necessary.

If the different criteria are compared then it is possible to construct a chart from the design tables [11.12] for a given cover, bar size, grade of concrete and grade of steel (see Fig. F5) showing the service moment M/bd^2 plotted against steel ratio ρ_c for different bar spacings. For limit state design the limit state of collapse is critical at low steel ratios whereas at higher steel ratios the limit state of cracking becomes critical with different curves for different bar spacings. The steel stresses under service loads permitted by the limit state of cracking, or collapse, are significantly higher than the permissible values permitted by the alternative design. In Fig. F5 the curve for f_{dst} = 100 MN/m^2 is drawn for $\lambda = d/h = 0.85$. A cover of 52 mm allows for 12 mm distribution bars on the outside of the main bars and a minimum cover of 40 mm.

Watertight construction for basements

The same principles as outlined for water-retaining structures can be applied to watertight construction for basements to buildings. The essential difference is that the crack widths which can be tolerated are significantly finer, since seepage of water which is only sufficient to cause a damp area can be objectionable. Design crack widths should therefore be somewhat less than 0.1 mm. Alternatively, 0.2 mm crack widths with an internally applied waterproof coating can be considered. See references [11.13] and [11.14] for further information on watertight construction.

References

11.1 Base, G.D., Reed J.D., Beeby, A.W., and Taylor H.P.J., An investigation of the crack control characteristics of various types of bar in reinforced concrete beams, Research Report 18 parts 1 and 2, Cement and Concrete Association, London, December, 1966.

11.2 Illston, J.M., and Stevens, R.F., 'Long-term cracking in reinforced concrete beams', Proc. Instn Civil Engrs, Part 2, vol. 53, December, 1972, pp. 445-59.

11.3 Beeby, A.W., Private communication

11.4 Evans, E.P., and Hughes, B.P., 'Shrinkage and thermal cracking in a reinforced concrete retaining wall', Proc. Instn Civil Engrs, vol. 39, January, 1968, pp. 111-25.

11.5 Hughes, B.P., and Miller M.M., 'Thermal cracking and movement in reinforced concrete walls', Proc. Instn Civil Engrs (supplement paper 7254S) vol. 45, February, 1970.

11.6 British Standards Institution CP 2007:1960, Design and constructio of reinforced and prestressed concrete structures for the storage of water and other aqueous liquids, London 50 pp.

11.7 Hughes, B.P., 'Designs to CP 2007 and BS 5337 for controlled cracking of water retaining structures', The Structural Engineer, vol. 54, (To be published) 1976.

11.8 British Standards Institution BS 5337 The structural use of concrete for retaining aqueous liquids, London 1976.

11.9 Large pours for RC Structures, Symposium of the Concrete Society, West Midlands Region, Birmingham, September, 1973, pp. 91.

11.10 Road Research Laboratory, A guide to structural design of pavements for new roads, Road Note 29 (3rd edn) London, HMSO, 1970, pp. 30.

11.11 Deacon, R. Colin., Concrete ground floors: their design constructi and finish, Cement and Concrete Association, London, 1970, pp.30.

11.12 Anchor, R.D., Hill, A.W., and Hughes, B.P., Explanatory handbook on BS 5337:1976 (To be published).
11.13 Design and construction of deep basements, Instn of Structural Engineers, London, August, 1975, pp.64.
11.14 Concrete - Can it hold water? The Concrete Society, West Midlands Region, Birmingham, September, 1974 pp.87

Exercises

1. (i) Show that the crack spacing, s, for a continuous length of reinforced concrete wall, fully restrained at the ends and without any contraction joints, is given by

$$\frac{{}_*f_{ct}}{f_b}\,\frac{\Phi}{2\rho_c} \geq s \geq \frac{{}_*f_{ct}}{f_b}\,\frac{\Phi}{4\rho_c}$$

when $\rho_c > {}_*f_{ct}(f_{yt} + f_{sc})$

and

Φ = diameter of steel bars
f_b = average bond strength between concrete and steel
${}_*f_{ct}$ = ultimate tensile strength of concrete
f_{sc} = stress for steel in compression
f_{yt} = yield strength of steel in tension
ρ_c = steel ratio

(ii) A wall contains 0.6% mild steel (f_{yt} = 250 MN/m^2) in the longitudinal direction. The peak hydration temperature was 30°C. It then fell to 10°C and then to -10°C a few months later, both with an effective coefficient of thermal expansion of half its normal value (5×10^{-6} strain/$^\circ$C). Calculate the mean crack width at -10°C if f_b and ${}_*f_{ct}$ both equal 1.3 MN/m^2, the shrinkage strain of concrete is 120×10^{-6}, ultimate tensile strain of concrete (elastic + creep) is 200×10^{-6}, coefficient of thermal expansion (steel or concrete) is 10×10^{-6} strain/$^\circ$C and Φ = 8 mm. (Modified from UB2 1966, 36 min)

2. A typical rib to a solid garage floor is as Section AA shown in Fig. A. The ribs occur at 1 m centres and the total longitudinal reinforcement in rib and slab for the 1 m wide section AA adjacent to

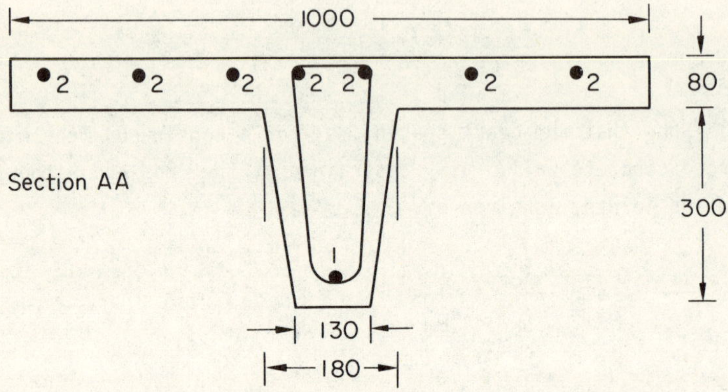

Ch.11 Fig.A Question 2

a support consists of 7 No. 7 mm bars type 'X' (Mark 2) and 1 No. 16 mm bar type 'Y' (Mark 1). However, the anchorage length for the 16 mm bar beyond section AA is only 450 mm.

Analyse section AA to determine, from first principles, the predicted crack width at the section (due to early thermal effects) by considering that

(a) the section contracts as a whole

(b) the 300 mm deep rib contracts independently of the slab.

Can you predict the spacing for (b) if the anchorage length for the 16 mm bar is doubled?

Assume the following:

Concrete tensile strength at 3 days = 1.3 MN/m^2

Total effective contraction due to early thermal and other effects = 300 microstrain

	Type 'X'	Type 'Y'
Characteristic yield strength of steel bars (MN/m^2)	485	425
Bond strength with concrete at early age (MN/m^2)	1.95	1.625

(UB3/H, 1974, 50 min)

3. The retaining wall for a large shallow reservoir has the cross-section shown in Fig. B.

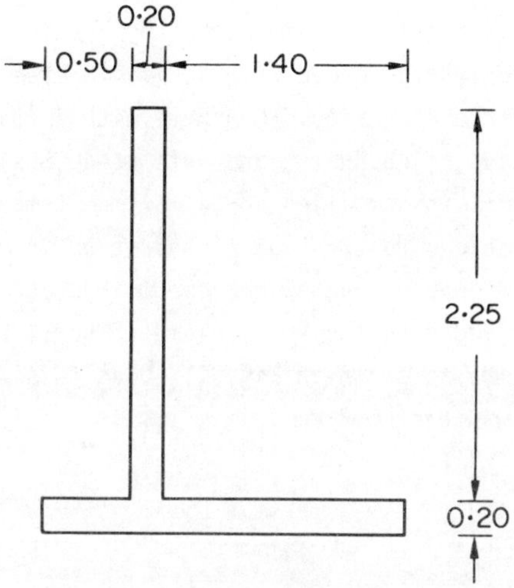

Ch.11 Fig.B Question 3

The horizontal reinforcement in the wall consists of 16 mm diameter plain round mild steel bars at 450 mm centres in each face.

(a) What is the maximum joint spacing for vertical contraction joints which would be satisfactory if no cracking is permitted?

Is the amount of horizontal steel satisfactory if partial contraction joints (i.e. discontinuity in the concrete only) are used in most cases? The tensile strength of the immature concrete, and the average bond strength between concrete and reinforcement can both be taken as 1.55 MN/m^2. The characteristic strength of the reinforcement is 250 MN/m^2.

(b) Extreme concern is caused by the discovery that the drainage of the site is not as assumed in the design. Land drains are therefore installed along the toe around the outside perimeter of the reservoir as an immediate expedient. Determine the stability of the wall against overturning about the toe for the limit state of local damage, if the upthrust on the base is given by a maximum head of 2.25 m of fresh water (weighing 9.8 kN/m^3) at the heel, and varying linearly to zero at the toe. (Take all partial safety factors for load as unity and assume that reinforced concrete weighs 23 kN/m^3.) (UB3/H 1972, 40 min)

4. A massive concrete foundation slab for a multi-storey building is required to have a maximum crack width of not more than 0.3 mm. The top face of the slab is heavily reinforced with bars in two layers in one direction but is only reinforced with 40 mm diameter bars at 800 mm centres in the transverse direction. Concrete with a characteristic cube strength at 28 days of 45 MN/m^2 (and estimated tensile strength at 3 days of 1.75 MN/m^2) is specified and a maximum temperature rise at the core of the slab due to the heat of hydration is estimated as 40°C. Calculate the probable crack width due to hydration if 6 mm bars at 400 mm centres are added to the 40 mm diameter bars, which have a concrete cover of 80 mm, as shown in Fig. C. Does this satisfy the limit state of crack width of 0.3 mm?

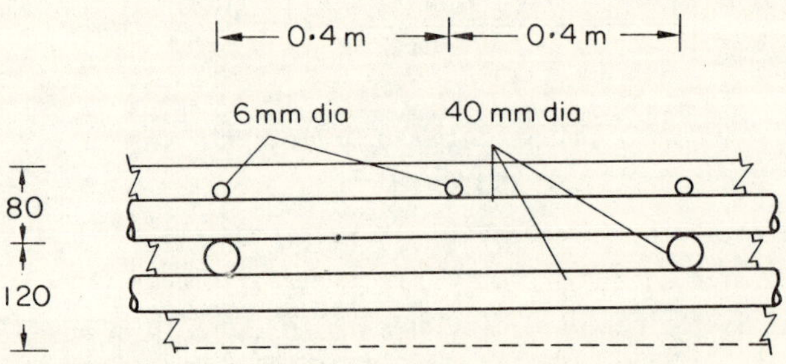

Ch.11 Fig.C Question 4

The yield strengths of the 40 mm and 6 mm diameter bars are 410 MN/m^2 and 460 MN/m^2 respectively. The ratio of bond strength to concrete tensile strength for all bars can be taken as 1.5. Assume a depth of surface zone of 0.2 m as shown (i.e. that the 40 mm bars are central in the surface zone of concrete) and that the concrete shrinkage strain equals one-half of the ultimate tensile strain of the concrete. The coefficient of thermal contraction of the immature concrete is 5 microstrain per deg. C after allowing for creep effects.

(UB3/H 1971, 40 min)

12 Torsion

Torsion is not, in general, of major importance in structures of conventional beam and slab construction and can often be ignored if sufficient flexural, axial and shear strength has been provided. This is because the torsional stiffness is generally small with respect to flexural stiffnesses and can often be ignored in assessing the force distribution in an indeterminate structure. Furthermore, small torsional moments cause no significant reduction in flexural capacity if nominal links are provided. However, if primary torsion occurs in a statically determinate structure, torsion should be both calculated and designed for. Even for an indeterminate system it may be desirable to design for torsion if the angle of twist occurs in a relatively short length and excessive cracking can occur. Thus beams cranked or curved on plan or with horizontal projections and main beams supporting secondary beams on one side only should generally be designed for torsion. Edge beams supporting the end span of a continuous slab need not necessarily be designed for torsion unless trimmer beams frame into the edge beam because of openings [5.4].

CP 110 considers torsion, like shear and bond, in terms of the limit state of collapse. The structures should be analysed on the basis of elastic theory and the flexural rigidity $E_c I_o$ should be based on the gross section. The torsional rigidity GC may be calculated taking G as $0.4E_c$ and the torsion constant, C, as one half the St Venant value calculated for the plain concrete section.

12.1 RECTANGULAR BEAMS WITHOUT TORSION REINFORCEMENT

A rectangular section subjected to pure torsion within the elastic range exhibits the greatest shear stress at the middle of the longer sides. The diagonal tension stress on one side of the web, due to torsion, therefore augments that due to transverse shear. It is therefore essential to consider the combination of transverse and torsional shear.

Elastic theory

The elastic axial torque for rectangular sections has been obtained by St Venant (e.g. see reference [12.1]) in terms of the angle of twist per unit length, θ, as follows:

$$T = \beta x^3 y G\theta \qquad (12.1.1)$$

where G = shear modulus of the material

x = shorter side of rectangle

y = longer side of rectangle

β = a function of x/y

The maximum elastic shear stress, v_t, occurs at the centre of the longer side, y, and is given by

$$v_t = \gamma x G\theta = T/\alpha x^2 y \qquad (12.1.2)$$

where γ is another function of x/y and $\alpha = \beta/\gamma$. Typical values of the functions α, β and γ are shown in Table 12.1.1. St Venant's torsional constant is given by $\beta x^3 y$

Table 12.1.1 Elastic torsion constants for rectangular sections

y/x	α	β	γ
1.0	0.208	0.141	0.675
1.2	0.219	0.166	0.758
1.4	0.227	0.187	0.822
1.6	0.234	0.204	0.872
1.8	0.240	0.217	0.904
2.0	0.246	0.229	0.930
2.5	0.258	0.249	0.965
3.0	0.267	0.264	0.989
5.0	0.292	0.291	0.999
∞	0.333	0.333	1.000

Plastic theory

The axial torque, for rectangular sections with a uniform shear stress over the whole section, is given by [12.3]:

$$v_t = T/\tfrac{1}{2}x^2(y - x/3) \qquad (12.1.3)$$

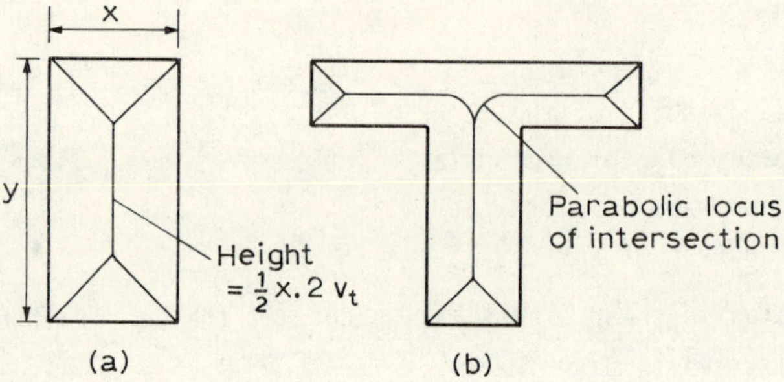

Fig.12.1.1 'Sand heap' analogy for sections in torsion. (a) Rectangular beam. (b) T-beam

This result can be obtained by use of the 'sand heap analogy', i.e. the fact that the torsional resistance of the cross-section is proportional to the volume of a sand heap of constant maximum slope on the section, the slope being equal to twice the constant shear stress. Reference to Fig. 12.1.1(a) indicates that the volume of the sand heap is given by

$$\{\tfrac{1}{2}x(y - x) + \tfrac{1}{3}x^2\}v_t x$$

Hence

$$T = \tfrac{1}{2}x^2(y - x/3)v_t$$

as given in Eqn (12.1.3).

Elasto-plastic behaviour

At collapse, a uniform shear stress probably exists over a considerable proportion of a concrete section since tests indicate that some stress redistribution occurs as the ultimate torque is approached. The extent to which plastic deformation occurs is somewhat uncertain. It has therefore been usual for codes of practice to adopt one of two extreme assumptions. Either:

(i) No plastic yield (i.e. elastic theory for which maximum stress nowhere exceeds a certain value), or

(ii) Extensive plastic yield (i.e. average stress for entire section does not exceed a particular value).

Accuracy of torsion calculations

The calculation of the torsional moments according to elastic theory depends upon the relative stiffnesses of the members of the structure. Such calculations are inevitably approximate. Design calculations for the torsional strengths of sections are also approximate and are based largely on experimental test results. A suitable choice of values for stress therefore enables the essentially different elastic and plastic theories to predict collapse loads which are reasonably consistent for a given set of experimental results.

Combined torsion and bending

Beams are rarely subjected to torsion only. Cowan has suggested that

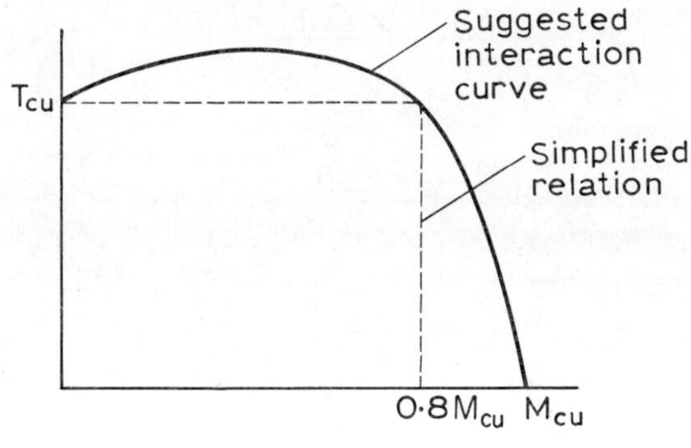

Fig.12.1.2 Relation between torque and bending moment at collapse (reference [12.4]).

the variation of the ultimate torsional moment with the ultimate flexural moment is indicated in Fig. 12.1.2(see reference [12.4] or [12.5]. The simplified, dotted, line therefore gives a safe value of the ultimate torsional moment provided that the bending moment is less than about 80% of the ultimate flexural value.

Visco-elastic limit

Cowan's suggested interaction curve is not strictly based on the collapse load, since failure in his tests was classified as the commencement of large-scale cracking, which was called the visco-elastic limit. This value can coincide with collapse although it is usually rather less and is therefore a safe estimate of the true ultimate torsional moment. This visco-elastic limit for a member subjected to pure torsion is defined as being the point at which the maximum tensile stress, which has the same value as the maximum shear stress, reaches the ultimate tensile strength of the concrete. Reinforcement has little effect on the magnitude of the torsional moment at which this cracking commences. If the beam is adequately reinforced, it sustains increased moments after cracking until eventually failure occurs by the steel yielding, with cracks opening up on three sides and crushing occurring on the fourth [12.6].

Torsional resistance of rectangular sections

CP 110 assumes a fully plastic section at collapse and therefore gives the following equation:

$$T = \tfrac{1}{2} v_t h_{min}^2 (h_{max} - h_{min}/3) \qquad (12.1.4)$$

where T = design torsional moment for collapse

v_t = torsional shear stress

h_{min} = x and h_{max} = y

Provided that the flexural moment does not exceed 80% of the collapse moment then Eqn (12.1.4) can be assumed to give a safe estimate of the torsional moment at collapse. Marshall [12.7] has confirmed, by analysing test results of many earlier investigators, that a reasonable estimate of the torque at cracking can be obtained from Eqn (12.1.4) by using the cylinder splitting strength for v_t.

Torsional resistance of concrete

No special provision of torsion reinforcement need be made if only nominal torsion is applied to the section, since nominal torsion reinforcement, in the form of minimum shear reinforcement and 'hanger bars', assists the torsional strength of the concrete. However, where the torsion shear stress, v_t, exceeds the value of v_{tmin} in Table CC2 special torsional reinforcement should be provided. Furthermore, in no case, even when reinforcement is provided as in section 12.2, should the sum of the shear stresses resulting from shear force and torsion, $v + v_t$, exceed the value of v_{tu} in Table CC2; nor, in the case of small sections ($y_1 < 550$ mm) should v_t exceed $v_{tu}y_1/550$ (see §3.3.7 Appendix CC). The values of f_{tu} in Table CC2 correspond with the maximum shear stress values of $0.75 \sqrt{(f_{cu})}$.

12.2 BEAMS WITH TORSION REINFORCEMENT

There is an analogy between shear and torsion for sections without shear or torsion reinforcement, since the CP 110 considers the principal tensile stress in the concrete in both cases. This similarity ceases, however, for sections with torsion reinforcement. This is because considerable rotation of the sections must occur before the torsion reinforcement becomes fully effective. Under these circumstances the torsional resistance of the cracked concrete must be ignored and the total torsion taken by the torsion reinforcement. Torsion reinforcement is therefore more analogous to flexural reinforcement in this respect.

Space truss analogy

Torsion reinforcement can be in the form of spirals in the appropriate direction. However, the most practical arrangement consists of longitudinal bars which are evenly distributed around the inside perimeter of the links as shown in Fig. 12.2.1. The links are vertical closed stirrups with the ends suitably anchored.

This arrangement lends itself to the space truss analogy in which the concrete between the cracks acts as compression diagonals. By considering the equilibrium of this truss Lampert and Collins [12.7] found that

$$T = x_1 y_1 \sqrt{\left\{\frac{A_{sl}\ f_{yl}}{x_1 + y_1}\ \frac{A_{sv}\ f_{yv}}{s_v}\right\}} \qquad (12.2.1)$$

where the symbols are as listed in Appendix CC (§3.3.7)

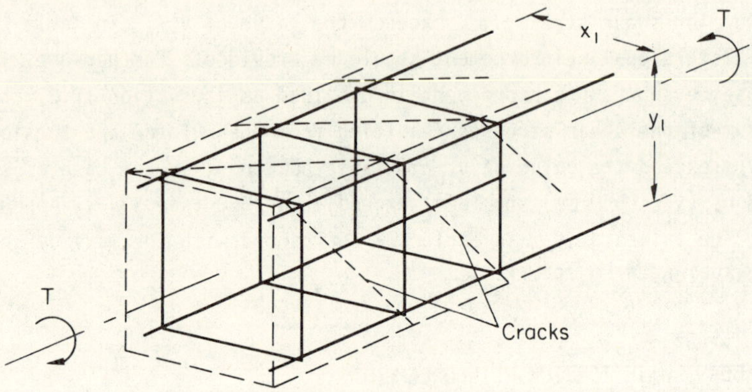

Fig.12.2.1 Space truss analogy for torsion

Considering a length s_v of the beam, the (steel volume) × (yield strength) products are respectively $A_{sl}\ f_{yl}\ s_v$ and $A_{sv}\ f_{yv}(x_1 + y_1)$. If the volume-strength products are made equal, the longitudinal bars and the links should yield simultaneously [12.4] i.e.

$$A_{sl}\ f_{yl}\ s_v = A_{sv}\ f_{yv}\ (x_1 + y_1) \qquad (12.2.2)$$

Substituting for $A_{sl}\ f_{yl}$ in Eqn (12.2.1) then

$$T = x_1 y_1 f_{yv}\ A_{sv}/s_v \qquad (12.2.3)$$

Lampert and Collins [12.8] showed experimentally that the torsional strength of a beam under-reinforced for torsion (i.e. the links and longitudinal reinforcement yield before the concrete crushes) was independent of the concrete strength. In practice torsionally over-reinforced beams should be avoided since they possess insufficient ductility when overloaded. This is ensured by not exceeding the v_{tu} values in Table CC2. They also showed that T varied as the area $x_1 y_1$ enclosed by the links and that the ratio x_1/y_1 was relatively unimportant. Notice that this implies that a box beam has the same torsional strength as a solid beam (see later).

Link reinforcement

If the required torsional resistance exceeds that for the concrete section, torsional reinforcement is necessary and is required by CP 110 to take the total torsion.

CP 110 also reduces the effective design stress in the steel by a factor of 0.80 as first obtained by Cowan for elastic theory [12.4] and because extensive concrete spalling occurs at the full design steel stress. The area of the closed links, which must be additional to any links required for shear due to normal loading, as given by CP 110 is therefore

$$A_{sv}/s_v = T/(0.8x_1y_1 \times 0.87f_{yv}) \qquad (12.2.4)$$

The Australian Code of Practice corresponding to CP 114 has a similar expression [12.10]. Eqn (12.2.1) effectively replaces 0.80/2 in the Australian Code with $0.80/\gamma_f\gamma_m$ which is approximately 0.46. CP 110 may appear less conservative compared with its Australian predecessor with respect to torsional reinforcement, except that the latter only requires the link steel to take the excess torsion.

Additional longitudinal steel

The area of longitudinal reinforcement, which is additional to that required for flexure, is given in CP 110 by the volume of steel which equals that for the links, suitably adjusted for any differences in yield strength. That is, exactly as in Eqn (12.2.2), where A_{sa} and f_{ya} refer to the additional longitudinal steel required. The longitudinal reinforcement should be distributed evenly round the inside perimeter of the links. At least four bars should be provided, one in each corner of the links. This is best effected by increasing the sizes of longitudinal bars at the levels of the tension and compression reinforcement above that which would be required for flexure alone.

If $y_1 > 0.3$ m, bars should be provided at points along the larger face such that the centre-to-centre distance between bars does not exceed 0.3 m. The torsional steel should extend a distance of at least h_{max} along the beam beyond points where such steel is theoretically necessary according to Tables CC1 and CC2.

Interaction curves for reinforced sections

Various interaction curves for torsion and bending for members reinforced for torsion have been proposed [12.8, 12.9] which are generally flatter than that given by Cowan [12.4] such that the point $M = 0.8M_u$, $T = T_u$ would fall outside the interaction curve. However, since the reinforcements for torsion and bending are calculated separately and then added together, Hsu and Kemp [12.9] have pointed out that such addition increases both the pure torsion and the pure bending strengths, as illustrated by points A and B in Fig. 12.2.2. Thus although the

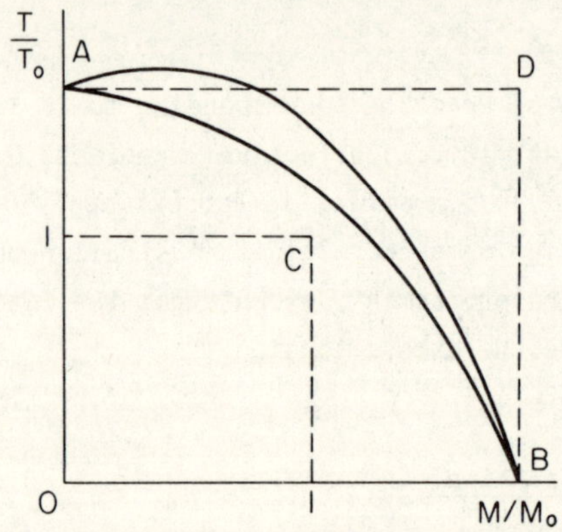

Fig.12.2.2 Typical interaction curves for torsion and bending concrete section [12.8]

interaction curve falls inside point D, it falls outside point C corresponding to the normal design procedure.

The behaviour of members in combined torsion, bending and shear is more complicated, requiring three-dimensional interaction diagrams and the normal design approach of calculating reinforcement requirements separately and then adding them together may be more conservative since, as Goode has indicated [12.6], the loading arrangement which gives the maximum torsional moment often does not give simultaneously the maximum bending moment and shear force.

Example 12.2.1

An RC beam, 0.30 m × 0.50 m overall and 0.45 m in effective depth, is subjected to a torsional moment of 20 kNm due to the characteristic imposed load.

(a) Calculate the design torsional moment for the concrete section without torsional reinforcement. Assume a design shear force of 27 kN, a longitudinal steel ratio of 1.0% and a characteristic cube strength

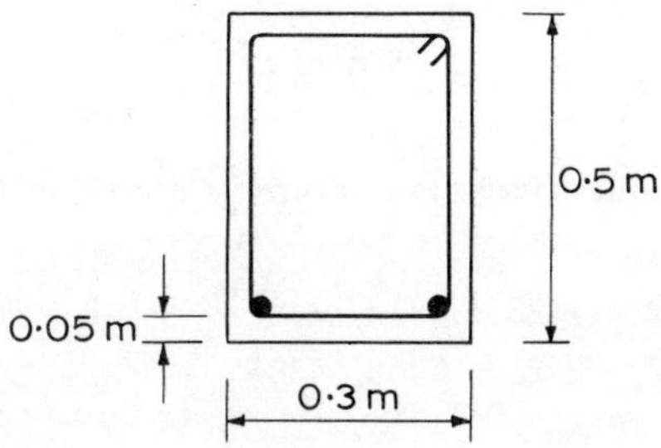

Fig.12.2.3 Example 12.2.1

of 30 MN/m^2 for the section concerned and assume that the torsional moment of resistance of the section with only nominal shear links and hanger bars is given by the total diagonal tension in the concrete due to both shear and torsion.

(b) Calculate the required spacing of 12 mm diameter links (if any) and additional area of longitudinal steel to resist the entire torsional moment as required by CP 110. Assume minimum cover of 25 mm, bar diameter of 12 mm for links, and characteristic yield strength for both links and main steel of 250 MN/m^2.

Solution

$T_k = 20$ kNm and $\gamma_f = 1.6$

Therefore

$T = 20 \times 1.6 \times 10^3 = 32 \times 10^3$

Substituting in Eqn (12.1.4) and transposing gives

$v_t = 2 \times 32 \times 10^3/0.3^2(0.5 - 0.1) = 1.78 \times 10^6$

From Tables CC1 and CC2 for $f_{cu} = 30 \times 10^6$:

$v_{tmin} = 0.37$, $v_{tu} = 4.10$ and

$v_c \quad = 0.70$ (all MN/m^2 units)

Hence $v_{tmin} < v_t < v_{tu}$. Also

$$v = 27 \times 10^3/0.30 \times 0.45 = 0.20 \times 10^6$$

Hence

$$v + v_t = (0.20 + 1.78) \times 10^6 = 1.98 \times 10^6 < v_{tu}$$

(a) Torsional resistance of concrete section is given by v_{tmin} i.e.

$$T_c = \tfrac{1}{2} \times 0.37 \times 0.3^2 \times 0.4 \times 10^6$$
$$= 6.66 \text{ kNm} \qquad \text{(Answer)}$$

(Hence notice the justification for only considering torsion when it is a major effect).

(b) $10^3 x_1 = 300 - 2(25 + 6) = 238$

$10^3 y_1 = 500 - 2(25 + 6) = 438$

and from Appendix CC

$$s_v \ngtr \tfrac{1}{2}y_1 = 0.217, \ngtr x_1 = 0.234 \text{ and } \ngtr 0.2$$

Also

$$A_{sv} = 2 \times \tfrac{1}{4}\pi \times 12^2 \times 10^{-6} = 226 \times 10^{-6}$$

Therefore

$$s_v = 0.8 x_1 y_1 \times 0.87 f_{yv} \times A_{sv}/T \ (\lambda = 0.80)$$
$$= 0.80 \times 0.238 \times 0.438 \times 0.87 \times 250 \times 226 \times 10^{6-6-3}/32$$
$$= 128 \times 10^{-3}$$
$$s_v = 125 \text{ mm say} \qquad \text{(Answer)}$$
$$A_{sl} = 226(0.238 + 0.438) \times 10^{-6}/0.125$$
$$= 1222 \text{ mm}^2 \qquad \text{(Answer)}$$

T-, L- and I-beams

Bach's approximation for elastic analysis of complex sections (e.g. see reference [12.2]) divides the section into its component rectangles, such that Eqn 12.1.1 becomes

$$T = G\theta \sum \beta x^3 y \qquad (12.2.5)$$

A similar approach could be used for plastic theory such that Eqn (12.1.3) becomes

$$T = \tfrac{1}{2} v_t \sum x^2 (y - x/3) \qquad (12.2.6)$$

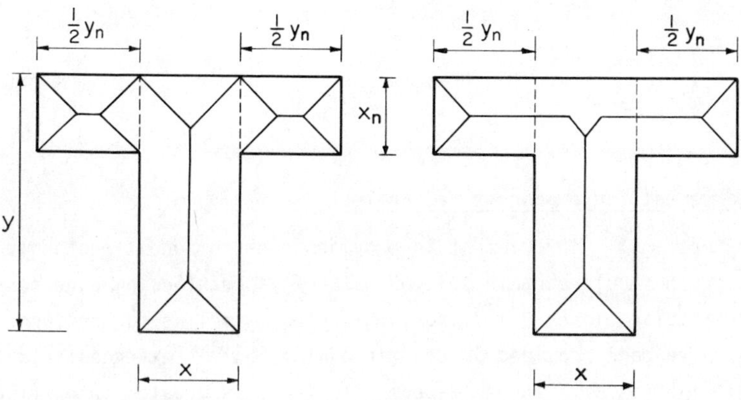

Fig.12.2.4 Approximations to 'sand heap' analogy. (a) Component rectangles. (b) Linear approximation

Reference to the 'sand heap' analogy in Fig. 12.1.1 and to Fig.12.2.4(a) shows, however, that this would be conservative. Fig.12.2.4(b) is a closer approximation in which the volume for the rectangle y_n by x_n is increased from

$$\tfrac{1}{2}v_t x^2(y - \tfrac{1}{3}x) \text{ to } \tfrac{1}{2}v_t x_n{}^2(y_n)$$

Hence the expression for all the additional rectangles should be $\frac{1}{2}v_t x_n{}^2 y_n$ if the expression for the central rectangle x_1 by y_1 remains as $\frac{1}{2}v_t x^2(y - \frac{1}{3}x)$, hence a close approximation is given by

$$T = \tfrac{1}{2}v_t x^2(y - \tfrac{1}{3}x) + \sum_{n=2}^{n=N} \tfrac{1}{2}v_t x_n{}^2 y_n \qquad (12.2.7)$$

where

$x > x_2$, etc.

CP 110 however considers the component rectangles for an elastic distribution and therefore suggests that

$\Sigma(h^3_{min}\ h_{max})$

can be made a maximum if the widest rectangle is made as long as possible. The torsion shear stress carried by each component rectangle may then be calculated as for a rectangular section subjected to a torsional moment of

$$T\left(\frac{h^3_{min}\ h_{max}}{\Sigma h^3_{min}\ h_{max}}\right) \tag{12.2.8}$$

Reinforcement for component rectangles

Where torsional reinforcement is provided each component rectangle of the section should contain all, or part of, a reinforcing cage consisting of rectangular closed links and longitudinal steel as for rectangular beams. The cage provided in the rib of the beam is extended into the flange (T-beam) or flanges (I-beam) if x(rib) is greater than x(flange). The proportion of the total torsion taken by each cage is given according to (12.2.8).

For example, for a T-beam there will be two component rectangles: the rib and the flange. If b_r exceeds h_f then the major rectangle is given by

$h \times b_r$

The second component rectangle should not duplicate any area in the first rectangle and is therefore given by

$(b - b_r) \times h_f$

provided that the reinforcement is continuous through the area common to both rectangles as shown in Fig. 12.2.5 .

The torsion constant C is given by

$C = \frac{1}{2}\beta_1 b_r^3 h + \frac{1}{2}\beta_2 h_f^3 (b - b_r)$

where β_1 is the St Venant constant from Table 12.1.1 for

$y/x = h/b_r$

and β_2 that for $y/x = (b - b_r)/h_f$

If $h_f > b_r$ then the rectangles would be

$b \times h_f$

and

$(h - h_f) \times b_r$

with

$C = \frac{1}{2}\beta_3 h_f^3 b + \frac{1}{2}\beta_4 b_r^3 (h - h_f)$

where β_3 is given by b/h_f and

β_4 is given by $(h - h_f)/b_r$

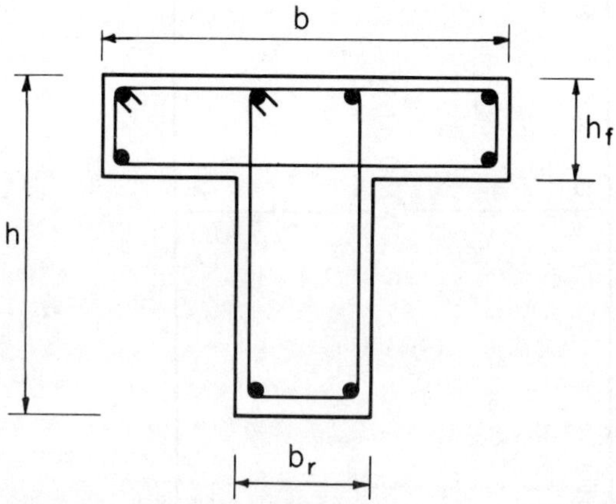

Fig.12.2.5 Reinforcement for component rectangles

Example 12.2.2

A tee beam section as shown in Fig. 12.2.6(a) is used for beams AB, CD and LM shown in Fig. 12.2.6(b). If the design load on LM at collapse is equivalent to a point load of 300 kN, determine the maximum torque applied to LB and design the torsion steel accordingly. The cover to the links is to be 30 mm, f_{cu} = 25 MN/m² and f_{yt} = 410 MN/m².

Solution

Taking first moments of area about the extreme top fibre

$300 \times 600 \times 300 + 1000 \times 200 \times 100 = (300 \times 600 + 1000 \times 200)\bar{y}$

or $\bar{y} = 195$

Therefore

$$I = \frac{1}{12}\ 300 \times 600\ + 300 \times 600(300\text{-}195)^2\ + \frac{1}{12}\ 1000 \times 200^3 + 1000 \times 200\ (195 - 100)^2\ \text{mm}^4$$

$= 9856 \times 10^{-6}\ \text{m}^4$

Now h_{max}/h_{min} = 600/300 and 1000/200, i.e. 2 and 5

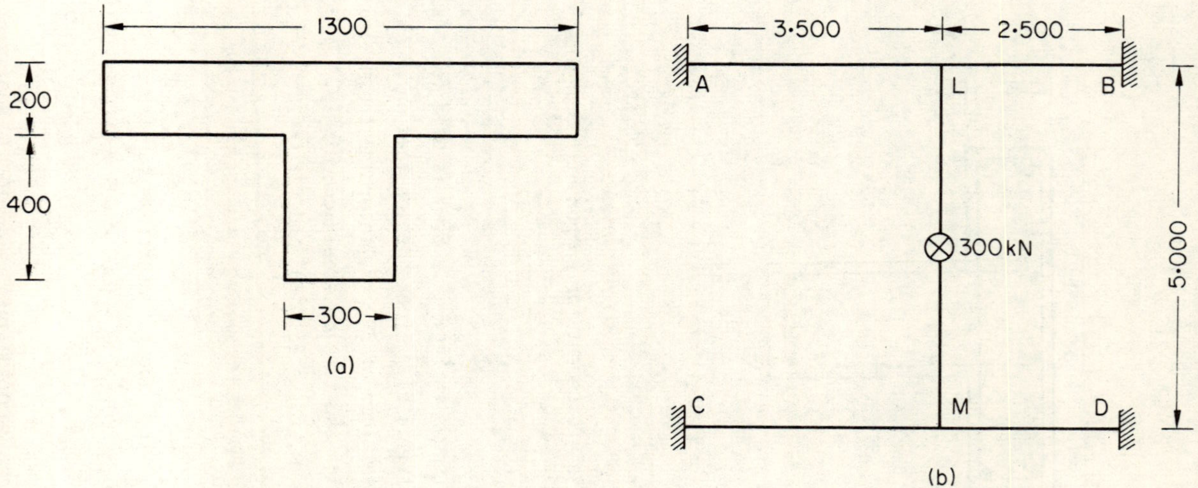

Fig.12.2.6 Example 12.2.2

Hence from Table 12.1.1, β = 0.229 and 0.291

$$C = \tfrac{1}{2} \times 0.229 \times 300^3 \times 600 + \tfrac{1}{2} \times 0.291 \times 200^3 \times 1000 \text{ mm}^4$$

$$= 1855 + 1164 = 3019 \times 10^{-6} \text{ m}^4$$

If a moment is applied at end L of beam LM then by slope deflection equations

$$M_{LM} = \frac{2EI}{L}(2\theta_L + \theta_M)$$

If the end M were assumed to be fixed (as normally the case for moment distribution) then $\theta_M = 0$ and $M_{LM}/\theta_L = 4EI/L_{LM}$

However, in this case the grid is symmetrical and $\theta_M = -\theta_L$

i.e. $M_{LM}/\theta_L = 2EI/L_{LM} = K_{LM}$

Hence

$$K_{LM} = \frac{2EI}{L_{LM}} = \frac{2 \times 9856 \times 10^{-6}}{5}E = 3942 \times 10^{-6}E$$

$$K_{TLA} = \frac{CG}{L_{LA}} = \frac{3019 \times 0.4 \times 10^{-6}}{3.5}E = 345 \times 10^{-6}E$$

$$K_{TLB} = \frac{CG}{L_{LB}} = \frac{3019 \times 0.4 \times 10^{-6}}{2.5}E = 483 \times 10^{-6}E$$

Therefore $\Sigma K = 4770 \times 10^{-6}E$

The distribution factors are as follows.

D_{LM} = 3942/4770 = 0.827

D_{LA} = 345/4770 = 0.072

D_{LB} = 483/4770 = 0.101

The fixed end moment at L due to point load

$$= \frac{300 \times 5.0}{8} = 188 \text{ kNm}$$

By moment redistribution; releasing joints L and M

M_{LM} = 188(1-0.827) = 32.4 kN m

T_{LA} = 188 × 0.072 = 13.5 kN m

T_{LB} = 188 × 0.101 = 18.9 kN m

These are the actual torque values since the carry over factors for torque are +1.0. Thus the torque applied to LB at collapse is 18.9 kN m.

The torques applied to the component rectanges are as follows.

Rib: $18.9 \times 10^3 \times 1855/3019$ = 11.6 kN m

Flange: $18.9 \times 10^3 \times 1164/3019$ = 7.3 kN m

The torsional shear stresses, from Eqn (12.1.4), are therefore as follows.

$$\text{Rib:} \quad \frac{2 \times 11.6 \times 10^3}{0.3^2(0.6 - 0.1)} = 0.52 \text{ MN/m}^2$$

$$\text{Flange:} \quad \frac{2 \times 7.3 \times 10^3}{0.2^2(1.0 - 0.2/3)} = 0.39 \text{ MN/m}^2$$

Thus in both cases from Table CC2,

$v_t > v_{tmin} = 0.33 \text{ MN/m}^2$ and $<< v_{tu} = 3.75 \text{ MN/m}^2$

Both component rectangles require reinforcement.

The cover to the links is 30 mm; hence the centre line dimensions, assuming ≯ 10 mm diameter links, are as follows.

Rib: $x_1 = 300 - 2 \times 35 = 230$ mm

$y_1 = 600 - 2 \times 35 = 530$ mm

Flange: $x_1 = 200 - 2 \times 35 = 130$ mm

$y_1 = 1000 - 2 \times 35 = 930$ mm

Hence from Eqn (12.2.4), the required links are given as follows.

$$\text{Rib:} \quad \frac{A_{sv}}{s_v} = \frac{11.6 \times 10^{3-6}}{0.8 \times 0.23 \times 0.53 \times 0.87 \times 410} = 0.333 \times 10^{-3}$$

$$\text{Flange:} \quad \frac{A_{sv}}{s_v} = \frac{7.3 \times 10^{3-6}}{0.8 \times 0.13 \times 0.93 \times 0.87 \times 410} = 0.212 \times 10^{-3}$$

The area for the rib should be added to the area required for normal shear. For the flange, consider 6 mm links,

$$A_{sv} = 2 \times \frac{\pi}{4} 6^2 \times 10^{-6} = 56.5 \times 10^{-6}$$

Therefore

$$s_v = 56.5 \times 10^{-6+3}/0.212 = 267 \text{ mm}$$

Maximum spacing (≯0.2m, x_1 or $\frac{1}{2}y_1$) = 130 mm

Therefore either use 6 mm links at 130 mm spacing in flange or redesign as a rectangular beam, i.e. if rib takes total shear then

$$\frac{A_{sv}}{s_v} = \frac{18.9 \times 10^{3-6}}{0.8 \times 0.23 \times 0.53 \times 0.87 \times 410} = 0.543 \times 10^{-3} \qquad \text{(Answer)}$$

Closed box sections

Closed hollow sections of circular or square shape are the most efficient for resisting torsion and should be used if the torsional moments are very large. Hsu and Kemp [12.9] have confirmed that box beams have similar strengths in torsion to solid beams, provided that the wall thickness is not less than one quarter the overall width of the box beam in the direction of measurement. Wall thicknesses down to one-tenth the overall width are suggested [12.9] if the reduction factor is taken as four times the ratio of wall thickness to member thickness. These conservative estimates for thin walls are generally desirable since thin walls could fail in compression in a brittle manner.

Other sections

The shear stresses due to torsion in multi-cellular sections may be calculated by using shear flow analysis. Each box should be reinforced to resist the torsion arising from its shear flow. Torsional stress for complex sections should be calculated assuming plastic distribution of shear stresses. Where torsional reinforcement in the form of non-rectangular reinforcing cages is required, the formulae for rectangular beams may be used except that x_1y_1 shall be replaced by the cage area, and $x_1 + y_1$ by half the cage perimeter.

References

12.1 Timoshenko, S., Strength of Materials, vol. 1, Macmillan, 1938.

12.2 Timoshenko, S., and Goodier, J.N., Theory of Elasticity McGraw-Hill, 1961.

12.3 Nadai, A., Theory of flow and fracture of solids, McGraw-Hill, 1950.

12.4 Cowan, H.J., Elastic theory for torsional strength of rectangular reinforced beams, Magazine of Concrete Research, vol. 2, 1950 pp 3-8.

12.5 Cowan, H.J., and Armstrong S., Experiments on the strength of reinforced and prestressed concrete beams and of concrete-encased steel joints in combined bending and torsion, Magazine of Concrete Research, vol. 7, 1955, pp 3-20.

12.6 Goode, C.D., Torsion—CP 110 and ACI 318-71 compared, Concrete, vol.8, March 1974, pp 36-40.

12.7 Marshall, W.T., Torsion in concrete and CP 110:1972, The Structural Engineer, vol. 52, March 1974, pp 83-8.

12.8 Lampert, P., and Collins, M.P., Torsion, bending and confusion—an attempt to establish the facts, Proc. American Concrete Institute, vol.69, August, 1972, pp 500-4.

12.9 Hsu, T.C., and Kemp, E.L., Background and practical application of tentative design criteria for torsion. Proc. American Concrete Institute, vol.66, January, 1969, pp 12-23.

12.10 Crawley, D.B., Commentary on the code for the use of concrete in buildings ca2:1963, Constructional Review Technical Supplement Australia, October, 1968.

13 Reinforced concrete slabs

A structural slab is essentially a thin, plane-sided member which carries transverse loads. Reinforced concrete slabs are eminently suitable for structural floors, bridge decks, retaining walls, etc., since they can offer a combination of strength, ductility and resistance to fire, corrosion, abrasion, or sound transmission, which is superior to that of other materials of similar cost. For solid slabs the strength in shear of the concrete is usually more than adequate, even for concentrated loads. In addition, since the minimum depth of the slab is usually governed by the limit states of deflection, the flexural compressive strength of the concrete is also more than adequate so that the section is appreciably under-reinforced. It can therefore be assumed that the rotational capacity of slabs is very high.

Many proprietary floors reduce self-weight by forming voids in the lower surface of the slab with hollow clay pots, lightweight permanent shuttering, removable shuttering with suitable inserts, waffle construction and similar techniques which are suitable for larger spans. Such floors can be designed by a combination of the methods used for T-beams with those for the uniform solid slabs considered in this chapter. Practical rules for the design of ribbed and hollow block slabs are given in Appendix CB (§3.7).

13.1 EQUILIBRIUM OF THIN PLATES IN BENDING

A two-way spanning slab is a redundant structure. The distribution of moments throughout the slab can therefore be determined most conveniently from either an elastic or a plastic analysis. In both cases, however, an exact solution must satisfy the equilibrium equations. Comprehensive discussions of plate theory are given elsewhere, e.g. references [13.1] and [13.2], so a brief summary is adequate for the present introduction to the subject. Plate theory for slabs considers thin uniform slabs which are subjected to loads and support reactions which are normal to the plane of the undeformed slab.

Notation and sign convention

Axes x and y are taken in the plane of the slab and the load acts at right angles along the positive direction of the z-axis. That is, the z-axis is vertically downwards for a horizontal slab. The notation which will be used is only applicable to plates and slabs and is shown in Fig. 13.1.1. It is widely used although it is a little confusing at first for those who are familiar with the general mathematician's notation (in which the vectorial direction of the moment also controls the notation, i.e. M_x would be a moment which acts clockwise about the positive x-axis direction and not as given below).

M_x = normal bending moment acting upon x-plane
M_y = normal bending moment acting upon y-plane
M_{xy} = torsional moment acting upon x-plane
M_{yx} = torsional moment acting upon y-plane

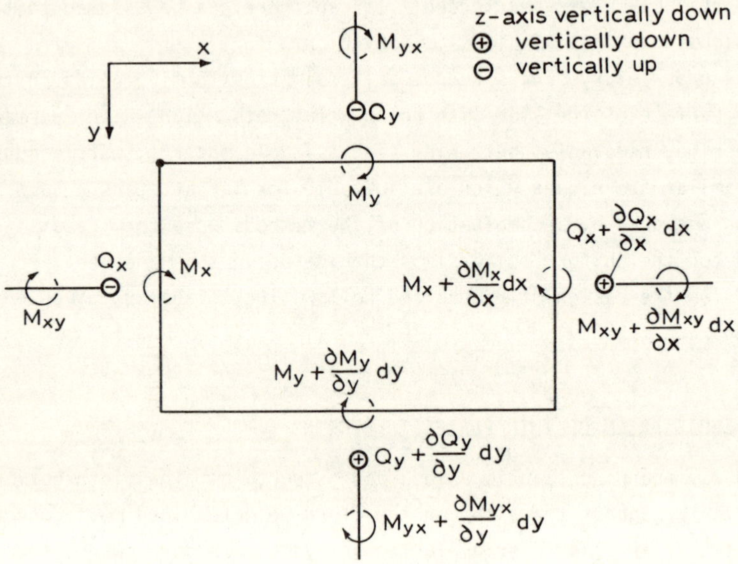

Fig.13.1.1 Element of slab (plan view) showing positive shear forces, bending moments and twisting moments

The first suffix gives the plane on which the moment acts: the 'x-plane' being a plane which is perpendicular to the x-axis. The second suffix (if different) indicates that it is a torsional moment.

Positive moments M_x, M_y, cause sagging curvature along the x- and y-directions respectively.

Positive shear forces Q_x, Q_y, act in the positive z-direction on planes whose outward normals are in the positive x- and y-directions respectively.

Positive torsional moments M_{xy}, M_{yz} act in a clockwise direction about the outward pointing normals.

Note: The x-plane is parallel to both the y- and z-axes and could therefore also be referred to as a plane parallel to the yz-plane.

Equilibrium equations

Consider the equilibrium of a small element, δx by δy, subjected to a uniformly distributed load per unit area, w. Let Q and M be the shear force per unit width of plate and moment per unit width of plate respectively. For example, Q_x is the shear force per unit length of plate along the y-direction and M_x the bending moment per unit length along the y-direction.

Resolving vertically (see Fig. 13.1.1),

$$\frac{\partial Q_x}{\partial x} dx\delta y + \frac{\partial Q_y}{\partial y} dy\delta x + w\delta x\delta y = 0$$

Taking moments about y-axis (and ignoring second-order terms):

$$\frac{\partial M_x}{\partial x} dx\delta y + \frac{\partial M_{yx}}{\partial y} dy\delta x - Q_x\delta y.\delta x = 0$$

Taking moments about x-axis (and ignoring second-order terms):

$$-\frac{\partial M_y}{\partial y} dy\delta x + \frac{\partial M_{xy}}{\partial x} dx\delta y + Q_y\delta x.\delta y = 0$$

Hence

$$\frac{\partial Q_x}{\partial x} + \frac{\partial Q_y}{\partial y} = -w \tag{13.1.1}$$

$$\frac{\partial M_x}{\partial x} + \frac{\partial M_{yx}}{\partial y} = Q_x \tag{13.1.2}$$

$$\frac{\partial M_y}{\partial y} - \frac{\partial M_{xy}}{\partial x} = Q_y \tag{13.1.3}$$

In addition, it is shown later that

$$M_{xy} = -M_{yx} \tag{13.1.4}$$

Hence by differentiating Eqns (13.1.2) and (13.1.3), w.r.t. x and y respectively, and substituting in Eqn (13.1.1), the basic equilibrium equation for a thin plate or slab is obtained as follows:

$$\frac{\partial^2 M_x}{\partial x^2} + \frac{\partial^2 M_y}{\partial y^2} - 2\frac{\partial^2 M_{xy}}{\partial x \partial y} = -w \tag{13.1.5}$$

Equilibrium of triangular element

Let a triangular element ABC be cut out of a thin plate, of thickness h, which is subjected to pure bending moments M_x and M_y in two perpendicular directions. Consider a thin triangular lamina abc which is in the z-plane as shown in Fig. 13.1.2. The stresses acting on the face ac can be resolved into a component σ_n along the normal direction and a shearing component τ_{nt} along the tangential direction. Summing the normal stresses, σ_n, for all laminas such as abc over the thickness of the plate gives the bending moment acting, per unit length, on the section AC of the plate; that is,

$$M_n = \int_{-\frac{1}{2}h}^{\frac{1}{2}h} \sigma_n z\,dz \tag{13.1.6}$$

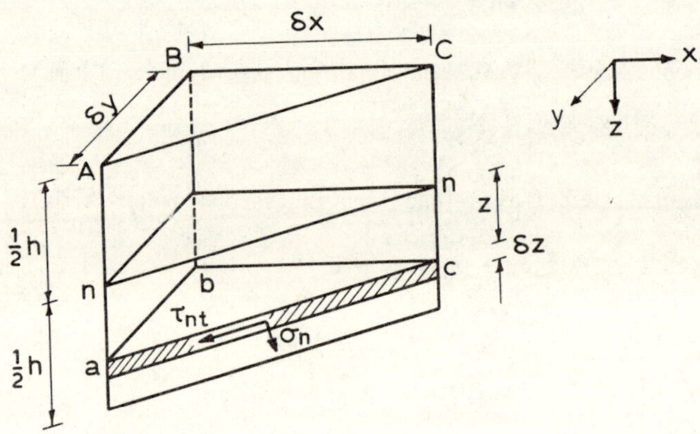

Fig.13.1.2 Triangular element in plate subjected to moments in two perpendicular directions

The shearing stresses τ_{nt} give the twisting moment acting, per unit length, on the section AC of the plate, that is

$$M_{nt} = -\int_{-\frac{1}{2}h}^{\frac{1}{2}h} \tau_{nt} z dz \qquad (13.1.7(a))$$

Thus it is evident that the twisting moment per unit length along a section is due to the horizontal shearing stresses continuously distributed along the section.

For the n-plane shown in Fig. 13.1.2, τ_{nt} acts in a direction at 90° clockwise from the n-axis. For the t-plane, however, τ_{tn} acts in a direction at 90° anticlockwise from the t-axis. Hence

$$M_{tn} = +\int_{-\frac{1}{2}h}^{+\frac{1}{2}h} \tau_{nt} z dz \qquad (13.1.7(b))$$

Hence since the complementary horizontal shear stresses are equal, i.e. $\tau_{tn} = \tau_{nt}$, it follows that $M_{tn} = -M_{nt}$. Similarly for the x- and y-axes, $M_{xy} = -M_{yx}$.

Note: Reference [13.1] has a different sign convention for shear stresses, such that $\tau_{nt} = -\tau_{tn}$. However, Eqn (13.1.5) is still obtained in the same form.

Moment vectors

Bending moments are vectorial quantities and can therefore be resolved in the same way as other vectors. The direction of the vector is given by the clockwise rotation of the moment about its axis, as shown in Fig. 13.1.3. It is essential to remember this, even though it is

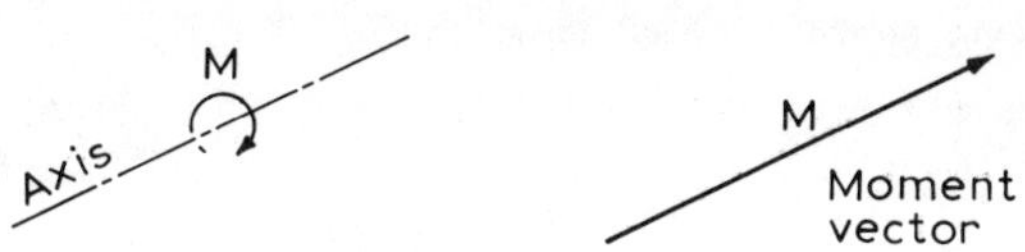

Fig.13.1.3 Vectorial representation of a moment

convenient, as shown above, to adopt a notation which relates moments to the planes on which they act, rather than to the direction of the moment axis. When drawing diagrams for which moment vectors are

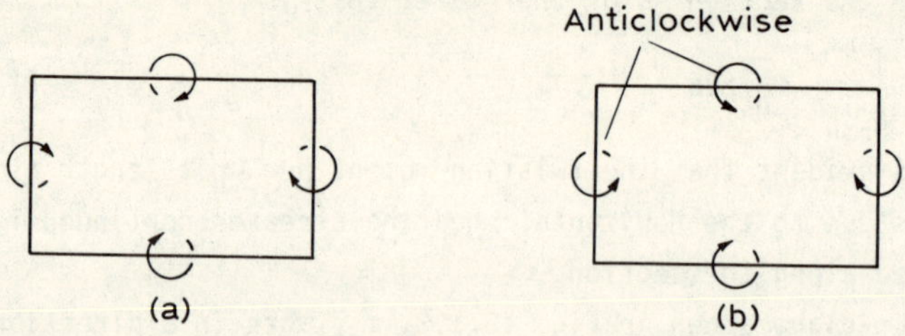

Fig.13.1.4 Methods of showing moments prior to vectorial representation
(a) Preferred. (b) Not recommended

considered it is desirable to show arrows at the clockwise extremity in all cases as in Fig. 13.1.4(a). Although sagging moments can be clearly indicated with arrows at either end, as in (b), errors in vectorial representation are more likely.

Resolution of moment vectors

The moments acting on a plane whose external normal makes some angle ϕ with the x-axis can be determined by considering the equilibrium of small triangular elements as shown in Fig. 13.1.5. Notice that a similar notation can be used for rectangular axes n, t, making angles of ϕ with the x-, y-axes respectively, as was done above. Clockwise rotation about the outward normals gives the positive directions of the torsion moments in all cases.

Resolving along t-axis for element (i):

$$M_n c = M_x b \cos\phi + M_y a \sin\phi - M_{xy} b \sin\phi + M_{yx} a \cos\phi$$

$$b/c = \cos\phi \quad a/c = \sin\phi \quad \text{and } M_{yx} = -M_{xy}$$

Hence

$$M_n = M_x \cos^2\phi + M_y \sin^2\phi - M_{xy} \sin 2\phi \tag{13.1.8}$$

Resolving along n-axis for element (ii), (or putting $\phi = 90 + \phi$ and writing M_t for M_n in above) gives

$$M_t = M_x \sin^2\phi + M_y \cos^2\phi + M_{xy} \sin 2\phi \tag{13.1.9}$$

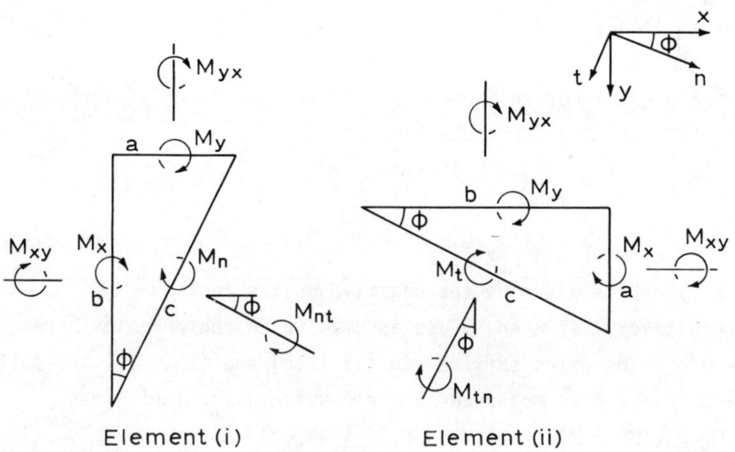

Fig.13.1.5 Triangular elements with faces having external normals along n and t axes

Resolving along n-axis for element (i):

$$M_{nt}c = M_x b \sin\phi - M_y a \cos\phi + M_{xy} b \cos\phi + M_{yx} a \sin\phi$$

Hence

$$M_{nt} = \tfrac{1}{2}(M_x - M_y)\sin 2\phi + M_{xy}\cos 2\phi \qquad (13.1.10)$$

Resolving along t-axis for element (ii), (or putting $\phi = 90 + \phi$ in above) gives

$$M_{tn} = -\tfrac{1}{2}(M_x - M_y)\sin 2\phi - M_{xy}\cos 2\phi \qquad (13.1.11)$$

Principal moments

The maximum and minimum bending moments act on planes at right angles to each other and can be found by differentiating Eqn (13.1.8) or (13.1.9)

$$dM/d\phi = - M_x \sin 2\phi + M_y \sin 2\phi - M_{xy}\, 2\cos 2\phi$$
$$= 0$$

or

$$\tan 2\phi = - 2M_{xy}/(M_x - M_y) \qquad (13.1.12)$$

Substituting for ϕ in Eqns(13.1.10) and (13.1.11) shows that $M_{nt} = 0$ and $M_{tn} = 0$, which confirms that no torsional moments act on the principal planes.

Substituting

$$\cos^2 \phi = \tfrac{1}{2}(1 + \cos 2\phi), \text{ etc.,}$$

in Eqn (13.1.8) and then substituting

$$\sin 2\phi = -2M_{xy}/[(M_x - M_y)^2 + 4M_{xy}^2]^{\frac{1}{2}}, \text{ etc.,}$$

gives

$$M_n = \tfrac{1}{2}(M_x + M_y) + \tfrac{1}{2}[(M_x - M_y)^2 + 4M_{xy}^2]^{\frac{1}{2}} \qquad (13.1.13)$$

Similarly, M_t is given by the negative square root.

Alternatively, if x and y are assumed to be the principal axes, then $M_{xy} = 0$. Transposing suffixes in (13.1.13) and (13.1.12) and putting $\phi = -\phi$ (since ϕ is measured in a clockwise direction) gives

$$M_x, M_y = \tfrac{1}{2}(M_n + M_t) \pm \tfrac{1}{2}[(M_n - M_t)^2 + 4M_{nt}^2]^{\frac{1}{2}} \qquad (13.1.14)$$

and

$$\tan 2\phi = 2M_{nt}/(M_n - M_t) \qquad (13.1.15)$$

Mohr's circle

It will be evident that Eqns(13.1.6) to (13.1.13) are similar to those for plane stress given in standard textbooks on structures and strength of materials. Mohr's circle construction is therefore equally applicabl as shown in Fig. 13.1.6.

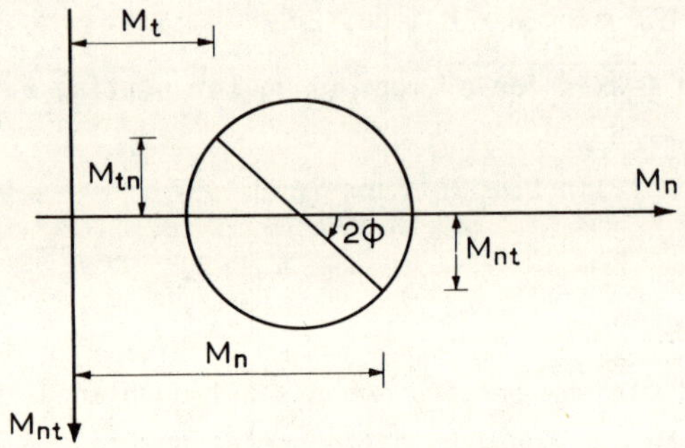

Fig.13.1.6 Mohr's circle construction for moments

Terminal forces

If a torsional moment M_{nt} per unit length acts on the diagonal of a slab ABCD as shown in Fig. 13.1.7(a) it can be effectively replaced by equal and opposite vertical shear forces of M_{nt} acting unit distance apart as shown in (b). This replacement of the distributed horizontal force by a statically equivalent system of two vertical forces cannot cause any appreciable disturbance in the plate, from St Venant's principle, except within a distance comparable with the plate thickness (which is assumed small). Similarly a torsional moment of $M_{nt} + \partial M_{nt}/\partial t$ per unit length can also be replaced as shown. The torsional moment

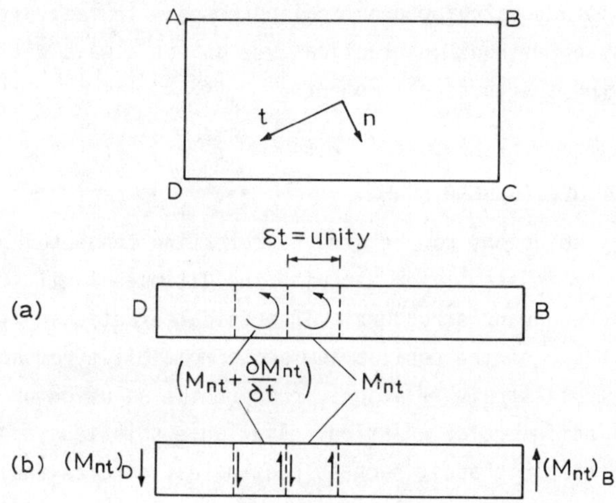

Fig.13.1.7 'Terminal' forces for one edge (or section) of a plate

along the diagonal section BD is therefore statically equivalent to a distributed vertical shear force $- \partial M_{nt}/\partial t$ per unit length together with concentrated 'terminal' forces M_{nt} at the ends B and D. Notice that if the sides AD and DC of the slab were subjected to torsional moments of M_{xy} and M_{yx} then since $M_{xy} = - M_{yx}$ there are unbalanced corner forces of magnitude $2M_{nt}$ at each corner. The corners of a loaded slab therefore tend to lift if the supports cannot provide adequate downwards restraint. If the corners are restrained then they should be provided with suitable torsion reinforcement. This is theoretically required along or perpendicular to the diagonal, according to whether it is placed in the bottom or the top of the slab respectively.

Boundary conditions at a free edge

Let the external reaction at a supported edge be V_n per unit length and the internal shear in the slab be Q_n per unit length. Hence from Fig. 13.1.7.

$$V_n - Q_n = -\partial M_{nt}/\partial t \tag{13.1.16}$$

However, $V_n = 0$ for a free edge, hence $Q_n = \partial M_{nt}/\partial t$. This implies that neither Q_n, $\partial M_{nt}/\partial t$ nor M_{nt} need be zero. This is because simple plate theory does not distinguish between stresses due to rate of change of torsional moment and those due to vertical shear. Although the edge conditions $M_{nt} = Q_n = 0$ cannot be specified completely in simple plate theory, this is adequate for practical purposes. In fact, it is shown later that design methods in practical use do not always satisfy other requirements for the torsional moments.

13.2 DESIGN OF RECTANGULAR SLABS

Elastic theory solutions can be used to determine the bending moments and shear forces in slabs, for use with the ultimate limit states, as for any other redundant structure. Theoretical elastic analysis for thin plates, based on the equilibrium and compatibility equations and the elastic stress-strain relations, require the solution of a fourth-order partial differential equation. Available solutions are therefore restricted to simple shapes, such as rectangles, circles and triangles. Such solutions, even when available, take no account of variations in the flexural rigidity of the slab due to cracking of the concrete or curtailment of the steel, or the effects of thermal movement, shrinkage and creep. In practice, therefore, a theoretical analysis could be adjusted according to experimental data, as was the case with Westergaard method. However, CP 110 has now replaced the coefficients for this method for restrained rectangular slabs with values based on yield line analysis.

Serviceability requirements for slabs

The limit state of cracking is deemed to be satisfied if the bar spacing rules given in Appendix CE and Table CE5 are satisfied. These rules can therefore be considered after the analysis and design for collapse. To estimate the slab thickness, however, the designer must consider the cover and the limit state of deflection. This will normally be satisfied from the span/effective depth ratios as used for beams, using the shorter span only and the reinforcement in that direction at the centre of the span. Notice, however, that as the steel ratio is usually low for slabs the ξ-multiplier for tension reinforcement will be higher, and hence a higher span/effective depth ratio is possible. Flat slabs require an additional multiplier: see Appendix CE (§3.3.8). The reader should be aware of the arbitrary rules in Appendix CE so that reference can be made as and when necessary. Notice, for example, that for steel ratios below 0.5% and no moment redistribution Table CE5 permits spacings (if $\ngtr$ 5d) of up to 600 mm with mild steel, whereas BS 5337 permits generally a maximum of 300 mm for either mild steel or high yield steel.

Simplified rules for curtailing reinforcement in solid, flat and ribbed slabs are given in Appendix CD. The minimum steel ratios are also given in Appendix CD (§3.11.4,5)

One-way slabs

Elastic theory for plates indicates that the effective flexural rigidity of a strip of unit width is $EI/(1 - \nu^2)$ and not simply EI as for a beam. This increase in stiffness is due to the transverse strain, which is associated with the longitudinal stress, being restrained by the adjacent slab. Poisson's ratio for concrete is only about 0.20 so that ν^2 is small and this 'Poisson's ratio effect' may be neglected. The effect may also be ignored for the steel reinforcement since this is in the form of discrete bars. One-way spanning reinforced concrete slabs can therefore be considered as beams, as shown in earlier chapters, provided that suitable provision is made for distribution reinforcement. Such reinforcement is required to distribute the effects of point loads over an adequate breadth and to control the cracking due to thermal and shrinkage effects.

Theoretical elastic solutions

Theoretical elastic analyses for thin plates, based on the equilibrium and compatibility equations and the elastic stress-strain relations, require the solution of a fourth-order partial differential equation. Available solutions are therefore restricted to simple shapes, such as rectangles, circles, triangles, etc., and are usually obtained by

Fourier series or finite difference techniques. Such a solution, even when available, takes no account of variations in the flexural rigidity of the slab due to cracking of the concrete or curtailment of the steel, or of the effects of thermal movement, shrinkage and creep. CP 110 therefore adopts alternative approaches for rectangular slabs, one based on a simplified elastic method and one on a yield-line theory.

Rectangular slabs simply supported on four sides

The Grashof-Rankine method (see e.g., reference [5.3]) assumes that a uniformly distributed load on a rectangular slab is shared between strips running in the two directions parallel to the sides of the slab. If all torsional moments in Eqn (13.1.5) are neglected, as well as any interaction between the x-strips and the y-strips, the strips can be designed as beams for which

$$d^2M_{sx}/dx^2 = -n_x, d^2M_{sy}/dy^2 = -n_y \text{ and } n_x + n_y = n$$

where n is the total design load per unit area for collapse, (i.e. $1.4g_k + 1.6q_k$).

The ratio of n_x to n_y is determined by a compatibility requirement that the midspan deflections of the central strips in the two directions should be equal:

$$\text{e.g.} \quad Kl_x^2 M_{sx}/EI = Kl_y^2 M_{sy}/EI \text{ or } M_{sx}/M_{sy} = l_y^2/l_x^2$$

Notice that this method neglects the effects of torsional moments and therefore gives relatively high bending moments for the strips. CP 110 states that for simply supported slabs, where adequate provision to resist torsion at the corners and to prevent the corners from lifting is not made, the bending moments per unit width at midspan should be taken as follows:

$$M_{sx} = \alpha_{sx} n l_x^2 \text{ and } M_{sy} = \alpha_{sy} n l_x^2$$

where l_x = length of shorter side

l_y = length of longer side (see Table 13.2.1)

α_{sx}, α_{sy} = coefficients as shown in Table 13.2.1.

Table 13.2.1 Bending moment coefficients for slabs spanning in two directions at right angles, simply supported on four sides

l_y/l_x	1.0	1.1	1.2	1.3	1.4	1.5	1.75	2.0	2.5	3.0
$10^3\alpha_{sx}$	62	74	84	93	99	104	113	118	122	124
$10^3\alpha_{sy}$	62	61	59	55	51	46	37	29	20	14

Notice that M_{sy} is given in terms of l_x. Notice also that $\alpha_{sx} = 0.124$ for $l_y/l_x = 3.0$ is virtually the same as for a one-way spanning slab (for which $\alpha_{sx} = 1/8$). At least 50% of the tension reinforcement provided at midspan should extend to the supports and the remaining 50% should extend to within $0.1l_x$ or $0.1l_y$ of the support, as appropriate.

Rectangular slabs restrained on four sides

CP 110 gives the coefficients reproduced in Table 13.3.2. for rectangular slabs with any combination of continuous or simply supported edges provided that all four corners are prevented from lifting and suitable provision is made for torsion.

The slabs are considered as being divided in each direction into middle strips and edge strips, as shown in Fig. 13.2.1. The maximum moments are calculated as follows and apply only to the middle strips.

$M_{sx} = \beta_{sx} n l_x^2$ and $M_{sy} = \beta_{sy} n l_x^2$

where M_{sx}, M_{sy} are the maximum moments at midspan on strips of unit width spanning l_x and l_y respectively,

n is the total ultimate load per unit area $(1.4g_k + 1.6q_k)$

β_{sx}, β_{sy} are coefficients (of l_x^2) given in Table 13.2.2

Where l_x/l_y exceeds 2, design slabs as spanning one way only.

Table 13.2.2 Bending moment coefficients for rectangular panels supported on four sides with provision for torsion at corners

Type of panel and position of negative or positive moment	$10^3 \times$ short-span coefficient β_x							
	Values of l_y/l_x							
	1.0	1.1	1.2	1.3	1.4	1.5	1.75	2.0+
Case 1. Interior panels								
Negative moment at continuous edge	32*	37	43	47	51	53	60	65
Positive moment at midspan	24*	28	32	36	39	41	45	49
Case 2. One short edge discontinuous								
Negative moment at continuous edge	37*	43	48	51	55	57	64	68
Positive moment at midspan	28*	32	36	39	41	44	48	52
Case 3. One long edge discontinuous								
Negative moment at continuous edge	37*	44	52	57	63	67	77	85
Positive moment at midspan	28*	33	39	44	47	51	59	65
Case 4. Two adjacent edges discontinuous								
Negative moment at continuous edge	47*	53	60	65	71	75	84	91
Positive moment at midspan	35*	40	45	49	53	56	63	69
Case 5. Two short edges discontinuous								
Negative moment at continuous edge	45	49	52	56	59	60	65	69
Positive moment at midspan	35*	37	40	43	44	45	49	52
Case 6. Two long edges discontinuous								
Negative moment at continuous edge	45*	-	-	-	-	-	-	-
Positive moment at midspan	35*	43	51	57	63	68	80	88
Case 7. Three edges discontinuous (one long edge continuous)								
Negative moment at continuous edge	57	64	71	76	80	84	91	97
Positive moment at midspan	43*	48	53	57	60	64	69	73
Case 8. Three edges discontinuous (one short edge continuous)								
Negative moment at continuous edge	57*	—	—	—	—	—	—	—
Positive moment at midspan	43*	51	59	65	71	76	87	96
Case 9. Four edges discontinuous								
Positive moment at midspan	56*	64	72	79	85	89	100	107

* Long-span coefficient, β_y, for all values of l_y/l_x is equal to the value of β_x for $l_y/l_x = 1.0$.

No redistribution should be made and tension reinforcement should be provided as indicated below for restrained slabs.

These coefficients are based on values derived by Taylor et al. [13.3] using yield-line analysis for a uniform distribution of steel. They have been mutiplied by 4/3 to allow for distribution over the middle strip only as defined in Fig. 13.2.1 [5.4].

Westergaard's method (see e.g. reference [5.3]) was based on theoretical elastic analyses supplemented by experimental data. The CP 110 coefficients are generally lower except for cases 8 and 9 and at high l_y/l_x ratios for cases 3, 4 and 6.

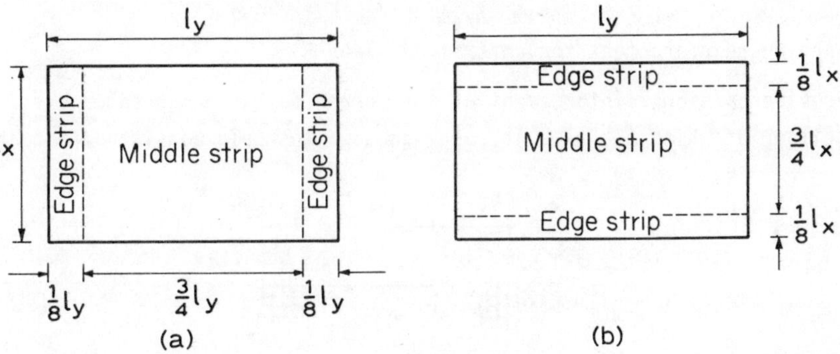

Fig.13.2.1 Division of slab into 'middle' and 'edge' strips. (a) For span l_x. (b) for span l_y

<u>Tension reinforcement for restrained slabs</u>

Provide tension reinforcement at midspan in a middle strip and extend ≮ 50% in the lower part of the slab to within 0.15l of a continuous edge, and within 50 mm of a discontinuous edge. Remainder should extend to within 0.25l of a continuous edge, or 0.15l of a discontinuous edge.

Provide tension reinforcement over the continuous edges of a middle strip and extend ≮ 50% in the upper part of the slab a distance of 0.3l. Remainder should extend 0.15l into the span.

Provide tension reinforcement at a discontinuous edge according to degree of fixity at the nominally free edge. 50% of that provided at midspan extending 0.1l into the span should generally be adequate.

Reinforcement parallel to the adjacent edges of the slab in the edge strips need only satisfy the minimum for shrinkage etc (see Appendix CD, §3.11.4) and for torsion (see below).

If the ends of the slabs are fully restrained then the minimum should be taken as ρ_{crit} (see chapter 11). This requirement occurs, for example, in cantilever slabs and especially canopies where temperature differences between the outside and the inside of the building are also present. Very unsightly blemishes can occur if crack control is ignored as shown in the frontispiece (Plate II).

<u>Torsion reinforcement for restrained slabs</u>

Provide torsion reinforcement at any corner contained by edges over neither of which the slab is continuous. It should consist of top and

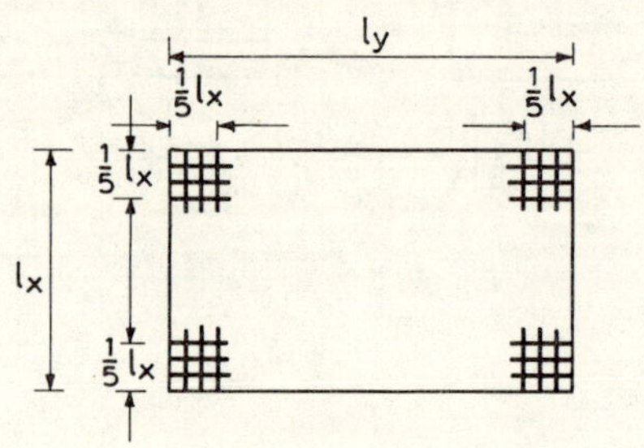

Fig.13.2.2 Torsion reinforcement for 'restrained' simply supported slab

bottom reinforcement, each level containing bars placed parallel to the sides of the slab and extending in these directions for a distance of one-fifth of the shorter span, as shown in Fig. 13.2.2. The area of the bars in each of the two layers at each level, per unit width of slab, should be 3/4 of the area required for the maximum midspan moment in the slab. Torsion steel equal to half the preceding amount should be provided at corners contained by edges over only one of which the slab is discontinuous. No torsion steel need be provided at corners contained by edges over both of which the slab is continuous.

Concentrated loads on slabs

The bending moments due to concentrated loads may be calculated from any methods based on elastic theory, such as Pigeaud or Westergaard (see, e.g. reference [5.3]). Alternatively, they may be calculated by the use of Johansen's yield-line method or Hillerborg's strip method, which are discussed later.

In the case of solid floors, or other slabs with approximately equal transverse and longitudinal flexural rigidities, the effective width may be taken as the sum of the load width and $2.4x(1 - x/l)$ where x is the distance from the nearer support to the section under consideration and l is the span. Where the concentrated load is nearer an unsupported edge of a slab (see Fig. 13.2.3) the effective width should not exceed the above, nor half that value plus the distance of the centre of the load from the unsupported edge [5.3].

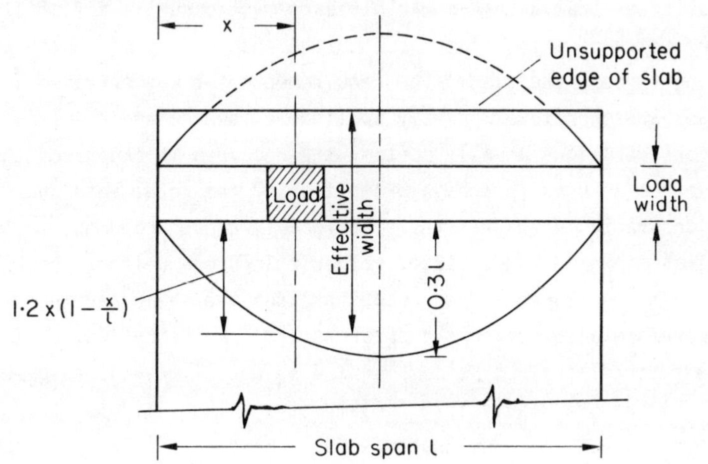

Fig.13.2.3 Effective width of solid slab carrying a concentrated load near an unsupported edge

In the case of precast floors constructed of narrow precast units such that loads can be transferred between the units the same methods may be employed. The width of floor assumed by the code to be effective in withstanding the loads must not exceed the width of the loaded area by more than:

(i) width of three precast units and joints,

(ii) width of four precast units and joints where there is a topping at least 30 mm thick.

Flat slabs

Normal beam and slab construction requires supporting the slab on all four sides on either beams or walls. Beams can seriously reduce the headroom and so impose restrictions on the use of the space beneath. This disadvantage can be avoided by supporting the slab directly on the columns. The term 'flat slab' means a reinforced concrete slab, with or without 'drops', supported by columns with or without flared column heads. As for a normal slab, a flat slab may be solid or may have recesses formed in the soffit to provide a series of ribs in two directions. The recesses may be formed by removable or permanent filler blocks.

It is clear that severe shear stresses must occur in a flat slab at the perimeter of each column, due to the possibility of the column punching through the slab. The requirements for concentrated loads on slabs are therefore directly applicable (see Appendix CC, §3.4.5.2). Such considerations usually control the required thickness of the slab. Self-weight is usually a high proportion of the total load and it is often desirable to thicken the slab locally around columns, to form a 'drop' as shown in Fig. 13.2.4, rather than use a uniform thickness throughout. A flared column head increases the shear perimeter and can therefore also reduce the shear stress. A theoretical elastic analysis for flat slabs is complex and so design methods as given in CP 110 divide the panels into 'column strips' and 'middle strips' as shown in Fig. 13.2.5 (see Appendix CB). However, yield-line theory, which is discussed in the next section, is equally suitable for either flat slabs or beam-and-column construction. Practical design of flat slabs using the continuous-frame method of analysis is outlined in Appendix CB (§3.6). CP 110 itself also contains an alternative empirical design method. All shear stresses should be reduced by 20% and calculated shear forces may need to be magnified—see Appendix CC (§3.6.2). There is an additional ζ_3 multiplier for span-effective depth ratios for deflection—see Appendix CE (§3.6.3). Reinforcement spacing rules are as for solid or ribbed slabs—see Appendix CE

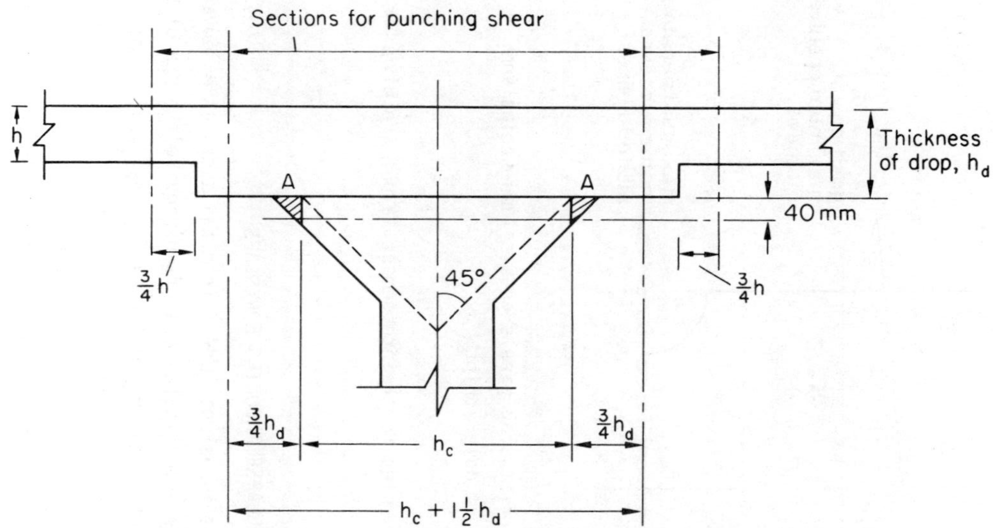

Fig. 13.2.4 Slab with drop and column with column head showing critical sections for punching shear.

The presence of any concrete in the shaded areas A does not affect the calculation. Critical sections for shear (diagonal tension) are $\frac{3}{4}h$ further out than the outer sections for punching shear.

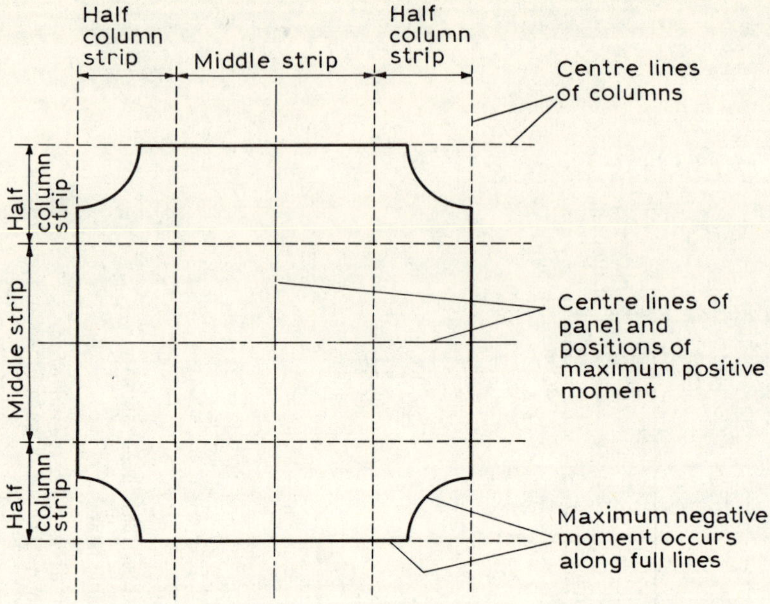

Fig.13.2.5 Division of flat slab panels into column strips and middle strips

(§3.6.4)—and simplified rules for reinforcement layout are given in Appendix CD (§3.6.5.3).

13.3 INTRODUCTION TO COLLAPSE THEORIES FOR SLABS

Elastic theory solutions for slabs suffer from two distinct disadvantage Theoretical solutions are complicated and often require the use of computers, electrical analogues or similar techniques, while semi-empiri methods are strictly limited in their applicability. Even for simple rectangular shapes CP 110 now gives solutions based on yield-line theory and recommends that bending moments for more complicated shapes may be determined by using Johansen's yield-line method or Hillerborg's strip method. These two methods are applicable to slabs of any shape, loading and edge conditions so that their use is preferred. CP 110 requires, however, that the ratio between support and span moments are similar to those obtained by the use of elastic theory: values between 1.0 and 1.5

being recommended. The rules for detailing of the reinforcement should also still follow the principles inherent in the elastic analysis of slabs in order to ensure that the limit state of cracking is satisfactory. However, it is recognized that, as for other structural members, very extensive redistribution of bending moments can occur before the collapse load is reached. The collapse load of a slab in flexure may be determined by either an upper-bound method, a lower-bound method, or by an exact solution.

Historical background

Yield-line theory was initiated by Ingerslev [13.4] and substantially pioneered by Johansen who produced his thesis, in Danish, in 1943. Extensive design experience had been gained in Scandinavian countries by the time Hognestad published a summary of yield-line theory, in English, in 1953. It had a very controversial introduction into English-speaking countries, due very largely to the difficult theoretical justification for Johansen's so-called 'equilibrium method'. Anomalies were apparent and extensive research continued following the publications, from 1961–2, by Wood [13.2], Jones [13.5] and the English translation of Johansen's thesis [13.6]. A special publication appeared in 1965 [13.7] and a book by Jones and Wood [13.8] in 1967 resolved the earlier lack of agreement.

Limit theory for slabs

The basic assumptions for limit theory, discussed in chapter 6, are that failures due to shear, bond and sudden flexural compression are prevented. That is, it is assumed implicitly that the structure has a long horizontal portion to its moment-rotation curve, otherwise a section might fail and lose its moment-carrying capacity before all the plastic hinges had formed. In the case of a slab this means that all yield lines must be capable of reaching the boundaries of the slab before any portion loses its moment-carrying capacity due to excessive rotation. RC slabs are ideally suited to limit theory because the need to limit deflections leads to thicknesses which require relatively low steel ratios for flexure and also low shear stresses. Hence slab sections are usually very under-reinforced so that the steel ratio is often considerably less than $0.227f_{cu}/f_{yt}$ (i.e. $x < \frac{1}{2}d$). It is now convenient to reconsider upper- and lower-bound solutions with special reference to slabs.

Upper-bound solutions

Upper-bound methods postulate a collapse mechanism and ensure that:

(i) The moments at the plastic hinges (which have been postulated) are not greater than the ultimate moments of resistance of the sections.

(ii) The collapse mechanism is compatible with the boundary conditions.

The elements of the slab between the plastic hinges are not examined to ensure that the moments there do not exceed the ultimate moments of resistance of the sections. This method therefore gives a collapse load which may be too high, and is therefore an upper limit or bound to the correct collapse load. If all the possible collapse mechanisms for the slab are examined, the one giving the lowest collapse load is the most likely and is therefore the correct mechanism. The ultimate moments of resistance between the plastic hinges are exceeded only if an incorrect collapse mechanism is postulated.

Lower-bound solutions

Lower-bound methods choose a particular distribution of moments at collapse and ensure that:

(i) The ultimate moments of resistance are not exceeded anywhere in the slab.

(ii) The equilibrium equations are satisfied at all positions (which includes compliance with the boundary conditions).

Lower bound methods give a collapse load which may be too low.

Yield-line theory

Yield-line theory gives an upper-bound solution so that it is essential to ensure that the most critical collapse mechanism has been selected if the load-carrying capacity of the slab is not to be over-estimated. The correct collapse mechanisms for most cases are now well known, however, so that the experienced designer is rarely faced with the possibility that a further (more critical) mechanism may exist. Yield-line theory using the virtual work method will be considered in the most detail in the present chapter.

Yield-line theory applies to almost any shape of slab, any load and any edge condition. However, it is restricted as a design method to slabs of constant thickness which are uniformly reinforced in each of two mutually perpendicular directions. Thus the steel ratio for the longitudinal direction is constant across the width of the slab and the steel ratio for the transverse direction is constant along the length of the slab. However, the steel ratios may be different in the two principal directions and also in the top and bottom. For such slabs the collapse moment of resistance per unit width has a constant value along any straight line in the plane of the slab. Yield-line theory gives no information on the best distribution of the steel within the slab, but can be used to analyse a slab where the steel has been distributed according to some other method (e.g. Hillerborg's method).

Hillerborg's method

Lower-bound solutions are much more difficult to obtain, as has been shown by Wood [13.2]. However, a method which has been developed by Hillerborg obtains the moment distribution for the slab by assuming strip action. It has been shown by Wood and Armer [13.9] that Hillerborg's method can provide an exact solution with an unlimited number of simultaneous modes. Hillerborg, however, originally intended it as a lower-bound rather than an exact solution. It is a powerful alternative to yield-line theory since it is essentially a design technique which provides information on the best distribution of the steel within the slab.

13.4 YIELD LINES AND YIELD-LINE MECHANISMS

The progressive development of yield lines is described in order to indicate how yield-line mechanisms are formed. Satisfactory prediction of the possible yield-line pattern is essential if the correct collapse load for the slab is to be calculated. The conditions which assist in determining yield-line mechanisms are outlined and typical examples are illustrated.

Inelastic deformation

A simply supported, two-way spanning slab, subjected to a uniformly distributed load of increasing intensity, initially deforms elastically. The concrete on the tension side of the slab cracks as the load increases causing the stiffness of the cracked section to be reduced and the distribution of moments in the slab to change. This redistribution produces, for equal increments of load, greater increases in moment at an uncracked section than before cracking occurred.

Yield-line propagation

The initial stages of the inelastic deformation produces yielding of the reinforcement in the central area of the slab, where the moment is greatest. Once the steel in an under-reinforced section has yielded, then although the section continues to deform its resistance moment does not increase by any appreciable amount. Consequently an even

Fig.13.4.1 Initial propagation of yield lines

greater redistribution of moments must take place after yielding of the steel commences. As further loading is applied the steel in the section adjacent to the central area also yields, since an increasing proportion of the moment has to be carried by these sections. Lines along which the steel has yielded are therefore propagated from the point at which yielding initially occurred. The yield lines at this stage could be as shown in Fig. 13.4.1.

Yield-line pattern

Even further increase in load causes the reinforcement in more and more sections to yield and so extend the yield lines. The yield lines eventually reach the boundaries of the slab, at which stage the resistan moments along the yield lines are almost at their ultimate values. Since the yield lines cannot extend any further, the slab is virtually carrying

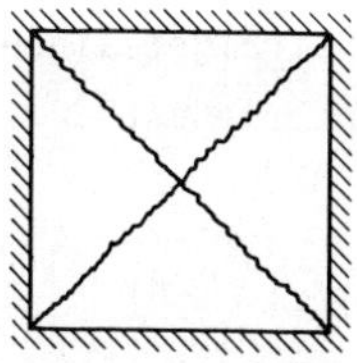

Fig.13.4.2 Final extent of yield lines

the maximum possible load. The system of yield lines (or fracture lines), such as those shown in Fig. 13.4.2, is called a yield-line pattern.

Collapse load for slab

A slight increase in load now causes a state of unstable equilibrium: the slab continues to deflect under this load until the curvature at some section along the yield lines is so great that the concrete crushes. This section then loses its ability to carry any moment. This increases the state of unstable equilibrium even more and causes crushing to occur along the whole length of the yield lines. Hence the condition when the yield lines just reach the boundary can be regarded as the collapse criterion of the slab.

Mode of deformation

The curvature of the slab along the yield lines where the steel is yielding is much larger than that for the areas in between, which are still behaving elastically. Consequently, it is reasonable to assume that the elastic deformations are negligible in comparison with the plastic deformations. This means that the slab elements between the yield lines can be assumed to remain plane and that all deformations take place in the yield lines.

Straight yield lines

Consider the slab elements in the deflected state. These elements are inclined planes and therefore intersect in straight lines. It follows that the yield lines are also straight since these are, in fact, the intersections between the plane elements.

Axes of rotation

The plane elements must rotate about certain axes if the slab is to deflect. In Fig. 13.4.3, element a rotates about AB, element b about BC, c about CD and d about DA. It is also a necessary condition of deformation that the yield line (being a line of intersection between two adjacent elements) passes through the intersection of the axes

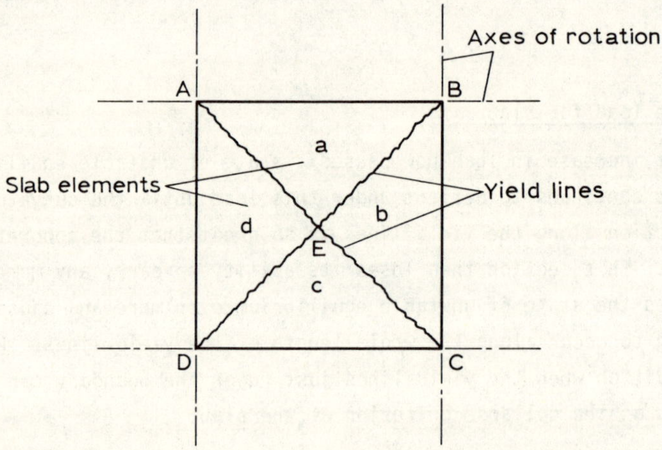

Fig.13.4.3 Fully developed yield-line pattern

of rotation of these plane slab elements: i.e. yield line EB must pass through point B, the intersection of AB and BC, EC through point C, etc.

Prediction of yield-line pattern

The foregoing can be summarized by the following five conditions which assist the prediction of the yield line pattern for any slab:

(i) Yield lines terminate at a slab boundary.

(ii) Yield lines are straight.*

(iii) Yield lines (produced if necessary) pass through the intersection of the axes of rotation of adjacent slab elements.

(iv) Axes of rotation generally lie along lines of supports and pass over columns.

(v) Any symmetry of the slab is maintained in the yield-line pattern.

Notation

The notation used is shown in Fig. 13.4.4.

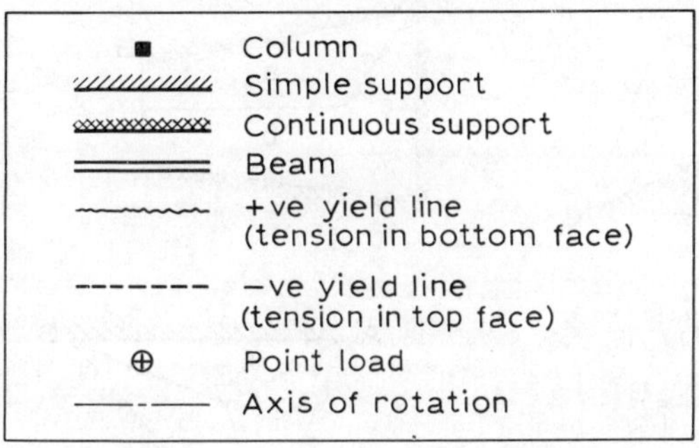

Fig.13.4.4 Standard notation for yield-line diagrams

Typical failure mechanisms

Typical yield-line patterns of various slabs subjected to uniformly distributed loads are shown in Fig. 13.4.5 while two examples of a point load are shown in Fig. 13.4.6. Notice that, in Fig. 13.4.5, the yield line EF between segments a and c, for a rectangular slab, passes through the intersection of the axes of rotation AB and DC at infinity. Similarly the other yield lines pass through the corners.

* It should be noted however that for some systems of point loads or circular supports a multiple system of yield lines can approximate to a single curved yield line, e.g. see Fig. 13.4.6(b).

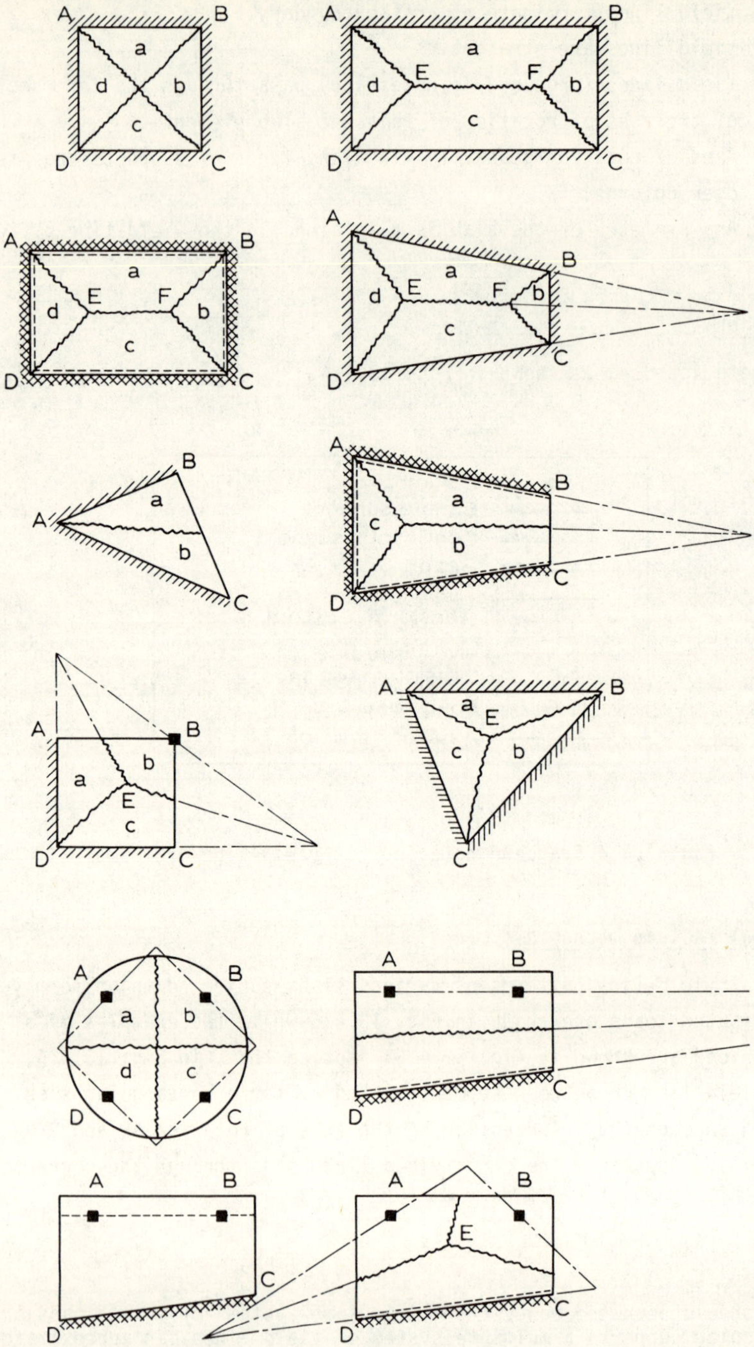

Fig.13.4.5 Various yield-line patterns for uniformly distributed loads

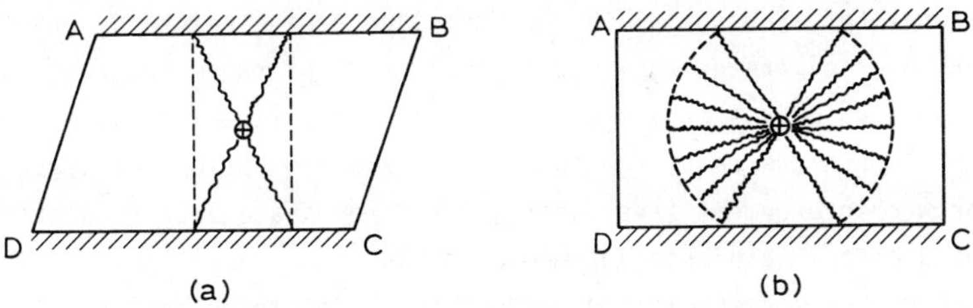

Fig.13.4.6 Typical yield-line patterns for point loads

The negative yield lines for the continuous rectangular slab must also form along the lines of support before the latter can act as axes of rotation and transform the slab into a mechanism. The more irregularly shaped slabs are seen to follow the same general principles. Notice that only the general patterns have been indicated in some cases. For example, the exact positions of E or F must be found in the process of analysis. Fortunately, however, the analysis can be commenced with only a qualitative assessment of the general yield-line pattern. In other cases it is necessary to consider more than one yield-line pattern. For example, consider the last three patterns in Fig. 13.4.5, which are all for the same slab layout.

13.5 NORMAL BENDING MOMENT CRITERION FOR YIELD-LINE THEORY

It is shown in this section that the criteria for failure on which yield-line theory is based are effectively considerations of the normal bending moments in the slab. This is consistent with an upper-bound approach. 'Classical' yield-line theory, however, ignores any partial kinking of the bars, which can increase the resistance moment by up to about 9% for an isotropically reinforced slab. The criterion normally used is therefore seen to be a 'lower-bound' for an upper-bound method'.

Co-ordinate axes

As seen previously it is often convenient to let the x- and y- axes be parallel to the directions of the reinforcement as shown in Fig. 13.5.1, even though the ultimate bending strength is in terms of moments per unit length of slab normal to the direction of the reinforcement. This convention facilitates easy location of parts of the slab, e.g. between corresponding points on actual and affine slabs (see later), although in cases where the reinforcement moment axes are not obvious, e.g. for skew reinforcement (see later), the co-ordinate axes should then be taken as the reinforcement moment axes.

Whichever axes are selected it must again be emphasized that the reinforcement moment axis gives the moment vector.

Yield principal moments

Let m_x be the yield principal moment for the x-direction steel, per unit length of the slab parallel to the y-direction. Similarly, let m_y be the yield principal moment for the y-direction steel, per unit length of slab parallel to the x-direction.

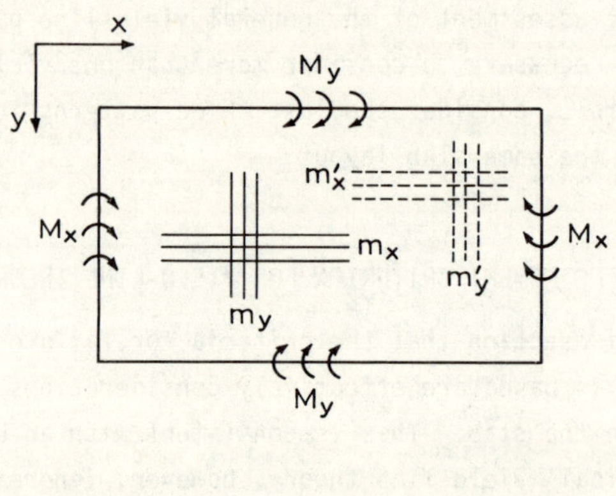

Fig.13.5.1 Bending moments and resistance moments per unit length

Fig. 13.5.1 indicates that the ultimate moment per unit length about a positive yield line parallel to the y-axis is m_x. If the steel is in the bottom face of the slab, giving positive (sagging) bending strength, the reinforcement lines are shown full and they provide resistance along a positive yield line. If the steel is in the top of the slab to give negative (hogging) strength, it is shown by broken lines and it provides resistance along a negative yield line.

Moments along a yield line

Consider a small element of slab subjected to principal moments M_x and M_y. If M_n and M_{nt} are the normal and twisting moments respectively, both per unit length, applied along a yield line whose normal makes an angle ϕ with the x-axis, then their values can be obtained by resolving vectors and putting $M_{xy} = 0$. Hence from Eqns (13.1.8) and (13.1.10)

$$M_n = M_x \cos^2\phi + M_y \sin^2\phi \qquad (13.5.1)$$

and

$$M_{nt} = \tfrac{1}{2}(M_x - M_y) \sin 2\phi \qquad (13.5.2)$$

Yield-line criterion

If it is assumed that all the reinforcement which crosses a yield line is yielding, then yield criteria can be obtained from Eqns (13.5.1) and (13.5.2) by putting

$$M_x = m_x = m \quad \text{and} \quad M_y = m_y = \mu m$$

The criterion for classical yield-line theory, however, considers only the normal bending moment as given in Eqn (13.5.1). That is,

$$m_n = m_x \cos^2\phi + m_y \sin^2\phi \qquad (13.5.3(a))$$

or

$$m_n = m \cos^2\phi + \mu m \sin^2\phi \qquad (13.5.3(b))$$

Lower-bound solutions

Lower-bound solutions would require that the twisting moment is also satisfied, that is

$$m_{nt} = m(1 - \mu)\sin\phi \cos\phi \qquad (13.5.4)$$

Wood has shown, however, that the assumption that all steel is yielding leads to yet a third condition for lower-bound solutions [13.2]. It is extremely difficult to find, in general, a statically admissible yield moment which gives a collapse load at all close to the upper-bound solutions. Thus true lower-bound solutions are very difficult to obtain. However, twisting moments at a yield line do little, if any, virtual work; hence although simple yield-line theory satisfies one basic condition only (and then in general not exactly), it has been found to give surprisingly satisfactory predictions of the collapse load.

Johansen's stepped yield criterion

Johansen's 'stepped' yield criterion is the yield criterion traditionally used for yield-line theory and can be reiterated as follows [13.8].

(a) The normal and twisting moments on a yield line can be obtained by considering each band of reinforcement in turn, the total effect being the addition of the individual effects.

(b) For each band of reinforcement taken on its own, the yield line may be considered to be divided into small steps parallel to, and at right angles to, the reinforcement.

(c) All reinforcement crossing the yield line is assumed to yield.

(d) All reinforcement is assumed to stay in its original straight line when the steel yields: i.e. there is no 'kinking', or change in horizontal direction of the steel in crossing the yield line.

(e) When each band of reinforcement is considered on its own, on the small steps at right angles to the reinforcement there is only a normal moment/unit length m whose value is given by the ultimate moment for that direction, whilst on the steps parallel to the reinforcement there is neither normal nor twisting moment.

(f) The values of normal and twisting moments on the yield line are such that they are equivalent to the components of the normal moment on the steps.

When applying the above it is also necessary to assume that only normal rotations θ_n are allowed as the strains.

Square yield criterion

For an isotropically reinforced slab it is possible to consider the simplified 'square' yield criterion due to Prager (see Fig. 13.8.6). For the orthotropic case, the yield moments change with change of axis, hence one diagram could only apply to one axis. However, for the isotropic case although the principal moments M_1 and M_2 may be in any direction, the yield moment for the slab, $\pm m$, does not alter with direction. $\mu = 1$ and Eqns (13.5.3) and (13.5.4) reduce to

$$m_n = m \qquad (13.5.5)$$

and

$$m_{nt} = 0 \qquad (13.5.6)$$

Plotting M_n against ϕ according to Eqn (13.5.1) provides the curve shown in Fig. 13.5.2. The slab will yield wherever the maximum principal moment equals the yield moment of resistance of the slab. For an isotropic slab the yield moment is constant for any direction, that is, for any value of ϕ. The lines for M_n and m touch at $\phi = \pm 90^0$

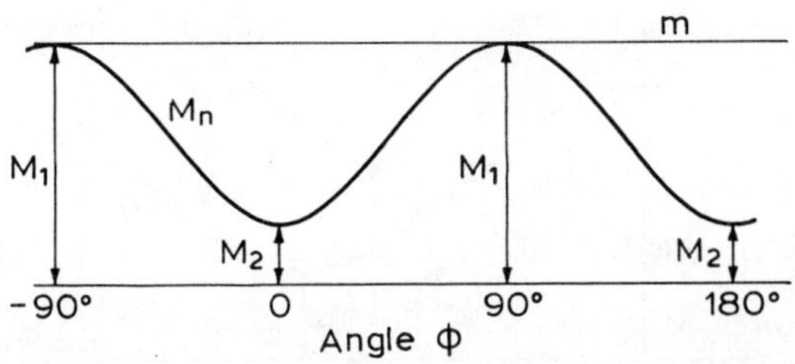

Fig.13.5.2 Graphical representation of normal moment

in Fig. 13.5.2, hence the slab yields in that direction. If there is reinforcement in the bottom face only, the yield criterion is given by:

$0 \le M_i \le m \quad (i = 1,2)$

Negative reinforcement

If isotropic reinforcement is also provided in the top face to produce a negative yield moment of resistance m' per unit length, then the yield criterion becomes

$m' \le M_i \le m \quad (i = 1,2)$

To summarize for an isotropic slab: the normal moment along a positive yield line is the maximum principal moment, i.e. ($M_1 = m$) and the twisting moment is zero. The normal moment along a negative yield line is the minimum principal moment (i.e. $M_2 = m'$) and the twisting moment is zero.

Orthotropic slabs

Kemp has shown that the yield criterion for orthotropically reinforced slabs is essentially an envelope of permissible normal bending moments [13.10]. If the plane of action of the applied principal moment, M_1, makes some clockwise angle, δ, with the x-direction reinforcement, it should be noticed that yield occurs where the two curves just touch, which corresponds to a yield line with a normal at $\phi = \delta + \psi$ measured

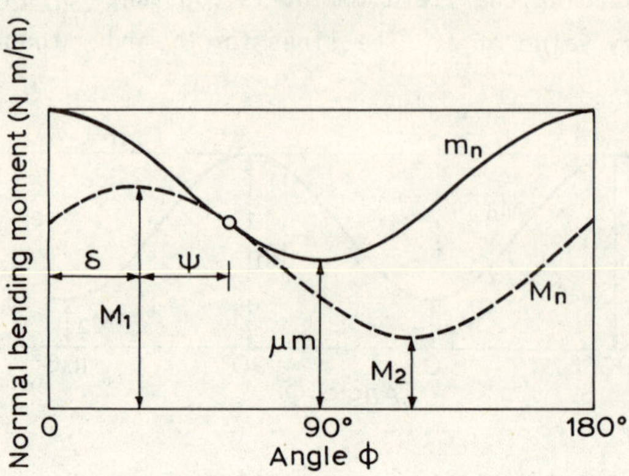

Fig.13.5.3 Positive yield in an orthotropically reinforced slab

clockwise from the x-direction, as shown in Fig. 13.5.3. In other words, the conditions satisfied by classical yield-line theory are:

$M_n = m_n$ and $dM_n/d\phi = dm_n/d\phi$

<u>Kinking of bars</u>

Eqn (13.5.3(a)) assumes that the reinforcing bars crossing the crack at a yield line remain straight as shown in Fig. 13.5.4(a). However, if

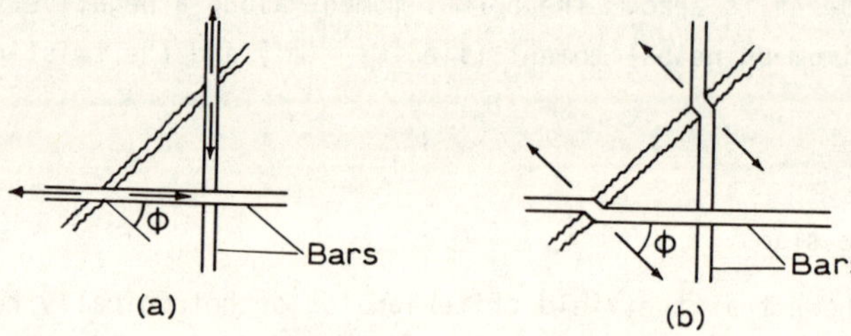

Fig.13.5.4 Bars with (a) no kinking and (b) full kinking
Note: the crack widths in this diagram are exaggerated.

the bars are kinked at the crack, as shown in Fig. 13.5.4(b), the yield moment is increased. Maximum possible kinking would occur if the bars were forced to cross the crack at right angles. If this is assumed it is evident that resolving moment vectors along the yield line, for unit length of yield line, now gives

$$m_n = m_x \cos\phi + m_y \sin\phi \qquad (13.5.7)$$

Notice that the complication due to kinking is present even for an isotropically reinforced slab. The maximum effect due to kinking occurs if the yield line is at 45° to the bars. Hence putting $m_x = m_y$ and $\phi = 45^o$ in Eqns (13.5.5) and (13.5.7) shows that:

m_n for full kinking = $(\sqrt{2}) \times m_n$ for no kinking

This is an increase of over 40%. However, full kinking cannot occur, because the high bearing stresses against the concrete tend to allow the bars to straighten and also because the crack width is in fact much smaller than the bar diameter. Although some increase in moment due to kinking is possible this has usually been ignored since it is on the safe side. Kemp and Prince have proposed a strain compatibility theory [13.11] and Mills [13.12] has indicated a maximum increase of about 9% at $\phi = 45^o$, after modifying their theory, confirming that this effect can usually be neglected in practical design.

13.6 METHODS OF ANALYSIS FOR YIELD-LINE THEORY

There are two well-known methods for yield-line theory:

(i) The virtual work method

(ii) Johansen's 'equilibrium' method.

Both methods are based on a postulated yield-line pattern. Johansen's 'equilibrium' method is therefore a misnomer in that it is an upper-bound solution and not, as its name implies, a lower-bound solution. It is necessary, for either method, to analyse alternative yield-line patterns in certain cases in order to determine the most critical collapse mechanism.

Virtual work method

When the yield-line pattern has formed completely, only an infinitesimal increase in load is necessary to cause the structure to deflect. By the principle of virtual work, the work done on the slab by the loads equals the work done against the yield lines as they rotate. Hence if a point on the slab is given a virtual deflection, δ, displacements in the form of rotations compatible with this deflection must take place along the yield lines. The internal work done on the slab is the sum of the rotations in the yield lines multiplied by the ultimate resistance moments. The external work done by the loads is the sum of the loads multiplied by their respective deflections. Equating the internal and external work gives the relation between the collapse resistance moments in the slab and the collapse load.

'Equilibrium' method

A yield-line pattern is postulated, as before, and then the equilibrium of each slab element is studied. The equilibrium equations obtained for each element, by taking moments and resolving vertically, are solved simultaneously to obtain the solution. The equilibrium method cannot be used, however, without further specialized knowledge. The reason for this is that moments cannot be taken about appropriate axes for each individual slab element until the magnitude and distribution of the shear forces (or reactions) between adjacent elements are known. Certain theorems due to K. W. Johansen enable the statical equivalents of these forces to be calculated, so the 'equilibrium' method, once understood, usually gives solutions more readily than the virtual work method.

Although the equilibrium method enables column reactions to be calculated it is still restricted, as a design method, to isotropically or orthotropically reinforced slabs. The reader is therefore advised not to consider this method until after studying both the virtual work method and Hillerborg's strip method. Reference can be made subsequently to more advanced texts (e.g. reference [13.8]).

Uniform steel distribution

Yield-line solutions are greatly simplified by using a mesh type of reinforcement which is constant throughout the slab. In other words, the steel ratio along a given line is constant. The steel ratio for different directions, however, may still vary. Both of the following commonly used types of slab have a constant steel ratio along any given line:

(i) Isotropically reinforced slabs. If the ultimate flexural moment about any two perpendicular lines in the plane of the slab is the same, then the slab is isotropically reinforced.

(ii) Orthotropically reinforced slabs. If the ultimate flexural moments about two perpendicular lines in the plane of the slab are different, then the slab is orthotropically reinforced.

13.7 YIELD-LINE ANALYSIS BY VIRTUAL WORK

Consider a rigid body which is in equilibrium under the action of a system of forces. If the body is given a small arbitrary displacement the sum of the work done by the forces (i.e. force times its corresponding displacement) is zero since the resultant force is zero. Hence the principle of virtual work can be stated as follows:

'If a rigid body which is in statical equilibrium under a system of forces is given a virtual displacement the sum of the virtual work done by the forces is zero'. The virtual displacement is a small arbitrary displacement and the virtual work is the work resulting from the displacement.

External work

Once a yield-line pattern at failure has been postulated, a convenient point in the slab is selected and given a virtual displacement δ. The deflections of all parts of the slab can then be expressed in terms of δ and the total work done by the external loads calculated from the relation

external work = $\iint w\delta \, dx dy$

which may be written as $\sum(W\delta)$ for convenience. Notice that the reactions at the supports do not contribute to the virtual work since their displacement is zero.

Internal work

The virtual work done on the internal actions at the yield lines is given by the yield moments only. The twisting moments and the shear forces on each side of a yield line are assumed to undergo no relative movement, hence the virtual work done on the twisting moments and the shear forces summed over the whole slab is zero. The work done on a yield line is therefore the product of the total normal moment along the yield line and the rotation at the yield line. Hence the total work done on all the yield lines is given by

$$\text{internal work} = \sum m_n l\theta = \sum(M\theta)$$

where m_n = resistance normal moment of yield line, per unit length along yield line

l = length of yield line

θ = rotation at the yield line.

Virtual work equation

The solution for the slab is obtained by equating the external work done by the loads to the internal work done against the yield line normal moments. That is:

$$\sum(W\delta) = \sum(M\theta) \tag{13.7.1}$$

Notice that the virtual work equation (i.e. Eqn (13.7.1) is applied once the yield-line pattern has been postulated, hence the loads and moments are at their maximum values, e.g. the expression for the external work done is $W\delta$ and not $\frac{1}{2}W\delta$ as for real (i.e. total) elastic deflections.

Example 13.7.1

A rectangular slab is simply supported along two opposite edges and subjected to a uniformly distributed load. Determine the collapse load per unit area, n, in terms of the resistance moment per unit width, m. Let L be the span.

Solution Assume a yield-line pattern as shown in Fig. 13.7.1 and let λL equal the length of each support. Then the average deflection of the slab elements is $\frac{1}{2}\delta$ if the maximum deflection is δ. The total rotation of the yield line is $(\theta_a + \theta_b) = 2(\delta/\frac{1}{2}L)$, hence from Eqn (13.7.1)

$$n.\lambda L.L.\tfrac{1}{2}\delta = m.\lambda L.2(\delta/\tfrac{1}{2}L)$$

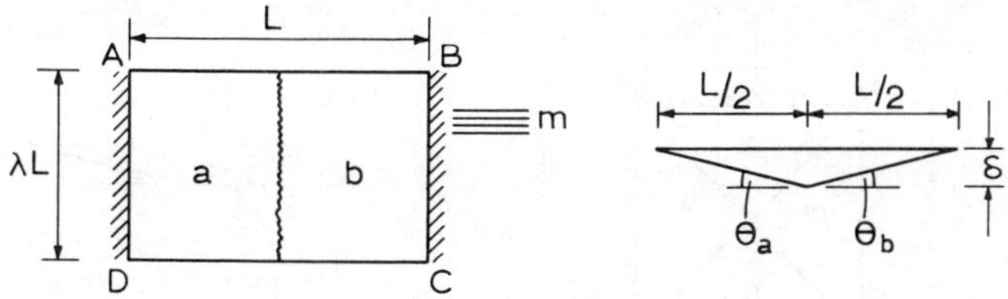

Fig.13.7.1 Example 13.7.1

Simplifying and rearranging gives

$m = nL^2/8$, (Answer)

a well-known result.

Note: The direction of the reinforcement in the figures is shown by four parallel lines. A single line should not be used unless it is intended as the moment key line representing the moment axis direction (in which case the single line should be drawn vertical in Fig.13.7.1). If a single line is shown it should always be assumed that this indicates the moment axis direction and not the reinforcement direction.

Example 13.7.2

A simply supported, isotropically reinforced, square slab is subjected to a uniformly distributed load of n per unit area, as shown in Fig. 13.7.2. Determine the resistance moment per unit width, m, in terms of n.

Solution Using Eqn (13.7.1) as before:

$$n.L^2.\delta/3 = 2(2mL\sqrt{2})(2\delta/L\sqrt{2})$$

Simplifying and rearranging gives

$m = nL^2/24$ (Answer)

Notice that since δ always cancels out, a unit deflection can be considered at the selected point if so desired. However, the advantage of considering δ is that it provides a partial check and the working is easier to follow.

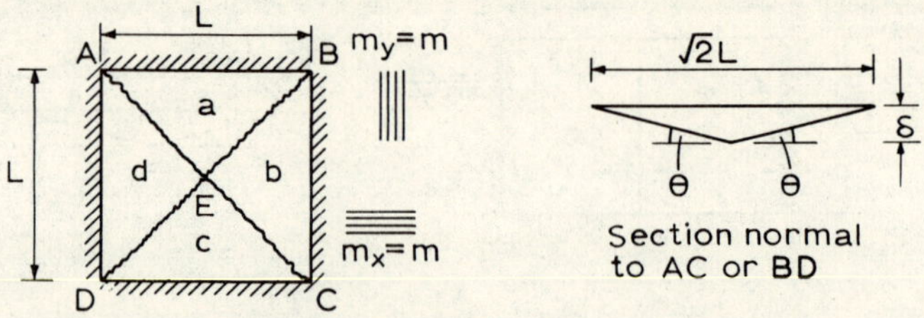

Fig.13.7.2 Example 13.7.2

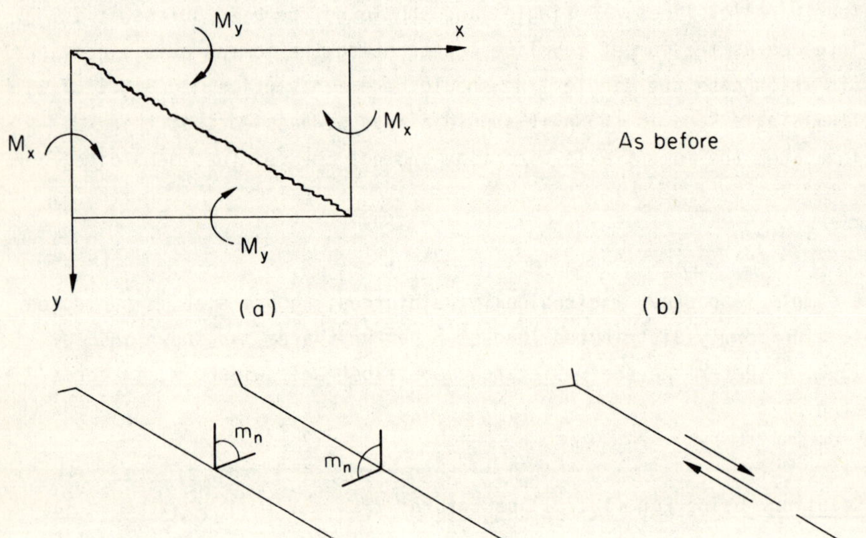

Fig.13.7.3 Components of the work done at a yield line

Components of work

Consider the external normal moment on a typical yield line split into its components as indicated in Fig. 13.7(a) and (b). The work done on a yield line is always positive, hence the moment vector should point in the same direction as the rotation vector if the product is to be positive. Assuming the right-hand screw rule for rotation, it is clear that a left-hand screw rule is necessary for the external moment vector if the vectors for rotation and moment for each element are to point in the same direction. Fig. 13.7.3(c) and (d) shows how the external normal moment acting on the yield line can be represented. Thus in (d) the vector arrow is drawn on the rigid element or region to which the external moment applies, in a direction such that the external moment acts anti-clockwise about the direction of the vector arrow.

Although the work done at a yield line is simply the product of the ultimate normal moment along the yield line and the true rotation of the yield line, it is usually easier to calculate the components of the work done using the following relation

$$m_n l\theta = m_y l_x \theta_y + m_x l_y \theta_x \qquad (13.7.2)$$

where m_x = collapse moment of x-direction steel per unit length normal to x-direction

l_x = projected length of the yield line on the x-axis

θ_x = rotation at yield line of x-direction steel

Similarly for m_y, l_y and θ_y.

If the notation follows the reinforcement moment axes, rather than the reinforcement directions, then the subscripts x and y are interchanged for both M and θ in Fig. 13.7.3 and Eq (13.7.2) can be written as

$$m_n l\theta = m_x l_x \theta_x + m_y l_y \theta_y$$

where m_x = collapse moment about **x**-axis (of y-direction steel) per unit length in x-direction

l_x = projected length of the yield line on the x-axis

θ_x = rotation at yield line about x-axis

Special reference is made in the text wherever this latter (mathematician's) notation is used.

Slab-element rotations

The rotation of a yield line is the sum of the rotations of the adjacent slab elements. When resolving along the axes it is convenient to consider the rotations of the elements instead of the yield lines. This is illustrated in Example 13.7.3, which reconsiders Example 13.7.2.

When considering the rotations of slab elements it is important in more complicated cases to check the signs of all terms since they need not all be positive. A term is negative if the rotation of the element is opposite to the direction of the applied moment. Hence although the total work is positive not all individual 'work' component terms need be positive (although they are in simple cases.)

Example 13.7.3 (same as Example 13.7.2)

Alternative solution Element a rotates about the x-axis (see Fig. 13.7.2). The portions of the yield lines which surround this element only are AE and EB. Hence by considering only the rotation of these yield lines which is due to the rotation of element a from the horizontal, it is evident that the work component for rotation about the x-axis is given by

$m_y L_x \theta_y$ where $m_y = m$ and $\theta_y = \delta/\tfrac{1}{2}L$

as shown in Fig. 13.7.4.

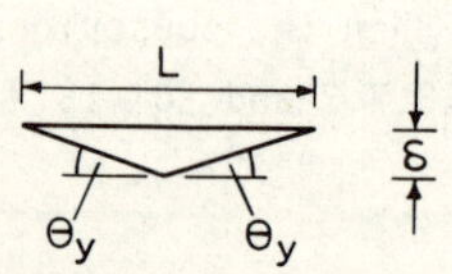

Fig.13.7.4 Example 13.7.3
Section is normal to AB of Fig.13.7.2

Work component for rotation about y-axis for element a is zero since $\theta_x = 0$. Hence work done by element a is

$mL.2\delta/L = m.2\delta$

Work done by element b is $m_x L_y \theta_x$ which in this case also equals $m.2\delta$. Hence applying the principle of virtual work to the whole slab:

$4(m.2\delta) = n.L^2.\delta/3$ or

$m = nL^2/24$ as before. (Answer)

Derivation of components of work for orthogonal reinforcement (after Jones and Wood [13.8])

The work done on a yield line is the scalar product of two vectors: $m_n l$ and θ_n where $m_n l$ is the resistance moment and θ_n is the angle of rotation. If the vectors $m_n l$ and θ_n are inclined to each other at some angle ϕ, then the scalar product is $m_n l\theta_n \cos\phi$.

Let θ_{na}, θ_{nb} be the normal rotations of the regions a and b on either side of the yield line respectively (e.g. see Fig. 13.7.1). The total rotation of the yield line, $(\theta_n)_{ab}$ is therefore given by

$$(\theta_n)_{ab} = \theta_{na} + \theta_{nb}$$

where the vector addition of θ_{na} and θ_{nb} is also the algebraic addition since the three vectors act in the same direction.

It is convenient to consider the work done solely by an individual element. Thus if D_1 is the work done on the yield line and $(D_a)_1$, $(D_b)_1$ represent the 'work done' solely by the rotation of elements a and b respectively then

$$D_1 = (D_a)_1 + (D_b)_1$$

i.e. $(m_n l\theta_n)_1 = (m_n l\theta_{na})_1 + (m_n l\theta_{nb})_1$

Notice that D_1 is always positive although a component $(D_a)_1$ or $(D_b)_1$ can be negative.

Consider the vectors along the a side only of the yield line and drop the suffix 1 (since only one line is being considered) as shown in Fig. 13.7.5(a). Notice that all moments within the plane of the element need to be included.

θ_{na} can be represented by any two vectors, according to vector algebra, hence let θ_{na} be replaced by θ_{nax} and θ_{nay}, as in Fig. 13.7.5(b) where θ_{nax} is along a selected x-axis and represents rotations of element a about this axis (Mathematician's notation). Similarly for θ_{nay} along the y-axis which is taken to be orthogonal to the x-axis. Consider θ_{nax} as shown in Fig.13.7.5(c), then the scalar products involving θ_{nax} are:

$$m_{nt} l\theta_{nax}\cos\left(\frac{\pi}{2} - \phi\right) \text{ and } m_n l\theta_{nax}\cos\phi$$

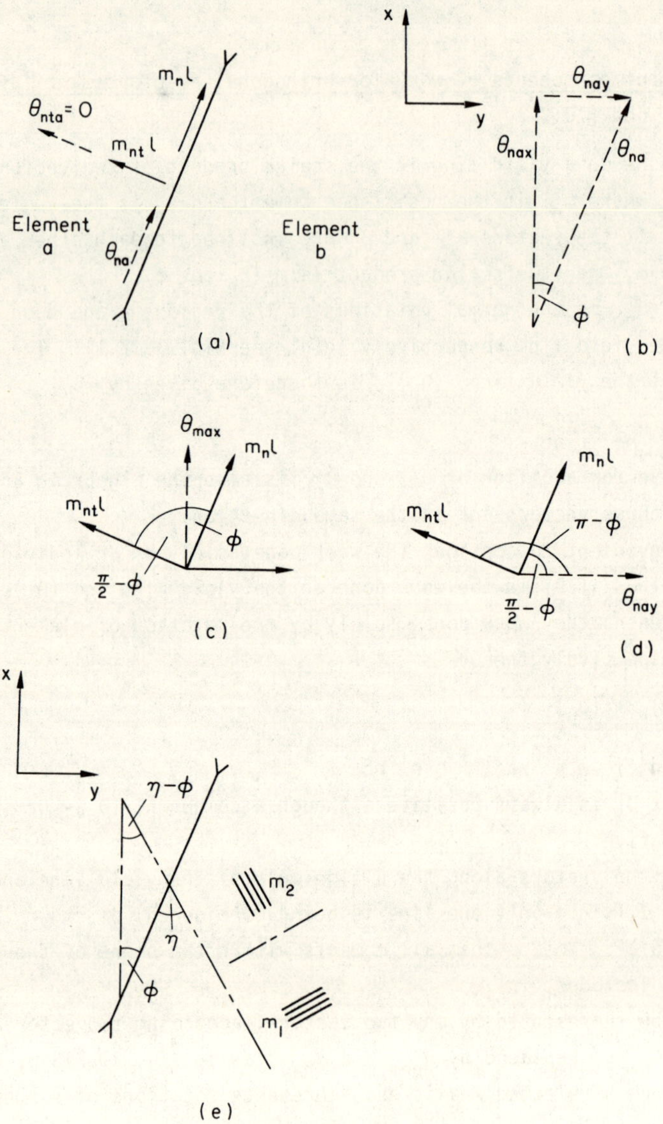

Fig.13.7.5 Component vectors for one element at a yield line

Similarly from Fig. 13.7.5(d) the scalar products involving θ_{nay} are:

$$m_{nt} l \theta_{nay} \cos(\pi - \phi) \text{ and } m_n l \theta_{nay} \cos(\frac{\pi}{2} - \phi)$$

Adding these four scalar products, simplifying and equating to $m_n l \theta_{na}$ gives

$$m_n l \theta_{na} = l \theta_{nax} \{m_{nt} \sin \phi + m_n \cos \phi\} + l \theta_{nay} \{-m_{nt} \cos \phi + m_n \sin \phi\}$$

If the projections on the x-axis and y-axis are l_x and l_y respectively, then $l_x = l \cos \phi$ and $l_y = l \sin \phi$ and substituting in the above gives

$$m_n l \theta_{na} = l_x \theta_{nax} \{m_n + m_{nt} \tan \phi\} + l_y \theta_{nay} \{m_n - m_{nt} \cos \phi\} \quad (13.7.4)$$

Comparing Eqn (13.7.4) with Eqn (13.7.3) shows that the above is valid only if

$$m_x = m_n + m_{nt} \tan \phi \quad (13.7.5)$$

and

$$m_y = m_n - m_{nt} \cos \phi \quad (13.7.6)$$

Consider the case when the moment key line m_1 (i.e. the moment axis for the m_1 reinforcement) is inclined at an angle η to the yield line as in Fig. 13.7.5(e). Putting $m = m_1$, $\mu m = m_2$ and $\phi = \eta$ in Eqns (13.5.3(b)) and (13.5.4) gives

$$m_n = m_1 \cos^2 \eta + m_2 \sin^2 \eta \quad (13.7.7)$$

$$m_{nt} = (m_1 - m_2) \sin \eta \cos \eta \quad (13.7.8)$$

Similarly putting $\phi = \eta - \phi$ gives m_x and $\phi = \frac{\pi}{2} - \eta + \phi$ gives m_y, i.e.

$$m_x = m_1 \cos^2(\eta - \phi) + m_2 \sin^2(\eta - \phi) \quad (13.7.9)$$

$$m_y = m_1 \sin^2(\eta - \phi) + m_2 \cos^2(\eta - \phi) \quad (13.7.10)$$

Substituting from Eqns (13.7.7) to (13.7.9) in Eqn (13.7.5) shows that Eqn (13.7.3) is only valid when

$$m_1 \cos^2(\eta - \phi) + m_2 \sin^2(\eta - \phi) = m_1 \cos^2 \eta + m_2 \sin^2 \eta + (m_1 - m_2) \sin \eta \cos \eta \tan \phi$$

Thus if $m_1 = m_2$ (i.e. for isotropic reinforcement) it is true for any value of η or ϕ.

Alternatively, if $m_1 \neq m_2$ (i.e. for orthotropic reinforcement) then the sum of the coefficients of m_1 and m_2 must each separately equal zero. That is

$$\cos^2 \eta + \sin \eta \cos \eta \tan \phi - \cos^2(\eta - \phi) = 0$$

i.e.

$$\cos(\eta - \phi)\{\cos \eta - \cos(\eta - \phi) \cos \phi\} = 0$$

Thus $\cos(\eta-\phi) = 0$ i.e. $\eta = (n\pi/2) + \phi$ where n is an integer is a valid solution.

Similarly, for the m_2 coefficients then $\sin(\eta-\phi) = 0$ but ϕ for the m_2 reinforcement differs by $\pi/2$ hence this leads to exactly the same solution. Again if Eqns (13.7.7), (13.7.8) and (13.7.10) are substituted into Eqn (13.7.6) the same condition satisfies. Thus if $m_1 \neq m_2$ (i.e. for orthotropic reinforcement) the chosen x- and y-axes must be in the direction of the reinforcement moment axes if Eqn (13.7.3) (or Eqn (13.7.2)) is to be valid.

Minimum load principle

While a general pattern of failure can often be drawn, the exact pattern cannot always be determined by inspection. Consider Fig.13.7.6 for example. Length EF is unknown for (a), the co-ordinates of E are unknown for (b), while α, β and ϕ are unknown for (c). Hence the general

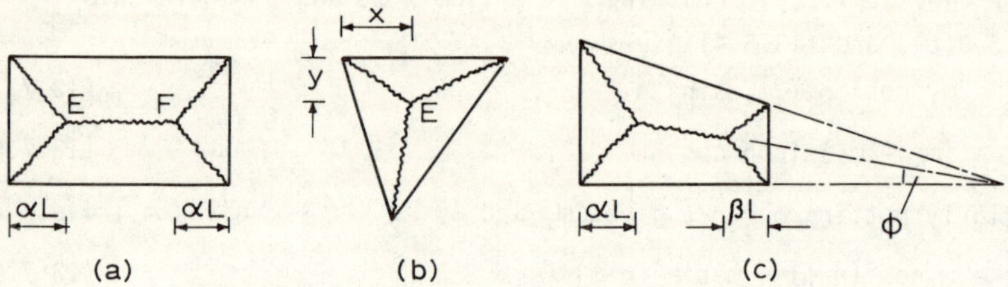

Fig.13.7.6 Yield patterns with different numbers of unknowns. (a) One unknown. (b) Two unknowns. (c) Three unknowns

pattern of yield lines is often defined by unknown parameters, such as ratios or angles, so that the virtual work equation is obtained in terms of these to give an equation of the form

$$m = nf(\alpha,\beta,\gamma\ldots\phi)$$

Only one equation relating m and n is obtained from virtual work hence j more equations are required for j parameters in order to solve for the collapse load. These are given by the turning values relative to each parameter, since m must be a maximum (or n a minimum) for the most probable collapse mechanism. That is,

$$\frac{\partial m}{\partial x} = \frac{\partial m}{\partial \beta} = \ldots = \frac{\partial m}{\partial \phi} = 0 \tag{13.7.3}$$

It is usually convenient to express m in the form

$m = nf_1(\alpha,\beta,\ldots)/f_2(\alpha,\beta,\ldots)$

A turning value is given by putting $\partial m/\partial x = 0$, that is

$$\frac{f_1(\alpha,\beta,..)}{f_2(\alpha,\beta,..)} = \frac{\partial f_1/\partial\alpha}{\partial f_2/\partial\alpha} = \frac{\partial f_1/\partial\beta}{\partial f_2/\partial\beta} = \ldots \text{ etc.} \tag{13.7.4}$$

Example 13.7.4

State clearly the conditions which help to predict the yield-line pattern for a reinforced concrete slab. An orthotropically reinforced rectangular slab is simply supported along all four edges and measures 4.8 m × 6 m. If the effective depth is 0.2 m, A_{st} is 381 and 296 mm^2 per metre width for the y- and x-direction reinforcement respectively, f_{cu} = 25 MN/m^2 and f_{yt} = 425 MN/m^2, determine the ultimate value of n. Derive general algebraic expressions where possible.

<u>Solution</u> For first part, see section 13.4.

Equating the steel and concrete forces for a metre width of slab gives

$(425/1.15) \times 296 \times 10^{+6-6} = 0.4 \times 25 \times x \times 10^6$

or

$x = 10.9 \times 10^{-3}$

Therefore

$m_x = (425/1.15) \times 296(200 - \frac{1}{2} \times 10.9) \times 10^{6-6-3} = 21300$

Similarly

$m_y = (425/1.15) \times 381(200 - \frac{1}{2} \times 14.1) \times 10^{-3} = 27200$

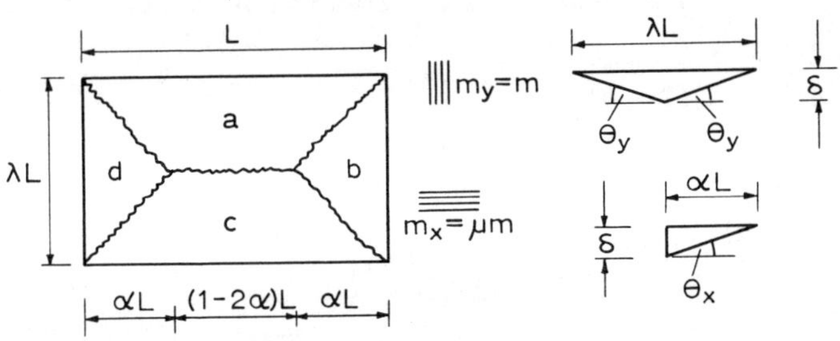

Fig.13.7.7 Example 13.7.4

Using $\sum W\delta = \sum M\theta$,

$$n \times \tfrac{1}{2}\delta[\lambda L(1 - 2\alpha)L] + n \times \tfrac{1}{3}\delta[\lambda L.2\alpha L] = 2(m_x\lambda L)(\delta/\alpha L) + 2(m_y L)(2\delta/\lambda L)$$

Let $m_y = m$ and $m_x = \mu m$ where $\mu = 21.3/27.2 = 0.783$. Then

$$n\lambda^2L^2/12 = (\mu\lambda^2 + 2\alpha)m/(3\alpha - 2\alpha^2)$$

$$dn/d\alpha = 0 \quad \text{when } (\mu\lambda^2 + 2\alpha)/(3\alpha - 2\alpha^2) = 2/(3 - 4\alpha)$$

or

$$-\alpha^2 - \mu\lambda^2\alpha + \tfrac{3}{4}\mu\lambda^2 = 0$$

$$\alpha = -\tfrac{1}{2}[\mu\lambda^2 \pm \sqrt{(\mu^2\lambda^4 + 3\mu\lambda^2)}]$$

If $\lambda = 0.8$ and $\mu = 0.783$ then $\alpha = 0.412$. Hence

$$n = \frac{12}{(4.8)^2} \times \frac{0.783 \times 0.64 + 0.824}{(3 - 0.824) \times 0.412} \times (27200) = 20940$$

$$= 20.94 \text{ kN/m}^2$$ (Answer)

Example 13.7.5

Sketch two possible modes of failure for the uniformly loaded, isotropically reinforced slab shown in Fig. 13.7.8(a). Using either of your assumed modes of failure calculate m in terms of the ultimate uniformly distributed load n if $\mu = \eta = 1$ and $m' = m$.

Solution The possible modes are shown in Fig. 13.7.8(b) and (c).

General case for mode A

$$n.\lambda L(1 - \alpha)L(\delta/3) + n\lambda L.\alpha L(\delta/2)$$
$$= (m + m')\lambda L\delta/(1 - \alpha)L + 2(\mu m + \eta m')L(2\delta/\lambda L)$$

or

$$(n\lambda^2L^2/6)(2 + \alpha)(1 - \alpha)$$
$$= \lambda^2(m + m') + 4(\mu m + \eta m')(1 - \alpha)$$

$dn/d\alpha = 0$ when

$$(2 + \alpha)(1 - \alpha)[-4(\mu m + \eta m')]$$
$$= [\lambda^2(m + m') + 4(\mu m + \eta m')(1 - \alpha)](-2\alpha - 1)$$

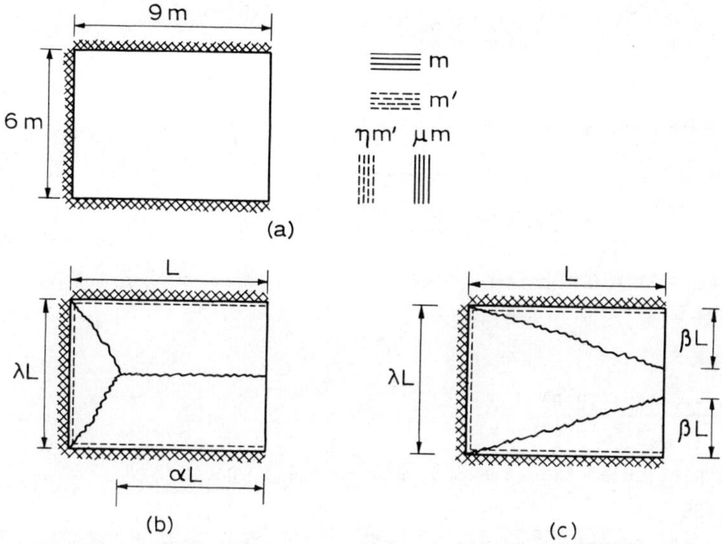

Fig.13.7.8 Example 13.7.5 (a) Slab layout. (b) Mode A (c) Mode B.

which simplifies to

$\frac{1}{2}\alpha^2 - (1 + A)\alpha + \frac{1}{2}(1 - A) = 0$ where

$A = \lambda^2(m + m')/4(\mu m + \eta m')$

Hence

$\alpha = (1 + A) \pm \sqrt{[(1 + A)^2 - (1 - A)]}$

For this particular slab $\lambda = 2/3$ and $A = 1/9$, hence α equals either 0.522 or a value greater than one. $\alpha = 0.522$ is a valid solution.

Hence, substituting

$\alpha = 0.522 \quad \lambda L = 6 \quad \mu = 1 \quad \eta = 1 \quad m' = m$

in the above equation for m and transposing gives

$m = 1.54n$ (Answer)

General case for mode B

$$n(\delta/3)(2\beta L^2) + n(\delta/2)L^2(\lambda - 2\beta)$$

$$= (2m\beta L + m'\lambda L)(\delta/L) + 2(\mu m + \eta m')L(\delta/\beta L)$$

or

$$\frac{nL^2}{6} = \frac{2m\beta^2 + \lambda m'\beta + 2(\mu m + \eta m')}{\beta(3\lambda - 2\beta)}$$

$dn/d\beta = 0$ when

$$[2m\beta^2 + \lambda m'\beta + 2(\mu m + \eta m')](3\lambda - 4\beta)$$
$$= \beta(3\lambda - 2\beta)(4m\beta + \lambda m')$$

or

$B\beta^2 + 4\beta - 3\lambda = 0$ where

$B = \lambda(3m + m')/(\mu m + \eta m')$

Hence

$$\beta = \frac{-4 \pm \sqrt{(16 + 12\lambda B)}}{2B}$$

For this particular slab $\lambda = 2/3$ and $B = 4/3$, hence

$\beta = 0.436$

But $\beta = 0.436$ is not valid since $0.436 > \frac{1}{2}\lambda$. Thus mode A is more likely than mode B.

Inclined axes of rotation

If the adjacent axes of rotation are at right angles to each other they can coincide with the directions of the reinforcement. However, if the adjacent axes of rotation are not at right angles the axes should be chosen to coincide with the reinforcement moment lines (i.e. perpendicular to the direction of the reinforcement itself) if the components of work are to be determined as previously.

Example 13.7.6

A triangular slab, ABC, carries a uniformly distributed load and is simply supported along two sides CA and CB at an angle θ. Length CA = a; length CB = b and the resistance moments per unit length for the reinforcement normal to and parallel to the side CA are m and μm respectively. If the assumed yield line pattern is as shown in Fig. 13.7.9, obtain an expression for the ultimate load n and hence determine the most probable value for the angle ϕ which the yield line makes with the side CA when $\mu = 1$.

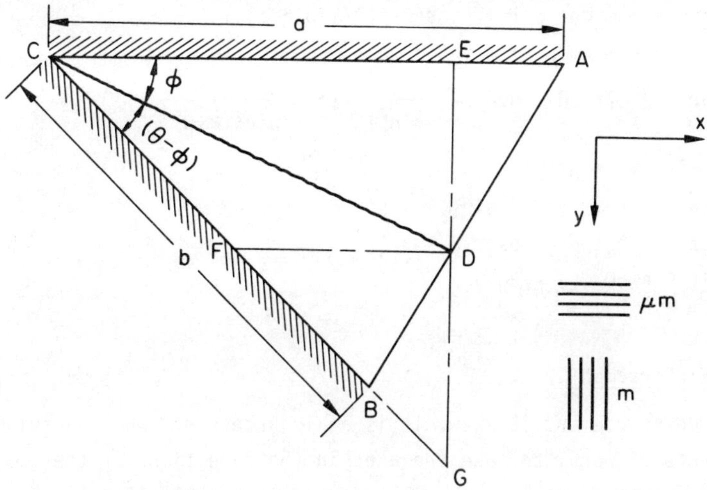

Fig.13.7.9 Example 13.7.6.

Solution

$\Sigma M\theta$ for element ACD

$$= m\ CD \cos\phi \left(\frac{\delta}{CD \sin\phi}\right) + \mu m\ CD \sin\phi\ (0)$$

$$= m\delta \cot\phi$$

$\Sigma M\theta$ for element BCD

$$= m\ CD \cos\phi \left(\frac{\delta}{DG}\right) + \mu m\ CD \sin\phi \left(\frac{\delta}{DF}\right)$$

Now $\dfrac{DF}{\sin(\theta-\phi)} = \dfrac{CD}{\sin(180-\theta)}$ and $\dfrac{DG}{\sin(\theta-\phi)} = \dfrac{CD}{\sin(90-\theta)}$

i.e. $DF = CD \sin(\theta-\phi)/\sin\theta$ and $DG = CD \sin(\theta-\phi)/\cos\theta$

Also $\Sigma W\delta = \frac{1}{2} nab \sin\theta \frac{\delta}{3} = \dfrac{nab\delta}{6} \sin\theta$

Therefore by virtual work $\Sigma W\delta = \Sigma M\theta$

i.e. $\dfrac{nab}{6} \sin\theta = m \cot\phi + m \cos\phi \cos\theta/\sin(\theta-\phi)$

$+ \mu m \sin\phi \sin\theta/\sin(\theta-\phi)$ (Answer)

If $\mu = 1$ then

$$\frac{nab}{6} \sin \theta = m \cot \phi + m \{\cos(\theta-\phi)\}/\sin(\theta-\phi)$$
$$= m \cot \phi + m \cot(\theta-\phi)$$

Therefore $\frac{\partial n}{\partial \phi} = 0$ when $-\frac{1}{\sin^2\phi} + \frac{1}{\sin^2(\theta-\phi)} = 0$

i.e. $\sin^2(\theta-\phi) = \sin^2\phi$

or $\sin\theta \cos\phi - \cos\theta \sin\phi = \pm \sin\phi$

or $\cot\phi = \cot\theta \pm \operatorname{cosec}\theta$ (Answer)

(e.g. if $\theta = 60^0$, $\phi = 30^0$)

Skew reinforcement

If the reinforcement is skew, it is again necessary, when deriving the components of work, to take the axes in the directions of the reinforcement moment lines as illustrated in Example 13.7.7.

Example 13.7.7

A skew slab is reinforced along the line of the slab, which makes an angle α with the supports. The resistance moments per unit width for the steel along the direction α and parallel to the supports are m and μm respectively. If the centre of the slab is subjected to a concentrated point load W, determine the most probable load at collapse and the width of slab which effectively resists the load if the yield line pattern is as shown in Fig. 13.7.10. The span perpendicular to the supports is l.

Solution

Let deflection at point load be δ and let θ_{APB}, θ_{APD} be the slopes of elements APB, APD respectively.

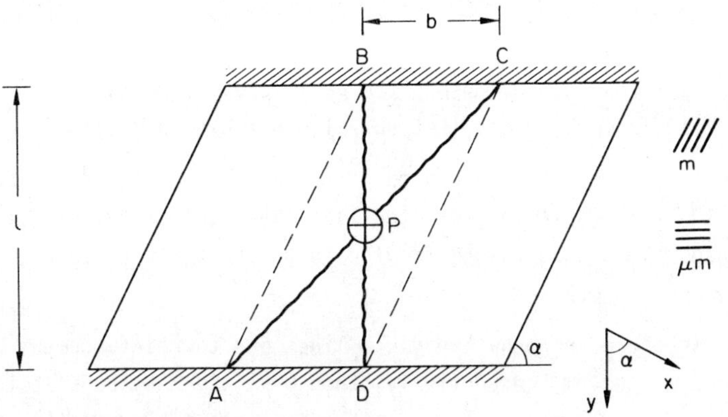

Fig.13.7.10 Example 13.7.7

Hence $\Sigma M\theta$

$= 2m$(projection of AD on x-axis) × (projection of θ_{APD} on x-axis)

$2\mu m$(projection of AD on y-axis) × (projection of θ_{APD} on y-axis)

$+ 2m$(projection of AB on x-axis) × (projection of θ_{APB} on x-axis)

$+ 2\mu m$(projection of AB on y-axis) × (projection of θ_{APB} on y-axis)

where projection of θ_{APD} on y-axis = slope of plane APD normal to y-axis, etc.

$= 2m(b \sin\alpha)(2\delta \sin\alpha/l) + 2\mu m(0)$

$+ 2m(0) + 2\mu m(l)(2\delta/b)$

$= 4\delta mb \sin^2\alpha/l + 4\delta\mu ml/b$

Hence using $\Sigma W\delta = \Sigma M\theta$ then

$W\delta = (mb \sin^2\alpha + \mu ml^2/b)4\delta/l$

or $\frac{1}{4}Wl = mb \sin^2\alpha + \mu ml^2/b$

Therefore $\dfrac{\partial W}{\partial b} = 0$ when $m \sin^2\alpha - \mu ml^2/b^2 = 0$

i.e. $b = +\sqrt{\mu}l/\sin\alpha$

Therefore $\frac{1}{4}Wl = \sqrt{\mu}lm \sin\alpha + \sqrt{\mu}lm \sin\alpha$

or $W = 8\sqrt{\mu}\, m \sin\alpha$ (Answer)

Note 1: The failure load for a yield-line pattern which is assumed to be rectangular as in Fig. 13.4.6 (rather than a parallelogram) is given by

$$W = 8\sqrt{\lambda}\ m \sin \alpha$$

where $\lambda = \mu + \cos^2 \alpha > \mu$. Hence the parallelogram pattern assumed in Fig. 13.7.10 is more likely (but see also Example 13.8.1).

Note 2: It is also of interest to note that the elastic distribution of $2.4x(1 - x/l)$ for a concentrated load (see earlier) gives $b = 0.61$ at midspan. This is obtained for the above when, for example, $\alpha = \pi/3$ and $\mu = 0.27$

Note 3: If there are negative yield lines due to reinforcement of resistance m' and $\nu m'$ respectively, then W is given by substituting $\mu m + \nu m'$ for μm and $m + m'$ for m in the initial working of the solution.

13.8 SPECIAL ASPECTS OF YIELD-LINE ANALYSIS

Since yield-line theory gives an upper-bound solution it is essential to select the most probable yield-line pattern if the lowest collapse load is to be determined. Alternative modes of failure which are possible due to point loads, curtailment of reinforcement, composite action of beams and slabs, corner levers, etc., therefore require special consideration.

Point loads

Consider a simply supported square slab as shown in Fig. 13.8.1(a).

Work done by load = $W\delta$

Internal work = $4mL.2\delta/L = 8m\delta$

Hence $W = 8m$

This is not the correct collapse load, however, since a lower value of W can be obtained if a yield-line mechanism in the form of a circular fan is considered, as shown in Fig. 13.8.1(b). The moment along OB or OA equals mR, since the slab is isotropically reinforced.

Hence

moment along AB = $mR.\delta\phi$

Rotation about AB = δ/R

Rotation about normal to AB = 0

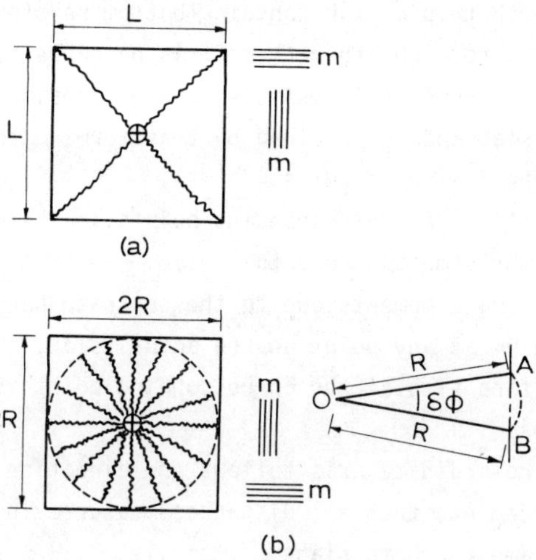

Fig.13.8.1 Central point load for square slab.(a) Diagonal yield lines.
(b) Radial yield lines

Therefore internal work is

$$\sum mR\delta\phi(\delta/R) = \delta\int_0^{2\pi} m\,d\phi = 2\pi m\delta$$

Work done by load = $W\delta$

Hence $W = 2\pi m$

If there is isotropic negative reinforcement with yield moment m':

Moment along AB = $m'(R.\delta\phi)$

$$\text{Total internal work} = \delta\int_0^{2\pi}(m + m')d\phi = 2\pi(m + m')\delta$$

Hence $W = 2\pi(m + m')$

Notice that this value is independent of the radius of the fan mechanism. For a uniformly distributed load n the result would be $n = 6(m + m')/r^2$ that is, the most critical fan mechanism is as large as possible and hence it would extend to a slab boundary. For a combination of distributed and point loads the fan would still tend to extend to a slab boundary.

Affine isotropic slab

If a skew or orthotropic slab contains bottom reinforcement of m and μm at an angle α to each other then it is possible to obtain an isotropic slab which, for purposes of analysis, can be considered to be equivalent to the actual slab and is obtained by transforming the shape and loading according to the following rules

(a) The deflections at corresponding points in the affine (isotropic) slab are the same as the actual slab.

(b) If the ultimate moments due to the separate bands of reinforcement are m and μm at any point on the actual slab, then the strength of the affine slab at the corresponding point is m in all directions.

(c) If the x co-ordinate axis follows the direction of the m-reinforcement, then all distances measured in the x-direction are the same for both slabs.

(d) If the y co-ordinate axis follows the direction of the μm reinforcement in the actual slab, it should be taken at right angles to the x-axis for the affine slab.

(e) All distances measured in the y-direction in the affine slab are obtained by dividing corresponding lengths in the actual slab by $\sqrt{\mu}$.

(f) All corresponding total loads for the affine slab are obtained by dividing the actual total loads by $\sqrt{\mu} \times \sin \alpha$.

Johansen conceived the concept of the 'affine' slab, which was then extended by Jones and Wood [13.8]. If top reinforcement is present it must be parallel to the bottom reinforcement and, furthermore, the ratio between the bands of reinforcement must be the same as that for the bottom reinforcement, i.e. m' and μm' and not some ratio other than μ.

Proof of affinity theorem

Let a point load W be located at a perpendicular distance l from the supported edge AB in Fig. 13.8.2. If the rotation about AB is θ then the load W deflects by lθ. Hence in terms of the skew co-ordinates l_x and l_y and slopes θ_x and θ_y in the directions x and y then

$$Wl\theta = W(l_x\theta_x + l_y\theta_y)$$

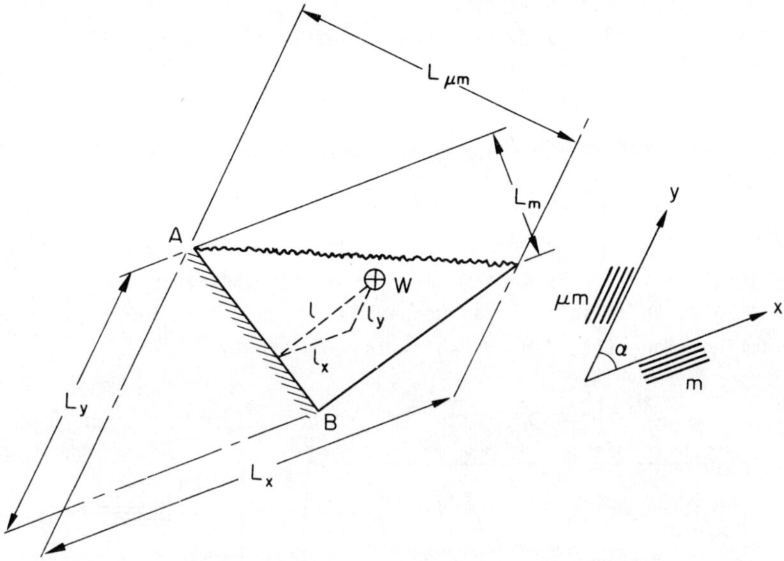

Fig.13.8.2 Slab element with skew areas

since element ABC is plane. The work done at the yield line, by the component vector technique is simply

$$mL_m\theta_m + \mu mL_{\mu m}\theta_{\mu m}$$

where L_m, $L_{\mu m}$ are the projections of the yield line on the directions of the m and μm moment axes respectively. If the total co-ordinate lengths of the yield line are L_x and L_y then $L_m = L_y \sin\alpha$; $L_{\mu m} = L_x \sin\alpha$.

Hence summing for the complete slab and substituting in the virtual work equation gives

$$m \sin\alpha\ \Sigma\{L_y\theta_m + \mu L_x\theta_{\mu m}\} = \Sigma W\{L_x\theta_x + L_y\theta_y\} \qquad (13.8.1)$$

Let the affine slab parameters be indicated by a prime ('). If deflections at corresponding points are unaltered and if, for the affine slab, the y-direction is perpendicular to the x-direction and y-co-ordinates are increased by a constant K_o then $L_y' = K_oL_y$. Thus it follows that

$$l_x' = l_x\ ;\quad \theta_x' = \theta_x\ ;\ \theta_m' = \theta_m$$

$$l_y' = K\ l_y;\ \theta_y' = \theta_y/K_o;\ \theta_{\mu m}' = \theta_{\mu m}/K_o$$

where $\theta_{\mu m}'$ means the rotation at right angles to θ_m'

Also if $W' = K_1\ W$ then substituting in Eqn (13.8.1) gives

$$m \sin\alpha\ \Sigma\{(L_y'/K_o)\theta_m' + \mu L_x'(K_o\theta_{\mu m}')\} = \Sigma(W'/K_1)\{l_x'\theta_x' + l_y'\theta_y'\} \qquad (13.8.2)$$

The corresponding equation obtained by applying the virtual work equation directly to the affine (isotropic) slab, remembering that the yield moment in any direction is m, is

$$m\,\Sigma\{L'_y\theta'_m + L'_x\theta'_{\mu m}\} = \Sigma W'\{l'_x\theta'_x + l'_y\theta'_y\} \quad (13.8.3)$$

Multiplying Eqn (13.8.2) by K_1 makes it identical with Eqn (13.8.3) if

$K_1 \sin\alpha/K_0 = 1$ and $K_1 \sin\alpha\,\mu\,K_0 = 1$

i.e. $K_0 = 1/\sqrt{\mu}$ $(= y'/y)$

and $K_1 = 1/(\sqrt{\mu}\sin\alpha)$ $(=W'/W)$

Example 13.8.1

Repeat Example 13.7.7 by considering the affine isotropic slab. Consider also an elliptical fan mechanism and show that for a concentrat[ed] load acting alone this fan mechanism is more likely.

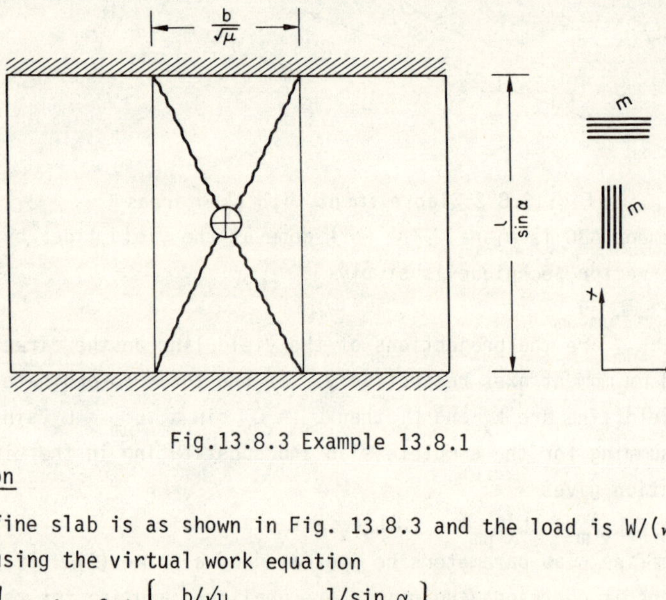

Fig.13.8.3 Example 13.8.1

Solution

The affine slab is as shown in Fig. 13.8.3 and the load is $W/(\sqrt{\mu}\sin\alpha)$. Hence using the virtual work equation

$$\frac{W}{\sqrt{\mu}\sin\alpha}\,\delta = 2m\delta\left\{\frac{b/\sqrt{\mu}}{\tfrac{1}{2}l/\sin\alpha} + \frac{l/\sin\alpha}{\tfrac{1}{2}b/\sqrt{\mu}}\right\}$$

or $\tfrac{1}{4}Wl = mb\sin^2\alpha + \mu ml^2/b$

hence etc. as in Example 13.7.7.

For a circular fan $W = 2\pi m$ (see Fig.13.8.1)

Therefore for the affine slab

$W/(\sqrt{\mu}\sin\alpha) = 2\pi m$

or $W = 2\pi\sqrt{\mu}\,m\sin\alpha$ for the actual elliptical fan.

This is always lower than $8\sqrt{\mu}\,m\sin\alpha$ obtained previously in Example 13.7.7. Thus for a concentrated load acting alone the elliptical fan mechanism is more likely. (QED)

Curtailment of reinforcement

Consider a rectangular slab continuously reinforced with top and bottom steel as shown in Fig. 13.8.4(a). If the top steel is curtailed as shown in Fig. 13.8.4(b) it is necessary to consider the alternative yield-line mechanism. This is illustrated by Example 13.8.1.

Example 13.8.2

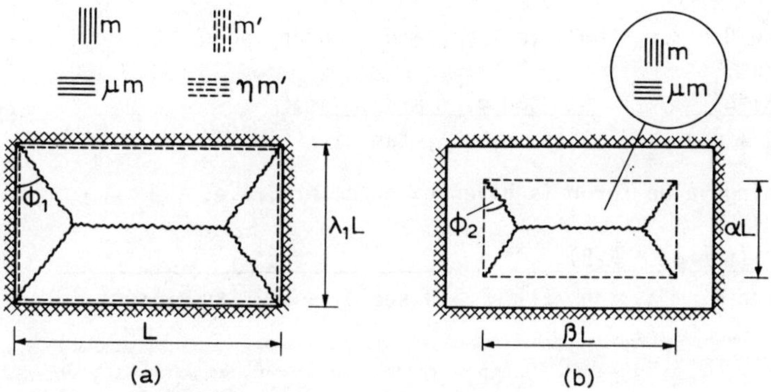

Fig.13.8.4 Example 13.8.2

Determine the uniformly distributed load, n at collapse of the rectangular slab, $\lambda_1 L \times L$, in terms of the angle ϕ_1 defining the yield-line mechanism shown in Fig. 13.8.4(a). Hence show that the most likely value for ϕ_1 is given by

$$\tan \phi_1 = \sqrt{(\lambda_1{}^2 B^2 + 3B)} - \lambda_1 B$$

where $B = (\mu m + \eta m')/(m + m')$. Hence show also that

$$n = 24(\mu m + \eta m')/\lambda_1{}^2 L^2 \tan^2 \phi_1$$

It is decided to curtail the top reinforcement in the central area, $\alpha L \times \beta L$, of the slab so that the alternative yield-line mechanism shown in Fig. 13.8.4(b) becomes possible. By comparison with the above result the alternative collapse load is given by

$$n = 24\mu m/(\alpha/\beta)^2(\beta L)^2 \tan^2\phi_2$$

Hence determine the maximum value of α if the curtailment of the top reinforcement is not to reduce the collapse load for the case when $\alpha = \lambda_1\beta$ and $\mu = \eta = 1$. (UB3/H 1969, 45 min)

Solution

$$\sum M\theta = 2(m + m')(L(2\delta/\lambda_1 L) + 2(\mu m + \eta m')\lambda_1 \, L(2\delta/\lambda_1 L \tan\phi_1)$$

$$\sum W\delta = n\lambda_1 L[\tfrac{1}{3}\delta(\lambda_1 L \tan\phi_1) + \tfrac{1}{2}\delta L(1 - \lambda_1 \tan\phi_1)]$$

Equating $\sum M\theta$ and $\sum W\delta$:

$$\frac{4(m + m')}{\lambda_1} + \frac{4(\mu m + \eta m')}{\tan\phi_1} = \frac{n\lambda_1 L^2}{6}\,(3 - \lambda_1 \tan\phi_1)$$

Putting $B = (\mu m + \eta m')/(m + m')$ and rearranging:

$$\frac{n\lambda_1{}^2 L^2}{24(m + m')} = \frac{\tan\phi_1 + \lambda_1 B}{3\tan\phi_1 - \lambda_1 \tan^2\phi_1} \qquad (1)$$

A turning value for n is given by $dn/d\phi = 0$, i.e.

$$\frac{(\tan\phi_1 + \lambda_1 B)}{(3\tan\phi_1 - \lambda_1 \tan^2\phi_1)} = \frac{\sec^2\phi_1}{(3\sec^2\phi_1 - 2\lambda_1 \tan\phi_1 \sec^2\phi_1)}$$

Hence

$$\tan^2\phi_1 + 2\lambda_1 B \tan\phi_1 - 3B = 0$$

$$\tan\phi_1 = \sqrt{(\lambda_1{}^2 B^2 + 3B)} - \lambda_1 B$$

Partial substitution for $\tan\phi_1$ in RHS of equation (1) gives

$$\frac{\sqrt{(\lambda_1^2 B^2 + 3B)}}{\tan\phi_1[3 + \lambda_1{}^2 B - \lambda_1\sqrt{(\lambda_1{}^2 B^2 + 3B)}]}$$

which simplifies to $B/\tan^2\phi_1$. Hence

$$n = 24(\mu m + \eta m')/\lambda_1{}^2 L^2 \tan^2\phi_1 \qquad \text{(QED)}$$

Alternative collapse loads are equal if

$$n = \frac{24(\mu m + \eta m')}{\lambda_1{}^2 L^2 \tan^2\phi_1} = \frac{24\mu m}{\alpha^2 L^2 \tan^2\phi_2}$$

or

$$\alpha^2 = [\mu m\lambda_1{}^2/(\mu m + \eta m')] \times (\tan^2\phi_1/\tan^2\phi_2)$$

If $\mu = \eta = 1$, then $B = 1$ and $\tan\phi_1 = \tan\phi_2$. Hence

$$\alpha^2 = m\lambda_1{}^2/(m + m_1) \qquad \text{(Answer)}$$

Composite action of beams and slabs

Consider an isotropic square slab, supported on flexible beams, and carrying a uniformly distributed load as shown in Fig. 13.8.5.

Collapse of slab alone (mode 1)occurs at $nL^2 = 24m$

Collapse of slab and beams together as given by mode 2 is calculated as follows:

Let M_B be the normal collapse moment for each beam at midspan. Resolving along the x- and y-axes in the normal way then, by addition,

$$\sum M\theta = 4(2M_B + mL)(2\delta/L)$$

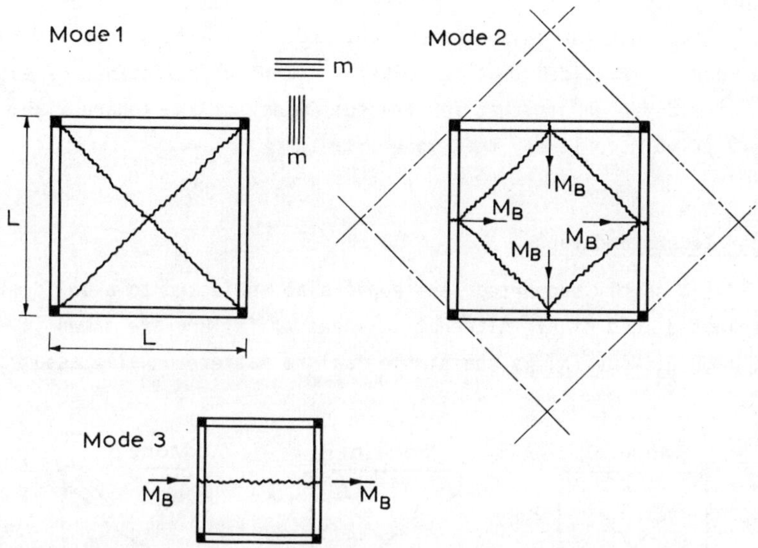

Fig.13.8.5 Alternative modes of failure for beam and slab construction

Notice that *each* beam has a total rotation of $2 \times (2\delta/L)$ at midspan. If the beams have a collapse moment of resistance at the supports due to their connection to the columns of M_C this should be included in $\sum M\theta$.

$$\sum W\delta = \tfrac{1}{2}nL^2(2\delta/3) + \tfrac{1}{2}nL^2(\delta) = nL^2(5\delta/6)$$

Hence

$$nL^2(5/6) = 8(m + 2M_B/L)$$

$$nL^2 = (48/5)(m + 2M_B/L)$$

A third mode is given by the normal mode of failure at midspan for either pair of beams. In this case there is simply a single yield line at midspan perpendicular to the pair of beams concerned. Hence, putting

$$W\delta = M\theta$$

for mode 3 gives

$$nL^2(\delta/2) = 2(mL + 2M_B)(2\delta/L)$$
$$nL^2 = 8(m + 2M_B/L)$$

If $M_B = mL$ the values of nL^2 for the three modes are as follows:

mode 1 24 m as before

mode 2 28.8m

mode 3 24m

It is clear that for this square slab, mode 2 is always less likely than mode 3. Provided that the plastic moment of resistance of each beam at midspan equals that for the full width of the square slab, mode 3 is no more likely than mode 1.

Corner levers and fans

Consider a simply supported, isotropic slab subjected to a uniformly distributed load of n. Alternative modes of failure are shown in Fig. 13.8.6. Mode (a) is the simple failure pattern usually assumed,

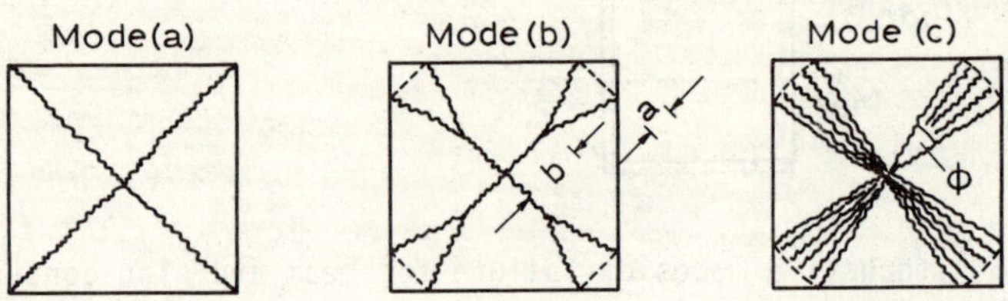

Fig.13.8.6 Alternative modes of failure for square slab

for which $nL^2 = 24m$. However, even if the corners are held down it is usually possible to obtain lower failure loads by assuming the presence of corner levers, as in mode (b), or corner fans, as in mode (c). An indication of the reduction in load is given in Table 13.8.1.

Table 13.8.1 Reductions for more critical modes of failure compared with the normally assumed mode

Mode	m'/m	nL^2/m	Reduction
(b) or (c)	1.0	24.0	—
(b)	0.5	23.0	4%
(b)	0.25	23.0	4%
(b)	0	22.0	8%
(c)	0	21.7	10%

Notice that the corner levers in mode (b) are eliminated (i.e. a = 0) if negative steel equal in magnitude to the positive steel is introduced at the corners. Good detailing procedure should therefore allow for this. Mode (c) gives the lowest value of all when m' = 0 and $\phi = 30^o$.

Acute angles with free edges

Corner levers can produce greater reductions in ultimate load for acute angles as shown in Fig. 13.8.7.

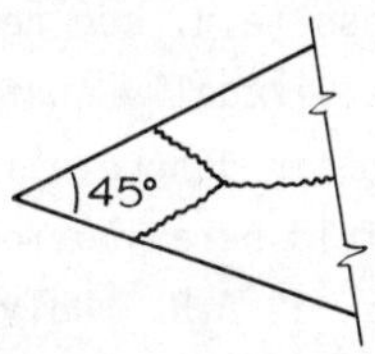

Fig.13.8.7 Corner lever at an acute angle

Table 13.8.2 Reductions due to corner levers for acute angles with free edges

Angle	m'/m	Reduction
45°	0	26%
45°	1	14%

Biaxial bending

The 'square yield criterion' which has been used for yield-line theory is shown by the broken line in Fig. 13.8.8. Yield in uniaxial bending is given by the four plotted points for which the moment equals m or m'.

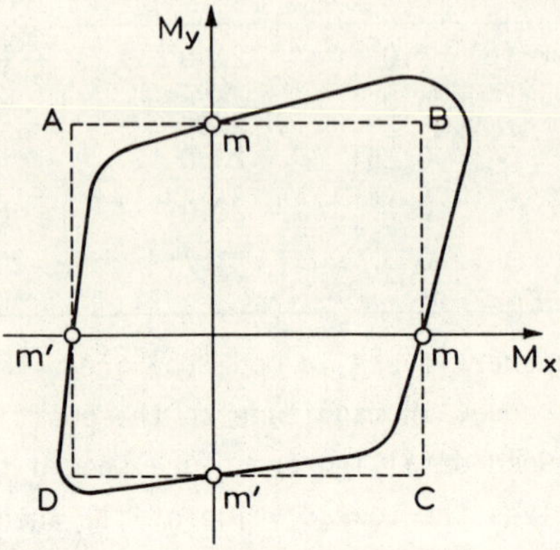

Fig.13.8.8 Yield criteria for a horizontally cast reinforced concrete slab

The behaviour of concrete under biaxial stress is dependent upon the magnitude of the lateral stress (e.g. see reference [13.13]). Hence although the reinforcement is virtually unaffected by the lateral stress the behaviour of the plain concrete is reflected (to a lesser and limited extent) in the yield behaviour of the reinforced slab as shown by the full line in Fig. 13.8.8. Only the compressive strength characteristics of the plain concrete are important in Fig. 13.8.8, hence the 'D' quadrant is affected by the compression/compression and not the tension/tension biaxial stress plot for plain concrete. Notice that there is a reserve of strength, using the square yield criterion, for biaxial moments of the same sign, but a deficiency of strength in the tension/compression quadrants (i.e. two moments of opposite sign). This deficiency of strength is made good by not curtailing the steel too rapidly. The use of the square yield criterion, which neglects interaction altogether, for yield-line analysis is satisfactory if slightly generous bond lengths in accordance with CP 110 are always provided (see chapter 5).

Compressive membrane action due to lateral restraint

Membrane action is the development of forces acting in the plane of the member. Such action can produce increases in the collapse load for slabs of up to ten times the value predicted by yield-line theory [13.2]. Arching action in deep beams has been discussed in chapter 5 (see Fig. 5.7.3). Although this simple form of arching must be negligible in simply supported slabs (due to the high span/depth ratio), similar arching action can be very pronounced if the edges of the slabs are restrained against lateral displacement. This is the case, for example, with interior panels of continuous slabs, where the horizontal thrust can be taken by the adjacent panels (instead of by the steel reinforcement) as shown in Fig. 13.8.9.

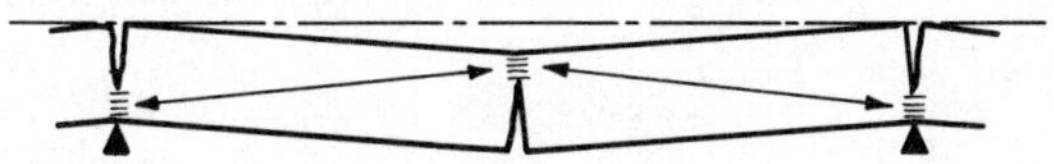

Fig.13.8.9 Compressive membrane (arching) action in laterally restrained slab

This action is made possible by the fact that slabs are under-reinforced so that the neutral axis depth is small. The centre of compression in the concrete at midspan can therefore be appreciably above that at the support.

Tensile membrane action

Tensile membrane action can also occur in restrained slabs at very large deflections as shown in Fig. 13.8.10, which shows a typical load-deflection curve for a uniformly loaded two-way rectangular slab with laterally restrained edges. Peak A represents the collapse load due to the fully developed yield-line pattern, enhanced by compressive membrane forces. Trough B occurs due to the reduction, with increasing deflections, in the compressive membrane action in the central region of the slab, and its replacement by tensile membrane forces. Beyond trough B the load is carried almost entirely by the reinforcement acting as a plastic tensile membrane with full depth cracking of the

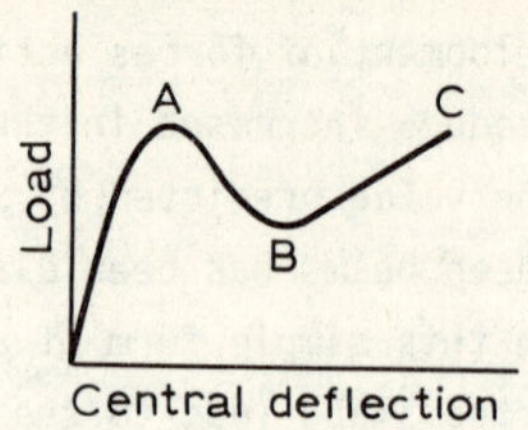

Fig.13.8.10 Typical load-deflection curve for a uniformly loaded, laterally restrained slab

concrete over the central region of the slab due to the extensive stretching of the slab surface. Fracture of the steel finally occurs at point C.

Membrane action in simply supported slabs

Tests on simply supported slabs have shown that arching action must be completely absent in this case since the cracks are seen to extend throughout the full depth of the slab as the deflections increase. Nevertheless moderate increases in the collapse load can occur at large deflections due to a modified form of membrane action. Consider a diagonal section AA across the slab shown in Fig. 13.8.11.

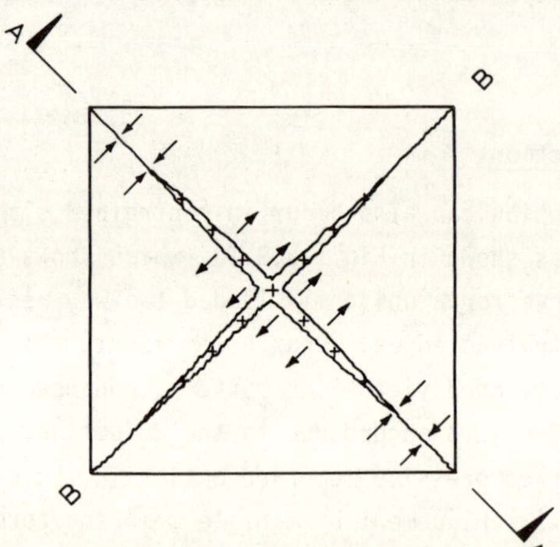

Fig.13.8.11 Membrane action in simply supported slab

The purely tensile forces across the centre at midspan must be balanced by additional compressive forces in the concrete (which is not separated in the flexural compression zone) near the edges of the slab. The same argument applies to the diagonal BB. The bending moment across AA exceeds the assumed yield value near the supports since the concrete can resist the superimposed compression, and the tension in the steel is relieved. This is illustrated by reference to the design curves for columns, which show increasing collapse moments for increasing axial loads when tension failures occur. The bending moment near midspan is only slightly reduced since the overstressed steel can continue to yield and provide a resistance moment due to the increasing inclination of the member. Tests indicate that the net result is a moderate increase in the collapse load.

Membrane action and design

These membrane effects are difficult to consider in design, either because of their interdependence with adjacent parts of the structure, due to lateral stiffness requirements or introduction of alternative modes of failure, or because of the extremely large deflections which are necessary. Nevertheless a knowledge of their existence increases the confidence which can be placed on upper-bound yield-line analysis.

13.9 HILLERBORG'S STRIP METHOD

Yield-line theory is essentially an upper-bound solution which is best suited for analysis. Hillerborg's strip method as modified by Wood & Armer [13.9], however, can be an exact solution, which is ideally suited for design. It gives, in effect, a great simplification of elastic design of slabs and the solution can then be checked by the virtual work method.

Simple strip method

The simple strip method assumes that the load at any point on a slab is carried to the nearest support by bending in the plane perpendicular to that support. For a rectangular slab the load would be allocated to the supporting beams by means of the lines of load discontinuity as shown in Fig. 13.9.1. This assumption in fact was used for determining

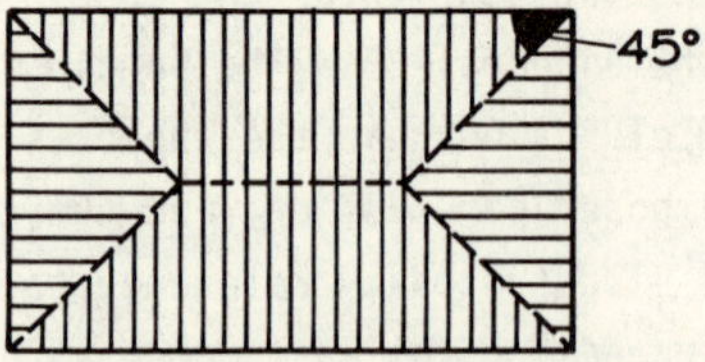

Fig.13.9.1 Lines of load discontinuity

the loads on beams supporting two-way slabs long before Hillerborg applied it to the slab itself. Although the elastic deflections at a given point would be dissimilar it can be assumed that plasticity enables the deflections for the two perpendicular spans to become compatible, before collapse occurs, and hence the design method is valid.

Equilibrium moments

The effect of distributing the load according to Fig. 13.9.1 on the strips XX and YY shown in Fig. 13.9.2 is to subject them to loading near their ends only. The bending moment diagrams are therefore parabolic near the ends and are constant over the central portions of the strips. A set of moments in equilibrium with the applied loading has therefore been formed. Reinforcement can now be provided to ensure that the yield condition is not violated anywhere in the slab. This requires reinforcement in the central portions of a strip even though it may not carry load locally. It is immediately apparent that the strip method attempts to design a slab with variable reinforcement, and not analyse a slab with given reinforcement.

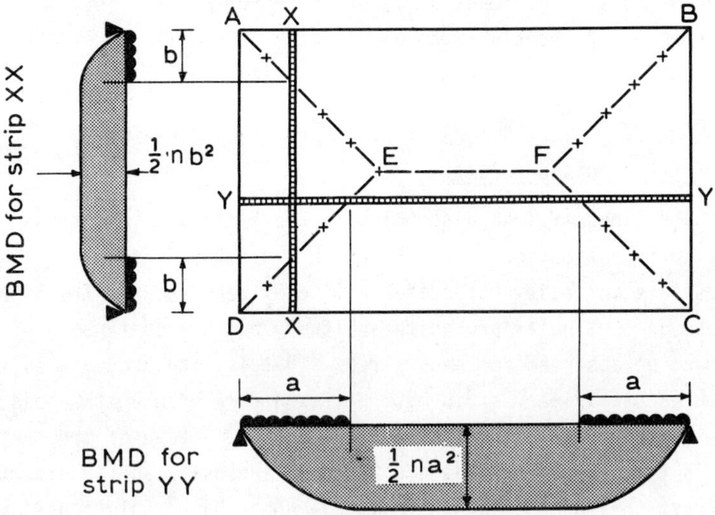

Fig.13.9.2 Bending moment diagrams for rectangular slab (after Hillerborg).

Equilibrium equation

Hillerborg intended his theory, in its simplest form, to provide a lower-bound solution. This would be valid only if the equilibrium equation was satisfied at all points, i.e:

$$\partial^2M_x/\partial x^2 + \partial^2M_y/\partial y^2 - 2\partial^2M_{xy}/\partial x\partial y = -n \qquad (13.9.1)$$

where moments M_x, M_y and twist M_{xy} follow the usual notation given in section 13.1.

It is clear from the above that the strip method attempts to make the stress field coincide with the field of resistance moments due to the reinforcement. This is a radical change in approach from yield-line theory methods and means that for the first time M_{xy} can be made zero, deliberately, so that the directions of the reinforcement (assumed to be along the x- and y-axes) coincide with the principal moment direction. Equation 13.9.1 can therefore be replaced by equations representing twistless beam-strip action:

$$\partial^2M_x/\partial x^2 = -\alpha n \quad \text{and} \quad \partial^2M_y/\partial y^2 = -(1-\alpha)n \qquad (13.9.2)$$

where α and $(1-\alpha)$ are the proportions of load taken in the strips in the x- and y-directions respectively. Usually α would be taken as

1 or 0, i.e. when $\alpha = 1$ all the load is dispersed in the x-direction and when $\alpha = 0$ all the load is dispersed in the y-direction. n could change throughout the slab without affecting the validity of the method, and point loads can be treated as local concentrations of n.

Lines of stress discontinuity

Most of the lines of load discontinuity are also lines of stress discontinuity but notice that line EF in Fig. 13.9.2 is not a line of stress discontinuity but rather a line of zero shear. The lines of stress discontinuity are quite arbitrary but the continuity of shear must be observed for each strip. That is, the accumulated shear force in each imaginary strip must be taken care of when the load dispersion changes. Notice that a very useful feature of the strip method is that the beam loads are known in intensity and in distribution. By contrast they are not known from true upper-bound solutions for yield-line theory.

Example 13.9.1

Use Hillerborg's strip method to determine the required values of m_x and m_y at any point of the rectangular slab shown in Example 13.7.4. Check the result by analysing the yield-line pattern which crosses every strip at a point of maximum bending moment (i.e. assume $\alpha = \lambda/2$).

Solution Refer to Fig. 13.9.3

Strip δx wide

Required resistance moments near centre

$$m_y \geq \tfrac{1}{2}nb^2$$

where b is the value of y at intersection of AE and strip.

Required resistance moments near edge

$$m_y \geq nby - \tfrac{1}{2}ny^2 = ny(b - \tfrac{1}{2}y) \qquad \text{(Answer)}$$

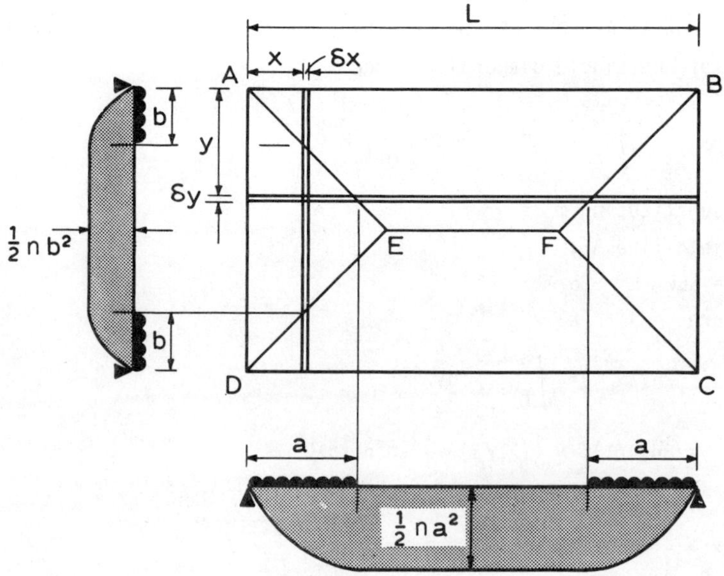

Fig.13.9.3 Example 13.9.1

Strip δy wide

Near centre

$m_x \geq \frac{1}{2}na^2$

Near edges

$m_x \geq nax - \frac{1}{2}nx^2 = nx(a - \frac{1}{2}x)$ (Answer)

where a is the value of x at intersection of AE and strip.

Hence maximum $m_x = \frac{1}{2}n(\alpha L)^2$ and maximum $m_y = \frac{1}{2}n(\frac{1}{2}\lambda L)^2$.

Yield-line analysis

If m_x and m_y are as given above only yield line EF has a constant resistance moment per unit length. The resistance moments of the diagonal yield lines AE, BF, CF and DE all vary. At any point along AE for example, m_x and m_y are obtained by putting a = x and b = y, i.e.

$m_x = \frac{1}{2}nx^2$ and $m_y = \frac{1}{2}ny^2$

The resistance moment of the x-direction steel at the yield line for the strip δy wide is $m_x \delta y$. Hence the total resistance moment of the x-direction steel for the slab shown in Example 13.7.4 is given by the four diagonals AE, BF, CF and DE, which equals four times the resistance moment of any single diagonal. Hence

$$\sum m_x \, \delta y = 4\int_0^{\frac{1}{2}\lambda L} m_x dy = 2n\int_0^{\frac{1}{2}\lambda L} x^2 dy$$

after substituting $m_x = \frac{1}{2}nx^2$.

On yield line AE,

$x/y = \alpha L/\frac{1}{2}\lambda L = 2\alpha/\lambda$

Therefore

$$\sum m_x \delta y = 2n(2\alpha/\lambda)^2 \int_0^{\frac{1}{2}\lambda L} y^2 dy$$

$$= 8n(\alpha/\lambda)^2 \times \tfrac{1}{3}(\lambda L/2)^3 = \tfrac{1}{3}n(\alpha L)^2\lambda L$$

Let m_X = maximum value of m_x, which is $\frac{1}{2}n(\alpha L)^2$. Hence

$\sum m_x \delta x = \frac{1}{3}(2\lambda L)m_X$

Similarly $\sum m_y \delta x$ for diagonals only equals $\frac{1}{3}(4\alpha L)m_Y$ where m_Y = maximum value of m_y, which is $\frac{1}{2}n(\frac{1}{2}\lambda L)^2$.

$\sum m_y \delta x$ for yield line EF is simply $(1 - 2\alpha)Lm_Y$, hence putting $\sum W\delta = \sum M\theta$ gives

$$n\lambda L^2[(1 - 2\alpha)/2 + 2\alpha/3]\delta$$
$$= 2(1 - 2\alpha)m_Y(2\delta/\lambda) + \tfrac{1}{3}(2\lambda)m_X(\delta/\alpha) + \tfrac{1}{3}(4\alpha)m_Y(2\delta/\lambda)$$

If $\alpha = \lambda/2$ then $m_X = m_Y$. Substituting in the above equation, putting $m_x = m_y = m$ and simplifying, gives

$m = n(\lambda L)^2/8$

which agrees with the previous value of maximum m_y (and m_x if $\alpha = \lambda/2$) given by Hillerborg's design.

When using Hillerborg's method it can be assumed that a load discontinuity at 45° is satisfactory for a 90° corner (and usually the most convenient) and that small variations from this distribution are not critical.

References

13.1 Timoshenko, S., and Wainowsky-Krieger, S., Theory of plates and shells, 2nd edn McGraw-Hill, New York, 1959 580 pp.

13.2 Wood R. H., Plastic and Elastic Design of Slabs and Plates, Thames & Hudson, London, 1961.

13.3 Taylor, R., Hayes B., and Mohamed Bhai, G. T. G., Coefficients for the design of slabs by the yield-line theory, Concrete, vol. 13, No. 5, May 1969, pp 171-2.

13.4 Ingerslev, A., The strength of rectangular slabs, Struct. Eng. vol. 1, No. 1, January, 1923, pp 3-14.

13.5 Jones, L. L., Ultimate load analysis of reinforced and prestressed concrete structures, Chatto & Windus, 1962.

13.6 Johansen, K. W., Yield-line theory, Cement and Concrete Association, London, 1962.

13.7 Kemp, K. O., Morley C. T., Nielson, M. P., Wood, R. H., and Jones, L. L., Recent developments in yield line theory, Magazine of Concrete Research special publication, London, May 1965, 74 pp.

13.8 Jones L. L., and Wood, R. H., Yield-line analysis of slabs, Thames & Hudson and Chatto & Windus, London, 1967.

13.9 Wood, R. H., and Armer, G. S. T., The theory of the strip method for design of slabs, Proc. Instn Civil Engrs, vol. 41, October, 1968, pp 285-311.

13.10 Kemp, K. O., The yield criterion for orthotropically reinforced concrete slabs, Int. J. Mech. Sci., vol 7, pp 737-46.

13.11 Prince, M. R., and Kemp K. O., A new approach to the yield criterion for isotropically reinforced concrete slabs, Magazine of Concrete Research, vol. 20, No. 62, March 1968, pp 13-20.

13.12 Mills, G. M., A partial kinking yield criterion for reinforced concrete slabs, Magazine of Concrete Research, vol 27, No. 90, March 1975, pp 13-22.

13.13 Newman, K., The structure and engineering properties of concrete, Theory of Arch Dams ed. J. R. Rydzewski, Pergamon, 1965, pp 683-712.

Exercises

1. A rectangular, orthotropically reinforced concrete slab measures 30 ft (9.15 m) by 20 ft (6.1 m). It is continuous over the shorter

sides and has ultimate resistance moments of 3000 lbf ft/ft(13.3 kNm/m) against both hogging and sagging parallel to these sides. It is simply supported on the longer sides and has an ultimate resistance moment of 6000 lbf ft/ft(26.6 kNm/m) against a sagging moment in this direction. Calculate the ultimate uniformly distributed load which can be carried by this slab. (UB3/G 1965, 36 min)

2. The triangular slab shown in Fig. A is isotropically reinforced to provide a yield moment of M per unit width, and carries a uniformly distributed load of n per unit area. Determine the relationship between M and n for the failure mode shown.

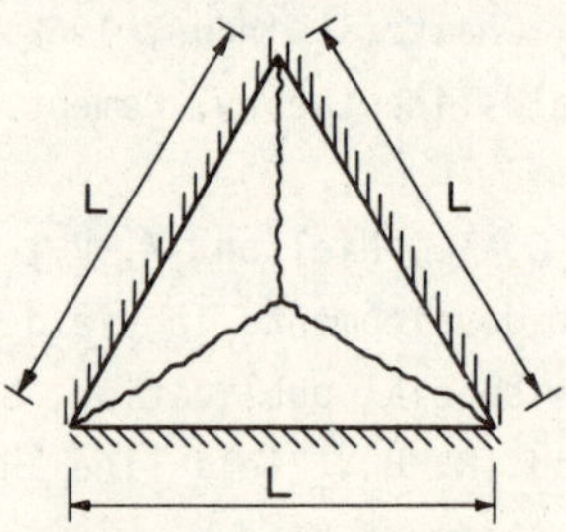

Ch.13 Fig.A Question 2

3. The rectangular slab shown in Fig. B is simply supported on two sides and built in on two sides. The reinforcement is isotropic

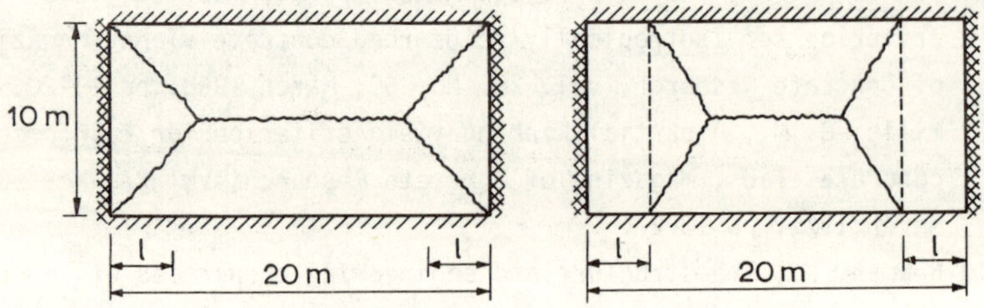

Ch.13 Fig.B Question 3

and provides sagging and hogging yield moments of +m and −m per unit width respectively. The top steel is curtailed at a distance l from the built-in supports. If the slab is uniformly loaded, two possible modes of failure are as shown. Determine the extent (l) of the top steel required for these two modes to occur at the same load, and determine the value of m required for this condition if the loading is 10 kN/m^2.

4. A reinforced concrete slab is subjected to a uniformly distributed load per unit area of n. The layout of the slab is as shown in Fig.C.

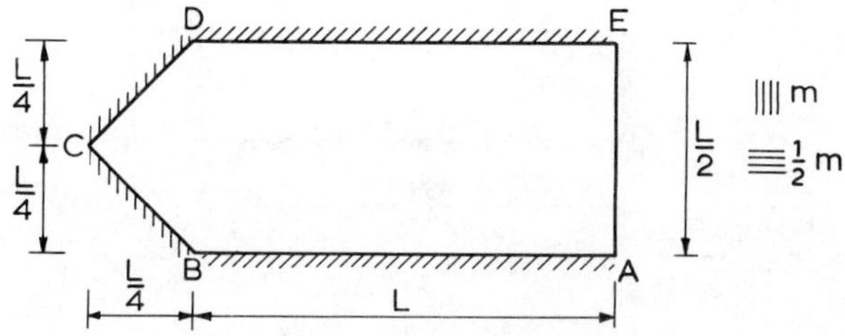

Ch.13 Fig.C Question 4

It is simply supported along sides AB, BC, CD and DE and unsupported along AE. The reinforcement in the direction parallel to AE produces an ultimate moment per unit length for a positive yield line parallel to AB of m. The reinforcement parallel to AB produces an ultimate moment per unit length of ½m. Determine the value of n if L = 40ft (12.2 m) and m = 6000 lbf ft/ft(26.7 kNm/m). (UB3/H 1968, 36 min)

5. A square reinforced concrete slab is rigidly built in at the edges and carries a central localized (point) load. It is isotropically reinforced at the bottom (m) and at the top (m'). A credible failure mechanism under overload would involve yield lines across the diagonals and around the sides. Alternatively, there might be numerous radial yield lines bounded by chords forming a multi-sided regular polygon contained within the boundaries of the slab. What is the percentage error in the failing load calculated by the less critical mode? What is the error if the load is uniformly distributed?

(UL3 1969, 36 min)

6. The reinforced concrete slab ABCD shown in Fig. D is to be designed for a uniformly distributed collapse load of 20 kN/m^2. Determine the required value of m_x' at E, point (0,2), and of m_x and m_y at F, point (2, 2), using Hillerborg's strip method. The co-ordinates are referred to A, point (0,0), as origin, m and m' are the resistance moments per unit width for positive and negative bending moments respectively and

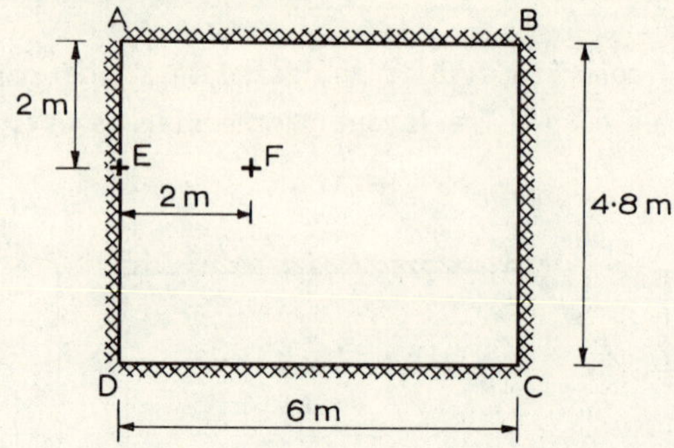

Ch.13 Fig.D Question 6

suffices x and y refer to the x-direction and y-direction reinforcement respectively.

The proposed Unified Code (i.e. CP 110—Ed.) recommends that the ratio of the support moment to the midspan moment should be between 1.5 and 2.0 for fixed slabs. Would a reduction of 15% in m_x' at E as obtained above be sufficient to satisfy this clause?

Comment, very briefly, on the essential differences in the disposition of the steel reinforcement for a design based on the strip method compared with one based solely upon the usual yield-line analysis approach. (UB3/H 1969, 45 min)

7. A rectangular orthotropically reinforced concrete slab is continuous over the longer sides and one shorter side as shown in Fig. E. The

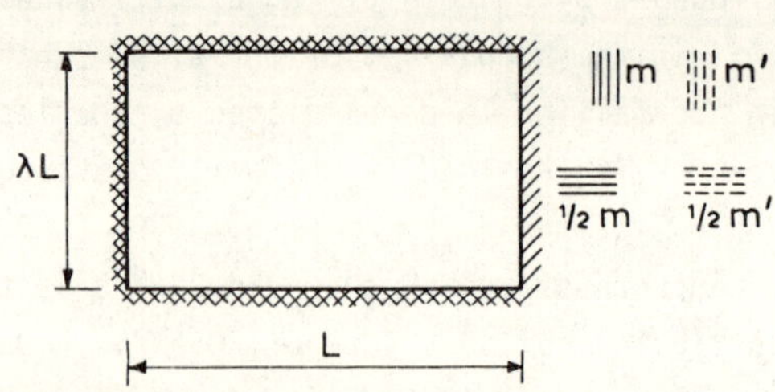

Ch.13 Fig.E Question 7

positive and negative resistance moments per unit length of the steel parallel to the longer side L are $\frac{1}{2}m$ and $\frac{1}{2}m'$ and the corresponding values for the steel parallel to the shorter side λL are m and m'.

Obtain a general expression for the ultimate uniformly distributed load n in terms of m and m'. Hence calculate n if L = 7.5 m, λ = 0.6 and m = m' = 40 kNm/m. (UB3/H 1970, 45 min)

14 Design calculations for reinforced concrete

The present chapter is based upon the (technical) report prepared by a working party of the Concrete Society [14.1] so that standard design calculations could be presented in as clear, consistent and economical manner as possible. Computers are also considered since they are now a very useful aid to design, and it is emphasized that design calculations even when computer-assisted, must be capable of being understood under all circumstances.

14.1 STANDARDIZED PROCEDURE FOR THE PRESENTATION OF CALCULATIONS

Calculations are prepared by an engineer for three main reasons:

(i) To satisfy himself as to the suitability of his design.

(ii) For use by the detailer preparing the detailed drawings.

(iii) For submission to other authorities for checking purposes to ensure conformity with Codes of Practice and Building Regulations currently in force.

It is evident that the results of the calculations must be clearly indicated so that the detailer can easily extract the reinforcement required. They must also be presented in such a way that they can be readily followed by a checking authority or by any engineer who may have cause to refer to them subsequently. The report considered that these requirements could best be achieved by utilizing:

(i) a consistent set of units and notation

(ii) a standard method of documentation

(iii) a logical sequence for the presentation of the calculations.

The units (SI) and the notation system should be consistent with those used in the appropriate code. Both the design sequence and the documentation are now considered.

Design process

The report considers two principal stages in the design process, 'preliminary design considerations' and 'analysis and detailed design', thus following generally the principles set out in the Institution of Structural Engineers' document *Aims of Structural Design*.

Preliminary design considerations

This stage commences with an appreciation of the client's requirements combined with the collection of facts on which the design will be based. Design decisions are taken and a preliminary design is carried out to determine sizes of structural elements for the information of other members of the design team. This design is usually produced with a minimum of analysis. The decisions and design data for this stage (1.0) are summarized on design information and structural summary sheets, as indicated later. A typical sequence for considering and presenting the design process and the resulting calculations is as follows:

1.0 PRELIMINARY DESIGN CONSIDERATIONS

1.1 USER REQUIREMENTS

1.1.1 Description of main requirements

e.g. nature of use—crane requirements etc.

1.2 CONDITIONS

1.2.1 Internal

Fire rating, heating requirements etc.

1.2.2 External

Wind, frost, temperature, adjoining buildings etc.

1.2.3 Subsoil

Soil conditions, chemical content, existing services, mining risk etc.

1.2.4 Others

Vibration, earthquakes, noise etc.

1.3 LOADING

Imposed, wind, inertial etc.

1.4 TYPE OF CONSTRUCTION

1.4.1 Superstructure

Framed, crosswall etc.

1.4.2 Foundations

Strip, raft etc.

1.5 STABILITY

1.5.1 Overall stability

1.5.2 Progressive collapse

1.6 MATERIALS

1.6.1 Structural

Concrete mix, reinforcement etc.

1.6.2 Cladding

Precast concrete, asbestos sheet etc.

1.6.3 Floor finishes

Lino-tile on screed, granolithic etc.

1.6.4 Partitions

Breeze block, plaster-board etc.

1.6.5 Jointing

Type of precast joint, expansion joint, contraction joint etc.

Factors influencing the choice of the above materials include cost, appearance, construction and workmanship, location and availability.

Structural analysis and detailed design

The second stage involves a full structural analysis to support the design decisions taken at the appraisal stage and to check the scheme for soundness in service and during erection. The structural analysis and detailed design (2.0) must be clearly presented on calculation sheets and the report outlines the extent of each item as follows:

2.0 ANALYSIS AND DETAILED DESIGN

A particular design could include additional items and exclude others. For example, prestressing losses and bursting reinforcement would only be considered in prestressed concrete design.

2.1 STRUCTURAL ANALYSIS

2.1.1 Geometry of structure and approximate element sizes

General sketch of the structural layout and choice of member sizes.

Output: frame and element dimensions.

2.1.2 Idealization of structure

Representation of frame as an analytical model.

2.1.3 Consideration of interaction of structural elements

Determination of degree of fixity at joints and degree of stiffness of members. Effect on analysis of local variations in member sizes.

2.1.4 Loading

An assessment of the dead, imposed, and wind loadings on the structu

Output: (characteristic) loads.

2.1.5 Overall stability

Ensure that the restoring moment is greater than the applied overturning moment multiplied by an appropriate safety factor.

2.1.6 Progressive collapse

A statement of assumptions made and method adopted.

2.1.7 Choice of critical limit state

The frame will generally be analysed for the limit state of collapse. Output: design loads.

2.1.8 Frame analysis

Choice of analysis is made and carried out for different loading combinations in order to determine the worst conditions in members, resulting in nodal bending moments, torsions, direct forces, shears, displacements and rotations. Output: bending moments etc.

2.2 ELEMENT DESIGN

2.2.1 Design data

This section will either refer to the pages giving the design data, concrete and steel strengths etc., or state the data if they are local to this part of the calculations.

2.2.2 Section and element geometry

The section size is confirmed. The effective span or height is calculated. The span/breadth, span/depth ratios etc. are calculated. Output: section size.

2.2.3 Forces and moments from structural analysis

The dead and imposed loadings are set out, together with the end forces and moments taken from the structural analysis. If there is no structural analysis, the end forces and moments are calculated in the next section. A loading diagram is very helpful.

2.2.4 Section design — forces and moments

This section sets out the calculations giving the maximum bending moments, shear forces, axial loads and torsions making allowance for redistribution etc. Output: Final bending moments, shear forces, axial loads and torsions.

2.2.5 Section design — concrete stresses

Check that the compressive, tensile and shear stresses in the concrete are within the upper limits. Output: concrete stresses.

2.2.6 Longitudinal reinforcement

The longitudinal bars required at the critical sections are

calculated. The minimum and maximum permissible quantities of reinforcement are checked. Output: Reinforcement number, type, size and pitch.

2.2.7 Transverse reinforcement

The quantities of stirrups, links, distribution reinforcement etc. are calculated from the shear torsion requirements. The minimum and maximum permissible quantities of reinforcement are checked. Output: reinforcement number, type, size and pitch.

2.2.8 Bond and anchorage

This section checks that suitable reinforcement is used to satisfy flexural bond and calculates the anchorage required for each bar and the stopping-off points for curtailed bars (including lap lengths). Output: anchorage and cut-off lengths.

2.2.9 Bearing stresses

Where bearing stresses may become high (e.g. beam seatings) a check calculation should be made.

2.2.10 Checks for other limit states

If necessary, check calculations should be given for deflection, local damage, vibration, etc.

2.2.11 Specific detailing problems

Check calculations should be made for problems specific to the job, e.g. stresses in concrete developed inside bends of reinforcement.

Documentation

The documentation based on the procedure given in the report should consist of four parts:

(i) design information sheet

(ii) structural summary sheet

(iii) index sheet

(iv) calculation sheets

Design information sheet

The design information normally includes many of the items in 1.0. A recommended design information sheet is used in Fig. 14.2.1 (see later). The sheet summarizes the information on which the engineer has based his design and includes those items which are of particular interest to the checking authority.

Structural summary sheet

This sheet supplements, as necessary, the information given on the design information sheet and illustrates the main structural elements, the form of the structure and the provisions for overall stability. Continuation sheets may be necessary to show all the relevant information for larger jobs. A recommended summary sheet is shown in Fig. 14.2.2 (see later).

Index

An index sheet should be prepared for each set of calculations to allow anyone who has to refer to them to appreciate readily the overall layout and find quickly a particular aspect they wish to examine. The index should be simple, consisting of a description of the item and the page number on which it starts. It is also advantageous to include at the end of the index the names of any reference books used. A special index sheet was not designed as it was considered that the standard calculation sheet or structural summary sheet could be used for this purpose.

Calculation sheets

Calculations should be written on a sheet of standard format which includes columns for member reference, the main calculation and the output of the calculation. 5-mm squares are used to assist in drawing sketches. Typical calculation sheets are used in Fig.14.2.3 (see later).

14.2 TYPICAL EXAMPLES OF DESIGN CALCULATIONS

Examples of designs of reinforced concrete members based on the recommended procedure are given in the Concrete Society Report [14.1] and by Higgins & Hollington [14.2]. Figs. 14.2.1 and 2 have been reproduced from the report and serve to illustrate the method of presentation. They show a design information page and a structural summary page. Figs.14.2.3 and 4 (after [14.2]) show calculations for an edge beam and a typical column. The reader should refer to [14.1] and [14.2] for more complete calculations.

It is important in the preparation of calculations that

(i) all headings are underlined

(ii) a horizontal line is drawn across the page after each stage in the calculation

(iii) the output of the calculation is clearly indicated in the right hand column.

14.3 APPLICATION OF SMALL COMPUTERS

Computers are an aid to the designer. They can relieve the designer of routine calculations and can offer alternative methods of solution which are not possible or practicable using a manual approach. Small computers can usually operate most economically by carrying out the computations involved in manual methods. Larger computers, however, can usually operate most satisfactorily using techniques which need not be suitable for solution by manual methods and are therefore considered separately. The final documentation for all computer calculations should follow the same general principles and be intelligible to the reader.

Documentation for computer calculations

The documentation should follow the same principles that have been indicated for manual calculations. The report realized that the type of computer which is used will have a considerable effect on the format of the calculations; nevertheless the following principles were considered important:

(i) The calculation must be completely self-explanatory and must not assume that the reader of the calculations has any knowledge of computers.

(ii) The calculations must state clearly the equations or process by which an answer is obtained (this can usually be satisfied simply by attaching preprinted calculation sheets to the computer answer sheets).

(iii) A description, and the magnitude, of all data and answers must be clearly stated.

(iv) In order to facilitate checking, intermediate answers should be given which will permit the calculations to be checked at an interim stage.

Operation of small computers

Small (desk-top) computers have very limited storage facilities and must therefore utilize familiar manual techniques, with human intervention often being required. For example, a series of programmes

A. D. Vance & Partners **Consulting Engineers** **123a Victoria Street** **London, S.W.1.**	Contract: Residential Block Creekgate Sub-section: Main building **DESIGN INFORMATION**	Job Ref A/704 Sheet No: 1 Made by: M.R.Smith Checked by: T.D.Brown Date: 15.4.70

Client: County Borough of Creekgate Architect: M. J. Cranthorne Esq. F.R.I.B.A. Borough [Architect. Chief Engineer County Borough of Creekgate	Engineer Responsible: K.R.E. Richards Building Regulation Authority or other
1. The Building Regulations, 1965 2. C.P. 114 Part 2, 1969, "The structural use of R.C. in [buildings" 3. C.P.3 Chap. V Loading, Part 1, 1967, "Dead & imposed loads." 4. Draft C.P.3 Chap. V Loading, Part 2, 1969, "Wind loading"	Relevant Building Regulations and Design Codes
45 individual study/bedrooms 3 dormitories with bathrooms etc. Dining hall and kitchen Recreation room Basement garage, boiler house & tank room.	Intended use of Structure
2 hours resistance	Fire resistance requirements
Roof — 1·5 kN/m² Bedrooms & dormitories — 1·5 kN/m² Dining & recreation rooms — 2·0 kN/m² Kitchen — 3·0 kN/m² Corridors — 4·0 kN/m² Finishes & partitions — 1·5 kN/m²	General loading conditions
Speed: 44 metres per second Factors: $S_1 = 1{\cdot}0$, $S_2 = 0{\cdot}86$, $S_3 = 1{\cdot}0$	Wind loading conditions
Normal atmospheric	Exposure conditions
Sulphate resistant cement required in foundations Buried C.E.G.B. high-tension cable across north end. Allowable bearing pressure – retaining walls = 250 kN/m² – end bearing on piles = 500 kN/m²	Subsoil conditions
Bored piles to accommodate extra loading for future extension. Retaining wall around basement area.	Foundation type
Concrete: Minimum cube crushing strengths at 28 days:– Columns — 30 MN/m² Beams & slabs — 25·5 MN/m² Foundations — 21 MN/m² Reinforcement: Maximum permissible working stresses. High tensile: p_{sc} = 175 MN/m², p_{st} = 210 MN/m² (if dia. > 20 mm) & 230 MN/m² (if dia. ≯ 20 mm) Mild steel: p_{sc} = 125 MN/m², p_{st} = 140 MN/m²	Material data
Mass density of reinforced concrete = 2400 kg/m³ Equivalent value in force units = 23·6 kN/m³	Other relevant information

Fig.14.2.1 Design information page [After 14.1]

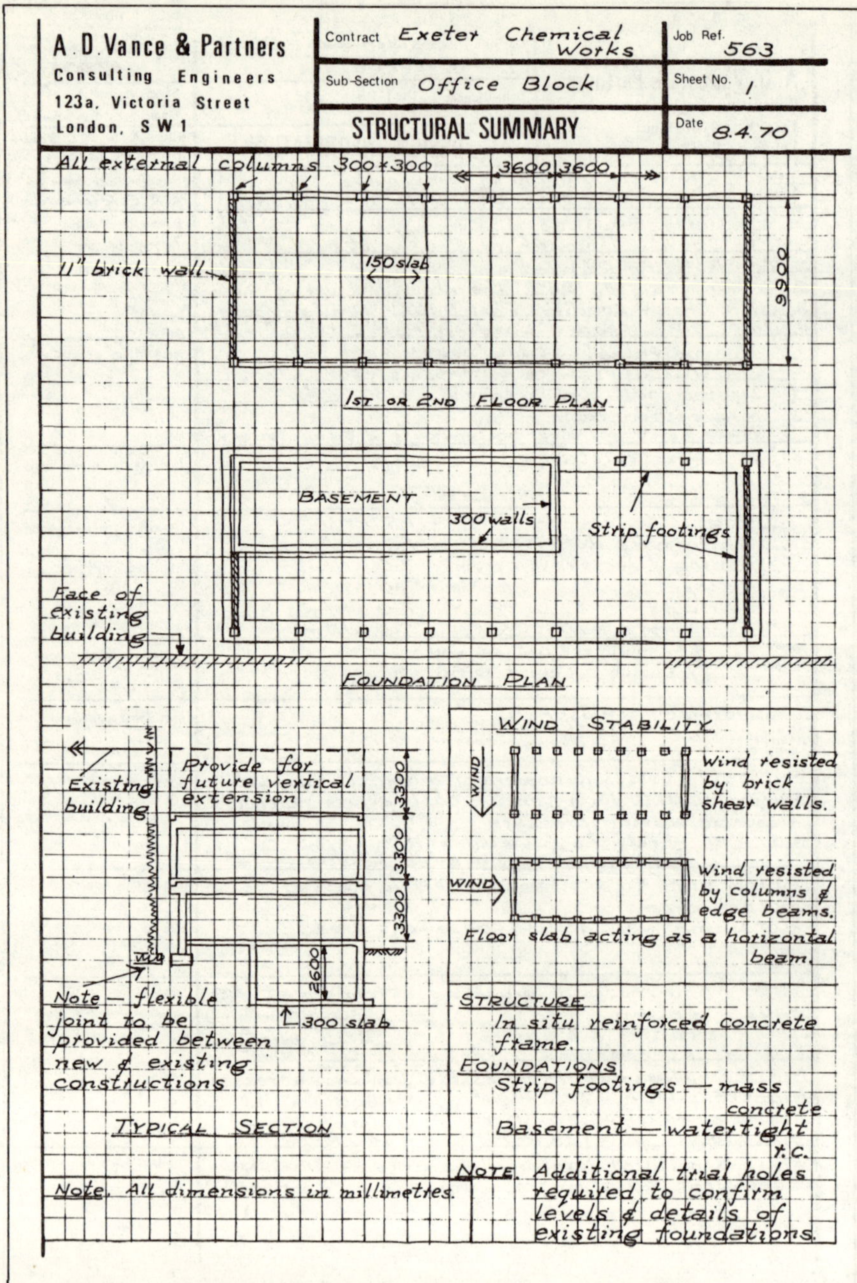

Fig.14.2.2 Structural summary page. [After 14.1]

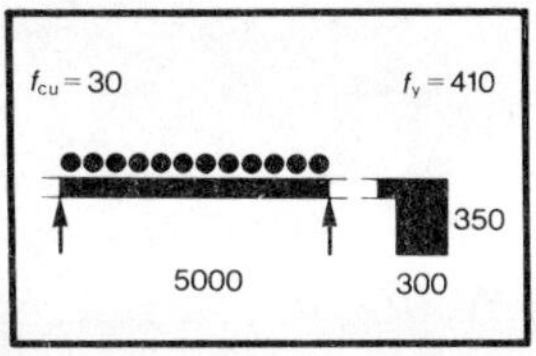

Edge beam

interior-span flanged beam

CP110 ref.	CALCULATIONS	OUTPUT
3.1.2.3 Table 19 Table 54	DURABILITY & FIRE RESISTANCE Nominal cover for moderate condition of exposure = 30 mm Fire resistance 300 wide beam with 40 av. cover to main reinft > 1 hr.	cover to links = 30 mm ∴ fire resistance OK
      2.3.3.1	LOADING Dead load from slab (1250 strip) = 28·2 × 1·25 = 35·2 Self-weight 0·35 × 0·3 × 23·6 × 5 = 12·4 Cladding @ 5 kN/m = 5 × 5 = 25·0 ∴ Characteristic dead load 72·6 kN " imposed " from slab 15 × 1·25 = 18·8 kN Design load = $1·4G_k + 1·6Q_k$ = 101·6 + 30·1 = 131·7 kN	 G_k = 72·6 kN Q_k = 18·8 kN F = 131·7 kN
3.3.4 Table 4	ULTIMATE B.M's Interior support: $\frac{Fl}{10}$ = 131·7 × $\frac{5}{10}$ = 65·9 kNm Mid-int. span: $\frac{Fl}{14}$ = 131·7 × $\frac{5}{14}$ = 47·2 kNm	
Design Chart No 2 3.3.1.2 Eqn 6 Table 4 Table 5 Table 6 Eqn 8 Eqn 9 3.11.4.3 3.3.6.1	REINFORCEMENT Interior support: $\frac{M}{bd^2} = \frac{65·9 \times 10^6}{300 \times 275^2}$ = 2·9 N/mm² ∴ A_s = 0·94% bd = 0·0094 × 300 × 275 = 775 mm² Mid-int. span: eff. flange width = 300 + ($\frac{1}{10}$ × 0·7 × 5000) = 650 mm ∴ $A_s = \frac{M}{0·87f_y(d - h_f/2)} = \frac{47·2 \times 10^6}{0·87 \times 410(300 - 175/2)}$ = 625 mm² Shear force; V = 0·55F = 0·55 × 131·7 = 72·4 kN Int. support A_s provided = 982 mm² = 1·2% bd ∴ v_c = 0·78 N/mm² $v = \frac{V}{bd} = \frac{72·4 \times 10^3}{300 \times 275}$ = 0·88 > 0·78 < 4·10 N/mm² $\frac{A_{sv}}{s_v} = \frac{b(v - v_c)}{0·87 f_{yv}} = \frac{300(0·88 - 0·78)}{0·87 \times 250}$ = 0·14 but min area (nominal) links, $\frac{A_{sv}}{s_v}$ = 0·002b_t = 0·002 × 300 = 0·60 and max s_v = 0·75d = 0·75 × 275 = 206 mm ∴ Provide R10 @ 200, ∴ $\frac{A_{sv}}{s_v} = \frac{157}{200}$ = 0·74 > 0·60	(982 mm²) 2Y25 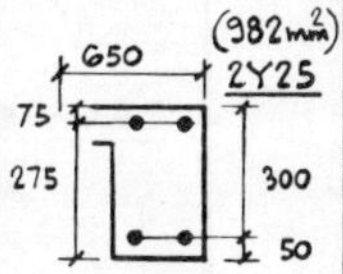 2Y20 (628 mm²)  R10 @ 200 throughout span
3.3.8 Table 10 3.3.8.2	DEFLECTION basic span/eff. depth ratio = 26 max. $100\frac{A_s}{bd} = \frac{100 \times 628}{650 \times 300}$ = 0·32, ∴ modification factor = 1·5 $\frac{b_w}{b} = \frac{300}{650}$ = 0·46 > 0·3, ∴ " " = 0·85 ∴ Allowable span/eff. depth ratio = 26 × 1·5 × 0·85 = 33 ∴ Max. span = 33 × 300 mm = 9900 > 5000	 ∴ l/d ratio OK
3.11.8.2(3) Table 24	CRACKING Bar mk 4, corner dist. = $\sqrt{75^2 + 60^2} + 13\sqrt{2} - 13$ = 101 mm Let $f_y = 410 \times \frac{775}{982} \simeq 330$ N/mm². ∴ Allowable dist = ½ × 208 = 104 > 101	∴ OK. 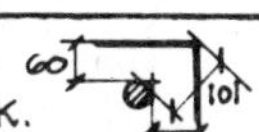
3.1.2.2	STABILITY PERIPHERAL tie F_t = 36 kN Min area continuity reinft = $\frac{36 \times 10^3}{410}$ = 88 mm²	TOP min 2Y12 OK (226 mm²)

Fig.14.2.3 Calculation page for edge beam. [After 14.2]

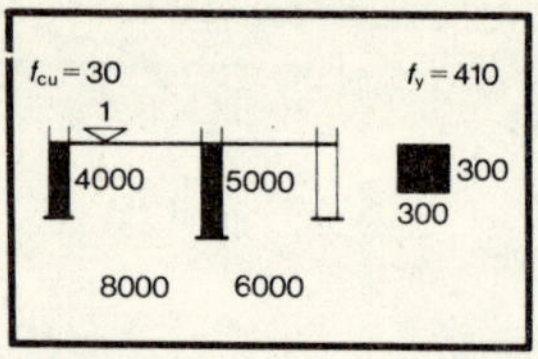

Columns

slender and short columns

CP110 ref.	CALCULATIONS	OUTPUT
3.5.1.3	All columns braced – lateral stability provided by walls	
3.1.2.3 Table 19 Table 59	DURABILITY & FIRE RESISTANCE Nominal cover to links for moderate condition of exposure = 30 mm Fire resistance of 300×300 column = 2 hrs > 1 hr reqd	∴ cover to main bars = 40 mm ∴ fire resistance OK.
2.4.3.1(1)	COLUMN STIFFNESSES $I = \frac{bh^3}{12} = \frac{300 \times 300^3}{12} = 675 \times 10^6\,mm^4$ Internal col: fdn → 1st, $\frac{I}{l} = \frac{675 \times 10^6}{5000} = 135 \times 10^3 mm^3$ " " 1st → 2nd, $\frac{I}{l} = \frac{675 \times 10^6}{3500} = 193 \times 10^3 mm^3$ External col: fdn → 1st, $\frac{I}{l} = \frac{675 \times 10^6}{4000} = 169 \times 10^3 mm^3$ " " 1st → 2nd, $\frac{I}{l} = \frac{675 \times 10^6}{3500} = 193 \times 10^3 mm^3$ BEAM STIFFNESSES $I = \frac{bh^3}{12} = \frac{300 \times 500^3}{12} = 3125 \times 10^6\,mm^4$ 8m span : $\frac{I}{l} = \frac{3125 \times 10^6}{8000} = 391 \times 10^3 mm^3$ 6m span : $\frac{I}{l} = \frac{3125 \times 10^6}{6000} = 521 \times 10^3 mm^3$	
3.5.2	MOMENTS IN COLUMNS 1st floor junctions, assume stiffnesses $(\frac{I}{l})$ as follows:	

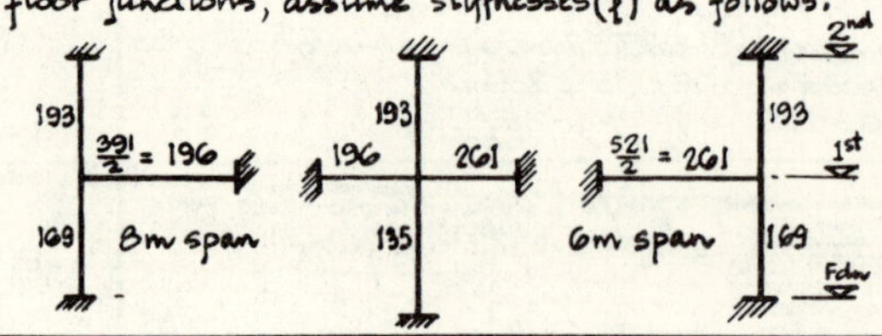

End of 8m span	Internal column	End of 6m span	
558 × $10^3 mm^3$	783 × $10^3 mm^3$	623 × $10^3 mm^3$	$\Sigma \frac{I}{l}$
356 kNm	264 kNm	200 kNm	Out of balance B.M. See page 10
Case I or III loading	Case I loading	Case II or III loading	
$356 \times \frac{193}{558} = 123$ kNm	$264 \times \frac{193}{785} = 65{\cdot}2$ kNm	$200 \times \frac{193}{623} = 62{\cdot}0$ kNm	B.M into upper col.
$356 \times \frac{169}{558} \doteq 108$ kNm	$264 \times \frac{135}{785} = 45{\cdot}4$ kNm	$200 \times \frac{169}{623} = 54{\cdot}6$ kNm	B.M into lower col.

Fig.14.2.4 Calculation pages for typical column. [After 14.2]

CP110 ref.	CALCULATIONS	OUTPUT
	INTERNAL COLUMN (foundation → 1st fl.)	
3.5.1.4	Effective height: $= l_e$	
	At foundation $\alpha_{c2} = 10$ (nominal moment connection)	
	At 1st floor $\alpha_{c1} = \frac{135+193}{391+521} = 0.36$ ($\alpha_{c\,min}$)	
Eqn 21	$l_e = l_o(0.85 + 0.05\alpha_{c\,min})$	
	$= 4500(0.85 + 0.05 \times 0.36) = 3900$ mm	$l_e = 3900$ mm
3.5.1.2	Design as slender col. since $\frac{l_e}{h} = \frac{3900}{12} = 13 > 12$	SLENDER COLUMN

AXIAL LOADING

Floor supported	Beam Load (see page 11) V kN	Column design loads kN Imposed $V \times \frac{1.6Q_k}{1.4G_k+1.6Q_k}$	Column design loads kN Dead $V \times \frac{1.4G_k}{1.4G_k+1.6Q_k}$
Roof	467	102	365
			self-wt 6
3rd	568	203	365
			" 6
2nd	568	203	365
			" 6
1st	568	203	365
			" 9
Sub-totals		711	1487
Reduced imposed load (CP3, Table 2) $711 \times 0.7 =$			500
∴ Total AXIAL load @ fdn.		=	1987 kN

CP110 ref.	CALCULATIONS	OUTPUT
		$N = 1987$ kN
	BENDING MOMENT	
	Moment into column = 45.4 kNm (see p.18)	
3.5.5	Nominal eccentricity B.M $= 0.05h_{min}N$	
	$= 0.05 \times 300 \times 1987 = 29.8 < 45.4$ kNm	
3.5.7.1	∴ $M_1 = 0$, $M_2 = 45.4$ kNm	
Eqn 34	$M_i = 0.4M_1 + 0.6M_2 = 0 + 0.6 \times 45.4 = 27.2$ kNm $> 0.4M_2$	
Eqn 33	$M_t = M_i + \frac{Nh}{1750}\left(\frac{l_e}{h}\right)^2\left(1 - 0.0035\frac{l_e}{h}\right)$	
	$= 27.2 + \left[\frac{1987 \times 300}{1750}(13)^2(1 - 0.0035 \times 13)\right]$	
	$= 27.2 + 55 = 82.2$ kNm $> M_2$	$M_t = 82.2$ kNm
	REINFORCEMENT	INTERNAL COLUMN 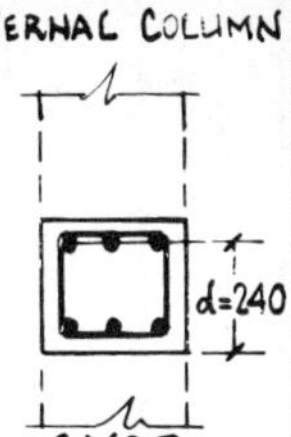
Design Chart Nº 66	$\frac{d}{h} = \frac{240}{300} = 0.8$, $\frac{N}{bh} = \frac{1987 \times 10^3}{300^2} = 22.1$ N/mm², $\frac{M}{bh^2} = \frac{82.2 \times 10^6}{300^3} = 3$ N/mm²	
	Adopt $A_{sc} = 5\%$ Then from simplified Design Chart, page 28	
Eqn 38	$K = \frac{N_{uz} - N}{N_{uz} - N_{bal}} = \frac{28.5 - 22.1}{28.5 - 7.5} = 0.31 < 1.0$	
Design Chart Nº 66	∴ reduced $M_t = 27.2 + (55 \times 0.31) = 44.3$ kNm. ∴ $\frac{M}{bh^2} = \frac{44.3 \times 10^6}{300^3} = 1.9$ N/mm²	6Y32
	∴ $A_{sc} = 5\%\,bh$ OK $= 0.05 \times 300^2 = 4500$ mm²	(4830 mm²)

Fig.14.2.4 Calculation pages for typical column. [After 14.2] (continued)

CP110 ref.	CALCULATIONS	OUTPUT
	EXTERNAL COLUMN { foundation → 1st fl. / end of 8m span.	
3.5.1.4	Effective height: $= l_e$	
	$\alpha_{c1} = \alpha_{c2} = 1.0$ (nominal moment connection)	
	$\therefore l_e = l_0 = 3500$ mm	$l_e = 3500$ mm
3.5.1.2	Design as short col. since $\frac{l_e}{h} = \frac{3500}{300} = 11.67 < 12$	SHORT COLUMN
	AXIAL LOADING	

Floor supported	Beam Loads (see pages 11 & 16) V kN	Column design loads kN: Imposed $V \times \frac{1.6Q_k}{1.4G_k+1.6Q_k}$	Column design loads kN: Dead $V \times \frac{1.4G_k}{1.4G_k+1.6Q_k}$
Roof	Main 184	40	144
	Edge 132	30	102
			self-wt. 6
3rd	Main 224	80	144
	Edge 132	30	102
			" 6
2nd	Main 224	80	144
	Edge 132	30	102
			" 6
1st	Main 224	80	144
	Edge 132	30	102
			" 7
Sub totals		400	1009

CP110 ref.	CALCULATIONS	OUTPUT
	Reduced imposed load (CP3, Table 2) $400 \times 0.7 = 280$	
	∴ Total AXIAL load @ fdn. = 1289 kN	N = 1289 kN
	BENDING MOMENT	
	Moment into column = 108 kNm (see p. 18)	
3.5.5	Nominal eccentricity B.M. $= 0.05h_{min}N$	
	$= 0.05 \times 300 \times 1289 = 19.4 < 108$ kNm	M = 108 kNm
	REINFORCEMENT	
Table 2	Age at loading = 6 months, ∴ conc. strength = 36 N/mm²	
Design Charts No 66 & No 70	$\frac{d}{h} = \frac{240}{300} = 0.8$	
	$\frac{N}{bh} = \frac{1289 \times 10^3}{300^2} = 14.3$ N/mm²	
	$\frac{M}{bh^2} = \frac{108 \times 10^6}{300^3} = 4.0$ N/mm²	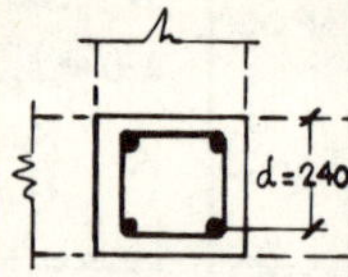
	From chart 66 ($f_{cu} = 30$) $A_{sc} = 4.5\%\,bh$	
	" " 70 ($f_{cu} = 40$) $A_{sc} = 3.1\%\,bh$	
	∴ interpolating, (conc. strength = 36) $A_{sc} = 3.6\%\,bh$	
	$= 0.036 \times 300^2 = 3240$ mm²	4 Y 32 (3220 mm²)

Fig.14.2.4 Calculation pages for typical column. [After 14.2] (continued)

may be necessary in order to obtain the required final output. In addition it may be necessary to use the same reasoning with the output from a particular program as is used in a manual approach. These two points, as well as the documentation, are illustrated by reference to work which has been carried out by GKN Reinforcements Limited [14.3].

This work demonstrates that not only can a small computer readily perform tedious arithmetic, it can also greatly reduce the time spent in writing out calculations if the system is also backed, as in the GKN Reinforcements' method, with pre-printed calculation sheets and answer sheets to which the computer output is attached. Such standard texts and setting-out of equations eliminate any need for further description. *

Frame and beam analysis by small computer

General framework analysis programmes covering various loading conditions for virtually any configuration of the members are readily available for large computers based on matrix methods. However, frameworks may also be analysed by the use of small computers. For example, a method of solution was developed by GKN Reinforcements Limited for use with the Olivetti Programma 101 desk-top computer. Because of the storage limitation of this machine, the problem was solved with a number of small related programs. The complete 'package' of programs determined both bending moments and shear forces at five points along each member of a frame or beam system of any size. The structural system could be subjected to any loading configuration but the stiffness of any member had to be constant along its length.

The technique used followed the same sequence as if analysis were being carried out manually using the moment-distribution method and is shown diagrammatically in Fig. 14.3.1. First the loading data was specified for each member, and 'program 20' used to calculate the fixed

*_Note:_ The development of computers is so rapid that reference to any individual machine is quickly outdated. The effect of this development is that pocket calculators with programmable facilities will move into the 'small computer' field while the desk-top computers will overtake the older medium size computers. Thus the GKN general type of approach will be applicable to even smaller computers with ever increasing capabilities.

end moments. Next the designer chose one from four available programs to carry out the moment-distribution process. The final member support moments determined by moment distribution were then added to the

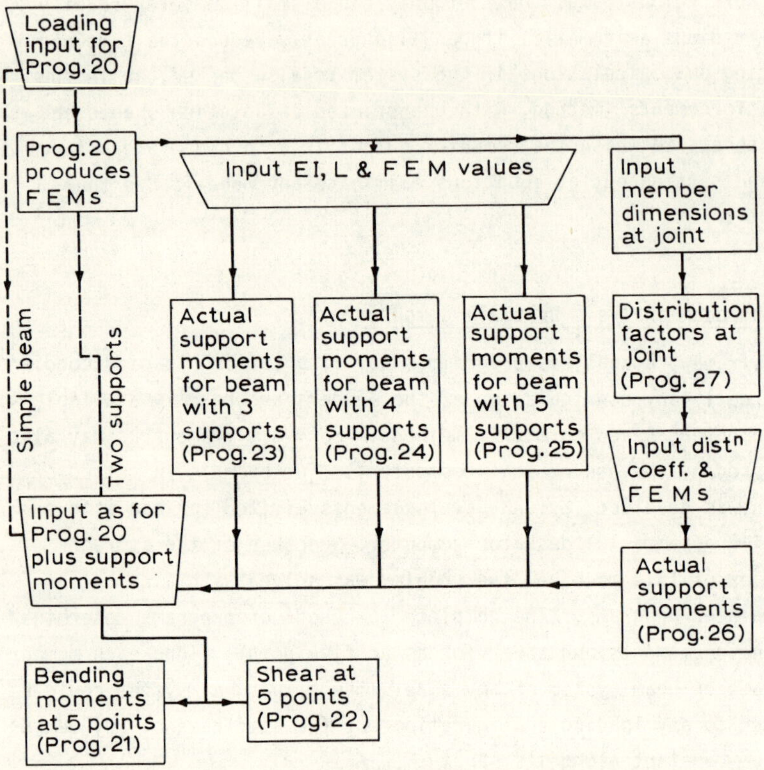

Fig.14.3.1 Example of GKN approach for small computer

original loading data and re-input to programs 21 and 22 to calculate the bending moments and shear forces.

The choice of moment-distribution program depended on the system being analysed. For beams on three, four or five supports, there were programs 23-5, which were quick in operation and required as data the values of EI, L and the fixed end moments for each span. For more general systems, program 26 was available, which required as data the distribution coefficients and fixed end moments for each member. This program was slower in operation and often a subsidiary program (program 27) was used first to determine the values of the distribution coefficients.

An explanation of the data sequence and input values for the GKN programs was normally available in the program manual and was condensed for the present description. After the program number, the first data required was the effective EI; thus three-quarters of the actual EI was needed for an outside span if the member was free to rotate over an end support (i.e. a pinned or projecting cantilever).

The spans were assumed continuous over internal supports but the user had to specify the type of end support. A zero marker was used for a fixed end. Although the programs were designed primarily for pinned or fixed-end outside supports, the effect of a projecting cantilever could be catered for by initially reducing the fixed end moment at the appropriate first internal support by half the value of the cantilever moment. In addition, because of built-in checks associated with a zero marker, the final outside support moment had then to be adjusted to minus the value of the cantilever moment value. The use of program 24 is demonstrated in Example 14.3.1.

Example 14.3.1

Determine the final support moments for the beam on four supports shown in Fig. 14.3.2 for the two load cases given below. The actual

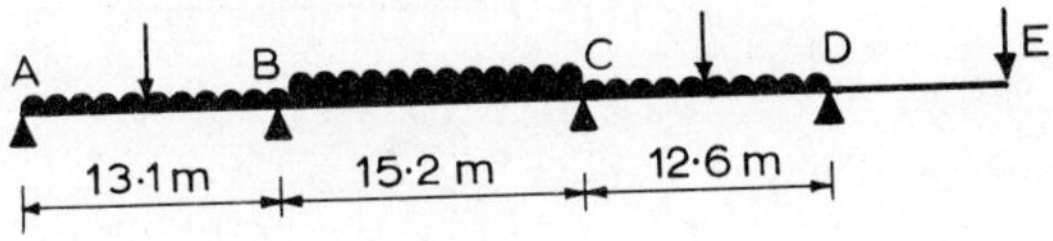

Fig.14.3.2 Example 14.3.1

I-values for spans AB, BC and CD are 200, 160 and 160 dm^4 respectively.

Case 1: Dead load on all spans and live load on spans AB, BC and cantilever DE.

Case 2: Dead load on all spans and live load on spans BC and DE only.

For case 1 the fixed end moments, for pinned supports at A and D, are M_{AB} = 0, M_{BA} = 960, M_{BC} = −750, M_{CB} = 750, M_{CD} = −570, M_{DC} = 0 and M_{DE} = 220 (all in kNm units).

For case 2 the fixed end moments, for pinned supports at A and D, are M_{AB} = 0, M_{BA} = 420, M_{BC} = −750, M_{CB} = 750, M_{CD} = −570, M_{DC} = 0 and M_{DE} = 220.

Solution The data for load case 1 are shown as the first 15 items listed on the left-hand side of the answer sheet in Fig. 14.3.3. It should be noted that three-quarters of the I value is in each case taken for spans AB and CD, and M_{CD} is reduced by one-half the value of the cantilever moment to give

$$-570 - \tfrac{1}{2}(-220) = -460$$

The next four items of data are the marker and fixed end moments at the end supports. The calculated support moments are the last six items listed and, as indicated by the note at the bottom of the sheet, the value of M_{DC} must be adjusted to 220. Similarly, the data and answers for case 2 are shown on the right-hand side of Fig. 14.3.3.

Note: It cannot be over-emphasized that the mechanization of a manual method such as moment distribution should only be considered for a small computer. Manual techniques are not normally used for more sophisticated computers.

Drawings and schedules using computers

Drawings which must be to scale require an X-Y plotter. However, if drawings need not be to scale it is possible to have a few standard drawings, e.g. one for a single-span beam, one for a two-span beam, one for a column, etc., and to do all the printing by relatively inexpensive automated processes, using a scheduling system for the standard structural units. Only general arrangement and non-standard members would then require to be done manually. Bar schedules lend themselves ideally to being computerized now that standard shape codes are available. The adoption of a shorthand, schematic or computer method of drawing and detailing will undoubtedly become more common in the future. Such scheduling and any other symbolism used must be clearly explained, if necessary on a special drawing prepared for the purpose, and the basic methods used should still conform with

MENT DISTRIBUTION FOR A BEAM WITH UR SUPPORTS	PROGRAM 24 ANSWERS	Job No.	*Ex. 14.3.1*
		Sheet No.	
		Initials	
		Date	

INPUT SPECIFICATIONS

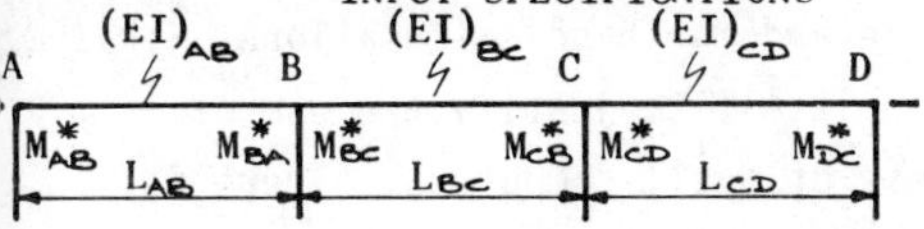

END SUPPORT MARKERS
Simple Support - 0
Fixed Support - 1

VE LOAD ON SPANS AB, BC, DE		ITEM	INPUT CHECK	*LIVE LOAD ON SPANS BC, DE*		ITEM	INPUT CHECK
		24		24	S	24	
24	S	$(EI)_{AB}$		150	S	$(EI)_{AB}$	
150	S	L_{AB}		13·1	S	L_{AB}	
13·1	S	$(EI)_{BC}$		160	S	$(EI)_{BC}$	
160	S	L_{BC}		15·2	S	L_{BC}	
15·2	S	$(EI)_{CD}$		120	S	$(EI)_{CD}$	
120	S	L_{CD}		12·6	S	L_{CD}	
12·6	S	M^*_{BA}		420	S	M^*_{BA}	
960	S	M^*_{BC}		-750	S	M^*_{BC}	
-750	S	M^*_{CB}		750	S	M^*_{CB}	
750	S	M^*_{CD}		-460	S	M^*_{CD}	
-460	S	Mark A		0	S	Mark A	
0	S	M^*_{AB}		0	S	M^*_{AB}	
0	S	Mark D		0	S	Mark D	
0	S	M^*_{DC}		0	S	M^*_{DC}	
0	S	RW			RW	RW	
	RW						
-0·00	A0	M^{**}_{AB}		0·00	A0	M^{**}_{AB}	
883·84	A0	M^{**}_{BA}		648·98	A0	M^{**}_{BA}	
-883·84	A0	M^{**}_{BC}		-648·98	A0	M^{**}_{BC}	
583·24	A0	M^{**}_{CB}		648·02	A0	M^{**}_{CB}	
-583·24	A0	M^{**}_{CD}		-648·02	A0	M^{**}_{CD}	
220 ~~-0·00~~	A0	M^{**}_{DC}		*220* ~~-0·00~~	A0	M^{**}_{DC}	

AM *ABCDE (FIG. 14.3.2)*	BEAM *ABCDE (FIG. 14.3.2)*
AD CASES *ONE*	LOAD CASES *TWO*
E.Ms ON PAGES *—*	F.E.Ms ON PAGES *—*
MENT UNITS *kN m*	MOMENT UNITS *kN m*
GN CONVENTION ↻+ CLOCKWISE	SIGN CONVENTION ↻+ CLOCKWISE

FINAL SUPPORT MOMENTS NOTATION

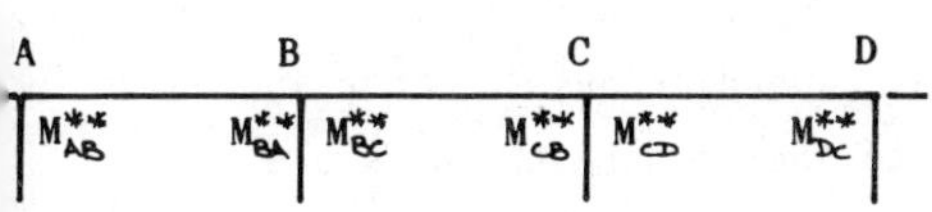

If there is a cantilever projecting over support A and/or D, the values of M^{**}_{AB} and/or M^{**}_{DC} must be adjusted for the cantilever moment value

Fig. 14.3.3

the principles of detailing given elsewhere (see Appendix H).

14.4 APPLICATION OF LARGE COMPUTERS

Computers are both fast and accurate and therefore have an ability to avoid the mistakes which characterize and mar hand calculations. This splendid efficiency is sufficient in itself to justify the use of machines, as evidenced by the success of small computers. There is, however, much greater justification for using large computers: first, their ability to make decisions and, second, the possibility they provide of utilizing techniques based on matrix methods. The large computer has had extensive influence on structural analysis and will also increasingly affect the process of design, which is largely concerned with decision-making. Although there is a complete range of computers with gradually increasing sophistication it is convenient in this section to consider only the largest digital machines.

Relative merits of computers

Large computers have large storage facilities and can therefore handle the basic input data and produce the final answers with a minimum of manual manipulation. They can accept comprehensive programs which can be made very general in application, again because of the large storage facilities. The advantages in structural analysis of large-computer techniques over manual methods can be briefly summarized as follows:

(i) Much larger frameworks can be contemplated for a single analysis.

(ii) Members of general orientation, in three dimensions if necessary, can be included without difficulty.

(iii) Axial, torsional and shear deformations can be taken into account as well as those due to bending.

(iv) A fast and accurate solution is readily obtainable.

(v) Several loading conditions can be analysed at once.

The disadvantage of large computers has been their extremely high initial cost and possible remoteness from the individual concerned with a particular problem. This has caused errors to occur in the input data and led to a reluctance to investigate the possible advantages of using any kind of computer.

Early development of computers

The first electronic computer was commissioned in 1946 at the University of Pennsylvania and the first automatic computer in the UK began operation at Cambridge in 1949. Early pioneers in structural computing realized that merely adapting existing techniques of manual calculation was not satisfactory for 'large' computers despite the relatively small—by modern standards—working store capacity which was then available. They saw very clearly the concept of a unified computer program which would expedite the complete process of structural analysis from raw engineering data to final print-out of stresses.

By the early 1950s Livesley had produced practical computer programs for the general analysis of skeletal frames by stiffness methods [14.4,5]. Rapid developments took place in computing which also stimulated renewed interest in matrix algebra. The development of matrix methods not only provided computer programmers with practical techniques for solving problems but also helped to provide a unified approach for all structures, which could replace the traditional diverse approaches for beams, frames, trusses, arches, shells, etc. These developments produced a better understanding of fundamental principles and a general realization that methods of structural analysis can generally be classified into either force methods or displacement methods, although techniques are now available which can combine the two approaches.

Matrix algebra

Algebra is a generalized form of arithmetic in which letters are symbols which represent the numerical data which can be substituted for each individual calculation. Matrices take the generalization a stage further by providing relationships which may hold between collections of numbers without the need to specify at the outset how many numbers are involved. Argyris emphasized in 1958 that matrix methods did not mean the application of matrix algebra merely to the solution of the linear equations in the unknowns. Little progress would have been achieved if the introduction of matrix operations had been confined to this essentially trivial task [14.4]. Thus matrix methods enable a cylindrical-shell roof and a pin-jointed frame, for example, to be analysed in the same general way. Furthermore, matrix methods are eminently suited to processing by computer, since computer programming is also concerned with separating the logical sequence of processes from the numerical data which are being processed. Generalized methods which enable structures with two, or two thousand, members subjected to one, or one thousand, loadings to be analysed with the same program represent an enormous economy of effort.

The first book setting out the detailed application of matrix techniques to problems in mechanics was written by Frazer, Duncan and Collar in 1937 following work mainly directed towards aircraft flutter. By 1947 Jenkins produced a book on shell structures which contained all the basic concepts that are currently used in the application of matrices to structures. By 1964 Livesley could produce a book with particularly outstanding coverage. (Reviews are given in references [14.4] and [14.5]).

Conventional elastic analysis of beam frameworks

Many different methods of solving problems associated with the deformation of frameworks under load evolved over the years as the theory of structures developed. The classical slope-deflection method systematically provides a series of simultaneous equations and effectively describes the elastic behaviour of statically indeterminate frameworks. However, the equations for equilibrium, compatibility and slope-deflection (load deformation) cannot be solved by hand using ordinary techniques for other than simple systems.

A significant advance was made by Hardy Cross in the early 1930s when he derived the method of moment distribution. This method can be regarded as providing a numerical solution to frame problems (which implicitly incorporates satisfaction of the slope-deflection requirements) which is suitable for hand computation. It has the advantage that the beam end moments which are of primary interest are obtained directly. It was only after the advent of the digital computer that matrix methods could be implemented for the solution of large numbers of simultaneous equations and thus enable much larger and more complex structures (not only frameworks) to be solved.

Historical outline of modern structural analysis

Both of the main methods of modern elastic structural analysis(the force, flexibility or compatibility method and the displacement, stiffness or equilibrium method)had been developed by the 1860s by Maxwell (UK) and Clebsch (Germany) respectively. Both methods were later used by Mohr. Books by Muller-Breslau in 1886 and Bendixen in 1914 again aroused the interest of German-speaking engineers in the force and displacement methods respectively. Although Bendixen's book was not translated into English, the displacement method as applied to beams and frames became well-known in the UK and America as the 'slope deflection' method [14.5]. In 1926 Ostenfield showed, for the first time in Germany, that the two methods were complementary, although it was not until the methods were subsequently expressed in matrix form that this was explained more completely.

The formal methods of structural analysis were little known in English-speaking countries until the late 1940s since only the 'conventional' methods were in general use. The moment distribution method, for example, was a technique of successive approximations of increasing accuracy for the analysis of frames which was, in fact, based on the displacement relationships but which had not been generally recognized as such. However, by 1959, for example, Morice's book could outline the bases of the force and displacement methods and their relationship to the energy approximation methods of the 1930s, and Argyris had already given an interesting review of developments in the analysis of complex elastic aeronautical structures (see e.g. reference [14.5]).

The advent of computers and matrix methods has encouraged development within the whole field of structural analysis including special techniques for continuous-plate and three-dimensional problems. The 'finite-element' technique uses the displacement method to find displacements at discrete points having idealized the body into a number of suitably chosen elements. These elements are usually selected to suit the particular problem concerned.

Computer operation

The computer program contains the decisions and associated lists of operations and is communicated to the computer through the medium of paper tape or punched cards. The program is stored in the machine and is obeyed automatically so that a long sequence of operations can be followed, and the appropriate decisions made, without human intervention during the computation. The operations are performed on data that are fed in and stored at the start and further storage is required for the answers as they are produced. A computer with the necessary storage and facilities may not always be available, and even if it is, the cost in time and money of preparing the program and using the machine may outweigh its other advantages over a small desk computer. In general, the problem must have more than a certain minimum degree of intricacy before it becomes economical to use a large computer for single or limited runs.

Process of computing

The process of computing involves three main stages once the problem to be solved has been clearly defined and the method or procedure for solving the problem has been decided. The identification of the problem and determination of the form of solution is sometimes known as 'systems analysis' although this term can embrace flow charting as well.

(a) *Flow charting* A flow chart is a visual display of the precise logical procedures to be used. It is independent of any particular computer or programming language. Flow charts are considered in more detail in section 14.5.

(b) *Coding* Coding (or computer programming) concerns the translation of the procedures given by the flow chart into a form which is acceptable to the computer. The programmer must have a good working knowledge of the appropriate computer language in order to write the instructions to the computer.

(c) *Operation of computing* The actual computing sequence is as follows:

(i) *Punching and input* A card punch (or tape punch) operator punches the information from the coding sheet prepared by the programmer.

(ii) *Compilation* The cards (or tape) are fed into the computer once the computer's memory has been cleared and the standard compiler program has been read into it from magnetic tape.

The computer is able to translate the information from the cards into machine language and punch further cards (or tape) corresponding to the original cards (or tape).

(iii) *Execution and output* The new cards (or tape), i.e. those in machine language, are put in the card reader of the computer, with added data if any, and it is these that are actually used in the solution of the problem.

(iv) *Development of the program* All necessary checks should be incorporated at this stage and any inconsistencies or errors eliminated (a process sometimes called "de-bugging").

(v) *Production runs* The program is now available for actual production runs.

Procedure-orientated languages

The above procedure assumes the use of a symbolic programming language such as FORTRAN or ALGOL. The idea of writing a program in normal English and mathematical symbols, and getting the computer itself to do the translation into the appropriate internal code, dates from the early 1950s or perhaps even earlier. FORTRAN became available in the USA in about 1956. It is clear that a good symbolic language leaves the programmer free to concentrate all his thoughts on the logic of his procedure and this is desirable whatever the level of his expertise. Symbolic programming languages are much easier to learn since they eliminate any need for a detailed knowledge of the operations peculiar to a particular machine. Programs written in such languages are both permanent and universal since they are then based essentially on ordinary human logic. Programs written in machine code tend to die with their machines, and many programmers have spent tedious hours rewriting programs as a result. A program in a symbolic language, however, can be run on any machine equipped with a compiler for that language. Although compiler programs are very complex they are written by teams of programmers employed by a computer manufacturer and are delivered as part of the 'software' with the computer.

14.5 FLOW CHARTS FOR COMPUTER PROGRAMMING

The more difficult a problem, the more complicated it is to prepare a computer program directly from the basic specification. However, the pattern of the computation can be clarified by drawing a flow chart, which is a visual statement in the form of a network, of the routes through the computation and of the conditions that govern the choices of route. A flow chart is also the best method of documentation, since the program itself is difficult to comprehend as it inevitably includes the symbolism and grammar of a particular programming language and a particular computer. The flow chart, on the other hand, is completely general and can be made as detailed as the circumstances require.

Flow chart notation

A flow chart consists of connected boxes each of which contains the description of an operation. The connecting lines have arrows to show the direction of flow through the network. The boxes are of various shapes, chosen to identify different types of operation. This can be done in a number of ways to suit the whim of the program writer, hence the simplest of choices and the use of four shapes only are considered here: ovals, trapeziums, diamonds and rectangles. Other shapes can be considered if desired [14.6,7].

Oval

Oval shapes are used to mark the beginning and the end of a program.

Trapezium

A trapezium indicates input or output. The starting-point of the computation is defined by the data available. As far as possible the program should be of a general nature so that the values of the variables are fed in as data rather than being built into the program itself. This is of particular importance when the program is intended (as it often would be in design work) for repeated use. The form of output is important too since it is desirable to arrange in the program for the answers to appear in the final form and so avoid a further stage of processing. Output includes presentation in the form of graphs and, increasingly, of engineering drawings, as well as ordered arrangements of numbers.

Diamond

A diamond indicates a decision. In most cases the decision is made by answering 'yes' or 'no' to a question in the form of a numerical comparison. For example, is M_u (the moment of resistance of a section of a beam) greater than M_a (the moment applied to the section)? There are thus alternative routes away from the decision box corresponding to the 'yes' and 'no' answers. Computers can readily be programmed to cycle; that is, to repeat a sequence of computations a number of times with a change in the data for each cycle. This is very useful in many forms of numerical analysis (such as integration by Simpson's rule) and it depends on a decision that controls the cycle by moving on from the sequence when the required number of repetitions has been completed.

Rectangle

A rectangle indicates a computation or an operation which is not a decision, an input or an output (e.g. start, stop). This is normally an arithmetical process such as the evaluation of a formula, but it can also be the setting of a variable to one of a pre-selected range of values. In writing the program the four rules of arithmetic are shown by simple symbols; most programming languages include built-in sequences for evaluating common functions such as trigonometrical ratios and square roots, which are called up by the use of a short name. Thus quite involved expressions can be written in a simple and short manner.

Example 14.5.1

Construct a suitable flow chart for determining the area of steel required in a number of sections for various rectangular reinforced concrete beams according to the analytical approach of Hognestad, Hanson and McHenry. The following data can vary for each section: the applied moment M_a at collapse; the beam breadth, b, and effective depth, d, the material strengths (f_{cu} for concrete, f_y for steel in tension and f_{yc} for steel in compression).

The concrete stress block parameters, $\lambda_1\lambda_3$ and λ_2, should be incorporated in the program, which should be able to process any chosen number, M, of beam sections in a single run.

Solution The flow chart will require two separate loops, one to consider the need for compression steel and one to enable all M sections to be processed, and not just one section. A suitable flow chart is shown in Fig. 14.5.1. It should be realized that Example 14.3.1 referred to a practical problem, whereas Example 14.5.1 is essentially an illustrative problem for teaching purposes.

Computer-orientated processes

The above example, although relatively simple, still demonstrates the use of processes which were inconceivable before the advent of computers. It is seen that an analytical technique, for example, can be used as a design method because the hand computations can be suitably mechanized on a computer. Other processes, such as optimization of an analysis, can be considered when using computers whereas with hand computation a design would only be repeated if it was clearly uneconomical. A computer, however, can give a series of solutions for the designer's consideration at very little extra cost.

Problem-orientated languages

The next stage in development now that procedure-orientated languages are becoming well established will be problem-orientated languages such as are used in systems like ASKA and GENESYS. Programs using such systems will require to be fed with even less information and, by calling up standard routines, be able to provide even more answers. For example, optimization *in design* should become a reality in the future. At present optimization is usually based on analysis *after* the overall layout, material strengths, etc., have been decided (often for quite arbitrary reasons). Optimization of a series of analyses can point towards optimization in design but should not be confused with the latter in its widest sense.

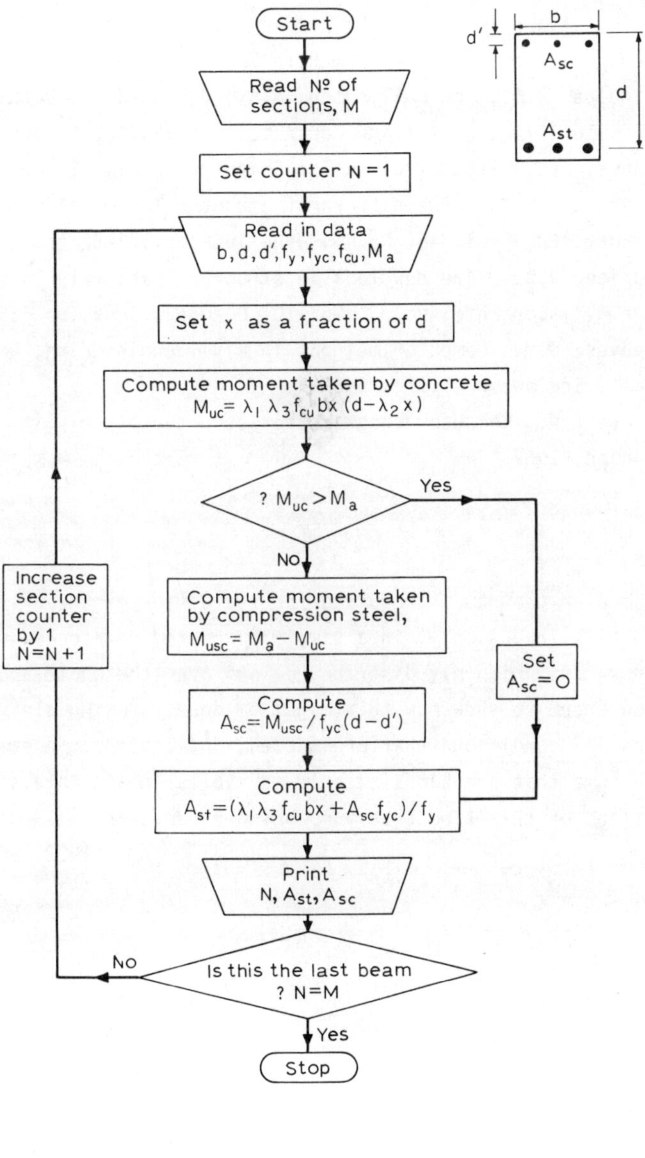

Fig.14.5.1 Example 14.5.1

References

14.1 Concrete Society, Model procedure for the presentation of calculations for reinforced and prestressed concrete structures, Report of Working Party Concrete Society, London, 1972.

14.2 Higgins, J.B., and Hollington, M.R., Designed and Detailed [CP 110:1972], Cement and Concrete Association, London, 1973, 28pp.

14.3 Jones, L.L., Structural Design Computer Weekly, March 20th 1969.

14.4 Livesley, R.K., 'The pattern of structural computing 1946-1966' Struct. Eng. vol. 46, No. 6, June 1968 pp 177-82.

14.5 Morice, P.B., 'The new look in structural analysis- a historical survey', Concrete, vol 3, No. 10, October, 1969, pp 415-7.

14.6 Fenves, S.J., Computer methods in civil engineering, Prentice-Hall, Englewood Cliffs, 1967.

14.7 Weaver, W., Computer programs for structural analysis, Van Nostrand, London, 1967.

Exercise

1. A computer program is required for the design of shear steel, in the form of stirrups only, in a simply supported reinforced concrete beam carrying a uniformly distributed load over the whole span. Draw up a flow chart to show how this might be done, showing clearly the relations and conditions that are needed, and listing the starting data. Assume that one bar size only is available and note any other restrictions of this kind that you consider necessary.

(UL2 1969, 36 min)

15 Concrete mix design and quality control

Good mix design is the selection of economical proportions of available materials to give concretes with properties which can be considered at least adequate in all major respects. Opinions can vary concerning the actual method of mix design and which properties should be considered of prime importance. For example, an experienced operator can carry out a series of trial mixes in the laboratory and by a process of trial and error obtain a good design after a relatively few trials— in much the same way that an experienced reinforced concrete designer can look at a beam section and know intuitively whether or not its size is about right. However, if the time spent in casting a preparatory trial mix is utilized in designing a mix based on extensive data which have been accumulated by others over many years, the effort can be well rewarded in terms of saving overall time and achieving the best results.

Two characteristics of concrete which are of vital importance for all structural concretes placed by vibration, are cube crushing strength and Vebe time. The method which is outlined in this chapter is therefore a development of rational concrete mix design [15.1, 15.2] which facilitates economical design of concretes for specified cube strengths and Vebe times, with or without air entrainment, and is suitable for use with locally available materials.

There are many good introductory and other texts on the properties of concrete and concrete materials (e.g. references [15.3] to [15.5]) so that only those properties directly concerned with the essential mix design procedure are considered here. The chapter concludes with a section on simple quality control procedures.

15.1 CUBE CRUSHING STRENGTH

The strength of concrete decreases with an increasing volume of voids. Only a small proportion of the water added initially actually combines with the cement — the remainder is necessary for adequate workability of the fresh concrete and results in water-filled voids. It is therefore common to show cube strength plotted against water cement ratio by weight, with different curves for different ages — the greater the age the higher the strength. However, for mix design purposes it is more convenient to consider volume fractions — i.e. the ratio of the volume of a constituent to the total volume of fully compacted concrete — and the inverse of water-cement ratio; since the latter tends to give more linear relationships. Thus the cement-water ratio, (by solid volumes) will be considered as the essential variable although the water-cement ratio by weight is also shown for reference (e.g. in Fig. 15.1.1).

Cube strength relation

Fig. 15.1.1 shows how the crushing strength of 100 mm cubes varies with c/w ratio (cement-water ratio by volume) for ordinary Portland cement at different ages, for various aggregates in a saturated and surface-dry condition prior to mixing (i.e. water in the aggregates in excess of the SSD condition is included in the volume fraction of free water w). Notice that there are two distinct linear ranges — one at high c/w ratios and one at moderate c/w ratios — and that the slopes of the two ranges are not necessarily related. The latter fact explains, for example, why both 'high-strength' and 'normal' designs were necessary for the Road Note No. 4 approach [15.6 and 15.7]. The 'Typical' curves correspond approximately with the data in Road Note No. 4 [15.6] and enable a clear comparison to be made. Notice that cements have continued to improve over the past twenty-five years so that the 'Typical' curves now only apply to moderately poor cements and aggregates. (Only the lightweight coarse aggregate S/F-28 curve is lower than the T-28 curve in Fig. 15.1.1. Even so, concrete strengths of 60 MN/m^2 are still possible with sintered pulverized fuel ash).

All aggregates in Fig. 15.1.1 are assumed to be presoaked and in a saturated and surface-dry condition. For most aggregates, absorption values are fairly low and readily estimated. However, lightweight aggregates are an exception, since absorption values can be high and difficult to assess accurately if the pore size is large. Such aggregates therefore often justify a special approach in mix design. It should also be realized that if initially dry pfa lightweight aggregate is used, the maximum strength will be less than the values indicated in Fig. 15.1.1 for pre-soaked aggregate.

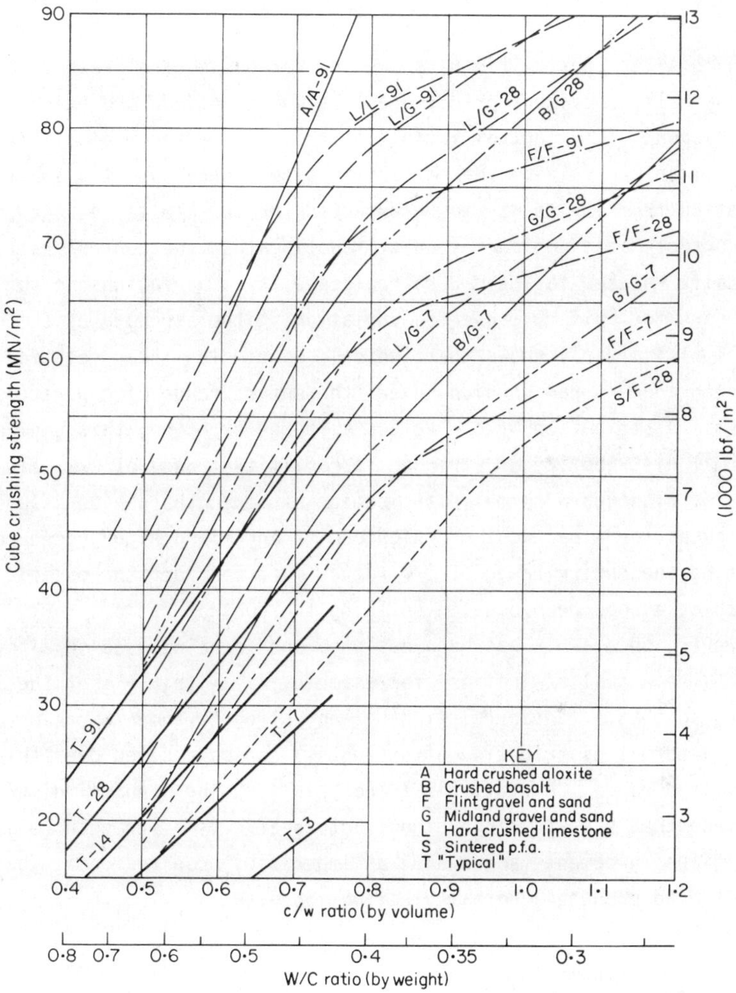

Fig.15.1.1 Cube crushing strength at different ages for concretes made with different aggregates and ordinary Portland cement

Standard of control

The standard of supervision and the degree of sophistication in batching which are economical depend on the quantity and the quality of the required concrete, as indicated in Table 15.1.1. Table 15.1.2, which is after Erntroy [15.8] shows the various control ratios which are typically obtained for different batching methods and standards of supervision. The 'control ratio' is the ratio of the minimum c/w value to the mean c/w value. The better the site control and the smaller the variation from day to day to be expected for the c/w value, the closer the control ratio approaches unity. Notice that the control ratio is logically related to the c/w ratio, since it is variations in the latter which are the primary cause for variation in the strength of the concrete Fig. 15.1.1 clearly shows that there is no single linear relation between strength and c/w ratio which covers the entire range of c/w ratios. Sinc a normal distribution of c/w values is to be expected, this can only give a normal distribution of strength for a limited range of c/w values. That is, although a normal distribution of strengths can be assumed for site cubes for a particular designed mix, any mix design procedures covering the entire range of c/w ratios must consider the basic variable, as has been done in defining the control ratio.

However, once a mix has been designed and is being used on site, over 68% of the actual c/w ratios, for example, lie within $\pm$ s of the mean value, so that a linear relation between strength and c/w ratio (and hence a normal distribution of strengths) can be assumed over this restricted range. A statistical analysis assuming a normal distribution of strengths is valid since other factors affecting strength (e.g. variations in cement, aggregate, or impurities) can reasonably be expected to produce a normal distribution also.

Current margins to CP 110

CP 110 requires (see Appendix A) that the current margin for grade 20 or above is not less than 7.5 MN/m^2 when based on at least 40 separate batches of concrete of nominally similar proportions of similar materials and produced over a period exceeding five days but not exceeding six months by the same plant under similar supervision. If less than 40 results are available, the current margin should be doubled to 15 MN/m^2 whereas if at least 100 results are available it may be halved to 3.75 MN/m^2.

Table 15.1.1 Economical control ratios

Concrete grade *	Contract size (m^3)			
	Very Large (over 50 000)	Large (over 10 000)	Medium (over 1000)	Small (under 1000)
30 or above	0.82+	0.82+	0.82	0.82
20-25	0.82	0.82	0.82-0.79	0.79
15	0.82	0.79	0.79	0.77
10	0.82	0.79	0.77	0.77
Unqualified	0.70	0.70	0.70	0.70

* e.g. Grade 30 concrete has a characteristic cube strength at 28 days of 30 MN/m^2.

'Unqualified' signifies an initial mix design where insufficient cube results will be available when permanent work commences.

Table 15.1.2 Suitable site facilities for different control ratios

Batching methods		Standards of supervision		
Cement	Aggregates	Poor	Normal	Good
Servo-operation (by weight)	Servo-operation (by weight)	0.78	0.82	0.86
By weight	By weight	0.75	0.79	0.83
By weight	By volume	0.72	0.77	0.82
By volume	By volume	0.70	0.75	0.80

Notice that for ready mixed concrete plants with normal supervision the assumed control ratio can be 0.82 for all the concrete grades except when commencing batching with new materials, when it is 0.70. Reference to Fig. 15.1.1 shows that even for the T-28 curve a control ratio of 0.70 ensures an initial margin of 15 MN/m^2 for grade 20 and over. CP 110 itself acknowledges that at strengths approaching the maximum possible this initial margin may be reduced—confirming the desirability of applying the control ratio to the c/w ratio for mix design purposes).

Example 15.5.1

Grade 25 concrete has been designed assuming a control ratio of 0.82 for materials providing the T-28 curve in Fig. 15.1.1. Would this design satisfy the CP 110 requirement (Appendix A, Table A) for either temporary works, or permanent works when 40 cube results would be available?

Solution

The required c/w ratio for a cube strength of 25 MN/m^2 is, from Fig. 15.1.1, equal to 0.495. Hence the mean c/w ratio is 0.493/0.82 = 0.601 which corresponds to a strength of 33.5 MN/m^2 and a margin of 8.5 MN/m^2. This design therefore satisfies the CP 110 requirement of a minimum margin of 7.5 MN/m^2.

Reduction in strength with air entrainment

If the concrete is to be air entrained there is a loss in compressive strength of up to 5% (or more) per 1% of entrained air for water cured concretes. When designing an air entrained mix the required strength should be increased by multiplying by the air voids factor. Appropriate values may be obtained from Table 15.1.3, which is based on data given by Klieger [15.9]. The original data indicated no loss (and gain up to 6.8%) in strength for lean concretes of 19 mm size. However, for design purposes the (cautious) values shown asterisked in Table 15.1.3 are recommended.

15.2 WORKABILITY OF CONCRETE AND THE VEBE CONSISTOMETER

If the concrete is not well compacted then the strength is reduced. The anticipated concrete strength will therefore not be achieved if the workability of the fresh concrete is not suitable for the available means of compaction and the method of placing. The meaning of the term 'workability' is outlined and reference is made to the compacting factor test and its limitations as well as to the Vebe consistometer.

Concrete workability

If no attempt is made to quantify the term 'workability' it can be used in a general way to embrace all the properties required of the fresh concrete, such as compactability, flowability, stability (i.e. resistance to segregation), finishability (i.e. ease with which surface can be trowelled), pumpability etc. Clearly, no single quantitative empirical test, such as slump, compacting factor or Vebe time, can give a satisfactory indication of all the above properties. The best that can be hoped for is to select the quantitative empirical test which measures that combination of properties which gives the closest mean approximation to the combinations most required in practice.

Table 15.1.3. Typical air voids factors for different entrained air contents and maximum sizes of aggregate

Cement content, c	< 7% (< 220 kg/m^3)		> 10% (> 310 kg/m^3)	
Maximum aggregate size, mm	19	38	19	38
Entrained air, n				
3%	1.00*	1.01	1.12	1.17
4%	1.05*	1.08	1.18	1.27
5%	1.12*	1.19	1.25	1.41
6%	1.16*	1.33	1.34	1.56

* Values considered suitable for design.

If the term 'workability' is to be used in a quantitative sense then it is preferable to consider fundamental properties as considered in the science of rheology (i.e. the science of the deformation and flow of matter). Tattersall [15.10] considered that fresh concrete approximates to the Bingham model so that workability can be described by two constants (i.e. two properties). He has developed a simple two-point test capable of providing the two constants which is essentially based on an ordinary food mixer equipped with a stirring hook that moves in planetary motion. The power requirements are measured and torque T is calculated as power divided by speed, N. Results indicate a relationship of the form

$$T = g + hN \qquad (15.2.1)$$

where N is the speed of the hook in revolutions per second and g, h are constants. The equation for the Bingham model is

$$\tau = \tau_o + \mu\dot{\gamma} \qquad (15.2.1)$$

where τ is the stress at a rate of shear $\dot{\gamma}$

τ_o is a constant known as the yield value and

μ is a constant known as the plastic viscosity

It can be shown [15.10] that g and h are proportional to τ_o and μ respectively. The two constants g and h can be determined by measuring power requirements at two speeds and typical results are shown in Fig. 15.2.1. Notice especially how the curves for two different mixes at either 0.6 or 0.7 water-cement ratio cross over at speeds of about 1 rev/s. Thus mix A which may appear less workable than mix B at slow speeds can appear more workable at high speeds.

Slump test

The slump test is a simple practical means of measuring the consistency of medium to high workability concretes. Changes in slump obtained during production of concrete may indicate changes in materials, in the water content or in the proportions of the mix, hence it is useful as a 'control test', i.e. in controlling the quality of the concrete produced.

The apparatus consists essentially of a truncated conical mould 100 mm (4 in) in diameter at the top, 200 mm (8 in) diameter at the bottom and 300 mm (12 in) high. The inside of the mould should be cleaned before each test and the mould placed on a hard, flat, impervious surface. It should then be filled in four layers (BS 812 test) of concrete of approximately equal depth. The rounded end of a steel tamping rod 16 mm ($\frac{5}{8}$ in) diameter and 600 mm (2 ft) long is used to rod each layer 25 times. After the top layer has been rodded the surface of the concrete is struck off level with the top of the mould with a trowel or the rod. After cleaning away any spillage from around the base of the mould, it is then lifted vertically from the concrete. The slump is the difference between the initial height of the concrete (i.e. the height of the mould) and the greatest height of the concrete after removal of the mould. If any specimen collapses or shears off laterally, the test should be repeated with another sample of the same concrete. If the repeat specimen also shears, the slump should be recorded together with the fact that the specimen sheared.

Compacting factor test

The compacting factor test is well described elsewhere [15.4, 15.5]. It consists essentially of allowing concrete to fall successively through two hoppers into a cylinder to determine the 'partially compacted weight' of concrete in the cylinder. The compacting factor (CF) is the ratio of the partially compacted weight of concrete to the fully compacted weight. It was clearly shown by Cusens in 1956 why the compacting factor could be very misleading for dry mixes suitable for vibration [15.11] . Reference to Tattersall's work in Fig. 15.2.1 shows that a measurement at low speed (and power) can reverse the comparison for two concretes measured at high speed (and power). Since most concretes are compacted by vibration it is important that any arbitrary single-point test should also use a relatively high speed, e.g. vibration. The compacting factor test can only be useful for assessing hand placed concretes and even in these cases it has limitations. For example, air entrained concrete which is very cohesive tends to stick in the hoppers and thus give CF values which are too low.

Vebe consistometer tests

The Vebe test was developed by V. Bahrner and is again well described elsewhere [15.4, 15.5]. It consists essentially of determining the time taken to remould concrete from a cone shape into a cylindrical shape on a standard vibrating table. Although it is again an arbitrary one-point workability test it has the distinct advantage that for concretes which are to be vibrated it uses vibration in the test, so that with reference to Fig. 15.2.1 the speed (and power) can be of the

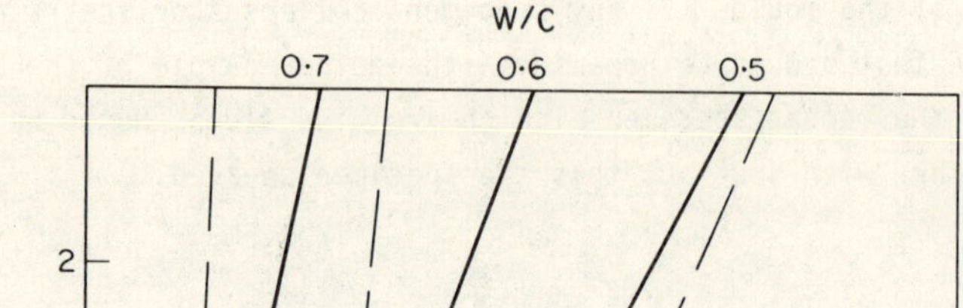

Fig.15.2.1 Typical results from two-point workability test after Tattersall [15.10]

right order of magnitude. Clearly, vibration can be very much more, or very much less, intense than that given by the standard vibrating table so that the same criticisms can be made as for the CF test. However, the differences in 'speed' are much less and can be reduced to reasonable limits by keeping within the practical limits of the test. For example, concretes requiring very intense vibration or mechanical compaction may never achieve an end point with the test, and Vebe times in excess of 40 s are best all classified as merely 40+ seconds. Similarly, concretes requiring little or no vibration cannot be judged satisfactorily either. To meet this latter need, a Vebe test with a reduced intensity of vibration can be used, as described elsewhere [15.2], which is referred to as the MVebe (modified Vebe) time and is approximately 2½ times the standard Vebe time. The Vebe and MVebe tests together can, with care, give a very good indication of the ease with which the vast majority of normal concretes can be placed on site [15.1, 15.13]. Table 15.2.1 gives guidance concerning the actual Vebe times which are appropriate to various conditions of placing and methods of compaction.

An important advantage of air entrained concrete is its improved cohesion which reduces segregation and internal water gain. Thus when considering Table 15.2.1 heavier vibration can be used for air entrained concrete without causing segregation—with a consequent reduction in the actual time of vibration required on site.

15.3 VEBE CHARTS

The cement content necessary to provide the required Vebe or MVebe time for a given c/w ratio can be obtained from Fig. 15.3.1 for various cements, shape and grading of fine aggregate, shape and grading (including maximum size) of coarse aggregate, air entrainment and (if necessary, as for lightweight or shielding concretes) specific gravity of concrete.

The advantage of these charts is that they can be used as accurately or as approximately as required. For example, if the properties of the aggregates and cement are not known, except that they are natural irregular materials, well-graded fine aggregate, maximum coarse aggregate size about 19 mm, and ordinary Portland cement, then all the adjustment charts can be taken as datum with no adjustment necessary to the main chart.

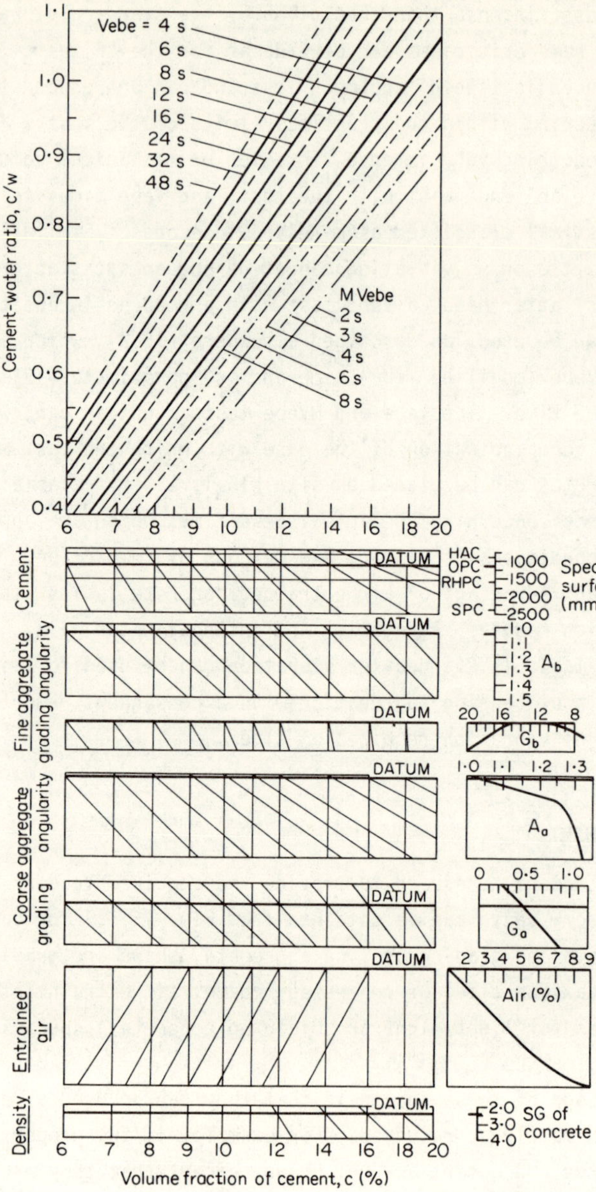

Fig.15.3.1 Datum and adjustment charts for Vebe and MVebe times

Cement fineness

An increase in the fineness of the cement reduces the workability of the concrete since the thickness of the 'lubricating film' of water decreases as the surface area of the cement increases. Typical values of specific surface (in mm^2/mm^3) are given in Fig. 15.3.1 for high alumina and OPC; a high value is shown for rapid-hardening Portland cement. A simple correction can therefore be made for any particular cement.

Table 15.2.1 Suggested mean values of workability for different applications

Section	Vibration	Air Entrained	MVebe	Vebe
Congested	None	No	6	-
"	Light immersion	Yes		
Reinforced	None	No	8	(3)
"	Light immersion	Yes		
Congested	Light immersion	No		
"	Immersion	Yes		
Reinforced	Light immersion	No	(10)	4
"	Immersion	Yes		
Congested	Immersion	No		
Pumped concrete	-	-		
Massive	Immersion	Yes	-	6
Reinforced	Immersion	No		
Pavements	Slip-form paver	Yes		
Reinforced	Heavy immersion	No	-	8
Massive	Heavy immersion	No	-	12
Pavements	Power-operated	Yes		
Heavy precast units	Table	No	-	16
Light precast units	Table	No	-	24
Precast units	Heavy table	No	-	32
Precast units	With pressure or jolting	No	-	40+

Aggregate angularity

It is well known that an increase in the angularity of the aggregate reduces the workability of the concrete. Estimates of the angularity of both the coarse and fine aggregates (A_a and A_b respectively) can be made, if necessary, by a visual comparison with Fig. 15.3.2, rather than by performing actual tests which are themselves fairly simple [15.1].

Aggregate grading

The aggregate gradings also affect the workability of the concrete: very coarse, or very fine, gradings of fine aggregate reduce the workability especially for leaner mixes. This is clearly shown by the G_b adjustment chart, since maximum workabilities are given by G_b values between about 10 and 15 mm^{-1}, and much larger adjustments are shown for low c values than for high c values. The grading modulus for the fine aggregate (G_b value) is obtained by multiplying the percentage retained on each sieve (from the sieve analysis of the aggregate) by the appropriate value given in Table 15.3.1. The sum of the products for all sieve sizes, divided by 100, gives G_b. The grading of the coarse aggregate, especially the maximum size, also affects workability. The larger the maximum size, the lower the value of G_a, the grading modulus of the coarse aggregate. G_a can be calculated from the sieve analysis in a similar way to G_b by using Table 15.3.1. Simple adjustments can therefore be made for the angularities and gradings of both aggregates.

The specific gravity adjustment chart is not normally required. However, if shielding concretes are being designed, their high density can reduce the effectiveness of vibration—depending on the volume of concrete, and size or type of vibrator. The adjustment chart for density, for the Vebe vibrator and sample size is as shown.

Many practical observations can be deduced from Fig. 15.3.1. For example, an increase in the maximum size of aggregate reduces the value of G_a and increases workability (i.e. reduces the cement content for a given Vebe time). Similarly, the omission of the 10 to 5 mm size group in a 20 mm maximum size continuous grading, to give a gap grading, also increases workability. Fig. 15.3.1 assumes that the most appropriat proportions of coarse and fine aggregate are used in each case. These are given by the recommended stone content, which is considered later.

Fig.15.3.2 Typical angularity factors

(a) Coarse Aggregate

1.00 rounded gravel; 1.03 irregular gravel

1.19 crushed aloxite; 1.24 crushed limestone

1.30 crushed basalt; 1.40 aglite

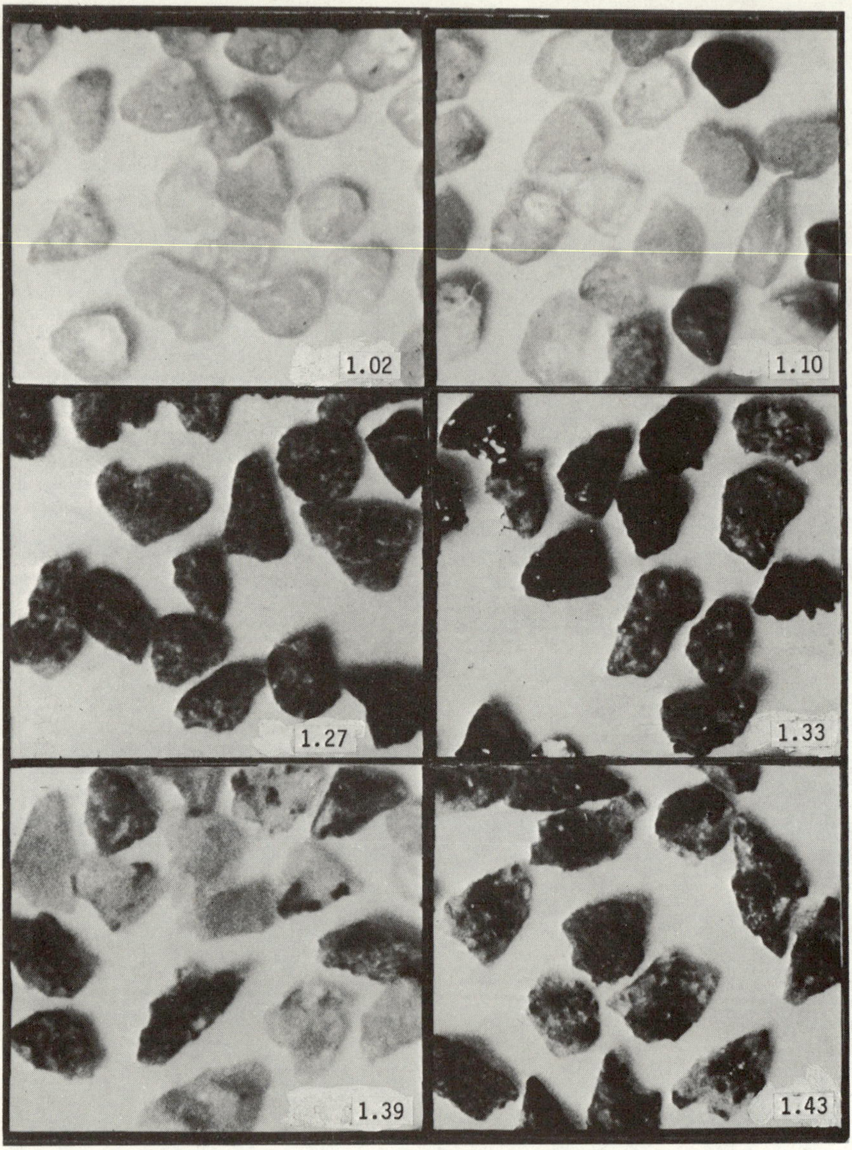

Fig.15.3.2. Typical angularity factors

(b) Fine Aggregate

1.02 rounded sand; 1.10 irregular sand

1.27 crushed limestone; 1.33 crushed silicon carbide

1.39 crushed granite; 1.43 crushed basalt

Table 15.3.1 Aggregate grading parameters

BS sieve (Old Ref.)	BS sieve aperture	Grading modulus (mm^{-1})	Equivalent mean diameter (mm)
	37.5 mm		
		0.191	
	26.5 mm	0.229	
		0.269	
$\frac{3}{4}$ in.	19.0 mm		
		0.381	
	13.2 mm	0.456	
		0.538	
$\frac{3}{8}$ in.	9.50 mm		
		0.756	8.10
	6.70 mm	0.911	7.13
		1.069	5.73
3/16 in.	4.75 mm		
		1.83	3.56
No. 7	2.36 mm		
		3.67	1.77
No. 14	1.18 mm		
		7.26	0.89
No. 25	600 μm		
		14.43	0.45
No. 52	300 μm		
		21.87	0.23
No. 100	150 μm		
		57.71	0.11
No. 200	75 μm		

Optimum coarse aggregate content

The optimum coarse aggregate content is that volume fraction of coarse aggregate, a, which enables maximum workability to be achieved with a given cement content and a given water content. (Alternatively, it enables a given workability and strength of concrete to be achieved

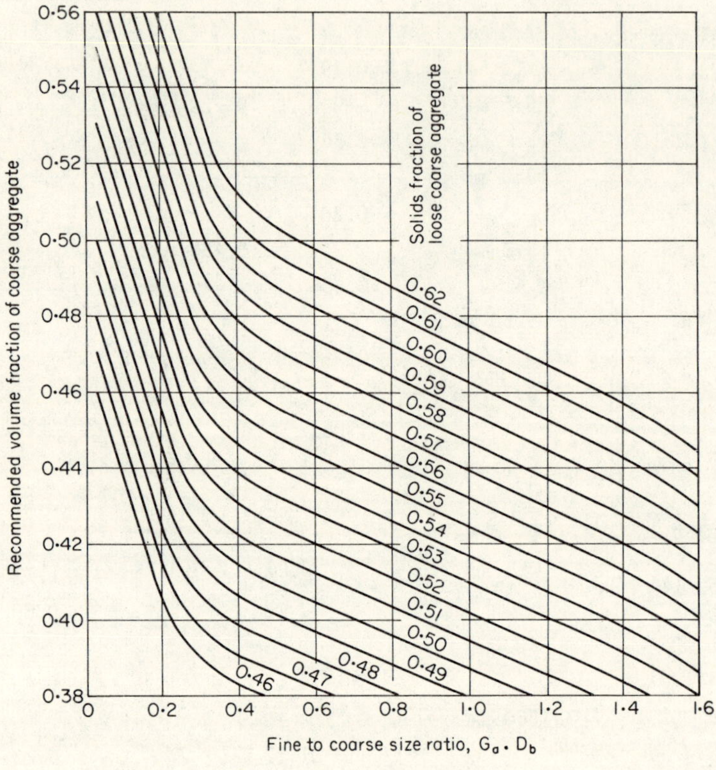

Fig. 15.3.3 The recommended stone content

with a minimum of cement.) In Fig. 15.3.3 this value has been amended (as indicated later) but is still given in terms of the solids fraction of *loose* coarse aggregate, a_B, and the relative sizes of the coarse and fine aggregate, d_a and d_b respectively. The optimum value of a reduces with increasing values of d_b/d_a, since the coarse aggregate particles must be pushed further apart in order to accommodate the larger fine aggregate particles. For reasons discussed elsewhere [15.1], the function G_aD_b expresses d_b/d_a most satisfactorily, the relation being effectively independent of both c and w.

The sieve analysis of the coarse and fine aggregates, together with Table 15.3.1, enable G_a (already used in Fig. 15.3.1) and the equivalent mean diameter of the fine aggregate, D_b, to be determined; hence G_aD_b. The loose bulk density test for the coarse aggregate should normally be carried out in a 300 mm by 150 mm diameter (12 inches by 6 inches) cylinder [15.1]. However, values of a_B can be assumed from those given in Table 15.3.2. Values vary with the actual gradings as well as with aggregate shape. Other sizes of similar range can be estimated: for example, 10-5 mm would be similar to 20-10 mm, and 40-10 mm similar to 20-5 mm.

Table 15.3.2 Solids fraction of loose coarse aggregate, and specific gravity on SSD basis, of some typical aggregates

Aggregate property	a_B		SG	
Size	20-10 mm	20-5 mm	Coarse	Fine
Irregular flint gravel	0.55	0.58	2.54	2.64
Irregular midland gravel	0.57	0.60	2.62	2.62
Crushed 'granite'	0.46	0.48	2.70	2.70
Hard crushed limestone	0.48	0.50	2.70	2.70
Crushed 'basalt'	0.45	0.47	2.84	2.90

Recommended stone content

The use of the optimum coarse aggregate content ensures a slight surplus of mortar. For example, if the point of a trowel is drawn across the surface of a compacted cube the mean depth of the incision should be about 3 mm ($\frac{1}{8}$ in). However, if approximations are made for a_B by simple estimates using Table 15.3.2, it is desirable to decrease the volume fraction of coarse aggregate in the concrete to allow for such errors. Furthermore, since a_B is based on the loose bulk density in a 150 mm (6 in) diameter cylinder, it follows that the surplus fines are only adequate when 150 mm diameter pipes are used for pumping concrete. If 100 mm (4 in) diameter pumps are to be used, the optimum value is reduced.

The simplest approach is to reduce the optimum value in all cases to allow for relatively large errors in estimating a_B as well as to eliminate the need for making corrections. Hence the recommended stone contents given in Fig. 15.3.3 are all 0.04 less than the corresponding optimum coarse aggregate contents. The increase in Vebe or MVebe times shown in Fig. 15.3.1 resulting from this adjustment may be neglected since workability varies only very slightly with coarse aggregate contents which are close to the optimum value.

Pumped concrete should have a Vebe time of about 4 s: Vebe times less than about 3 s or more than about 6 s can both cause difficulties. The use of an admixture consisting of a suitable surface-active agent is usually advisable both to facilitate sliding and to reduce any tendency towards segregation in the pipeline. Low cement contents, certain gradings of fine aggregate and angular coarse aggregates should generally be avoided.

It should be realized that the volume fractions used in gap-graded mixes, such as those advocated by Stewart [15.14], are based on the vibrated bulk density and are therefore much higher than the values given by Fig. 15.3.3. For example, the solid volumes of loose and vibrated 20 to 10 mm ($\frac{3}{4}$ to $\frac{3}{8}$ in) irregular flint gravel are 0.555 and 0.610 respectively, a difference of 0.055. These undersanded mixes (as well as Road Note No. 4's oversanded mixes) therefore have reduced workabilities. The harsh mixes advocated by Stewart tend to be unpopular on site, since they allow no margin for errors in batching, are prone to wet segregation if used for 'normal-strength' concretes, and are difficult to finish, since the aggregate tends to stand proud of the top surface.

15.5 MIX DESIGN PROCEDURE

The three requirements, cube crushing strength, Vebe time and recommended stone content have been considered in detail, and suitable relationships are available indicating how they vary with the properties and proportions of the basic concrete constituents — coarse aggregate, fine aggregate, cement and water. These proportions provide three independent variables which can therefore be used to satisfy the above requirements. Admixtures or additives such as entrained air should be regarded as further constituents which modify the relationships for cube crushing strength, etc., but which do not provide any further completely independent variables.

Volume fractions

The volume fraction of each constituent is the ratio of the 'solid' volume of the constituent to the total volume of fully compacted concrete. If a, b, c, n and w are the volume fractions of coarse aggregate, fine aggregate, cement, entrained air and water respectively, it follows that

$$a + b + c + n + w = 1.$$

Design procedure

The design procedure for determining the volume fractions, and hence the batch weights, of the concrete for the required strength and workability is as follows.

(1) Select the c/w ratio giving the specified minimum cube strength from the appropriate curve in Fig. 15.1.1.

(2) If the durability conditions are severe, or high workability without segregation is required, use air entrainment.

(3) If air entrainment is used, multiply strength by air voids factor, using Table 15.1.3 as a guide, before obtaining c/w ratio in (1).

(4) Select an economical control ratio from Table 15.1.1 and check from Table 15.1.2 that the equipment and supervision that can be made available are suitable.

(5) Divide the c/w ratio, from (1), by the control ratio, from (4) to obtain the mean c/w ratio which allows for variations and errors in site batching (and air entrainment if applicable).

(6) Determine G_a and D_b from Table 15.3.1 and the sieve analysis of the coarse and fine aggregates respectively and calculate G_aD_b. Also determine G_b.

(7) Calculate a_B from the specific gravity and loose bulk density of the coarse aggregate. Alternatively, estimate a_B from Table 15.3.2.

(8) Determine, from Fig. 15.3.3, the recommended stone content, a, for G_aD_b from (6) and a_B from (7).

(9) Select the required Vebe time from Table 15.2.1 for the given site conditions.

(10) Obtain c from Fig. 15.3.1 for the Vebe time, from (9), and the mean c/w ratio, from (5), making adjustments if necessary for the cement (specific surface), fine aggregate (A_b and G_b), coarse aggregate (A_a and G_a) and concrete density. A_a and A_b can be estimated from Fig. 15.3.2.

(11) The values of c/w, c and a are now known; hence b and w can be obtained (since $a + b + c + n + w = 1$). For air entrained concrete n is not zero and should be considered as replacing an equal volume fraction of fine aggregate.

(12) The batch weights for a mixer of 1 m^3 capacity and saturated surface-dry aggregates are obtained directly from the volume fractions and appropriate specific gravities. Due allowance can then, if necessary, be made for the actual moisture contents of the aggregates.

Trial mixes

Trial mixes in the laboratory can be unnecessary in some cases since the design specifically considers quality (as given by required cube strength), workability (as given by Vebe time), and the recommended stone content. However, full-scale site trials to check other factors, such as resistance to segregation under actual site conditions, are still very desirable. Failing this, then visual observation of a confirmatory trial mix in the laboratory can give some indication of the stability of the mix. Furthermore, confirmation of the actual strength characteristics of the aggregates and cement are necessary if these have not been obtained previously—especially if air entrainment is also used.

Segregation

The recommended stone content ensures that segregation of the stone from the mortar is virtually impossible, since the 'surface fines' will not exceed about 5 mm even after prolonged vibration. However, wet segregation (segregation of water and cement fines from the concrete) can occur for concretes of high workability and low cement content (i.e. medium to low-grade concrete). Wet segregation and excessive losses in strength due to bleeding and water gain are best avoided by using air entrainment. The net loss in strength due to the entrained air voids is then minimal and, of course, durability, impermeability and workability are all improved.

A possible alternative solution for wet segregation is to reduce the stone content and correspondingly increase the fine aggregate. However, the design method ensures a slightly oversanded mix; hence any further reductions in stone content will certainly reduce the workability. Assuming that reductions in workability cannot be tolerated, then increases in cement and water contents are necessary to offset this reduction. Other possible solutions for wet segregation are admixtures such as fly ash, or additives such as plasticizers which enable the same Vebe time to be achieved with a lower water content. If such admixtures or additives are used then trial mixes are essential unless data is available to enable the effects on Vebe time and strength to be estimated. It should be noted that air entraining agents which are also surface active agents can be expected to give greater improvements in workability for a given air content than those shown in Fig. 15.3.1.

Discussion

The advent of superplasticizers emphasizes the trend towards higher workability, high stability concretes. Since air entraining agents can do this, albeit to a more limited extent but at less cost, it is expected that they will be used more in the future with the availability of improved dispensers and modern dual-piston, positive displacement pumps which can cope with the compressibility of the entrained air. There is also an increasing awareness of the need for concrete with better weathering and durability than that provided by non-air-entrained concrete.

Examples of the application of the method to both an individual mix design and to a particular mix design chart, similar to those advocated by Owens [15.15], are given below. Particular mix design charts, appropriate to mixes designed from any given source of aggregates, are particularly useful for ready mixed concrete plants.

The basic approach of the rational method is direct and essentially simple. It presents sufficient data for both strength and workability to enable concrete mixes to be designed economically from the available aggregates in all cases and for the relative significance of various factors, such as a change in aggregate source or grading, to be readily assessed.

The method is well suited to ready mixed concrete plants where the aggregate properties are often already known but many mixes are required Once determined for particular coarse and fine aggregates, the recommended stone content is essentially constant irrespective of the grade of concrete and use of air entrainment or other additives or admixtures such as plasticisers or fly ash.

Example 15.4.1

Air entrained concrete having a Vebe time of 6 seconds and characteristi 7 -day cube strength of 20 MN/m^2 is required for concrete road slabs laid by a slip-form paver. 20 mm nominal single size crushed basalt coarse aggregate and irregular sand are available having the following properties.

Coarse

Sieve Size	19.0mm	13.2mm	9.50mm	6.70mm
% passing	100	48	22	0
Angularity	1.28			
Loose bulk density	1380 kg/m^3			
Specific gravity	2.80			

Fine

Sieve Size	4.75mm	2.36mm	1.18mm	600μm	300μm	150μm	75μm
% passing	100	91	79	60	25	5	0
Angularity	1.10						
Specific gravity	2.62						

Cement

Ordinary Portland Cement — specific gravity 3.14.
Determine the SSD batch weights for a 1 m^3 servo-operated mixer and 5% entrained air.

(UB2 1975, 50 mins Modified)

Solution

From Table 15.3.1 and the sieve analysis for the coarse aggregate, G_a is given by:

19.0 - 13.2 mm	0.52×0.381	=	0.198
13.2 - 9.5 mm	0.26×0.538	=	0.140
9.5 - 6.7 mm	$\underline{0.22} \times 0.756$	=	$\underline{0.166}$
	1.00		$0.504 = G_a$

Similarly, from the sieve analysis for the fine aggregate:

$$D_b = 0.911 \text{ and } G_b = 14.295$$

$$\text{Hence } G_a \times D_b = 0.504 \times 0.911 = 0.459$$

The solids fraction of loose coarse aggregate:

$$a_B = 1380/2800 = 0.493$$

Hence from Fig. 15.3.3 the recommended stone content is given by

$$a = \underline{0.406}$$

Assume a control ratio of 0.82 from Table 15.1.1 and an air voids factor of 1.25 for 5% air from Table 15.1.3.
Minimum c/w for curve B/G-7 from Fig. 15.1.1 for cube strength of $1.25 \times 20 = 25$ MN/m^2

$$= 0.545$$

Hence required mean c/w

$$= 0.545/0.82 = \underline{0.665}$$

From Fig. 15.3.1

for Vebe = 6 s

c = 9.5% before adjustments

11.5% after normal adjustments

and 10.1% after 5% air adjustment $\underline{c = 0.101}$

Hence w = 0.101/0.665 = 0.152 and batch weights for SSD aggregates are as follows.

	Volume fractions	Specific gravities		Batch weights
a	0.406	2.80	1.137	1137 kg
b	0.291	2.62	0.762	762 kg
c	0.101	3.14	0.317	317 kg
w	0.152	1.00	0.152	152 kg
n	0.050	—	—	—
	1.000		2.368	

Example 15.4.2

The sources of aggregate and cement for a ready mixed concrete plant are as indicated in Example 15.4.1. Draw up a suitable Particular Mix Design Chart for 20 mm maximum size air entrained concrete with workabilities of 4, 6 and 8 Vebe seconds and for 28-day strengths.

Solution

Mix designs can be conveniently carried out for cement contents of 200, 300, 400 and 500 kg/m^3 for each Vebe time. A typical calculation for 400 kg/m^3 would be as follows.

$c = 0.127$ hence progressing upwards through Fig. 15.3.1

$c/w = 0.745$, 0.785 and 0.810 for Vebe = 4, 6 and 8 s respectively. Allowing for air voids factor of 1.25 then the corresponding strengths from Fig. 15.1.1 are 51.8, 54.3 and 55.8 MN/m^2 respectively. Volume fractions w and b can be determined and the batch weights for 1 m^3 summarized in the required Particular Mix Design Chart as shown in Fig. 15.4.1.

15.5 QUALITY CONTROL PROCEDURES

The basic requirement for the characteristic strength, as discussed in chapter 1, is that there shall be not more than 5% defectives. Thus the target mean strength must exceed the characteristic strength by the current margin. Statistically from chapter 1 this equals 1.64s but additionally CP 110 stipulates certain minimum values for the current margin (see Table A6). However, in practice the sampling and testing of concrete is not generally carried out at a sufficiently rapid rate for the characteristic strength to be estimated in good time to establish compliance of small sections of a structure. CP 110 therefore gives a testing plan which aims to give a reasonable

Cube Crushing Strength (MN/m^2)

Age. 28 ..days

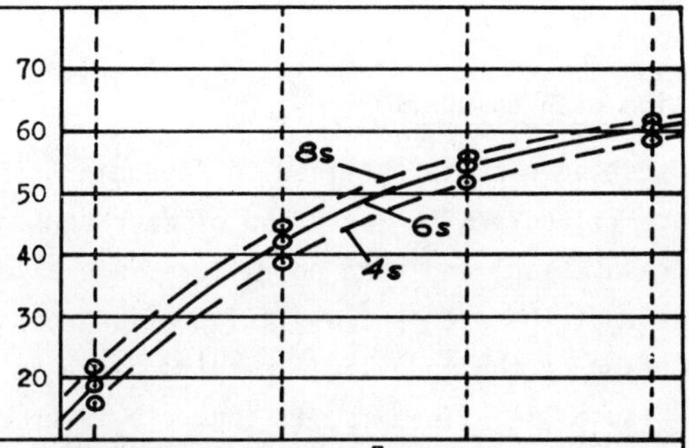

Vebe/ ~~MVebe~~		Batch Weights (kg/m^3)			
	Cement	200	300	400	500
.4.s	Water	160	157	170	183
	Fine Aggregate	838 *	764	647	529
.6.s	Water	151	148	162	175
	Fine Aggregate	862	787	668	550
.8.s	Water	142	143	156	170
	Fine Aggregate	886	800	684	563
Coarse Aggregate. 20 .mm		1137* - - - - - - - -		1137 - - - - -	
.....mm		* - - - - - - - -		- - - - -	

*Check cohesion of lowest Vebe/low cement proportions with trial mix and if necessary increase b (and decrease a accordingly).

AGGREGATE SOURCE. Local crushed basalt & local sand

AGGREGATE CONDITION. SSD .... CEMENT. OPC & 5% air ......

Fig.15.4.1 Example 15.4.2.

assurance that the quality of the concrete is at least as good as that specified based on a small number of results. Suitable rates of sampling are given in Table A8 and the testing plan is discussed below. Quality control procedures as recommended by BRMCA (British Ready Mixed Concrete Association) are also considered in this section.

Testing plan to CP 110

The testing plan is outlined in Appendix A (§6.8.2.2). Notice that clause (1) refers to any group of four consecutive test cubes (i.e. groups overlap) and that no failures are allowed. However Eqn (1.4.6) shows that the standard deviation of a group of 4 = $s/\sqrt{4} = \frac{1}{2} \times$ standard deviation of the individual results—hence 5% failures can be expected for clause (1). That is, if concrete is produced to the characteristic strength requirement (i.e. 5% defectives) the testing plan nevertheless will on average require rejection of the concrete in 5% of cases.

The effect of clause (2) is to encourage better quality control in the same way as the BS 4449 and BS 4461 clauses do (see Example 2.1.2).

Initial target mean strength

BRMCA [15.16] suggests that target mean strengths not less than 27 MN/m^2 should be calculated, assuming an **increased** margin of 2s (rather than 1.64s) i.e. that

$$\text{target mean strength} = f_k + 2s$$

This corresponds to 2.28% defectives (rather than 5%) and therefore greatly reduces the occasions when the concrete can be at risk due to non-compliance with the testing plan clauses of CP 110 (§6.8.2.2).

Producer's risk and consumer's risk

With any testing plan, the producer has a risk of good quality material being judged as non-complying—the producer's risk, and the consumer has a risk of accepting material which is less than the quality specified—the consumer's risk.

It was seen above that if concrete is produced which is exactly of the quality specified, there is a producer's risk of 5%. With the same testing plan, however, if the producer makes concrete with 20% defectives the producer's risk only increases to about 50% (with an ideal testing plan it would increase to 100%). In order to improve the consumer's risk of detecting concrete with a high percentage of defectives, CP 110 has stipulated the use of overlapping averages of any consecutive tests of four [5.4]. Computer simulation suggests that the probable producer's risk (at reasonably low values), on the other hand, remains sensibly constant when either overlapping or non-overlapping averages are used [5.4].

Initial standard deviation

Where less than 40 cube results are available it is necessary to assume an initial value of standard deviation approved for mix design purposes and BRMCA [15.16] suggest 6 MN/m^2. Where 40 or preferably 100 results are available the standard deviation s can be calculated and expressed to the nearest 0.5 MN/m^2.

Estimated and actual 28-day results

If pairs of cubes are cast from each sample then one cube can be used for an early accelerated curing method and one for the normal 28-day curing. Careful and consistent control measures are essential for 3-day normally cured, or 1-day accelerated cubes, if a reasonable correlation is to be established with the 28-day cube results for each type and combination of aggregate and cement used at the batching plant. The results from the early accelerated curing method are then used to estimate the 28-day results, the estimated values being examined continuously to detect changes from the initial standard deviation and target mean strength values. Whenever a change is detected, new values of the standard deviation and target mean strength are determined and appropriate alterations made in the concrete mix proportions to provide concrete of the required quality. The possible cause of the variation should be investigated and the actual 28-day results on the second cube of each pair used to confirm that the control method is functioning correctly.

Reference and approved mixes

One only of the concrete mixes tested should be designated as the 'reference' mix and used as the basis for the analysis of test results.

Test results from other mixes may be used in the analysis and called 'approved' mixes provided that they can be correlated with the reference mix. The target mean strengths of both reference and approved mixes should not be less than 27 MN/m^2 and the target mean strength of approved mixes should preferably be within 7.5 MN/m^2 of that for the reference mix [15.16].

Correlation charts should be established between 28-day cube strength and cement content in kg/m^3, preferably from at least 100 related sets of results for each type of aggregate and workability. Such charts can then be used to estimate the strengths of (approved) mixes with cement contents higher or lower than the reference mix. If such data is not available a cement content of 6 kg/m^3 can be taken as equivalent to changes of 1 MN/m^2 in the target mean strength [15.16].

Concrete of different workabilities may be included in the above correlation and subsequently in the quality control analysis provided that adjustments are made as follows. To convert the cement contents to an equivalent mix of 'medium' workability—i.e. 75 mm slump—add or subtract 10 kg/m^3 to the cement content for every 25 mm in slump below or above 75 mm respectively. For example, subtract 20 kg/m^3 from cement content to convert a mix of 125 mm slump to an equivalent mix of medium workability [15.16]. Concrete with slumps in excess of 150 mm should not be included. Mixes containing 10 mm aggregates should not be included, but mixes containing 40 mm and 20 mm nominal maximum sized aggregate may be treated as the same and included in the analysis without separate modification [15.16].

Separate correlation charts should be established for mixes containing

(a) admixtures and

(b) cements other than ordinary Portland or sulphate resisting Portland cement.

Clearly, where low-slump or no-slump concrete is involved, it is necessary to carry out a Vebe test (or possibly a CF test) as the control test instead of slump, and to obtain correlation charts between correction to cement content and variations in workability as well as between 28-day cube strength and cement content in kg/m^3.

Example 15.5.1

A large ready mixed concrete depot has a high demand for low-slump concrete and they are able to use the Vebe as a control test for such concrete. If their extra-reference mix for low-slump concrete has a Vebe time of 8 seconds and a cement content of 315 kg/m^3 (i.e. $c \simeq 0.1$) determine the required corrections to the cement contents of extra-approved mixes of different Vebe times, using the standard mix design data, when the Vebe = 4s and 16s respectively.

Solution

From Fig. 15.3.1 when c/w ratio (and hence strength) is constant at value for Vebe = 8s and c = 10%, then when Vebe = 4s, c = 11.2% and when Vebe = 16s, c = 9.1%.

Therefore for Vebe = 4s, decrease cement content by c = 1.2%, i.e. by 38 kg/m^3

and for Vebe = 16s, increase cement content by c = 0.9%, i.e. by 28 kg/m^3

(Answer)

Note: It would be necessary to calculate intermediate values of, say, Vebe = 6s and 12s and then plot a graph of correction to cement content (kg/m^3) against Vebe time which could then be used to convert approved mixes of various workabilities to an equivalent mix of low-slump workability—i.e. 8s Vebe.

Initial cement content

The initial cement content corresponding to the target mean strength should be obtained from the 28-day cube strength versus cement content data. Where insufficient data are available to establish such correlation charts, the initial cement contents should be those given in the appropriate code of practice (i.e. for CP 110 as for prescribed mixes given in Table A7). The initial cement content should be used for production until the subsequent control procedures establish that a change is justified.

Subsequent procedures

Weighing equipment should be calibrated regularly (e.g. see [15.16]). Cement, aggregate and water should be weighed to a tolerance of ± 2% and admixtures measured to a tolerance of ± 5%. Rates of sampling for sites should be as Table A8 whereas for ready mixed concrete depots at least one pair of cubes should be made from current mixes approved for quality control purposes on at least 80% of the full working days during which concrete is produced [15.16]. At least two pairs of cubes should be made from different batches for depots producing over 150 m^3 of concrete in any one day of testing.

As results become available, the early age and 28-day cube results should be examined continuously by a suitable method of analysis to detect changes from the plant standard deviation and target mean strength. The 28-day cube results should be used to revise or confirm the strength/cement content correlation data at least once per month and to confirm the relationship assumed in the analysis between early age and 28-day strength.

When concrete produced by a plant is restricted for a period to a mix which cannot be correlated with the reference mix, an extra-reference mix is necessary and the results of cubes made from this concrete should be assessed separately by a suitable analysis.

All test results should be included in the analysis unless there is definite evidence of faulty sampling and testing. If significant alterations in materials, equipment or technical control are introduced at the depot, a new analysis may be started from that time.

Where significant changes are observed, the adoption of new values of standard deviation and target mean strength and cement content can be made, as described for the method of analysis, except that standard deviations smaller than 3.75 MN/m^2 (plants to CP 110—see Table A6) or 3.5 MN/m^2 (depots to reference [15.16]) should not be used.

Mathematical method of analysis

One method of quality control referred to as the mathematical method of analysis [15.16] is as follows.

(1) The 28-day cube and earlier age cube results are recorded regularly in tabular form together with the estimated or predicted 28-day strengths obtained from the appropriate correlation chart.

(2) The mean and range of each group of four consecutive, but not overlapping, estimated 28-day results are calculated and plotted on continuous control charts on which are drawn the lines indicating the average values and the limits outside which 1 in 10

and 1 in 40 points may be expected to fall. Appropriate values for the limits are included in Table 15.5.1.

(3) A significant change from the initial standard deviation and target mean strength should be assumed if:

(a) Three individual estimated 28-day cube results in any 40 consecutive results fall below the specified strength.

(b) For the means or ranges of groups of four estimated 28-day results, three values in any ten consecutive values fall outside the 1 in 10 limits, or two values in any twenty consecutive values fall outside the 1 in 40 limits.

(4) When a significant change is indicated, by either the mean or range of groups of four results, assume that the situation can be restored by an increase, or decrease as appropriate, of 10 kg/m^3 in the cement content of the reference mix. Until sufficient results after the change are available assume corresponding changes in target mean strength and standard deviation of 1.6 MN/m^2 and 0.8 MN/m^2 respectively (if corresponding values from correlation charts are not available). Also adjust the cement content of mixes other than the reference mix accordingly, using the appropriate strength—cement content correlation chart.

Table 15.5.1. Limit lines for mean and range charts (groups of 4 estimated 28-day results), also mean and standard deviation charts for correlation checks between estimated and actual 28-day results (groups of 10 actual 28-day results).

Chart	Average or mean value	Limit lines	
		1 in 10	1 in 40
Mean of groups of 4	Target mean strength T	T ± 0.64s	T ± 0.98s
Range of groups of 4	2.06s	0.98s 3.24s	0.59s 3.98s
Mean of groups of 10	Target mean strength T	T ± 0.405s	T ± 0.62s
Standard deviation of groups of 10	0.92s	Not used	0.52s 1.38s

(5) If the indicated change is in the mean of groups of four results, assume that the alteration in cement content restores the mean to its previous level. If the change is in the range, however, calculate the new limits and draw on the control charts on the basis of the provisional new values of the target mean strength and standard deviation.

(6) Lines should be drawn on the charts to indicate any change in mix proportions. Analysis is then started afresh, and continued until either a further change is indicated or 100 consecutive individual 28-day results are available for re-assessment of the plant standard deviation and target mean strength.

(7) The actual 28-day cube results are used to check that the control method is functioning correctly. Calculate the mean and standard deviation of groups of ten consecutive, but not overlapping, actual 28-day results. Continuous control charts, similar to those in (2), should be plotted. Appropriate values for the limits 1 in 10 and 1 in 40 are also included in Table 15.5.1. If, on the basis of the statistical checks given above, these charts indicate a change which was not shown by the estimated 28-day results, an investigation should be made of the accuracy of the correlation charts used to estimate the 28-day results and of the cube making and testing procedures.

<u>Causes of variation</u> [15.16]

(1) Whenever a significant change has been detected and the concrete mix proportions adjusted, the cause of the change should be investigated, and wherever possible and appropriate, action taken to correct it. If the change has been corrected then the analysis of test results subsequent to the correction should indicate that another change has occurred.

(2) Significant changes in the mean of groups of four consecutive cubes can be due to variations in the quality of the materials used, especially the cement. Examine the cement test certificates to assess whether the alterations in cement quality are the cause of the variation.

(3) Significant changes in the range of groups of four consecutive cubes are usually the results of variations in the manufacturing or

testing procedures and of any unusual variations in the moisture contents and gradings of the fine and coarse aggregates.

(4) Whenever significant changes occur in the correlation between the estimated and actual 28-day results, an investigation should be made into the whole procedure of control testing and analysis especially the making and testing of cubes.

To assess the testing variation a third cube can be made every fourth test, the extra cube being tested alternately at 28 days and at the earlier age. If the difference between successive pairs of cubes is calculated it may be expressed as a percentage of the average strength of the pair. A value of 3% represents good control values, while values over 5% are excessive.

Example 15.5.2

The initial approved standard deviation of a depot is 6 MN/m^2 and the required target mean strength is 33 MN/m^2 for a specified characteristic strength of 21 MN/m^2 at 28 days. The initial cement content for the reference mix is 280 kg/m^3. The early test results are obtained by an accelerated technique, from which the 28-day strength can be estimated on the day after the cubes are cast. Test results are included from two mixes in addition to the reference mix. The first is a prescribed mix with a fixed cement content of 300 kg/m^3 and the second is a designed mix with a cement content of 250 kg/m^3.

The data are given in columns (1), (2), (3), (4) and (11) of Table 15.5.2 (This example forms part of Example 2 in reference [15.16] to which the reader is referred for more details.)

Solution

Column (5) of Table 15.5.2 can be filled in after reference to the correlation chart between early results and estimated 28-day results. Similarly, column (6) requires the correlation chart between 28-day strength correction and cement content. Column (7) is simply the sum of columns (5) and (6). Columns (8), (9) and (10) are calculated for successive groups of four. Column (12) is simply the sum of columns (11) and (6). Columns (13) to (16) are obtained for successive groups of ten in column (12), to calculate the standard deviation by using

Table 15.5.2 (Example

Cube Serial Number	Date Cast	Cement content of mix tested	Early result (Accelerated) for mix tested	Estimated 28-day result for mix tested	Correction for difference in cement content from reference mix (21.0/28)	Equivalent estimated 28-day result for reference mix (5) + (6)
(1)	(2)	(3)	(4)	(5)	(6)	(7)
1	Feb.5	300	27.2	48.0	-3.0	45.0
2	8	280	14.7	29.0	0	29.0
3	9	280	17.2	33.0	0	33.0
4	9	300	16.8	32.5	-3.0	29.5
5	10	250	15.0	29.5	+5.0	34.5
6	11	280	16.5	32.0	0	32.0
7	12	280	17.8	34.0	0	34.0
8	15	280	12.2	25.0	0	25.0
9	16	250	21.0	39.0	+5.0	44.0
10	17	280	12.0	24.5	0	24.5
11	17	280	12.2	25.0	0	25.0
12	19	280	18.2	34.5	0	34.5
13	22	280	17.2	33.0	0	33.0
14	23	300	13.8	27.5	-3.0	24.5
15	24	250	11.6	24.0	+5.0	29.0
16	25	300	19.8	37.0	-3.0	34.0
17	26	280	15.5	30.5	0	30.5
18	Mar.1	280	18.5	35.0	0	35.0
19	2	250	13.2	26.5	+5.0	31.5
20	2	280	17.2	33.0	0	33.0

15.5.2) Courtesy:BRMCA

Equivalent estimated 28-day results			Actual 28-day result for mix tested	Equivalent actual 28-day result for reference mix (X)	Equivalent actual 28-day results			
Groups of 4					Groups of 10			
Sum	Mean	Range			Sum of X	X^2	Sum of X^2	Standard deviation
(8)	(9)	(10)	(11)	(12)	(13)	(14)	(15)	(16)
			47.0	44.0		1936		
			26.0	26.0		676		
			32.0	32.0		1024		
136.5	34.1	16.0	32.5	29.5		870		
			29.5	34.5		1190		
			30.0	30.0		900		
			38.5	38.5		1482		
125.5	31.4	9.5	28.5	28.5		812		
			39.5	44.5		1980		
			23.5	23.5	331.0	552	11423	7.2
			21.0	21.0		441		
128.0	32.0	19.0	34.5	34.5		1190		
			31.5	31.5		992		
			28.0	25.0		625		
			21.5	26.5		702		
120.5	30.1	9.5	35.5	32.5		1056		
			29.5	29.5		870		
			37.5	37.5		1406		
			29.0	34.0		1156		
130.0	32.5	4.5	32.0	32.0	304.0	1024	9463	5.0

equation (1.3.5) in the form

$$s = \sqrt{\frac{\Sigma x^2 - (\Sigma x)^2/n}{n - 1}}$$

Fig. 15.5.1 shows the control charts for mean and range of groups of four results for cubes 1-20 (shown in Table 15.5.2) and for a further 20 cubes (given as cubes 21-40 in reference [15.16] only).

Figure 15.5.1 shows that the 6th, 7th and 10th groups of four results are below the 1 in 10 limit, i.e. cube result No. 40 completes the third group of four results, in any ten consecutive groups, with a mean strength lower than the 1 in 10 limit. This indicates that a significant decrease in the target mean strength has occurred. Thus remedial action is required, i.e. of increasing the cement content of the mix by 10 kg to 490 kg/m^3. This should increase the mean strength by 1.5 MN/m^2 to about the former target mean strength of 33 MN/m^2. As the standard deviation remains unchanged, the limits on the control charts (Fig. 15.5.1) are not altered.

The 28-day results can also be added, as they become available, to charts similar to those shown in Fig. 15.5.1. Cube No. 20 completes two groups of 10 results (see Table 15.5.2). Comparisons between the mean and standard deviation of the groups of ten actual 28-day results and the appropriate control limits indicate that no change has occurred. That is: none of the individual equivalent actual 28-day results falls below the characteristic strength of 21 MN/m^2 at 28 days. Also $T \pm 0.405s = 33 \pm 2.43$ MN/m^2, for the 1 in 10 limits for the mean of groups of ten, just excludes one of the two results and 0.52s to 1.38s = 3.12 to 8.28 MN/m^2 for the 1 in 40 limits for standard deviation of groups of ten contains both results. Thus only if a further two values in the next eight consecutive results failed the 1 in 10 limits for the mean of groups of ten, would a change be indicated by the actual 28-day results. However, since a further change (increasing the cement content by 10 kg) would have already been instigated, this could well not occur. The main purpose of the actual 28-day results is to check that they do not indicate a change which was not shown by the estimated 28-day results. If they do, then an investigation should be made of the accuracy of the correlation charts used to estimate the 28-day results and of the cube making and testing procedures.

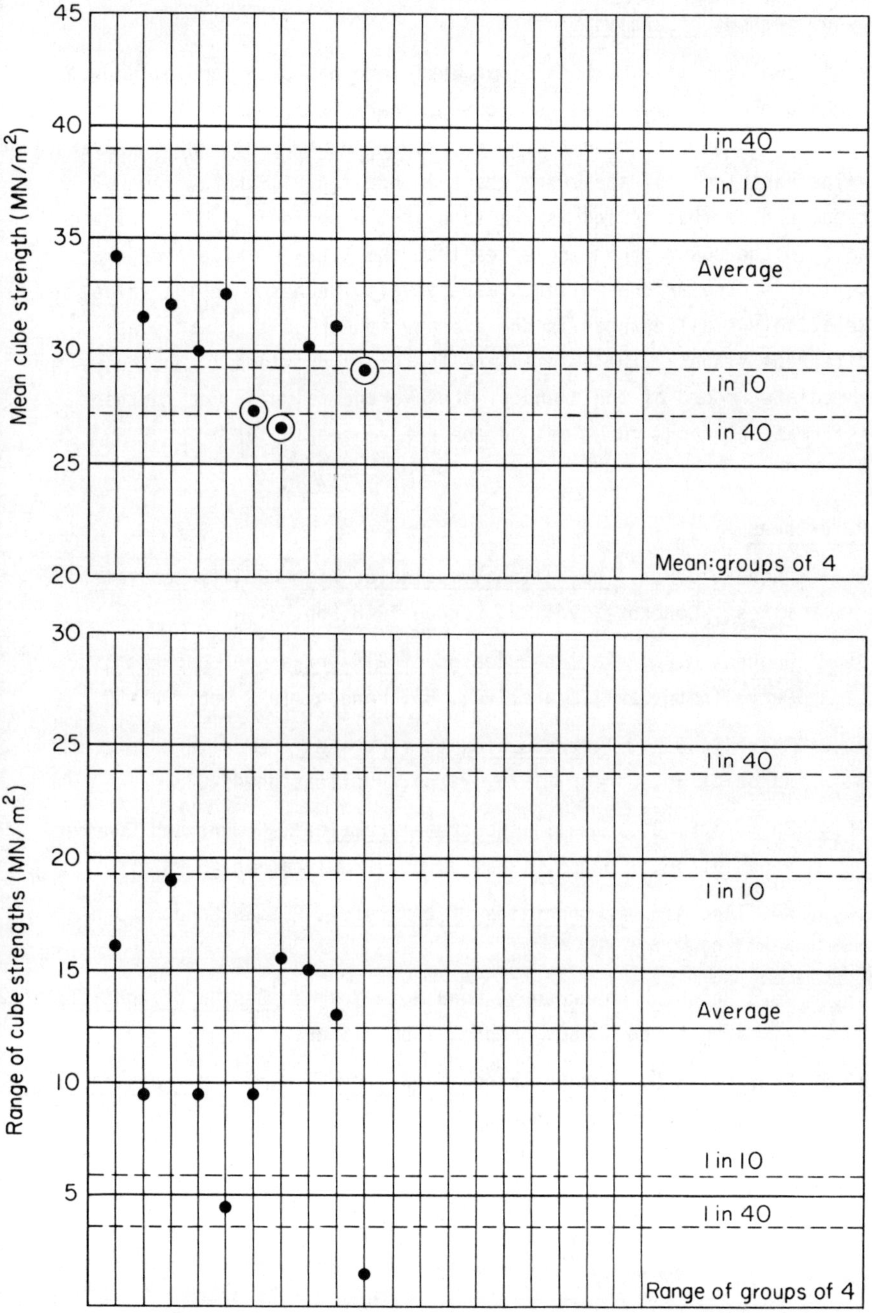

Fig.15.5.1 Control charts for mean range of groups of four results (Example 15.5.2)

Cusum methods of analysis

Various alternative methods of analysis are based on cumulative sum techniques. A cumulative sum (cusum) chart is drawn by plotting the accumulated sum of the differences between each result and the average value expected. If the cusum chart is horizontal, the average of the results over that period is the same as the design value. A rise or fall in the cusum graph indicates that the average value over that period is greater or smaller, respectively, than the design value. Relatively small changes in the average appear quite clearly as different slopes. The cusum chart therefore provides a continuous and immediate record of the trend of the average values. For further information of one cusum method see reference [15.16].

References

15.1 Hughes, B.P., 'The rational design of high quality concrete mixes', Concrete, vol. 2, London 1968, pp. 212-22.

15.2 Hughes, B.P., 'Rational design of air entrained concrete', Proc. Instn Civil Engrs, vol. 55, London, 1973, pp. 841-53.

15.3 Shirley, D.E., 'Introduction to concrete', Cement and Concrete Association, London, 1975, 24 pp.(Publication 45.028).

15.4 Barnbrook, G., et al., 'Concrete Practice', Cement and Concrete Association, London, 1975, 63 pp. (Publication 48.037).

15.5 Neville, A.M., 'Properties of concrete', Pitman London, 3rd edn, 1976, 702 pp.

15.6 Road Research Laboratory, Road Note No. 4: 'Design of concrete mixes', 2nd edn, HMSO, London, 1950, 16 pp.

15.7 Erntroy, H.C., and Shacklock, B.W., Proc. Symp. 'Mix design and quality control of concrete', Cement and Concrete Association, London, 1954, pp. 55-73.

15.8 Erntroy, H.C., Int. Ass. Bridge Struct. Engineering, Sixth Congr., Stockholm 1960, IV b3 p. 689.

15.9 Klieger, P., 'Effect of entrained air on strength and durability of concrete made with various maximum sizes of aggregate', Proc. Highway Research Board, vol. 31, Washington, D.C., 1952, pp. 177-201.

15.10 Tattersall, G.H., 'Fresh concrete and the workability problem', Conf. on Ready Mixed Concrete, Dundee University, 1975, Theme 3, 21 pp.

15.11 Cusens, A.R., 'The measurement of the workability of dry concrete mixes', Magazine of Concrete Research, vol. 8, London, March 1956, pp. 23-30.

15.12 Hughes, B.P., 'The workability of concrete subjected to vibration', Cement and Concrete Research, vol. 2, Pergamon Press, Oxford, 1972, pp. 101-22.

15.13 Hughes, B.P., 'The Vebe test and the effect of aggregate and cement properties on concrete workability', Rilem Seminar: Fresh Concrete, Leeds 1973, pp. 4.3, 1-12.

15.14 Stewart, D.A., 'The design and placing of high-quality concrete', Spon, 2nd edn, London, 1962. 162 pp.

15.15 Owens, P.L., 'Basic mix method', Cement and Concrete Association Training Centre, TDH 3529, London, 1973.

15.16 BRMCA Code for Ready Mixed Concrete, British Ready Mixed Concrete Association, London, May 1975, 56 pp.

Exercises

1. Ten consecutive cube results for Grade 25 concrete produced under actual site conditions gave the following seven values (in MN/m^2) for the running means of four results for the testing plan in CP 110.

   37.750, 36.375, 36.875, 34.875, 30.000, 31.375, 31.750

   Determine the mean value and the standard deviation for these seven running mean values.

   What are the characteristic cube strength values as predicted by (a) the seven running means and (b) the ten individual results for which the mean strength and standard deviation are 34.15 MN/m^2 and 5.53 MN/m^2 respectively? Are these predicted values satisfactory for the required grade when considering:

   (i) basic statistical principles

   (ii) the CP 110 requirement that the mean strength determined from any group of four consecutive test cubes should exceed the specified characteristic strength by not less than 0.5

times the current margin.

If the concrete was designed using curve T-28 (see Fig. 15.1.1), with a control ratio of 0.82, what would be the corresponding standard deviation?

(UB2 1975, 50 min)

2. Concrete suitable for pumping is to have a characteristic cube strength at 7 days of 25 MN/m^2. Good supervision with weight batching (control ratio = 0.83) is to be used with materials having the following characteristics:

Coarse aggregate	Angularity	1.25
	Loose bulk density	1300 kg/m^3
	Specific gravity	2.70
	Sieve analysis	
	$\frac{3}{16}$ in. - ½ in.	503 gm
	½ in. - ⅜ in.	310 gm
	⅜ in. - ¼ in.	187 gm
Fine aggregate	Angularity	1.20
	Specific gravity	2.62
	Sieve analysis	
	$\frac{3}{16}$ in. - 7	90 gm
	7 - 14	105 gm
	14 - 25	210 gm
	25 - 52	392 gm
	52 - 100	165 gm
	100 - 200	38 gm
Cement	OPC. Specific gravity	3.14

Use curve L/G - 7 for strength vs c/w ratio (i.e. see Fig. 15.1.1)

Design a suitable concrete mix and determine the required batch weights for a 1 m^3 mixer assuming the corrected absorption of the coarse aggregate is ½% and the free moisture content of the sand is 1%.

It is realized subsequently that this concrete cannot first be used for temporary works to obtain 40 test cube results under operating conditions. It is therefore specified that the minimum margin (i.e. difference between mean and characteristic strength) is to be

15 MN/m^2. What would be the corresponding maximum control ratio which could be assumed for the initial design?

(UB2 1974, 50 min)

Appendix A
Concrete grades, properties and specifications

(Extracts from CP 110:1972 and BS 5337:1976. Cross references to the codes are shown in brackets).

CONCRETE GRADES AND PROPERTIES (CP 110 §APP. D, BS 5337 §6.3.2)

Table A1 Concrete grades for different applications and short-term elastic moduli

Grade*	Cube strength at 28 days (MN/m^2)	Concrete modulus Static E_c (GN/m^2)	Dynamic E_{cq} (GN/m^2)	Possible applications are shown against the lowest suitable grade*
7	7	-	-	Plain concrete—BS12[1] or BS146[2] (CP 110)
10	10	-	-	-
15	15	-	-	Reinforced lightweight concrete (CP 110) Blinding concrete—BS12[1] or BS146[2] (BS 5337)
20	20	25 ± 4	35 ± 4	Reinforced concrete—dense aggregate (CP 110)
25	25	26 ± 4	36 ± 4	Reinforced concrete (BS 5337) Blinding concrete—BS 4027[3] or BS 4248[4] (BS 5337)
30	30	28 ± 5	38 ± 5	Prestressed concrete post-tensioned tendons
40	40	31 ± 5	40 ± 5	Prestressed concrete pre-tensioned tendons
50	50	34 ± 6	42 ± 6	
60	60	36 ± 6	44 ± 6	

* The grade also designates the characteristic cube strength at 28 days e.g.
(1) Portland or (2) Portland blast furnace cement (3) Sulphate-resisting Portland or (4) Supersulphated cement

Design may be based on the characteristic strength or, if appropriate, the strength given in Table A2 for the age at loading.

Table A2 Cube strengths of RC grades at various ages

	Grade				
Age	20	25	30	40	50
7 days	13.5	16.5	20	28	36
2 months	22	27.5	33	44	54
3 months	23	29	35	45.5	55.5
6 months	24	30	36	47.5	57.5
1 year	25	31	37	50	60

Static modulus can be estimated from dynamic modulus from

$$E_c = 1.25\ E_{cq} - 19$$

Poisson's ratio for concrete for the serviceability limit states may be taken as 0.2.

CONSTITUENT MATERIALS FOR CONCRETE (CP 110 §6.2, BS 5337 §6.2)

Cements (§6.2.1)

For BS 5337 use cements to BS 12 (ordinary Portland, rapid hardening Portland, white Portland, coloured Portland), BS 146 (Portland blastfurnace), BS 1370 (low-heat Portland), BS 4027 (sulphate-resisting Portland), BS 4246 (low-heat Portland blastfurnace) and BS 4248 (supersulphated cement).

For CP 110, cements with additives but otherwise to BS 12 (ultra high early strength Portland, water-repellent Portland and hydrophobic

Portland) may also be used.

Aggregates (§6.2.2)

For BS 5337 use coarse and fine aggregates to BS 882 or blastfurnace slag coarse aggregate to BS 1047. Aggregate absorption to BS 812 generally $\ngtr$ 3%. Grading and shape should be such as to faciliate well compacted, dense impervious concrete with the specified proportions and consistency without segregation or use of an excessive water content. Preferred nominal maximum size of coarse aggregate: 40 mm and 20 mm. Maximum size of coarse aggregate $\ngtr \frac{1}{4} \times$ minimum thickness of member.

For CP 110, aggregates to BS 877 (foamed or expanded blastfurnace slag lightweight aggregate) BS 3797 (lightweight aggregates) and BS 4619 (heavy aggregates) may also be used. Other aggregates, including other types or gradings, are not excluded if satisfactory data on the properties of concrete made with them is available. Aggregates having high drying shrinkage or other properties likely to have an unusual effect on the concrete strength, density, shrinkage, moisture movement, thermal properties, creep, modulus of elasticity or durability may be used, provided that special measures or account is taken in the structural design and in the workmanship requirements.

Preferred nominal maximum sizes also include 14 mm and 10 mm.

Marine aggregates (§6.2.2.3)

For BS 5337 marine aggregates may be used in reinforced concrete (but not prestressed concrete) provided that:

(a) Sodium chloride content of fine aggregate $\ngtr$ 0.10% and of coarse aggregate $\ngtr$ 0.03% by weight of dry aggregate
or total sodium chloride concentration for aggregates $\ngtr$ 0.32% by weight of cement in the mix.

(b) Hollow shells generally of negligible amount and total shell content as calcium carbonate $\ngtr$ 2, 5, 10, 15 and 30% by weight of dry aggregate for nominal maximum aggregate size of 40 mm, 20 mm, 14 mm, 10 mm and fine aggregate respectively.

For CP 110 sodium chloride $\ngtr$ 1.0% by weight of cement and $\ngtr$ 0.1% for prestressed concrete if calcium chloride admixtures are not permitted. Also, if sodium chloride is taken as equal to an equivalent anhydrous calcium chloride content and if concrete contains embedded metal then:

equivalent (calcium + sodium) anhydrous calcium chloride $\not>$ 1.5% by weight of cement

No actual limits are given for hollow and flat shells and CP 110 suggests that inclusion of an appropriate admixture may ease some workability problems.

Admixtures (§6.2.3)

For BS 5337 calcium chloride or admixtures containing chlorides should not be used. Other admixtures and constituents only with the engineer's approval.

For CP 110 suitable admixtures may be used for 'special structural concrete' with the prior approval of the engineer, who should be informed of the following:

(1) the typical dosage and adverse effects of under and overdosage
(2) the chemical name(s) of the main active ingredient(s)
(3) the chloride content (if any) expressed as equivalent anhydrous calcium chloride by weight of admixture
(4) the entrainment of air (if any) when used at the manufacturer's recommended dosage.

Calcium chloride (CP 110 only)

Calcium chloride must not be used with sulphate-resisting Portland cement or with supersulphated cement, or in pre-tensioned prestressed concrete or in the main concrete of post-tensioned prestressed concrete unless there is an impermeable durable and continuous barrier, in addition to any grout, protecting the tendons. Calcium chloride can be obtained as flake or solid (undesirable form), or in liquid grade, or included in a compound admixture or interground in extra rapid hardening Portland cement. The corrosion risk for any embedded metal is greatly increased by inadequate mixing, curing at elevated temperatures and subsequent exposure to warm moist conditions.

Pulverised fuel ash

The fineness zone and the maximum sulphate content of PFA should be specified in accordance with BS 3892. The total sulphate content (cement to BS 4550, aggregate to BS 1047 or BS 3797 if appropriate and PFA to BS 3892) of the concrete expressed as sulphuric anhydride (SO_3) $\ngtr$ 4% by weight of the cement
PFA should not be used with sulphate-resisting Portland cement to BS 4027 in concrete required to be sulphate resistant.

Air entraining agents

The AEA admixture should enable the air content to be maintained within the required limits irrespective of extension of mixing time to 30 min.

REQUIREMENTS FOR HARDENED CONCRETE (CP 110 §6.3 & BS 5337 §6.3)

The engineer should decide the minimum requirements for strength and durability of the concrete in the hardened state. If, in addition, a special property or a particular surface finish is required, these minimum requirements may have to be considerably exceeded. See Table A1 for grade designation, Table A3 for nominal cover and conditions of exposure (see also Appendix CW §2.1.9) and Tables A4 and A5 for minimum and maximum cement contents. The type of aggregate may affect the fire resistance, creep, shrinkage, durability etc. of the hardened concrete.

Table A3 Nominal cover to reinforcement (§3.11.2)

Conditions of exposure	Nominal cover for concrete grade of:				
	20	25	30	40	≮50
Mild: e.g. completely protected from weather etc. except for brief period during construction.	25mm	20mm	15mm	15mm	15mm
Moderate: e.g. sheltered from severe rain and freezing whilst saturated. Concrete continuously under water or buried.	-	40	30†	25	20
Severe: e.g. exposed to driving rain, alternate wetting and drying and to freezing whilst wet, or heavy condensation or corrosive fumes.	-	50	40	30	25
Very severe: e.g. exposed to sea or moorland water and with abrasion	-	-	-	60	50
Subject to de-icing salt	-	-	50*	40*	25

* applicable only if concrete is air entrained.

† BS 5337 specifies a minimum of 40 mm nominal cover in all cases (§4.10)

Table A4 Minimum cement contents in Portland cement concretes for durability and impermeability (§6.3.3 or §22.3)

Exposure	Reinforced concrete					Plain concrete				
	Nominal maximum size of aggregate				Maximum free W/C ratio*	Nominal maximum size of aggregate				Maximum free W/C ratio
	40	20	14	10		40	20	14	10	
Mild	220 -20	250 -20	270 -20	290 -30	0.65	200 -20	220 -20	250 -30	270 -30	0.70
Moderate (or Class B†)	260 -20	290 -30	320 -30	340 -30	0.55	220 -20	250 -20	280 -30	300 -30	0.60
Severe (or Class A†)	320 -30	360 -30	390 -40	410 -40	0.45	270 -30	310 -30	330 -30	360 -40	0.50
De-icing salts	260 -20	290 -30	320 -30	340 -30	0.55	240 -20	280 -30	310 -30	330 -30	0.55

* When the maximum free water/cement ratio can be strictly controlled to values shown and required concrete workability is still obtainable then the cement contents can be reduced by the values shown as negative. Minimum cement contents for prestressed concrete are 300 kg/m^3 or as for reinforced concrete if higher (CP 110 and BS 5337).

† Class A and Class B exposures (to BS 5337) apply to 40 mm and 20 mm nominal maximum size of aggregate only, and (since strict control is assumed) no reductions are permitted for stricter control of water/cement ratio.

Table A5 Maximum cement contents in Portland cement concretes for early thermal contraction (§6.3.4 or §22.4)

CP 110	BS 5337	Maximum cement content (kg/m^3)
-	Reinforced concrete	400
All concrete grades*	Reinforced concrete	550

* Higher cement contents may be used for high grade concretes to CP 110 (including lightweight aggregates) if special consideration is given to the increased risk of cracking due to early thermal stresses and shrinkage.

Sulphate attack (§6.3.3)

CP 110 itself (§Table 49) shows how resistance of concrete to attack by sulphates depends on the type of cement used.

Entrained air (§6.3.7)

Special concretes of grade 40, or lower, required to resist the effects of salt used for de-icing should use an air entraining agent and have average air contents as follows:

4%, 5%, 6% and 7% for natural aggregates of 40 mm, 20 mm, 14 mm and 10 mm nominal maximum size.

SPECIFICATION OF CONCRETE (CP 110 §6.1)

Ordinary and special structural concretes (§6.1.2)

Ordinary structural concrete of any grade contains only materials as follows:

(i) Cements to BS 12, BS 146 or BS 4027 only

(ii) Aggregates from natural sources for concrete to BS 882, BS 1201 or air-cooled blastfurnace slag coarse aggregate for concrete to BS 1047

(iii) Water which is clean and free from harmful matter.

'Special' structural concrete contains admixtures or materials other than the above.

Designed and prescribed mixes (§6.1.2)

For a 'designed mix' the contractor selects the mix proportions in accordance with §6.5 to achieve the required strength and workability, while the engineer is responsible for specifying the minimum cement content and any other properties required to ensure durability.

For a 'prescribed mix' the engineer specifies the mix proportions and is responsible for ensuring that these properties will provide the required strength and durability. The contractor undertakes to provide a properly mixed concrete containing the constituents in the specified proportions in accordance with the appropriate provisions of §6.6.

Designed mix for ordinary structural concrete (§6.1.3.2)

The engineer need only specify the items shown below and, unless specified otherwise, the contractor decides the workability of the fresh concrete.

(a) Designed mix for ordinary concrete
(b) Grade designation (affects cover: see Table A3 and fire resistance)
(c) Minimum cement content (see Table A4)
(d) Nominal maximum size of aggregate (40, 20, 14 or 10 mm preferred)
(e) Type of cement (see (i) above).

Prescribed mix for ordinary structural concrete (§6.1.3.3)

If the engineer adopts one of the prescribed mixes given in Table A7, which are relatively rich in cement and are intended for use where the cost of testing is not justified, then since the materials are limited as indicated below (§6.6.1) the engineer need only specify:

(a) Prescribed mix for ordinary structural concrete
(b), (d) and (e) only as above.

If the engineer adopts a prescribed mix other than those in Table A7 then he should fully describe the limitations on constituent materials and the mix proportions in the specification, and assure himself of their suitability. Admixtures may not be specified or used but strength tests can be carried out during the progress of the work (as for a designed mix) to ensure that the engineer's specified mix proportions are suitable. The specification should include:

(a) Prescribed mix for ordinary structural concrete not in Table A7
(b), (c) and (d) as above
(e) Types of cement which may be used (see CP 110 §6.2.1)
(f) Types of aggregate which may be used (see CP 110 §6.2.2)
(g) Proportion of cement: fine aggregate: coarse aggregate
(h) Required workability (see CP 110 §6.4.2).

<u>Designed mix for special structural concrete</u> (§6.1.3.4)

This broad class of concrete includes those containing admixtures, cements other than Portland, lightweight or heavy aggregates, etc. as well as those designed to have a special property or to produce a particular surface finish. Specify the requirements in detail, giving reasons for special requirements, and give any of the following information, where appropriate, being careful to ensure that no requirements conflict with each other.

(a) Design mix for special structural concrete
(b), (c), (d) and (h) as before
(e) Required brand or type of cement (see CP 110 §6.3.5 for sulphate attack etc.)
(f) Required source or special type of aggregate (e.g. for fire resistance, low creep, etc.)
(g) Maximum cement content
(i) Required admixture (see CP 110 §6.2.4)
(j) Air content of fresh concrete (§6.3.7)
(k) Maximum or minimum temperature of fresh concrete (e.g. in massive sections or in cold weather)
(l) Rate of sampling or testing (§6.8.2.1)
(m) Other requirements (e.g. control of colour, see CP 110 §6.8.1 and §6.10.2)

See §6.5 for requirements for designed mixes.

<u>Prescribed mix for special structural concrete</u> (§6.1.3.5)

Only specify mixes in this class when the engineer is assured that the required concrete can be readily achieved; for example, a special finish may be of prime importance, and strength only a secondary consideration. Admixtures in an otherwise ordinary structural concrete render them 'special'. The mix proportions in Table A7 are only suitable as prescribed mixes for special structural concrete if the admixture concerned is calcium chloride.

REQUIREMENTS FOR DESIGNED MIXES (CP 110 §6.5)

Target mean strength (§6.5.1)

Design the concrete mix to have at least the required minimum cement content and a mean strength

$\nless f_k$ + (current margin)

where the current margin depends on the availability of cube results on separate batches of concrete of nominally similar proportions and materials produced over certain periods by the same plant under similar supervision, and may be taken as the smaller of the values given in Table A6.

Table A6 Minimum values for current margins

Cube results available	$\nless$ 100	$\nless$ 40	< 40
Period tested	$\ngtr$ 12 months	> 5 days & $\ngtr$ 6 months	–
Grades 7, 10 and 15	$\nless \frac{1}{6} f_k$ & $\nless$ 1.64s	$\nless \frac{1}{3} f_k$ & $\nless$ 1.64s	$\nless \frac{2}{3} f_k$
Grades 20 and above	$\nless$ 3.75 MN/m² & $\nless$ 1.64s	$\nless$ 7.5 MN/m² & $\nless$ 1.64s	$\nless$ 15* MN/

* At high strengths a smaller margin (but $\nless$ 7.5 MN/m²) may have to be permitted for initial mix design

Evidence of suitability of proposed mix proportions (§6.5.2)

The contractor should provide the following information before supplying any designed mix.

(1) Nature and source of each material

(2) *Either* (a) appropriate existing data as evidence of satisfactory

previous performance for target mean strength, current margins and, if required, workability and W/C ratio

or (b) full details of tests on trial mixes

or (c) for ordinary structural concrete a statement that initially the appropriate mix proportions as for a prescribed mix (see Table A7) will be used

(3) Proposed quantities of each ingredient per cubic metre of fully compacted concrete

Subsequently, the contractor should declare any change in sources of materials and any change exceeding 20 kg/m^3 in cement content last declared.

Trial mixes (§6.5.3)

If trial mixes are required, make three separate batches of concrete using materials typical of the proposed supply and preferably under full scale production conditions, and sampling and testing to BS 1881. Determine the workability of each trial batch and cast 3 cubes for test at 28 days (and 3 for earlier test if desired). Trial mix properties are satisfactory if mean strength of the nine cubes tested at 28 days (or predicted mean from nine earlier tests)

$\nless$ specified f_k + (current margin - 3.5 MN/m^2)

If trial mixes are required to demonstrate that the maximum free water/cement ratio is not exceeded (see Table A4), make two batches in a laboratory using surface dry aggregates typical of the source of supply. Both batches should demonstrate that at the proposed degree of workability and at the correct cement content a free water/cement ratio below the maximum specified is appropriate. Existing laboratory test reports may be accepted only if the materials to be used are likely to be similar to those used in earlier tests.

Additional trial mixes may be necessary during production if a substantial change is made in the materials or the proportions. They are not necessary for normal control when merely adjusting mix proportions so as to approach more closely the target mean strength.

REQUIREMENTS FOR PRESCRIBED MIXES (§6.6)

Unless specified otherwise the mix proportions should be selected from Table A7 and the constituent materials selected from the following:

Grades 7, 10 and 15

Cement to BS 12 or BS 146, aggregate to BS 882 and coarse aggregate to BS 1047 (see Table A1 and §6.1.2). Higher sand contents as given in Table A7 are preferable for all-in aggregate to BS 882.

Grades 20, 25 and 30

Cement to BS 12, BS 146 or BS 4027, coarse aggregate to BS 882 or BS 1047 and sand to grading zones 1, 2 or 3 of BS 882.
No admixtures. Inform engineer of the nature and source of each material and of any changes.

Table A7 shows the total weights of dry aggregate to produce approximately 1 m³ of concrete. Slight adjustments may be necessary to the fine aggregate proportions or to the total aggregate quantities to produce this volume of concrete having the required workability, strength and cement content.

Where single-sized coarse aggregates are used, adjust the proportion to produce a combined grading to the graded coarse aggregate limits of BS 882 or BS 1047. Allow for a moisture content typical of the aggregate being used and calculate the actual batch weights to suit the size of the mixer. Aggregates for grades 7, 10 and 15 may be batched by volume assuming a bulk density of the damp (all-in) aggregate of 1500 kg/m³ and a whole bag of cement as weighing 50 kg.

PRODUCTION OF CONCRETE (CP 110 §6.7)

Protect cement before use and prevent accidental mixing of different types. Cement manufacturers' certificates should be available. Provide separate storage facilities with adequate provision for drainage for each different size of aggregate used. Inspect all aggregate deliveries and regularly test at least a proportion of them. Handle and store so as to minimize segregation and contamination. For grades 7, 10 and 15 may use all-in aggregate. For grades 20 and above, use separate fine and coarse aggregates and determine grading of each aggregate size from each source of supply at least once weekly. Report the test results to the engineer and check for any change from samples used to establish the batch weights. The results of routine control tests carried out by the aggregate producer should be accepted for this purpose.

Table A7 Prescribed mixes for ordinary structural concrete

Concrete grade	Nom. max. size of aggregate (mm)	40		20		14		10	
	Typical range of slump (mm)	50-100	100-150	25-75	75-125	10-50	50-100	10-25	25-50
7	Cement (kg)	180	200	210	230	-	-	-	-
	Total aggregate (kg)	1950	1850	1900	1800	-	-	-	-
10	Cement (kg)	210	230	240	260	-	-	-	-
	Total aggregate (kg)	1900	1850	1850	1800	-	-	-	-
15	Cement (kg)	250	270	280	310	-	-	-	-
	Total aggregate (kg)	1850	1800	1800	1750	-	-	-	-
7,10,15	Fine aggregate (%)	30-45	30-45	35-50	35-50	-	-	-	-
20	Cement (kg)	300	320	320	350	340	380	360	410
	Total aggregate (kg)	1850	1750	1800	1750	1750	1700	1750	1650
25	Cement (kg)	340	360	360	390	380	420	400	450
	Total aggregate (kg)	1800	1750	1750	1700	1700	1650	1700	1600
30	Cement (kg)	370	390	400	430	430	470	460	510
	Total aggregate (kg)	1750	1700	1700	1650	1700	1600	1650	1550
20,25,30	Sand, Zone 1 (%)	35	40	40	45	45	50	50	55
	Sand, Zone 2 (%)	30	35	35	40	40	45	45	50
	Sand, Zone 3 (%)	30	30	30	35	35	40	40	45

Appendix A

Batching and mixing (§6.7.4,5)

Batch cement, fine aggregate and various sizes of coarse aggregate by weight except that volume batching may be used for:

(1) Grades 7, 10 and 15 using dense aggregate

(2) Grades 7 and 10 using lightweight aggregate

(3) Grades 15 and above: lightweight coarse aggregate component only.

Weigh cement by separate weighing device or use whole number of bags in each batch. Measure water by volume or by weight. Measure any solid admixtures by weight except that liquid or paste admixtures may be measured by volume or weight. Adjust the batch weights of aggregate to allow for a moisture content typical of the aggregates being used. Maintain all measuring equipment in a clean, serviceable condition to batch to an accuracy within ± 3% for cement, water or total aggregates and ± 5% for any admixture.

The mixer should comply with BS 1305 or BS 4251 (batch type or truck type concrete mixers) if appropriate. The mixing time should be not less than that used by the manufacturer in assessing the mixer performance. This may not be adequate for low workability or high cement content mixes, hence comparison of the strengths of samples mixed for different times may be advisable.

Adjust the water content of each batch of concrete so as to produce a concrete of the workability required by the trial mixes or by Table A7 as appropriate.

Control of strength of designed mixes (§6.7.6)

Adjustments of mix proportions during production will normally be made in order to minimize the variability of strength and to approach more closely the target mean strength as part of the proper control of production—but maintain the specified limits of minimum cement content and maximum water/cement ratio. Declare any changes in cement content over 20 kg/m^3 (see §6.5.2). Such adjustments to mix proportions should not be taken to imply any change to the current margin.

Change of current margin (§6.7.6.2)

When a sufficiently large number of tests show that the previously established margin is significantly too large or too small, a change in the current margin used for judging compliance with the specified characteristic strength becomes appropriate as follows.

Recalculate margin to §6.5.1 and compare with existing margin. Adopt new margin if difference is:

$\not>$ 18% when based on tests on 40 separate batches
or $\not>$ 11% when based on tests on 100 separate batches
or $\not>$ 5% when based on tests on 500 separate batches

On the adoption of a recalculated margin it becomes the current margin for the judgement of compliance (§6.8.2.2) with the specified characteristic strength of concrete produced subsequently to the change.

COMPLIANCE WITH SPECIFIED REQUIREMENTS (§6.8)

If materials satisfy the specification and §6.2 and the methods of manufacture (§6.7) and practices of handling the raw materials and concrete are satisfactory, compliance of concrete mixes should be judged as follows.

(1) *Designed mix for ordinary structural concrete:* compare strength of hardened concrete with specified characteristic strength (§6.8.2) also compare cement content with specified minimum cement content (§6.8.3).
(2) *Prescribed mix for ordinary structural concrete:* judge on the basis of the specified mix proportions and required workability, unless specified otherwise. Also, concrete should be likely to have sufficient strength (§6.1.3.3).
(3) *Designed mix for special structural concrete:* as (1) above except that in addition, if appropriate, any special requirements as set out in §6.8.3, 4, 5, and 6 can be specified.
(4) *Prescribed mix for special structural concrete:* as (2) above.

If other requirements not specified in detail in CP 110 are specified (e.g. density or modulus of elasticity of concrete), a detailed description of the method of test, with tolerances which take appropriate account of variability due to manufacture, sampling and testing, should be given.

All sampling and testing of constituent materials should be to the appropriate British Standard, and all sampling and testing of fresh and hardened concrete should be to BS 1181 unless such provision is at variance with CP 110.

Appendix A

Strength compliance (§6.8.2)

The characteristic strength of concrete is that 28-day cube strength below which $\ngtr$ 5% of test results may be expected to fall. Judge compliance by 28-day cube tests or, if there is satisfactory evidence that a particular testing regime can predict the 28-day strength at an earlier age, on these earlier tests alone.

The minimum rate of sampling should be decided by the engineer using Table A8 as a guide.

Table A8 Rates of sampling and testing

	Average volume represented	Concrete at risk	Suitable applications
Rate 1	$\ngtr$ 10m^3 and $\ngtr$ 10 batches	$\ngtr$ 30m^3	Highly-stressed structural elements
Rate 2	$\ngtr$ 20m^3 and $\ngtr$ 20 batches	$\ngtr$ 60m^3	Ordinary structural work
Rate 3	$\ngtr$ 50m^3 and $\ngtr$ 50 batches	$\ngtr$ 150m^3	Mass concrete at low stresses

Each sample should be taken from one batch selected randomly to represent the appropriate volume of concrete. Higher rates of sampling and testing would be appropriate at the start of the work, to establish the level of quality quickly, or during periods of production when quality is in doubt. Conversely, rates may be reduced when high quality has been established. At least one sample should be taken on each day that concrete of that grade is used.

Testing plan (§6.8.2.2)

Make each cube from a single sample taken from a randomly selected batch of concrete, preferably at the point of discharge from the mixer, or, for ready-mixed concrete, at the point of discharge from the delivery vehicle. Assume compliance with the specified characteristic strength if:

(1) the average strength determined from any group of four consecutive test cubes

$\nless$ specified f_k + 0.5 × current margin and

(2) each individual test result

> 85% of specified f_k

Take the current margin as:

$\frac{2}{3}$ × specified f_k for Grades 7, 10 or 15 and

15 MN/m^2 for Grade 20 and above

unless under §6.5.1 or §6.7.6.2 a smaller margin has been established.

If only one cube results fails (2) then that result may be considered to represent only the particular batch of concrete from which that concrete was taken, provided that the average strength of the group satisfies (1).

If more than one cube result fails (2), or if the average strength of any group of four consecutive test cubes fails (1), then all the concrete in all the batches represented by all such cubes shall be deemed not to comply with the strength requirements. For the purposes of this subclause the batches of concrete represented by a group of four consecutive test cubes shall include the batches from which samples were taken to make the first and the last cubes in the group of four, together with all the intervening batches.

<u>Non-compliance with the testing plan</u> (§6.8.2.3)

When the average strength of four consecutive test cubes fails (1) of §6.8.2.2, the mix proportions of subsequent batches of concrete should be modified to increase the strength.

The action to be taken for non-compliance may range from qualified acceptance in less severe cases, to rejection and removal of the concrete which is represented in the most severe cases. In determining the action to be taken, the engineer should have due regard to the technical consequences of the kind and degree of non-compliance, and to the economic consequences of alternative remedial measures either to replace the substandard concrete or to ensure the integrity of any structure in which the concrete has been placed.

Wherever possible, establish the following in estimating the quality of the substandard concrete and in determining the action to be taken.

(1) The validity of the test results, and confirmation that specimen sampling and testing has been to BS 1881.

(2) The mix proportions actually used in the concrete under investigation. These may affect durability.

(3) The actual section of the structure represented by the test cube(s).

(4) The possible influence of any reduction in concrete quality on the strength and durability of this section of the structure.

The engineer may require tests on the hardened concrete in the structure, including non-destructive methods or the taking of cored samples (CP 110, §9). The results of any such tests should not nullify the establishment of non-compliance with the testing plan provided that this was based on valid cube test results.

Cement content (§6.8.3)

The cement content of any batch of concrete should be

$\nless$ 0.95 × specified minimum

and $\ngtr$ 1.05 × specified maximum

One way of ensuring this is to observe the batching in relation to the indicated batch weights. Alternatively determine cement content from samples representative of any batch of concrete provided that a suitable testing regime (including errors due to sampling) is used to measure the cement content of fresh concrete to ±5% of the actual value with a confidence of 95%.

Water–cement ratio (§6.8.5)

If a maximum water/cement ratio has been specified the ability to comply, at a suitable level of workability, will have been determined from trial mixes. If the constituent materials and mix proportions are substantiall[y] the same as for the trial mixes, a maximum water/cement ratio may be judged from workability tests. Alternatively determine the water–cement ratio from samples and a suitable testing regime within the same limits as for cement content above.

Air content of fresh concrete (§6.8.6)

The percentage air content determined from individual samples taken at the point of placing the concrete and representative of any given batch of concrete should be within ±1.5 of the required value. The average value from any four consecutive determinations from separate batches should be within ±1.0 of the required value. It should be noted that the method of measuring air content in BS 1881 does not necessarily apply to lightweight aggregate concrete.

Additional tests on concrete for special purposes (§6.9)

Make and test any additional cubes to BS 1181 but the methods of sampling and the conditions under which the cubes are stored should be varied according to the purpose for which they are required. For determining the cube strength before transfer of prestress, or of concrete before striking formwork etc., sampling should preferably be at the point of placing, and the cubes should be stored as far as possible under the same conditions as the concrete in the members. The extra cubes should not be used for normal quality control or compliance procedures and should be suitably identifiable.

SURFACE FINISHES FROM FORMWORK OR MOULDS (CP 110 §6.10.6, BS 5337 §6.5)

For general reference see 'Recommendations for the production of high quality concrete surfaces' CACA Technical Advisory Series 47.019. BS 5337 normally requires Type D finish for water storage, but Type B might be acceptable where algal growth is unlikely.

Type A finish

Properly designed formwork or moulds of closely-jointed sawn boards imprinting the joint pattern and grain on the concrete surface. In addition, small blemishes due to entrapped air or water may be expected, but surface should be free of large blemishes, voids or honeycombing. The finish is not necessarily suitable to receive in situ concrete or applied finishes, nor does it include special board-marked finishes. Consideration should be given to the type of connection, or applied finish, and if necessary other types of treatment specified.

Type B finish

Properly designed forms of closely-jointed wrought boards, steel or other suitable material imprinting the joint pattern and slight grain on the concrete surface. Otherwise as Type A.

Type C finish

Requires high quality concrete and properly designed forms having a hard, smooth surface. Smooth concrete surfaces with true, clean arrisses and only very minor surface blemishes and no staining or discoloration from the release agent.

Type D finish

Requires high quality concrete with Type B finish which is then improved by carefully removing all fins and other projections, thoroughly washing down and then filling the most noticeable surface blemishes with a cement and fine aggregate paste matching the colour of the concrete. The release agent should not permanently stain or discolour the concrete surface.

Type E finish

Requires a Type C finish for which, while the concrete is still green, all surface blemishes are filled with a fresh, specially prepared cement and fine aggregate paste of matching colour. After proper curing the faces are rubbed down where necessary to produce a smooth and even surface.

Other types of finish

Each case should be fully specified and related, if possible, to samples which are readily available for comparison. Includes any finish requiring permanently exposed coarse aggregate, special forms or linings, different concrete mixes near the surface, grinding, bush-hammering or other treatment.

CONCRETE BLINDING LAYER (BS 5337 §6.6)

Screeded layer of plain site concrete:

$\nless$ 75 mm thick for walls and floors founded on the ground.

For clay ground, hand excavate final layer (few cm) immediately before concreting. Concrete is usually weaker than remainder of structure, but not weaker than shown in Table A1. If necessary use concrete not weaker than Grade 25 and sulphate-resisting or other special cement if injurious soils or aggressive ground water is expected.

Appendix B
Practical guides for limit state design

(Extracts from CP 110:1972 and BS 5337:1976. Cross-references to the codes are shown in brackets. Conditions that are not as given by the codes are indicated by footnotes)

LIMIT STATES (§2.2)

(1) Ultimate limit states

*Collapse** Static loading acting on the structure as designed causing rupture of one or more critical sections. Also, overturning, buckling, etc.

*Impact** Static loading acting on the structure after damage by an exceptional event, and causing any of the above.

(2) Serviceability limit states

Deflection

Vibration

Local damage — Cracking of concrete or excessive strain in steel.

(3) Other limit states and considerations

Fatigue, durability, fire resistance, lightning or other relevant limit states as necessary.

Design on basis of most critical limit state and then check that the remaining limit states are satisfied.

* Terms not used in CP 110 but included here for clarity. The γ_f factors from CP 110 in Table B1 are generally for the limit state first defined for collapse or impact in each case.

DESIGN LOADS (§2.3.1)

The characteristic load should ideally be determined from the mean load and its standard deviation from the mean. However, the characteristic loads should be taken as equal to the loadings given in CP 3: chapter V: Part 1 for dead and imposed loads and Part 2 for wind loads (see Appendixes V, W and X). Thus:

F_k = working load given in CP 3: chapter V or other British Standard.

Design load

Design load $F = F_k \times \gamma_f$ e.g. $Q = Q_k \times \gamma_f$

Weight of reinforced concrete

Take as 24 kN/m³ under standard gravity conditions

DESIGN STRENGTHS (§2.3.2)

The characteristic strength is the strength below which not more than 5% of the test results fall. Thus:

f_k = strength for which there are $\not>$ 5% failures

Design strength

Design strength ${}_*f = f_k/\gamma_m$ e.g. ${}_*f_y = f_y/\gamma_m$

PARTIAL SAFETY FACTORS (§2.3)

The partial safety factors for load, γ_f, and for material strength, γ_m, are listed in Table B1.

Appendix B

Table B1 Specified γ_f and γ_m values for different limit states

Limit states and load combination	Partial safety factors γ_f Maximum	Minimum[1]	γ_m (& $1/\gamma_m$) Steel	Concrete
Ultimate (Collapse)				
(1) Dead and imposed	$1.4G_k+1.6Q_k$	$1.0G_k$	1.15[2] (0.87)	1.5 (0.67)
(2) Dead and wind	$1.4(G_k+W_k)$	$0.9G_k+1.4W_k$	"	"
(3) Dead, imposed and wind	$1.2(G_k+Q_k+W_k)$	–	"	"
(4) Excessive [3]	$1.05(G_k+Q_k+F_{exc})$	–	1.0	1.3 (0.77)
Ultimate (Impact) [4]	$1.05(G_k+Q_k+F_{im})$	–	"	"
Serviceability (Deflection)				
(1) Dead and imposed [5]	$1.0(G_k+Q_k)$	$1.0G_k$	1.0	1.0
(2) Dead and wind	$1.0(G_k+W_k)$	$1.0(G_k+W_k)$	"	"
(3) Dead, imposed and wind	$1.0G_k+0.8(Q_k+W_k)$	–	"	"
Serviceability (Cracking)				
As for deflection	As for deflection[6]	As for deflection	1.0	1.3 (0.77)

Notes

(1) The minimum intensities apply when a reduction in load provides a more adverse condition: e.g. alternate spans in continuous systems for load combinations (1) and (2).

(2) Steel in tension. γ_m varies for steel in compression as follows: $1.15 + f_y/2000$(beams), 1.5 or 1.67 (axially loaded columns).

(3) The excessive load and any other loads which act simultaneously (e.g. Q_k may represent only the permanent fraction of the total imposed load).

(4) CP 110 comments as follows: 'when considering the continued stability of a structure after it has sustained localized damage' but does not use the term 'impact'.

(5) These design loads apply when estimating the immediate deflections. When estimating the additional time-dependent deflections due to creep, shrinkage and temperature, consider dead load and permanent fraction of the total imposed load. Consider also temperature gradients if necessary and give special attention to any imposed loads which are predominantly cyclic in character.

(6) External applied loads affect cracking of hardened concrete. Early thermal cracking arises because of restraints to thermal movement (and shrinkage) of the concrete.

ANALYSIS (§2.4)

Analysis of structure (all limit states)(§2.4.3.1 and 2.4.4.1)

Force distributions should be determined from material properties associated with characteristic strengths. Elastic analysis may be used with relative stiffnesses based on any one (i.e. one only) of the following cross-sections throughout.

(a) Concrete section—given by bh
(b) Gross (i.e. equivalent uncracked) section—given by bh + $\alpha_e A_s$
(c) Transformed (i.e. equivalent cracked) section—given by $bd_n + \alpha_e A_s$ or by $bx + \alpha_e A_s$ as convenient.

For prestressed concrete gross section only should be used.

Moment redistribution is permitted for limit state of collapse only:

$\not>$ 30% if members possess adequate ductility

$\not>$ 10% for structures with $\not<$ 4 storeys and relying on columns for stability

For slabs, yield line or other appropriate plastic theory may be used.

Analysis of sections (§2.4.3.2 and 2.4.4.2)

Material properties associated with the design strengths appropriate to the limit states being considered should be used as follows.

(a) *Ultimate*—inelastic analysis based on short term stress-strain curves (see Appendix D).

(b) *Serviceability*—elastic analysis based on short-term stress-strain curves (e.g. instantaneous deflection) or elastic analysis allowing for creep and shrinkage (e.g. long-term deflection of uncracked section) or partially inelastic analysis allowing for creep, shrinkage and cracking (e.g. long-term deflection of pseudo-cracked section).

Properties of Materials (§2.4.2)

Concrete grades, elastic moduli, etc., are given in Appendix A. Short-term stress-strain curves are given in Appendix D.

Short-term stress-strain curves may require modification for sustained loading as follows.

Reinforcement:	none
Prestressing tendons:	due allowance for relaxation
Concrete:	due allowance for shrinkage and creep

Model analysis and testing (§2.4.5)

A design may be deemed satisfactory on the basis of model tests coupled with model analysis predicting the behaviour of the actual structure, if suitable expertise and equipment is available.

Experimental development (§2.4.6)

Actual tests of prototype units and similar structures may be used to justify the analytical or empirical basis of a design based on novel methods.

LIMIT STATE REQUIREMENTS (§2.2)

Ultimate (§2.2.2)

Ensure adequate resistance to:

(a) bending, shear and axial loads at any critical section

(b) buckling of any member

(c) overturning of the structure

(d) progressive collapse following an exceptional event.

Deflection (§2.2.3.1)

Horizontal members (deflection below horizontal):
Final long-term deflection $\ngtr$ span/250 (generally).

Additional deflection after erecting partitions and applying finishes $\ngtr$ span/350 and $\ngtr$ 20 mm unless provision is made for movement. Reduce for brittle finishes.

Precamber (above horizontal) $\ngtr$ span/300 for prestressed units with finishes and non-uniform camber.

Cracking (§2.2.3.2)

(a) Crack width at surface $\ngtr$ 0.3 mm (RC generally)
$\ngtr$ 0.2 mm (Prestressed Class 3)
No cracks (Prestressed Class 1 and 2) and no tension (Class 1).

(b) Crack width at surface nearest to main reinforcement for an aggressive environment $\ngtr$ 0.004 × nominal cover (RC) and $\ngtr$ 0.1 mm (Prestressed Class 3).

(c) Limit compressive stresses both finally and at transfer to prevent (i) excessive losses in prestress (prestressed concrete only) and (ii) spalling of concrete in compression zone.

(d) Maximum width of surface cracks for different classes of exposure for water retaining structures are given in Appendix CW (§2.2)*

(e) Provide movement joints both within the structure and between the structure and cladding, as necessary, in tall or long buildings unless the effects of temperature, creep and shrinkage are otherwise adequately catered for.

* BS 5337:1976

Appendix C
Practical guides for design

(Extracts from CP 110:1972 'The Structural Use of Concrete' (CA,CB,CC, CD,CE,CL,CU and CX) and BS 5337:1976 'The Structural Use of Concrete for Retaining Liquids' (CW and CX). Cross-references to the codes are shown in brackets. Conditions that are not as given by the codes are indicated by footnotes.)

The ultimate limit states (collapse) are usually critical for RC structures. Thus the guides are arranged as follows:

Appendix CA - General requirements
CB - Structural members at ultimate
CC - Shear and torsion
CD - Reinforcement, bond and anchorage
CE - Limit states of serviceability
CL - Lightweight concrete
CU - Plain walls
CW - Water-retaining structures
CX - Joints

APPENDIX CA
GENERAL REQUIREMENTS

Methods of analysis and design for reinforced concrete structures are given which are, in general, the most suitable for satisfying the requirements given in Appendix B. Use other methods only if they can be shown to be satisfactory for the particular type of structure or member concerned.

Stability (§3.1.2.2)

The layout of the structure in plan and the interaction between the structural members should ensure a robust and stable design.

Recommended precautions are as follows.

(a) Horizontal design force for collapse
$\nless$ 1½% total characteristic dead load above any level. Assume force acts at the height of the centre of gravity of the dead load and is shared by the structural members according to stiffness and strength.

(b) Consider impact for members vital for stability, i.e. either shield member from source of danger (e.g. provide bollards to prevent vehicle impacts) or design to sustain damage.

(c) Provide effective continuous vertical tie from foundations to roof level in all structures of five or more storeys of: $\nless$ minimum for vertical reinforcement with lap lengths for tension in all RC columns and walls.

(d) Provide effective horizontal ties both peripherally and internally. Ensure adequate anchorage or effectiveness at all positions, including re-entrant corners, so as to resist, by catenary, cantilever or other action, the results of damage due to an exceptional event.

Tie forces (§3.1.2.2)

The horizontal tie forces to be resisted at each floor level are as follows.

Basic tie force, $F_t = (20 + 4n_s)$ or 60 whichever is least, where n_s is the number of storeys.

(a) *Peripheral tie force* $\nless F_t$ kN
Locate in the perimeter wall or within 1.2 m of the edge of the building.

(b) *Internal tie force* $\nless F_t (g_k + q_k)\ell/37.5$ kN/m and $\nless F_t$ kN/m
where $(g_k + q_k)$ is the sum of the average characteristic dead and imposed load per unit area of the floor in kN/m^2
and ℓ in metres is the total distance in the direction concerned between the centres of columns or walls etc., (i.e. total distance whether spanned by single slab or by a system of beams and slabs) *or* 5 × clear storey height (under beams, if any) whichever is least.

For floors spanning on to walls in one direction only, the internal tie force for direction parallel to the walls $\nless F_t$ kN/m

(c) *External column or wall tie force*

$$\nless F_t \frac{\ell_o}{2.5} \text{ kN/m (for walls) or } \nless F_t \frac{\ell_o}{2.5} \text{ kN (for each column)}$$

and $\nless$ 3% of total vertical design load at collapse in column or wall at that floor level

where ℓ_o is 5 or the floor to ceiling height (in metres) whichever is least.

Provide internal ties at each floor and roof level in two directions approximately at right angles. Make effectively continuous throughout their length and either anchor directly to the peripheral tie at both ends or continue as column or wall ties. Space at intervals within the slabs, beams or walls, or spread evenly over the width of the structure. Ties in a wall to be within 0.5 m of the near face of the floor slab.

Anchor all external load-bearing vertical members into the structure with the column and wall ties, which may consist partly or wholly of the same reinforcement as that provided for the peripheral or internal ties. Tie corner columns in two directions approximately at right angles.

Ignore any forces other than the above tie forces and assume that reinforcement acts at its characteristic strength (i.e. for limit state of impact $\gamma_m = 1.0$).

Regard reinforcement provided for other purposes as providing part or all that required for these ties.

<u>Construction details</u> (§3.11.1)

The sizes of members should normally be in terms of the nominal overall size. Give preference to the dimensions given in BS 2539. Specify a degree of tolerance as large as possible consistent with the finished structure or member being acceptable for its intended use and function.

The partial safety factor for loads allows for normal tolerances on nominal dimensions. PD 6440 and Appendix H gives guidance on tolerances and the design can normally assume that the reinforcement is in its nominal position. If large tolerances are specified for small highly stressed members it may exceptionally be necessary to base the design on the minimum net dimensions after allowing for the maximum specified tolerance; for example, if the reinforcement cages are positioned accurately on one face such that an accumulation of tolerances affect the position of highly stressed reinforcement at the opposite face, the consequent possible reduction in effective depth could exceed the allowance included in the normal value for the partial safety for loads. In such cases it may be necessary to adjust the effective depth assumed after considering tolerances appropriate to cover blocks, spacers and/or chairs (including movement bearing surfaces), dimensions and straightness, etc., of reinforcement, planeness and dimensions of formwork and size of member.

See also: Appendix CX.

APPENDIX CB

STRUCTURAL MEMBERS AT ULTIMATE

STRUCTURAL FRAMES (§3.2.2)

Framed structures may be idealized as frameworks with member lengths equal to the distance between the intersections of the centre lines framing into the joints. See Appendix B for assumptions in analysis of complete structures and frameworks. See below and Fig. CB1 for alternative analyses for simplified sub-frameworks.

Design load arrangements

Use arrangement causing most severe effect. For all framed structures under uniformly distributed imposed loading assume:

(a) *Beams* and *slabs:* (i) alternate spans loaded;
(ii) any two adjacent spans loaded.

(b) *Columns:* maximum out-of-balance moment at either end.

(c) *Wind:* sideways movement of columns produces reactions on all beams or slabs.

The maximum and minimum loads for vertical loads only are therefore $(1.4G_k + 1.6Q_k)$ and $1.0G_k$ respectively.

Appendix CB

Single-span continuous frame (§3.2.1)

If ratio of beam span to supporting column height > 3
and steel ratio for beam < 1%
then *either* use transformed (i.e. equivalent cracked) section only for analysis of structure
or ensure that top reinforcement in the beam at the support yields (under overload) before the concrete on the inside of the column crushes.

Frame supporting vertical loads only (§3.2.2.1)

Consider a series of sub-frames, each with beams at one level together with the columns above and below assumed to be fixed at their ends remote from the given level. See Fig. CB 1 (ii).

Alternatively, consider a simplified sub-frame for each loaded beam in which the ends of any beams and columns remote from the two joints at each end of the loaded span are assumed fixed, and the outer beams possess half their actual stiffness. This simplified sub-frame can also be used to determine the moments in an individual column, provided that the longer of the two spans framing into the given column is the central beam in the sub-frame. See Fig. CB 1 (iii).

Frame supporting vertical and lateral loads (§3.2.2.2)

In framed structures which rely on the columns for lateral stability, base design of the individual beams and columns on the larger moments, loads and shear forces obtained from either consideration of vertical loads only, as above, or from the sum of (a) and (b) below.

(a) An elastic analysis of a series of sub-frames of beams at each level, as for vertical loading only. Ignore lateral loads but load all spans with the full vertical design load for load combination (III), i.e. $1.2G_k + 1.2Q_k$.

(b) An elastic analysis of the complete frame, assuming points of contraflexure at the centres of all beams and columns, considering only the design wind load for load combination (III), i.e. $1.2W_k$, on the structure.

If necessary consider also the effects of sway (if columns are slender) and, in certain cases, the effects of load combination (II), i.e.

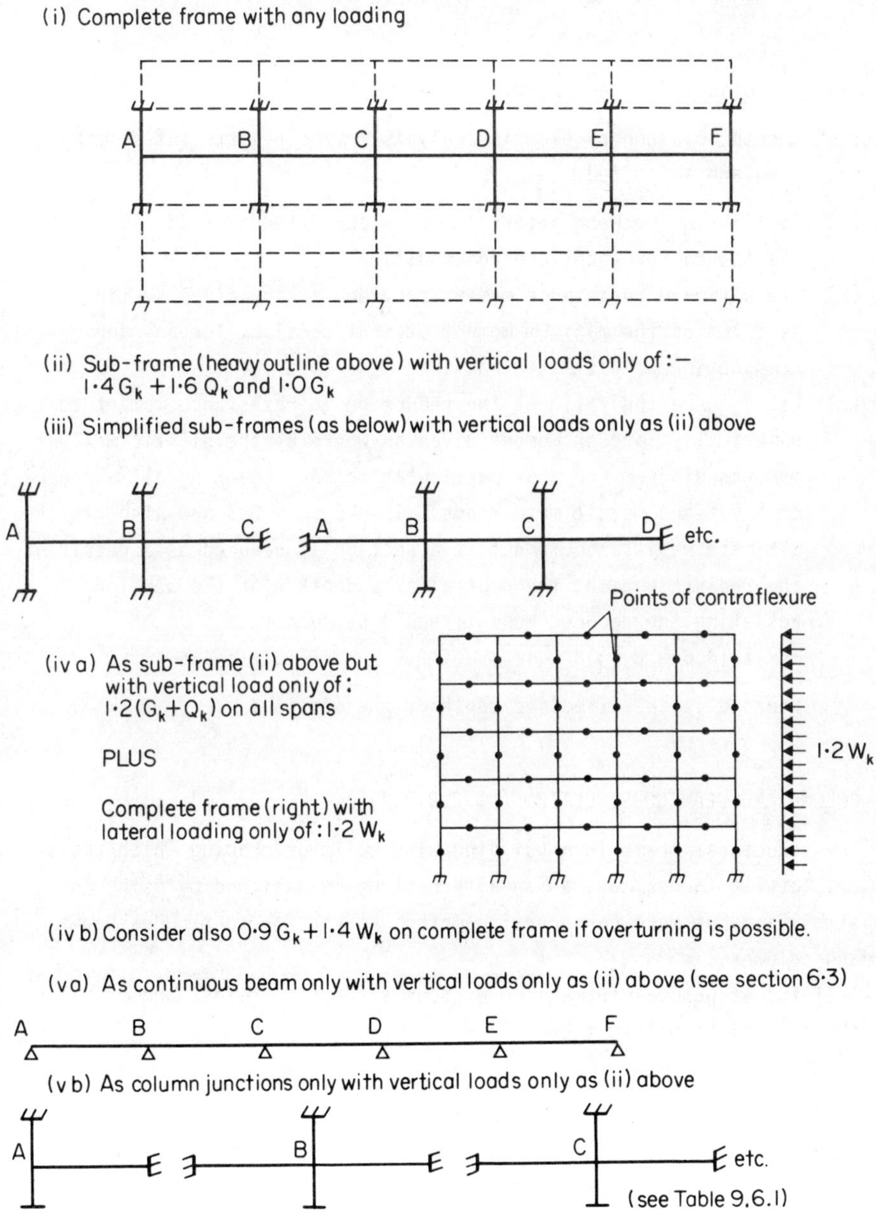

Fig.CB1 Alternative simplifications of a frame for analysis of beams and columns

$0.9G_k + 1.4W_k$. (See Fig. CB1 (iva) and (ivb)).

Redistribution of moments (§3.2.2.3)

The moments obtained by elastic analysis may be redistributed if the following are satisfied.

(a) Equilibrium between internal forces and external loads is maintained for each loading condition.

(b) The ultimate resistance moment provided at any given section is $\nless$ 70% of the maximum moment at that section, for any appropriate combination of ultimate loads.

(c) Let β_{red} be the ratio of the reduction in resistance moment to the numerically largest moment given anywhere by the elastic maximum moments diagram for that particular member, covering all appropriate combinations of ultimate loads. Then $\beta_{red} \ngtr 0.3$ and also, if the ultimate resistance moment at a section is reduced as a result of the redistribution, the neutral axis depth x of the section resisting the reduced moment should be such that

$$x \ngtr (0.6 - \beta_{red})d$$

where d is the effective depth of the section.

COLUMN AND BEAM CONSTRUCTION (§3.2.2.3,4)

Any structural frame in a building with walls or bracing which resist all lateral forces may, as an alternative, be designed by using the simplified but more conservative methods given for continuous beams and columns.

If the structural frame is over four storeys in height and provides the lateral stability then

$$\beta_{red} \ngtr 0.1$$

Notice that the condition for x generally rules out, for columns, the possibility of a redistribution involving a reduction in moment, unless the axial column load is small.

BEAMS (§3.3)

Continuous beams may be analysed in accordance with Appendix B by using the design load arrangements as for beams and slabs in framed structures.

Analyse either as part of a frame or as continuous over its supports and capable of free rotation about them. Redistribution of moments may be considered for either method above, but not for the simplified moments and shear forces given below.

Alternatively, simplified moments and shear forces may be used for beams which support substantially uniformly distributed loads over three or more spans which do not differ by more than 15% of the longest. See RC text (Table 6.3.1). If $Q_k \ngtr G_k$ Table I4 may be used.

<u>Analysis of beam sections</u> (§3.3.5)

See RC text.

<u>Effective span of beams</u> (ℓ) (§3.3.1.1)

$\ell \ngtr$ distance between centres of bearings (all spans)
and $\ngtr$ clear span plus effective depth (simple spans only)
and $\ngtr$ length to centre of support (cantilever continuous with beam only)
and $\ngtr$ length to support face plus half effective depth (built-in cantilever only).

<u>Effective width of flanged beams</u> (§3.3.1.2)

$b \ngtr$ actual flange width
and $\ngtr$ web width + ξa
where ξ = 0.1 for L-beam and 0.2 for T-beam
a = distance between points of zero moment (0.7 × effective span for a continuous beam).

<u>Slenderness limits for beams</u> (§3.3.1.3)

(a) *Simply supported and continuous beams*

$L/b_c \ngtr 60$, and $Ld/b_c^2 \ngtr 250$ in all cases
where L = clear distance between lateral restraints
b_c = breadth of compression face midway between restraints.

(b) *Cantilevers*

$L/b_c \ngtr 25$, and $Ld/b_c^2 \ngtr 100$ in all cases, where L = clear distance from face of support to end of cantilever.

See also: Appendixes CC,CD,CE,CL and CW.

SOLID SLABS (§3.4)

The recommendations given for beams apply, in general, to solid slabs. Two-way spanning slabs may be analysed using either elastic (e.g. Pigeaud and Westergaard) or plastic (e.g. yieldline or Hillerborg's) theories provided that the ratios of support to span moments are between 1.0 and 1.5 and similar to those obtained by elastic theory (see RC text).

The loads on beams supporting solid two-way slabs with uniformly distributed loads may be assumed to be given by lines of load discontinuity making an angle of 45° at each right-angled corner (see RC text, Fig. 13.9.1).

See also: Appendixes CC,CD,CE,CL and CW.

COLUMNS (§3.5)

Consider durability and fire resistance first since both the column size and the position of the bars can be affected. A column which acts as a very short length of plain concrete wall may be designed accordingly (see Appendix CU). Moments in columns may be obtained by using a simplified sub-frame where column and beam ends remote from the junction under consideration are fixed and the beams on either side possess half their actual stiffness.

Short columns (§3.5.1.2)

Column is short if $\ell_{ex}/h \ngtr 12$ and $\ell_{ey}/b \ngtr 12$. Otherwise consider as slender

where ℓ_{ex}, ℓ_{ey} = effective height in respect of the major, and minor axis respectively; and

h, b = depth in respect of the major, and minor, axis respectively.

Braced columns (§3.5.1.3)

A column is braced in a given plane if lateral stability to the structure as a whole is provided by walls or bracing designed to resist all lateral forces in that plane.

Effective height of columns (ℓ_e) (§3.5.1.4)

Braced column properly restrained in direction at both ends

$\ell_e = 0.75\ell_o$

Braced column imperfectly restrained in direction at one or both ends

$0.75\ell_o \le \ell_e \le \ell_o$

Unbraced or partially braced column, properly restrained in direction at one end, imperfectly restrained in direction at the other

$\ell_o \le \ell_e \le 2\ell_o$

where ℓ_o is the clear height between end restraints in the plane considered.

Alternatively, for framed structures, obtain ℓ_e as follows.

(a) *Braced column*

$\ell_e \not> \ell_o$

and $\not> \ell_o \{0.70 + 0.05 (\alpha_{c_1} + \alpha_{c_2})\}$

and $\not> \ell_o (0.85 + 0.05 \alpha_{cmin})$

(b) *Unbraced column*

$\ell_e \not> \ell_o \{1.0 + 0.15 (\alpha_{c_1} + \alpha_{c_2})\}$

and $\not> \ell_o (2.0 + 0.3 \alpha_{cmin})$

where $\alpha_c = \Sigma \dfrac{EI_c}{\ell_c} / \Sigma \dfrac{EI_b}{\ell_b}$ (for plane considered) i.e. the ratio of the sum of the column stiffnesses to the sum of the beam stiffnesses at appropriate end of the column. Only members properly framed into the end in the appropriate plane of bending should be considered.

Subscripts 1 and 2 refer to upper and lower ends of column

α_{cmin} = lesser of α_{c_1} and α_{c_2}

ℓ_b, ℓ_c = length of beam and column respectively along centreline from centre to centre of joints

I_b, I_c = second moment of area of concrete section of beam and column respectively.

If beams are designed as simply supported: $\alpha_c = 10$

(c) *Column cast onto footing*

If base is designed to resist moment. $\alpha_c = 10$

If base is not designed to resist moment: $\alpha_c = 10$

Slenderness limits for columns (§3.5.1.5)

(a) *All columns* $\ell_o/b \not> 60$

(b) *Unbraced columns only*

For cantilever columns, or if unrestrained in one plane at one end:

$\ell_o h/b^2 \not> 100$ and consider also deflection,

where h is the depth of cross-section in the plane considered

b is width perpendicular to h

Short columns resisting moments (§3.5.5,6)

Column sections subjected to combined bending and axial load may be analysed as for beam sections. For predominantly axial loads make nominal allowance for eccentricity due to construction tolerances of $\not< 0.05h_{min}$

where h_{min} is the minimum depth of cross-section.

Need generally consider only single axis bending but if necessary then for biaxial bending:

$(M_x/M_{ux})^{\alpha_n} + (M_y/M_{uy})^{\alpha_n} \not> 1.0$

where M_x, M_y are the moments about the major and minor axes

M_{ux} is the maximum moment capacity, assuming ultimate axial load N and bending about major axis only.

M_{uy} is the maximum moment capacity, assuming ultimate axial load N and bending about minor axis only.

α_n is given below in terms of N/N_{uz}

and $N_{uz} = 0.45\, f_{cu}A_c + 0.75\, f_y A_{sc}$

Table CB1 Values of α_n for biaxial bending condition

N/N_{uz}	≤0.2	0.4	0.6	≥0.8
α_n	1.0	1.33	1.67	2.0

Slender columns (§3.5.7)

Additional moments are induced in a slender column (due to its deflection) and should be taken into account in the design. Suitable values for uniform rectangular columns with a symmetrical arrangement of reinforcement are given by the second term, in each case, in the following expressions for M_t.

(a) *Bending about both axes*

Design for the ultimate axial load N together with the moments M_{tx} and M_{ty} about the major and minor axes respectively as given by:

$$M_{tx} = M_{ix} + \beta_{cx}{}^2 (1 - 0.0035\ \beta_{cx})\ Nh/1750$$

$$M_{ty} = M_{iy} + \beta_{cy}{}^2 (1 - 0.0035\ \beta_{cy})\ Nb/1750$$

where M_t is total moment due to ultimate loads

M_i is the maximum initial moment due to ultimate loads (i.e. primary moment by simple elastic analysis ignoring deflection) and is $\nless 0.05$ Nh

β_c is the effective height to depth ratio for the column in the plane considered.
(i.e. $\beta_{cx} = \ell_{ex}/h$; $\beta_{cy} = \ell_{ey}/b$)

b is the width of the column (i.e. depth normal to the minor axis)

h is the depth normal to the major axis

additional subscripts x and y refer to the major and minor axes respectively.

(b) *Bending about minor axis only*

In this case the width b is the overall depth of the cross-section in the plane of bending; hence M_t and M_i are given by M_{ty} and M_{iy} respectively, except that ℓ_e for obtaining β_{cy} is given by the larger of ℓ_{ex} and ℓ_{ey}.

For a braced column where no transverse loads occur in its height M_i may be reduced to

$$M_i = 0.4M_1 + 0.6M_2$$

For all cases $M_i \nless 0.4M_2$ and $M_t \nless M_2$

where M_1 is the smaller initial end moment due to ultimate loads (assumed negative if column is bent in double curvature) and M_2 is the larger initial end moment due to ultimate loads (assumed positive).

(c) *Bending about major axis only*

If $h < 3b$ assume M_t is given by

$$M_t = M_{ix} + \beta_{cy}^2 (1 - 0.0035\beta_{cy})Nh/1750$$

except that ℓ_e for obtaining ℓ_{cy} is given by larger of ℓ_{ex} and ℓ_{ey}.

Alternatively, treat column as biaxially loaded with no initial moment about the minor axis.

(d) *Adjustments to slender columns*

The additional moments, i.e. the second term of the equations in (a), (b) and (c), may in all cases be reduced by the factor K given by

$$K = (N_{uz} - N)/(N_{uz} - N_{bal}) \leq 1$$

where N is the ultimate axial load;

N_{uz} is the capacity under axial load only, for corresponding short column ($= 0.45f_{cu}A_c + 0.75f_yA_{sc}$);

N_{bal} is the axial load corresponding to 'balanced' condition (i.e. maximum concrete strain of 0.0035 and tensile strain in outermost layer of tension steel of 0.002)

Unbraced columns, at any given level or storey, subject to lateral load are usually constrained to deflect sideways by the same amount and may be assumed to have the average value of ℓ_e/h (or ℓ_e/b etc. as appropriate) for all the columns acting in the same direction.

If the cross-section is not uniform throughout the column length see §3.5.7.5 of CP 110 itself.

See also: Appendixes CD, CE and CL.

FLAT SLAB CONSTRUCTION (§3.6)

(For definition and description of flat slabs, see RC text).

Column heads (§3.6.1.3)

Any column heads should, for analysis purposes, satisfy:

(a) Angle of greatest slope of head ≯ 45^0 from vertical

(b) h_c ≯ $\frac{1}{4}\ell_m$ and ≯ diameter at level of 40 mm below soffit of slab (or drop if present),
where h_c = diameter of circle of same area as cross-section of head,
$\ell_m = \frac{1}{2}(\ell_1 + \ell_2)$ and ℓ_1, ℓ_2 are the panel length and width respectively (centre to centre of columns).

Division of panels (§3.6.1.4)

Width of column strip = $\frac{1}{2}\ell_m$ (if without drops)

or = width of drop (if present)

Width of middle strip = width of panel - width of column strip

Slab thickness (§3.6.1.5)

Slab thickness ≮ 125 mm in all cases (see also: Appendix CE)

Openings in panels (§3.6.1.6)

If the reduced section can resist the redistributed total shear and moments (see analysis, below) the sizes of openings can be as follows:

(a) *Within area common to two intersecting middle strips:*
maximum dimension parallel to a centre-line of panel ≯ $0.4\ell_m$;

(b) *Within area common to two intersecting column strips:*
aggregate length or width ≯ $\frac{1}{10}$ × width of column strip (perimeter for calculating shear stress may also require reduction);

(c) *Within areas common to one column strip and one middle strip:*
aggregate length or width ≯ $\frac{1}{4}$ × width of column strip.

Otherwise openings must not encroach upon a column head and must be completely framed on all sides with beams to carry the loads to the columns.

Appendix CB

Analysis-continuous frame method (§3.6.5)

(a) Divide structure longitudinally and transversely along centre-lines of panels to provide strips of slab which form frames with the rows of the columns. Assume slab stiffness is given by full width generally for vertical loading but only one-half the slab width for horizontal loading.

(b) Use elastic analysis of each frame in its entirety. *Or*, for vertical loads only, consider each strip of floor (or roof) separately, assuming that the far ends of the columns above and below are rigidly fixed.

(c) Use gross concrete cross-section to determine relative stiffnesses of members. Ignore stiffening effect of drops and flared column heads; also ignore solid part of recessed or coffered slabs if the solid part extends $\ngtr 0.15\ell$ into span from centre-line of columns.

(d) Consider maximum bending moments due to design load arrangements for normal slabs (see 'Structural Frames'). If the sections possess adequate ductility (i.e. as for any framed structure) reduce negative moments by $\nless$ 15% only.

(e) Assume maximum support moments occur at a distance of $\frac{1}{2}h_c$ from column centre-line. Divide the total moment (appropriate to the panel of the flat slab forming the continuous frame) between column and middle strips as indicated in Table CB2.

Panels with marginal beams or walls (§3.6.5.4)

If supported by marginal beam with $h > 1.5h_s$ or by wall then:

(a) Beam or wall carries, in addition to those loads which act directly on beam or wall, a uniformly distributed load of $\frac{1}{4}$ × total load on panel.

(b) Reduce bending moments on half column strip adjacent to beam or wall by 50% (i.e. $\frac{1}{4}$ × moments obtained by continuous frame analysis).

Table CB2 Distribution of moments in panels of flat slabs designed as continuous frames

Strip	Width	Apportionment of total moment	
		Support	Span
Column	$\frac{1}{2}\ell$	75%	55%
Middle	$\frac{1}{2}\ell$	25%	45%
Column	$<\frac{1}{2}\ell$*	$<75\%$*	$\ngtr 55\%$*
Middle	$>\frac{1}{2}\ell$*	$>25\%$*	$\nless 45\%$*

* Where the column strip is taken as equal to the drop width, increase moment for middle strip in proportion to increased width and reduce moment for column strip by the same amount, i.e. ensure that total negative and total positive moments resisted by column plus middle strip are not reduced after adjustment. In all cases, maximum positive moment plus average of negative moments used in design of any one span of slab, for whole panel width, should be $\nless (\ell_1 - 2h_c/3)^2 n\ell_2/8$, where n is the total ultimate load per unit area of panel. If not, increase negative moments.

Columns in flat slab construction (§3.6.5.5)

Design according to continuous frame procedures: i.e. take I_b as the second moment of area of width extending over column strip.

Analysis-empirical method (§3.6.6)

Refer to Code itself.
See also: Appendixes CC, CD and CE.

RIBBED, HOLLOW BLOCK OR VOIDED CONSTRUCTION (§3.7)

Ribbed, hollow block or voided construction refers to in situ slabs constructed as a series of concrete ribs cast in situ in one of the following ways.

(a) Between blocks which remain part of the completed structure. The tops of the ribs may be connected by a topping of concrete:
   (i) of the same strength as in the ribs, or
   (ii) which is not used to contribute to the structural strength.

For (ii):

The blocks should contribute to the structural strength (see later).

Overall thickness of block (and topping, if any):

$\nless \frac{1}{5} \times$ distance between ribs

Thickness of material above any void in the block:

$\nless$ 20 mm and

$\nless 0.1 \times$ dimension of the void measured transversely to the ribs.

Topping of crushing strength < 14 MN/m^2 should be ignored.

(b) With a topping cast on forms which may be removed subsequently.

(c) With a continuous top and bottom face but containing voids of rectangular, oval or other shape.

<u>Analysis of structure</u> (§3.7.2)

Analysis as for solid slabs. *Alternatively,* if not exposed to weather or corrosive conditions, consider as a series of simply supported spans (usually desirable to provide some nominal 'anti-crack' steel at supports.)

<u>Ribs</u> (§3.7.1.3)

The size and position of in situ ribs should be as follows:

(a) Rib width $\nless$ 65 mm

(b) Rib depth, excluding any topping, $\ngtr 4 \times$ rib width

(c) Rib spacing $\ngtr$ 1.5 m

Ribs should generally be formed along each edge parallel to the span of one-way slabs. They should be at least as wide as the bearing if the edge is built into a wall or rests on a beam.

<u>Thickness of topping</u> (§3.7.1.2)

If the topping contributes to the structural strength then its thickness, after allowance for effect of wear if necessary, should be as follows.

(a) Permanent blocks (see below) with clear distance between ribs of not more than 500 mm: $\nless$ 30 mm.

(b) As (a) but with each row of blocks jointed with a 1:3 cement-sand mortar of 11 MN/m^2 minimum crushing strength: $\nless$ 25 mm.

(c) Other slabs with permanent blocks:
$\nless$ 40 mm and $\nless$ 0.1 × clear distance between ribs.

(d) Slabs without permanent blocks:
$\nless$ 50 mm and $\nless$ 0.1 × clear distance between ribs.

<u>Blocks and formers</u> (§3.7.1.4)

Blocks and formers may be of any suitable material, but if contributing to the structural strength of the slab they should (a) consist of *either* concrete *or* burnt clay to BS 3921:'Bricks and Blocks of Fixed Brickearth, Clay or Shale', and (b) crushing strength of $\nless$ 14 MN/m^2 on net section when axially loaded in direction of compressive stress in the slab.

<u>Analysis of sections</u> (§3.7.3)

As general assumptions for beams. The compressive stresses in burnt clay blocks under design collapse loads are either
$\ngtr$ 0.25 × strength given by (b) in §3.7.1.4,
or $\ngtr$ 0.30 × strength if 95% of blocks exceed specified strength.

See also: Appendixes CC,CD and CE.

REINFORCED CONCRETE WALLS (§3.8)

A reinforced wall is a vertical load-bearing concrete member such that its greatest lateral dimension $\nless$ 4 × its least lateral dimension, and the reinforcement is considered as contributing to its strength. A plain wall (Appendix CU) has nominal reinforcement only. Consider reinforced walls generally as columns. A wall is braced if at right angles to the plane of the wall, lateral stability to the structure as a whole is provided by walls or other suitable bracing designed to resist all lateral forces in that direction. The overall stability of a multi-storey building should not, in any direction, depend on unbraced walls alone.

Effective height of reinforced walls (§3.8.1.2)

For walls constructed monolithically with the adjacent construction assess as for columns (§3.5.1.4). If the construction transmitting the load to the reinforced wall is assumed to be simply supported, assess as for a plain wall (§5.5.1.2, Appendix CU).

Slenderness limits for reinforced walls (§3.8.1.3)

Unbraced:		$\ell_e/h \ngtr 30$
Braced:	Vertical $\rho_c < 1\%$:	$\ell_e/h \ngtr 40$
	Vertical $\rho_c \nless 1\%$:	$\ell_e/h \ngtr 45$

ℓ_e/h = slenderness ratio, i.e. ratio of effective height of wall to its thickness.

Analysis of structure (§3.8.2)

The axial force in a reinforced wall may be calculated by assuming that the beam and slabs transmitting forces into it are simply supported. The bending moments are obtained as follows.

Bending moments in plane at right angles to wall

(a) Use elastic analysis if members frame monolithically at right angles to the wall.

(b) Assume resultant moment is given by eccentricity as assessed for plain walls (§5.5.4, Appendix CU) if construction is designed to be simply supported by the wall.

(c) In other cases (but see short braced axially loaded walls below) moment at right angles to wall/unit length $\nless 0.05\, n_w h$ where
n_w is the axial load per unit length of wall and
h is the thickness (i.e. depth in plane of bending).

Significant additional moments may also be induced in a slender wall-consider as slender columns bent about the minor axis (§3.8.6).

Bending moments in plane of wall

Use elastic analysis and assume no tension in the concrete, i.e.

calculate moments for a single wall due to horizontal forces by statics alone. If horizontal force is resisted by several walls, distribute in proportion to the wall stiffnesses. Any shear connections necessary between vertical edges of adjacent walls must transmit the forces calculated from the elastic analysis.

Short braced axially loaded reinforced walls (§3.8.3,4)

If the nature of the construction precludes the application of significant moments consider as for short braced columns (see RC text) either axially loaded or supporting an approximately symmetrical arrangement of beams (for columns) and slabs (for walls), as appropriate.

Short reinforced walls resisting moments and axial forces (§3.8.5)

Consider first the bending moments in the plane of the wall (if any), assuming elastic theory and no tension in the concrete. This establishes the distribution of tension and compression along the length of the wall: an axial load may be assumed to be distributed uniformly along its length.

Consider the bending moments in the plane at right angles to the wall, and check the section for this moment and the compression or tension per unit length at various sections along the wall, applying the assumptions used for beam sections.

Slender reinforced walls (§3.8.6)

Determine the distribution of axial load along a slender reinforced wall as for a short wall. Assess the additional moment for the portion of wall subject to the highest intensity of axial load as a slender column of unit width bent about the minor axis (§3.5.7).

See also: Appendixes CD,CE,CL,CU and CW.

STAIRCASES (§3.9)

Design generally as for beams and slabs (for all limit states).

Appendix CB

Loading (§3.9.1.1)

Assume uniformly distributed over plan area. If two spans cross at right angles (e.g. around open wells) divide load on common area equally between two spans. If staircases or landings are built at least 110 mm into walls along part of their length, reduce loaded area by 150 mm wide strip adjacent to wall.

Effective breadth of staircase (§3.9.1.2)

Effective breadth of staircase without stringer beams generally equals actual breadth. However, for any parts along span which are built into a wall, include $\frac{2}{3}$ × embedded breadth (up to 80 mm maximum) in effective breadth.

Effective span (ℓ_e)(§3.9.1.3)

Effective span of staircase without stringer beams (including, if appropriate, sections of landing which span in same direction as, and are continuous with, the stair flight) is as follows.

(a) *Simply supported*

ℓ_e = horizontal distance between centre lines of supports.

(b) *Ends built-in* monolithically to structural member spanning in direction perpendicular to span of staircase:

ℓ_e = clear horizontal distance between supporting members plus ½ × breadths of supporting members (900 mm maximum addition at each end).

Depth of section (§3.9.1.4)

Depth of section = minimum thickness perpendicular to soffit of staircase.

BASES (§3.10)

Consider pad footings and pile caps generally as for beams and solid slabs. Assume moments and forces in bases to be given by uniformly distributed (axial loaded) or linearly varying (eccentrically loaded) ground reactions(see RC text)unless more accurate assessments can be made.

Consider the effects on strength of any pockets left for precast members unless cement mortar used for grouting up is ≮ compressive strength of concrete.

May ignore deflection but see also: Appendix CE (local damage).

APPENDIX CC

SHEAR AND TORSION

SHEAR (§3.3.6,3.4.5)

Calculate nominal design shear stress from

$v = V/bd$

where V = design shear force for collapse

b = breadth of beam (or web for T-beams)

d = effective depth

<u>Minimum (i.e. nominal) shear reinforcement</u> (§3.4.5.2)

Minimum shear reinforcement in form of links must always satisfy

$(0.87f_{yv})(A_{sv}/b_t s)(\sin\theta + \cos\theta) \nless 0.40^*$ (MN/m² units)

where f_{yv} = characteristic strength of links (taken as ≯ 425 MN/m²)

A_{sv} = cross-sectional area of links (e.g. both legs for 2-leg links)

b_t = breadth of beam at level of tension reinforcement

θ = inclination of links to longitudinal axis of member (≮ 45°)

s_v = link spacing along longitudinal axis of member, ≯ $\frac{3}{4}$d and ≯ clear web depth (I-beams)

Lateral spacing of legs of links ≯ $\frac{3}{4}$d.

Links should be anchored at both ends by passing round the main reinforcement.

* Approximately 0.44 for links in beams.

Appendix CC

Minimum (i.e. nominal) links for beams (§3.11.4.3)

Increase constant 0.40 above to about 0.44 and put $\theta = 90^o$ for minimum vertical links in beams:

$A_{sv}/b_t s_v \nless 0.002$ (mild steel) and $\nless 0.0012$ (high yield)

Links should enclose all tension reinforcement.

The minimum (i.e. nominal) links need not be provided in beams if $v < \frac{1}{2}v_c$ where v_c is given by Table CC1.

Shear reinforcement for beams (§3.3.6.1)

Table CC1 Ultimate concrete shear stress v_c (MN/m^2 units)

$\frac{100A_{st}}{bd}$	Characteristic cube strength (MN/m^2)			
	20	25	30	40 or more
0.25	0.35	0.35	0.35	0.35
0.50	0.45	0.50	0.55	0.55
1.00	0.60	0.65	0.70	0.75
2.0	0.80	0.85	0.90	0.95
3.0	0.85	0.90	0.95	1.00

Note: A_{st} refers to area of longitudinal tension reinforcement which continues beyond the section considered for $\nless$ d (generally) or $\nless$ minimum anchorage (at supports).

If $v > v_c$ provide reinforcement in form of links, or bent-up bars and links (minimum links as above and $\nless \frac{1}{2}(v - v_c)$) such that*

$$(0.87f_{yv})(A_{sv}/bs_v)(\sin\theta + \cos\theta) + (0.87f_{yb})(A_{sb}/bs_b)(\sin\theta_b + \cos\theta_b)$$

$$\nless v - v_c$$

where f_{yb}, A_{sb}, θ_b and s_b refer to bent-up bars.

Maximum shear stress for beams (§3.3.6.1)

In all cases $v \ngtr 0.75\sqrt{f_{cu}}$

i.e. $v \ngtr$ 3.35, 3.75, 4.10 and 4.75 MN/m^2 for

Grades 20, 25, 30 and 40 (or more) respectively

* CP110 considers only $\theta = 90^o$.

Concentrated loads on corbels and near beam supports (§3.3.6.2)

If the principal load is applied within 2d from support, continue main bars beyond face of support with anchorage length of $\nless 20\Phi$. Also if $v \ngtr (2d/a_v)v_c$ and $\ngtr 0.75\ \sqrt{f_{cu}}$ only nominal links are necessary. Otherwise, provide shear reinforcement for the surplus: i.e. $v - (2d/a_v)v_c$ but in all cases $v \ngtr \frac{3}{4}\sqrt{f_{cu}}$ where a_v = distance from nearest edge of the principal load to face of supporting member for corbels, cantilevers, continuous or simply supported beams.

Principal loads should be $\nless$ 70% of total shear force at support. If a_v for a corbel $\ngtr 0.6d$ give special consideration to the anchorage of the main tension steel and the provision of horizontal stirrups. Alternatively, design continuous concrete nibs, cast in situ or precast, as indicated in §5.2.4.2 (see CP 110 itself).

Minimum links for slabs (§3.4.5.1)

Unnecessary if $v < \xi_s v_c$

where ξ_s = 1.0, 1.10, 1.20 and 1.30 for overall slab depth

of $\nless$ 300, 250, 200 and $\ngtr$ 150 mm respectively

Maximum shear stress in slabs < 200 mm thick must be $\ngtr \xi_s v_c$

For slabs $\nless$ 200 mm thick shear reinforcement may be provided as for beams except that link spacing may be $\ngtr d$ and $\xi_s v_c$ substituted for v_c in the formula for beams (§3.3.6.1)

Maximum shear stress values at collapse

$\ngtr \frac{1}{2} \times$ values for beams i.e. $\ngtr 0.375\ \sqrt{f_{cu}}$

Concentrated loads on slabs (punching shear) (§3.4.5.2)

Take the critical section for calculating shear on a perimeter distance 1.5h from the boundary of the loaded area, where h is the overall slab thickness. Assume the normal shear stress v to have a constant value throughout the effective depth and length of this perimeter.

For the value of 100 A_{st}/bd in Table CC1 use the average percentage from all the tension steel that passes through the critical perimeter, i.e. include all tension reinforcement within the width in either direction given by the loaded width plus 3h.

If slab thickness < 200 mm, then v must be $\ngtr \xi_s v_c$

If $v < \xi_s v_c$ no shear reinforcement is necessary.

If $v > \xi_s v_c$ and if $h \nless 200$ mm and $v \ngtr 0.375\ \sqrt{f_{cu}}$ provide reinforcement as for beams with the beam breadth b replaced by u_{crit}, the length of the critical perimeter, and $v - v_c$ taken as $\nless 0.4$. Vertical or inclined links, or bent-up bars, may be used.

Lateral spacing of legs of links $\ngtr \frac{3}{4}h$

The perimeter distance $\frac{3}{4}h$ from boundary of loaded area should be reinforced for shear exactly as for critical perimeter distance $1\frac{1}{2}h$ from boundary of loaded area considered above. Also perimeters progressively $\frac{3}{4}h$ further out should be considered (i.e. u at $2\frac{1}{4}h$ etc.,) and if $\xi_s v_c$ continues to be exceeded for the appropriate values of u, further shear reinforcement should be provided.

Holes in slabs (§3.4.5.2)

If the distance of hole to edge of concentrated load < 6h then the part of the perimeter of the critical section which is enclosed by radial projections of the openings to the centroid of the loaded area as shown in Fig. CC1 shall be considered ineffective. Where one hole is adjacent to the column ignore the presence of the hole if the greatest hole width < ¼ column side and < ½h.

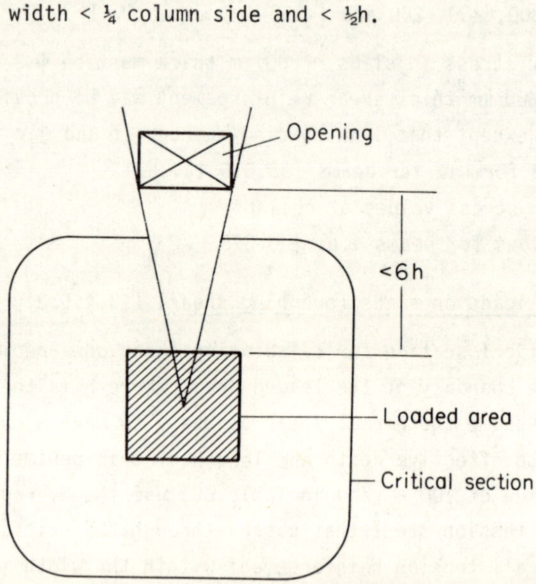

Fig.CC1 Openings in slabs

<u>Flat slab construction</u> (§3.6.2)

If ratio between adjacent spans > 1.25 it is necessary to consider various arrangements of ultimate imposed loads and hence various combinations of M and V and to modify shear force as below, except for edge columns being bent at right angles to the edge, or for corner columns.

(a) *Braced columns with ratios between adjacent spans $\nmid$ 1.25*
Shear stress $\nmid 0.8\xi v_c$ (i.e. reduce v_c values in Table CC1 by 20%)
Assume ultimate imposed loads are applied to all panels adjacent to column. Take design shear force as 1.25 times shear force calculated as above to allow for non-symmetrical distribution of shear round the column, and design for shear in accordance with 'concentrated loads on slabs (§3.4.5.2)'

(b) *Unbraced columns and columns with ratios between adjacent spans > 1.25*
Reduce v_c values and increase design shear force as (a) above. In addition, multiply the shear force by the following further magnification factor:

$$1 + 12.5\ M/V\ell$$

where M is the total moment transmitted to the slab at each column-slab connection
ℓ is the shorter of the two spans in the direction in which bending is being considered.

(c) *Shear reinforcement*
As for solid slabs.

<u>Ribbed construction</u> (§3.7.4)

Take V as shear force on width of slab equal to the centre to centre distance between ribs, and b as the width of the rib, b_w. Otherwise consider as for solid slabs and, if reinforcement is necessary, consider each rib as a beam. Allowance for shear strength of permanent hollow blocks and jointing mortar or concrete which contribute to structural strength can be made by increasing the width of the concrete rib by thickness of one wall of block, or of the narrow precast unit by width of jointing mortar or concrete.

Appendix CC
TORSION (§3.3.7)

Nominal shear reinforcement generally provides adequate control of any torsional cracking and the analysis of the structure need not take into account the torsional resistance or stiffness of members. However, if the latter need to be considered the following assumptions may be made.

Analysis (§3.3.7)

Assume that torsional rigidity is given by GC
where C = torsional constant = ½ × St Venant constant for the plain concrete section
and G = shear modulus = 0.4E

Rectangular beams (§3.3.7)

The torsional shear stress, v_t, at any cross-section is given by

$$T = \tfrac{1}{2} v_t x^2 (y - x/3)$$

where T = design torsional moment for collapse
x = length of shorter side
y = length of longer side

Torsional reinforcement (§3.3.7)

Provide torsional reinforcement in the form of closed links (similar to Code 74 of BS 4466) with extra longitudinal reinforcement, wherever torsion is a significant effect, i.e. if $v_t > v_{tmin}$ given in Table CC2.
In all cases $(v + v_t) \ngtr$ appropriate v_{tu} values in Table CC2.
If $y_1 < 0.55$ m then $v_t \ngtr v_{tu}\, y_1/0.55$.

Table CC2 Limiting design stresses for sections reinforced for torsion (MN/m² units)

	Characteristic cube strength			
	20	25	30	≮40
v_{tmin}	0.30	0.33	0.37	0.42
v_{tu}	3.35	3.75	4.10	4.75

Areas given by

$T \ngtr \lambda x_1 y_1 (0.87 f_{yv}) \times A_{sv}/s_v$ (Put $\lambda = 0.80$)

$A_{sl} \nless (A_{sv}/s_v)(f_{yv}/f_{yl})(x_1 + y_1)$

where A_{sv} = cross-sectional area of the legs of closed links at a section

f_{yv} = characteristic strength of links ($\ngtr$ 425 MN/m^2)

A_{sl} = area of longitudinal reinforcement provided for torsion (distributed evenly inside links with 4 bars minimum, one at each corner of the links)

f_{yl} = characteristic strength of longitudinal steel ($\ngtr$ 425 MN/m^2)

s_v = spacing of closed links ($\ngtr$ 0.2 m, x_1 or $\frac{1}{2}y_1$)

x_1 = smaller centre line dimension of links

y_1 = larger centre line dimension of links

Clear spacing between longitudinal bars $\ngtr$ 300 mm.

Extend longitudinal steel a distance $\nless$ y beyond point where torsion steel is no longer required.

T- and L-beams (§3.3.7)

Consider T, L and I-beams by dividing them into their component rectangles and calculating the shear stress for each component rectangle as for rectangular beams.

The function $\Sigma x_n^3 y_n$ is normally maximized by making the widest rectangle as long as possible and the reinforcement cages, consisting of rectangular links and longitudinal steel, should interlock and tie the component rectangles of the section together. Obtain the torsional moment carried by each component rectangle from:

$T(x_n^3 y_n / \Sigma x_n^3 y_n)$

where x_n, y_n = smaller and larger dimensions respectively of component rectangle being considered.

No torsion reinforcement need be provided if the torsion shear stress in a component rectangle is $< v_{tmin}$

Appendix CD

Closed box-sections†

As rectangular sections except that

$$v_t = T_c/2Ab$$

where A = area enclosed by median line of walls of box-section

b = minimum wall thickness.

If wall thickness > ¼ overall thickness (for both the x and y directions) treat as solid rectangular section.

Non-rectangular cages†

As rectangular cages except that x_1y_1 is replaced by cage area, and $2(x_1 + y_1)$ by cage perimeter.

APPENDIX CD

REINFORCEMENT, BOND AND ANCHORAGE

Sizes of reinforcement

6, 8, 10, 12, 15, 20, 25, 32, 40 and 50 mm

Specified strength and minimum elongation of reinforcement (§3.1.4.3)

Table CD1 Specified strengths and minimum elongations of reinforcement

Type of reinforcement	BS	Sizes (mm)	Strength (MN/m^2) f_y	Strength (MN/m^2) Ultimate	Minimum elongation on $5.65\sqrt{(S_o)}$
Mild steel	4449	All sizes	250	$\not< 1.15f_y$	22%
Hot rolled high yield	4449	All sizes	410*	$\not< 1.15f_y$	14%
Cold worked high yield	4461	> 16	425*	$\not< 1.10f_y$	14%
		$\not>$ 16	460*		12%
Hard drawn steel wire	4482	$\not>$ 12	485*	$\not< 1.10f_y$	

S_o = original cross-sectional area of test piece

* 0.2% proof stress may be assumed to be given by 0.4% total strain.

† Not as CP 110.

Bar schedule dimensions (§3.11.3.2)

Schedule bars in accordance with BS 4466 (see Appendix H). Where steel is to be bent (i.e. links and other bent steel) or cut to fit between two concrete faces, the dimension on the bending schedule is given by the nominal dimension of the concrete, less the nominal cover on each face, and less a deduction (for tolerance on member size and for bending) given in Table H1.

Cover to reinforcement (excluding finishes) (§3.11.2)

Nominal cover $\not<$ 15 mm and $\not<$ bar diameter in all cases (e.g. even for Grade 30 concrete under 'Mild' exposure)
The 'bar' diameter for bundled bars (three or more) is the size of a single bar of similar total area. Increase cover, if necessary, for envisaged conditions of exposure and required durability (e.g. see Table A3 and §6) fire resistance (see reference [1.3]), calcium chloride, lightweight (§3.12) and porous aggregates, surface treatment (e.g. bush hammering) and large-size aggregate.

Bundled bars (§3.11.3.1, 3.11.6.3)

Bars may be arranged in pairs, or groups of three or four parallel bars in contact, provided that:

(a) Effective perimeter for bond stresses for bundled bars
$\not>$ $\lambda \times$ perimeter for single bar
where λ = 1.6, 1.8 and 1.6 for bundles of 2, 3 and 4 bars respectively.

(b) $\not>$ 4 bars occur at any cross-section (including laps).

(c) They terminate at different points with a stagger of $\not<$ 40 bar diameters, except at a support.

(d) Minimum links are provided in all cases.

Maximum and minimum steel areas (§3.11.4,5)

(a) *Beams*

(i) Main tension steel: $\rho \not< 0.15\%\ b_t d$ (high yield)

and $\not< 0.25\%\ b_t d$ (mild steel)

For box, T or I-beams b_t is given by average breadth below the upper flange.

(ii) A_{st} or $A_{sc} \ngtr 0.04bh$ for calculation purposes

where b_h = gross cross-sectional area of the concrete.

(iii) Links: see Appendix CC

Laterally restrain compression bars as for columns.

(iv) Transverse reinforcement across full effective width of flange of T- or L-beam near top surface:

$\rho_c \nless 0.3\%$ of longitudinal cross-sectional area of flange.

(b) *Solid slabs*

Main steel: as for beams

Secondary steel: $\rho_c \nless 0.12\%$ (high yield) and $\nless 0.15\%$ (mild steel)*

In all cases distance between bars $\ngtr 5d$.

(c) *Columns*

(i) Longitudinal steel (based on gross area of column)

$\rho_c \nless 1\%$ or $\nless 0.15N/f_y$ (if only lightly loaded);

$\ngtr 6\%$ (cast vertically) and $\ngtr 8\%$ (cast horizontally)

and $\ngtr 10\%$ at laps (cast vertically or horizontally)

where N = ultimate axial design load for collapse

(ii) Number of longitudinal bars $\nless 4$ if rectangular

and $\nless 6$ if circular

(iii) Longitudinal bar diameter $\nless 12$ mm

(iv) Transverse bar diameter (links)

$\nless 6$ mm

and $\nless \frac{1}{4} \times$ diameter of any longitudinal compression bar

(v) Pitch of transverse links or stirrups

$\ngtr 12 \times$ diameter of any longitudinal compression bar

(vi) Transverse steel must change direction by $\nless 45^{\circ}$ at each corner and at each alternate longitudinal bar (or bundle) in an outer layer. Intermediate longitudinal bars $\ngtr 150$ mm clear distance from bars laterally supported as above do not require further change in direction of stirrup. Alternatively, for circular columns, a circular tie may be used.

* As given in CP 110, but higher values of ρ_c are necessary to control early thermal cracking if slabs are restrained.

(d) *Reinforced concrete walls* (i.e. walls acting as columns for which greatest lateral dimension $\nless$ 4 × least lateral dimension).

(i) Vertical reinforcement (in one or two layers)

$\rho_c \nless 0.4\%$ and $\ngtr 4\%$

Note: for fire resisting purposes a wall is classed as reinforced only if $\rho_c \nless 1.0\%$

(ii) Horizontal reinforcement

$\rho_c \nless 0.3\%$ (mild steel)* or $\nless 0.25\%$ (high yield)

(iii) Diameter of horizontal reinforcement

$\nless$ 6 mm

and $\nless$ ¼ × vertical bar diameter

(iv) If ρ_c (vertical) > 2% provide transverse links, of diameter as for (iii), and with pitch (in either direction) $\ngtr$ 2 × wall thickness and (vertical direction only) $\ngtr 16\Phi$

(v) Any vertical compression bar not enclosed by a link should be $\ngtr$ 0.2 m from a restrained bar if ρ_c(vertical) > 2%.

(e) *'Plain' concrete walls* (§5.5.9)

Vertical or horizontal reinforcement

(i) Cast in situ walls > 2 m long (except no-fines)

(a) External: $\rho_c \nless 0.3\%$ (mild steel)* and $\nless 0.25\%$ (high yield). Use small diameter bars (with adequate cover only) near the exposed face (or faces)

(b) Internal at junctions with floors and beams: As (a) but dispersed half near each face

(c) Around all openings: nominal

(ii) If eccentricity of loading for collapse in the plane of the wall $> \frac{1}{3}$ × total length

and length in tension under service loads > 0.1 × total length:

provide vertical reinforcement in tensile region of wall as for (1)(b) above.

Minimum distance between bars (§3.11.8.1)

Leave sufficient room for immersion vibrators to be inserted and

* As given in CP 110, but higher values of ρ_c are often necessary to control early thermal cracking (except no-fines and low-grade concretes)

concrete to achieve proper compaction. Minimum spacings are best based on experience and suitable site evidence, but may be taken as follows.

(a) Gaps between single bars: all cases $\nless$ bar diameter

horizontally: $\nless h_{agg} + 5$ mm and

vertically: $\nless \frac{2}{3}h_{agg}$ (if horizontal gaps are vertically in line)

where h_{agg} = maximum aggregate size

(b) Pairs of bars:

If pair has one bar above the other: as for single bars

If pair has bars side by side: gaps $\nless h_{agg} + 5$ mm both ways

(c) Bundled bars:

gaps $\nless h_{agg} + 15$ mm both ways (provided that horizontal gaps are vertically in line).

See also: Appendix CE (Local damage).

<u>Maximum distance between bars in tension</u> (§3.11.8.2)

See: Appendix CE (Local damage).

<u>Hooks and bends</u> (§3.11.6.4,7)

Use U or L-types to BS 4466 and only where necessary.

(a) Minimum internal radius $\nless \xi_{min} \times$ bar diameter

where ξ_{min} = 2(mild steel) or 3(high yield steel)

(b) Anchorage value of bends (including length forming anchorage)

$\ngtr$ 24 bar diameters

and $\ngtr \psi r$

where r = internal radius of hook or bend

and ψ = 4 (for 90° bend) or 8 (for 180° hook)

(c) Length beyond end of curve

$\nless$ 4 bar diameters (generally)

and $\nless$ 8 bar diameters (links with 90° bend)

(d) Minimum distance beyond (and inside) support face before start of hook $\nless$ 4 bar diameters.

Local bearing stress inside bends (§3.11.6.8)

Local bearing stress, $F_{st}/r\Phi$, should be $\ngtr 1.5f_{cu}/(1 + 2\Phi/a_b)$

where a_b $\ngtr$ centre to centre distance between bars (or bundles) perpendicular to the plane of the bend

and $\ngtr$ cover plus Φ (for bar or bundle adjacent to face of member)

also r = internal radius of bend

Φ = bar diameter (or equivalent bar size for a bundle)

and F_{bt} = design tensile force for collapse in bar or bundle at section.

Anchorage (average) bond stress (§3.11.6.2)

Given by

$$\frac{\text{Force in bar}}{\text{Anchorage length} \times \text{perimeter}}$$

The design values at collapse for any bar at any section $\ngtr$ anchorage bond stresses in Table CD2.

Table CD2 Maximum anchorage bond stress (MN/m² units)

	Characteristic cube strength			
Bar type & stress	20	25	30	$\nless 40$
Plain				
tension	1.2	1.4	1.5	1.9
compression	1.5	1.7	1.9	2.3
Deformed bars Type 1				
tension	1.7	1.9	2.2	2.6
compression	2.1	2.4	2.7	3.2
Deformed bars Type 2				
tension	2.2*	2.5*	2.9*	3.4*
compression	2.7*	3.1*	3.5*	4.2*

* Equal to 1.3 × Type 1 values.

Appendix CD

Local (flexural) bond (§3.11.6.1)

$$\text{Local bond stress} = \frac{V \pm (M/d)\tan\theta_s}{\Sigma u_s d} \ngtr \text{values in Table CD3}$$

where Σu_s = sum of perimeters of tension bars (or effective perimeters for bundles)

V = design shear force at section for collapse

M = design bending moment at section for collapse

θ_s = angle between compressive face and tensile steel (= 0 usually)

Use minus sign if M increases numerically in same direction as d

Critical sections for maximum local bond stress are:

(a) Faces of supports (simple spans)

(b) Where tensile bars are curtailed and at points of contraflexure (only if anchorage bond stresses > 0.8 × values in Table CD2)

Table CD3 Maximum local (flexural) bond stress (MN/m² units)

	Characteristic strength of concrete			
Bar type	20	25	30	$\nless$40
Plain	1.7	2.0	2.2	2.7
Deformed Type 1	2.1	2.5	2.8	3.4
Deformed Type 2	2.5	3.0	3.4	4.1

Laps and joints in bars (§3.11.6.5)

Lap length $\nless$ anchorage length for smaller bar generally but increase by 25% for deformed bars in tension.

Also lap length $\nless$ 25 diameters + 150 mm (tension)

or $\nless$ 20 diameters + 150 mm (compression)

Laps, welds or mechanical devices should generally be staggered and avoid high stress locations.

Compressive stresses in bars may be transferred by end bearing of square-sawn cut ends held in concentric contact by a suitable sleeve or mechanical device (see §5.3.2.3 of CP 110 itself)

Welded joints in bars (§3.11.6.6)

Welded joints may be used if the imposed load is not predominantly cyclic in nature. Avoid welded joints at bends in reinforcement and, for parallel bars, stagger in the longitudinal direction by $\nless$ end anchorage length for the bar. (See also §7.6 of CP 110 itself.)

If tests show that weld strength is $\nless$ strength of parent bar, then the strength of welded joints in tension or compression may be based on:

(a) $0.8 \times f_y/\gamma_m$ (i.e. $0.7f_y$ in tension)

or (b) f_y/γ_m for strict supervision and if $\ngtr$ 20% of tensile reinforcement at any section is welded.

For welds at a lapped joint,

$$l_{weld} \times t_{weld} \times v_{weld} \nless 0.87f_y \times A_{st}$$

where l_{weld} is the length of a run of weld

t_{weld} is the throat thickness of the weld

v_{weld} is the shear strength of filler material (taken as 0.38 × yield or proof stress given in appropriate British Standard)

If $l_{weld} > 5\Phi$ use separate runs of $\ngtr 5\Phi$ with spaces between runs $\nless 5\Phi$.

Curtailment of bars (§3.11.7.1)

Minimum anchorage generally $\nless$ d and $\nless$ 12 bar diameters from reference section (i.e. first section at which the curtailed bars are no longer required for flexure). In addition satisfy one of the following for all arrangements of load at collapse for tension bars:

(a) Anchorage length for design strength of $0.87f_y$ is provided from reference section for cut-off bars,

or (b) Shear capacity where bars stop $\nless$ 2 × shear force,

or (c) Continuing bars where bars stop, provide 2 × steel area for flexure,

or (d) A simplified design procedure is used (e.g. see below for beams, solid slabs, flat slabs or ribbed slabs).

Minimum anchorage at simple end supports must satisfy either:

(a) $\nless$ 12 bar diameters beyond centre line of support, or

(b) $\nless$ 12 bar diameters + ½d from face of support provided that straight length $\nless$ ½d, or

(c) If local bond stress $\ngtr$ ½ maximum values, straight length beyond centre line of support $\nless$ 30 mm and $\nless \frac{1}{3} \times$ support width.

Reinforcement for beams-simplified rules (§3.11.7.2)

The following simplified rules may be used if beams support uniformly distributed loads and, for (c) only, the spans are approximately equal and the characteristic imposed load $\ngtr$ characteristic dead load.

(a) *Simply supported beams:* $\nless$ 50% of tension reinforcement at midspan should extend beyond centre of supports by $\nless 12\Phi$
Remainder should extend within 0.08ℓ of interior supports.

(b) *Cantilever beams:* $\nless$ 50% of tension reinforcement over supports should extend to end of cantilever
Remainder to extend $\nless 0.5\ell$ and $\nless 45\Phi$ from support

(c) *Continuous beams:*

(i) Tension reinforcement over supports
$\nless$ 20% effectively continuous throughout spans
$\nless$ further 40% to extend $\nless 0.25\ell$ from support
Remainder to extend $\nless 0.15\ell$ and $\nless 45\Phi$ from support

(ii) Tension reinforcement at midspan
$\nless$ 30% should extend to the supports
Remainder should extend to within 0.15ℓ of interior supports and within 0.1ℓ of exterior supports

Consider simply supported end as for (a) above.

Reinforcement for solid slabs—simplified rules (§3.11.7.3)

(a) *Simply supported one-way slabs*—as for beams

(b) *Cantilever slabs*—as for beams

(c) *Continuous one-way slabs*—loading conditions as for beams

(i) Tension reinforcement over supports
$\nless$ 50% should extend 0.3ℓ into span
Remainder should extend $\nless 0.1\ell$ and $\nless 45\Phi$ from support
At end support with monolithic connections, provide for

suitable negative moment e.g. provide for half the moment at midspan and extend $\nless 0.1\ell$ and $\nless 45\Phi$ into span.

(ii) Tension reinforcement at midspan

$\nless$ 50% should extend into the support

Remainder should extend to within 0.2ℓ of interior supports and within 0.1ℓ of exterior supports.

For two-way slabs, see RC text.

<u>Reinforcement for flat slabs-simplified rules</u> (§3.6.5.3)

Satisfy following conditions, where $\ell = \ell_1$ or ℓ_2 as appropriate

(a) Each strip is reinforced over its full width, and slab is reinforced in both directions

(b) All positive span steel should extend $\nless 0.3\ell$ from centre line of panel and $\nless$ 40% should extend to within $\ell/8$ of centre line of columns

(c) All negative support steel should extend $\nless 0.2\ell$ from centre line of columns. Also, average distance for all negative steel $\nless 0.25\ell$

(d) If $h_c < 0.15\ell$, two-thirds of the negative steel in column strip should be placed centrally over column and within the central half-width of the column strip

(e) At all discontinuous edges, all steel should extend to within 70 mm of the edge and be provided with right angle bends.

<u>Reinforcement for ribbed slabs-simplified rules</u> (§3.7.6)

(a) Satisfy general requirements for distance between bars including maximum distance (as for beams) for areas of solid concrete. In all toppings, consider providing single layer of mesh of $\nless$ 0.12% of area of topping, in each direction, and wire spacing $\ngtr \frac{1}{2}$ rib spacing.

(b) Carry through $\nless$ 50% of positive span steel in bottom of rib to supports and satisfy general requirements for curtailment of bars (§3.11.7.1).

(c) 'Anti-crack' steel over supports in continuous slabs designed as simply supported $\nless \frac{1}{4} \times$ maximum area of steel in adjoining spans and should extend $\nless 0.1\ell$ into adjoining bays.

(d) Concrete cover ≮ 10 mm either laterally (if permanent blocks are present), or underneath (if slip tiles are present). Otherwise increase cover as for general requirements.

APPENDIX CE

LIMIT STATES OF SERVICEABILITY

DEFLECTION (§3.3.8)

Maximum deflections for rectangular beams need not normally be calculated if:

span/effective-depth ratio ≯ $\xi\psi\zeta(\ell/d)$

where (ℓ/d) = basic (or special) ratio of effective span to effective depth

ξ,ψ,ζ = various multipliers (see Tables CE1-CE3 below)

(See §A.2 if calculated deflections are required)

Table CE1 Basic and special ℓ/d ratios for rectangular beams

	Basic ℓ/d Values	Special ℓ/d values for spans with finishes				
Deflection:	≯$\ell/250$ *			<$\ell/250$		
Span (metres)	≯10	12	14	16	18	20
Cantilever	7	-†	-†	-†	-†	-†
Simply supported	20	18	16	14	12	10
Continuous	26	23	21	18	16	13

* Deflection of $\ell/250$ usually unacceptable for spans > 10 m since additional deflection (after finishes etc) can exceed 20 mm.

† Values to be justified by calculation.

ξ- Multiplier for tension reinforcement

The effect of tension reinforcement is shown in Table CE2.

Table CE2 ξ-multipliers for tension reinforcement

Service stress $f_s = 0.58f_y$ (MN/m^2)	Yield strength f_y (MN/m^2)	Tension steel ratio, ρ (per cent)							
		0.25	0.50	0.75	1.00	1.50	2.00	2.50	$\nless 3.0$
145	250	2.0	1.98	1.62	1.44	1.24	1.13	1.06	1.01
150	-	2.0	1.91	1.58	1.41	1.22	1.11	1.04	0.99
200	-	2.0	1.46	1.26	1.15	1.02	0.94	0.89	0.85
238	410	1.60	1.23	1.09	1.00	0.90	0.84	0.80	0.77
246	425	1.55	1.20	1.06	0.98	0.88	0.83	0.79	0.76
250	-	1.52	1.18	1.05	0.97	0.87	0.82	0.78	0.75
267	460	1.41	1.11	0.99	0.92	0.84	0.78	0.75	0.72
290	500	1.27	1.03	0.92	0.86	0.79	0.74	0.71	0.68
300	-	1.22	0.99	0.90	0.84	0.77	0.72	0.69	0.67

$\rho = A_{st}/bd$ at midspan (beams) or support (cantilevers)

$f_s = 0.58 f_y \beta_A / \beta_b$

where β_A = the ratio of the area of tension reinforcement required to resist the moment due to ultimate loads at midspan (or at support for a cantilever), to the area actually provided.

β_b = the ratio of the resistance moment at midspan obtained from the redistributed maximum moments diagram to that obtained from the maximum elastic moment at midspan before redistribution of moments.

Take f_s as $0.58f_y$ if the percentage redistribution is unknown, but the design ultimate moment is clearly $\nless$ elastic ultimate moment.

ψ-Multiplier for compression reinforcement

The influence of compression reinforcement on the long-term deflection can be obtained from Table CE3.

Table CE3 ψ-multipliers for compression reinforcement

Compression-steel ratio:	0.25	0.50	0.75	1.0	1.5	2.0	≥3.0
Factor:	1.07	1.14	1.20	1.25	1.33	1.40	1.50

Compression steel used for Table CE3 (and not considered for collapse) need not be tied in with links.

ζ_1-Multiplier for high shrinkage or creep

ζ_1 = 1.0 for normal creep and shrinkage deflection
ζ_1 = 0.85 for shrinkage coefficient > 0.75×10^{-3} or
creep coefficient > 4.0 or other abnormally adverse conditions.

ζ_2-Multiplier for flanged beams

For flanged beams use a further multiplier ζ_2 which varies according to the web breadth as follows.

$\zeta_2 = 0.9$ if $b_w = 0.65 \times$ effective flange width
$\zeta_2 = 0.8$ if $b_w \leq 0.30 \times$ effective flange width

Steel ratios should be based on effective flange width multiplied by d.

Solid slabs (§3.4.6)

Consider only the reinforcement at the centre of the span and determine span/effective depth ratios as for beams. Base ratio for a two-way spanning slab on the shorter span and the amount of reinforcement in that direction.

Columns (§3.5.8)

For all braced column construction, deflections are satisfactory if the slenderness limits for columns are satisfied.

For unbraced columns the average value of ℓ_e/h for all columns at each level acting in the same direction should satisfy

$\ell_e/h \not> 30$

Unbraced columns in single storey construction, where no finishes susceptible to damage are present, need only satisfy the slenderness limits for collapse (§3.5.15).

Flat slabs (§3.6.3)

Additional multiplier ζ_3 for use with Tables CE1 to CE3 equals 0.9 generally, or 1.0 if total width of drop $\nless \frac{1}{3} \times$ span for appropriate direction.

Ribbed slabs (§3.7.5)

As for solid slabs. Make allowance for permanent hollow blocks, if appropriate (e.g. when determining ζ_2-multiplier for web width). Take the effective web width as the mean web width below the upper flange.

Reinforced walls (§3.8.7)

Satisfy general and reinforcement requirements for reinforced walls. For a cantilever shear wall: total height $\ngtr 12 \times$ length

See also: Appendix CU

CALCULATED DEFLECTIONS (§A.2)

Calculated deflections of reinforced concrete members can be inaccurate as follows.

(1) Estimates of the restraints provided by supports are based on simplified and often inaccurate assumptions.

(2) The precise loading, or that portion which is of long duration, is unknown.

(3) The working load for a lightly reinforced member is often close to the cracking load and considerable differences in deflection occur according to whether the member has or has not cracked.

(4) The effects of the deflection on finishes and partitions are difficult to assess and are often neglected.

CP 110 approach

One approach is to assess the curvatures of sections under the appropriate moments and then calculate the deflections from the curvatures. Reasonable results are possible both for short-term and long-term loadings if the above are correctly assessed.

The moments and forces in members for serviceability limit states are conveniently obtained by an elastic analysis with relative stiffnesses based on the cracked, transformed section properties of the members at midspan or midheight. Take the concrete modulus as half the instantaneous value given in Table A1 and due account may be taken of the effects of any axial loads in assessing the second moments of area.

Curvature of pseudo*-cracked section

Calculate curvature r_A^{-1} of cracked section assuming the following:

(i) Plane sections remain plane (on average)

(ii) The reinforcement in tension and compression is elastic with E_c = 200 GN/m²

(iii) Concrete in compression is elastic with short-term E_c as in Table A1 and long-term modulus of $E_c/(1 + \phi)$ where ϕ is the appropriate creep coefficient.

(iv) Concrete in tension is inelastic and has a triangular stress distribution of zero at the neutral axis and, at the centroid of the tension steel, 1 MN/m² instantaneously,reducing to 0.55 MN/m² in the long term.

Curvature of uncracked section

Calculate curvature r_B^{-1} of uncracked section assuming a fully elastic section in compression and tension, i.e. assume (i), (ii) and (iii) above and (iv) below:

(iv) Concrete in tension is fully elastic with the same concrete modulus in tension as in compression.

Long-term curvature

Compare the curvatures obtained for the pseudo-cracked and uncracked

* Term not used in CP 110.

sections and consider the case which gives the larger curvature.

Assess the total long term curvature of a section as follows:

(1) Calculate instantaneous curvatures for (a) total load and (b) permanent load

(2) Calculate long-term curvature under permanent load

(3) Calculate the shrinkage curvature r_{cs}^{-1}, from

$$r_{cs}^{-1} = \rho_o \varepsilon_{cs}/d$$

where ε_{cs} is the free shrinkage strain making allowance for the shape of the section

ρ_o is a coefficient depending upon the percentages of tension and compression steel in the section as given by Table CE4.

(4) Determine the total long-term curvature from:

(1a) - (1b) + (2) + (3)

Table CE4 Values of ρ_o for calculation of shrinkage curvatures

$100 A_{st}/bd$	$100 A_{sc}/bd$								
	0.00	0.25	0.50	0.75	1.00	1.25	1.50	1.75	2.00
0.25	0.44	0.31	0.26	0.22	0.20	0.18	0.17	0.16	0.15
0.50	0.56	0.31	0.26	0.22	0.20	0.18	0.17	0.16	0.15
0.75	0.64	0.45	0.26	0.22	0.20	0.18	0.17	0.16	0.15
1.00	0.70	0.55	0.39	0.22	0.20	0.18	0.17	0.16	0.15
1.50	0.80	0.69	0.57	0.45	0.32	0.18	0.17	0.16	0.15
2.00	0.88	0.79	0.69	0.60	0.49	0.39	0.28	0.16	0.15
2.50	0.95	0.87	0.79	0.70	0.62	0.53	0.44	0.35	0.25
3.00	1.00	0.94	0.86	0.79	0.72	0.64	0.57	0.49	0.40
3.50	1.00	1.00	0.93	0.87	0.80	0.74	0.67	0.60	0.52
4.00	1.00	1.00	1.00	0.93	0.87	0.81	0.75	0.69	0.62

Calculation of deflection from curvatures

The deflection, a, may be calculated by using either the deflected shape directly or by using a simplified approach based on the equation

$$a = K\ell^2 r_b^{-1}$$

where ℓ is the effective span of the member

r_b^{-1} is the curvature at midspan or, for cantilevers, at the support section

and K is a constant which depends on the shape of the bending moment diagram.

LOCAL DAMAGE

Comply with limit states of collapse together with the following additional requirements for crack control to satisfy local damage for beams and slabs.

Crack width limit states (§A.3)*

The estimated maximum crack width for mature concrete in flexural tension may be calculated as below provided that steel stress $\not> 0.8f_y/E_s$

$$w_{max} = 3a_{cr}\varepsilon_m/K_1$$

where $K_1 = 1 + 2(a_{cr} - c_{min})/(h - d_n)$

$$K_2 = (1.2 \times 10^{-3})b_t hy/A_s(h - d_n)f_y$$

f_y above may be expressed in terms of the service stress (see Table CE2) and the other symbols are as given in Appendix CW.

In assessing the strains, take concrete modulus as half the instantaneous value. If shrinkage strain ε_{cs} exceeds 0.06% increase ε_m by adding $0.5\varepsilon_{cs}$

Beams in flexure (§3.3.9 and §3.11.8.2)

Satisfy either crack widths in Appendix B by using above calculations or the following.

(a) Clear horizontal distance between longitudinal bars in tensile face $\not>$ values in Table CE5.

* See Appendix CW for crack widths due to direct tension in immature concrete.

(b) Clear distance from the corner of a beam or rib to surface of nearest longitudinal bar $\ngtr \frac{1}{2} \times$ value in Table CE5.

Note: For (a) and (b), bar diameter $\nless 0.45 \times$ maximum bar diameter at section.

(c) For 'deep' beams (overall depth > 0.75 m) spacing of longitudinal bars near side faces $\ngtr$ 0.25 m for distance from tension face of $\nless \frac{2}{3}$ overall depth. Bar diameter $\nless \sqrt{(s_b b_w / f_y)}$ where s_b = spacing of bars and b_w = breadth at level considered.

(d) For particularly aggressive environment, take value f_y for collapse as $\ngtr$ 300 MN/m².

Table CE5 Maximum clear distance between bars (mm)

	Percentage redistribution to or from section considered										
f_y	-30	-25	-20	-15	-10	0	+10	+15	+20	+25	+30
250	215	230	245	260	275	300	300	300	300	300	300
410	130	140	150	155	165	185	205	215	220	230	240
425	125	135	145	155	160	180	200	210	215	225	235
460	115	125	130	140	150	165	180	190	200	205	215
500	105	115	120	130	135	150	165	175	180	190	195

<u>Solid slabs in flexure</u> (§3.11.8.2)

Either calculate crack widths for flexure and also, if fully restrained, for shrinkage and temperature effects (as for walls), and check compliance with values in Appendix B.

Or, satisfy one only of the following

(a) $h \ngtr 0.25$ m, $f_y \ngtr 425$ MN/m² and bar spacing $\ngtr 3d$

(b) $h \ngtr 0.20$ m, $f_y \ngtr 500$ MN/m² and bar spacing $\ngtr 3d$

(c) If $0.5\% < \rho_c < 1\%$
clear distance between bars $\ngtr$ value in Table CE5 $\times 1/100\rho_c$
where ρ_c = tensile steel ratio on gross concrete area

(d) If $\rho_c < 0.5\%$, clear distance between bars $\ngtr 5d$ (§3.11.4.2) and $\ngtr 2 \times$ value in Table CE5

(e) Otherwise, use rules as for beams.

Note: For slabs, if amount of redistribution is not known, assume -15% for support moments and zero for span moments.

Columns (§3.5.9)

Consider lightly loaded columns subjected to bending as for beams. If ultimate axial load > $0.2f_{cu}A_c$ no cracks are likely to occur and no check is required.

Flat and ribbed slabs (§3.6.4)

Satisfy reinforcement requirements for flat or ribbed slabs as appropriate. If blocks are present and contribute to structural strength, assume elastic modulus of block material equals that of surrounding concrete. Otherwise as for solid slabs.

Reinforced walls (§3.8.8)

Satisfy general and reinforcement requirements for reinforced walls. In any part of a wall where tension develops at collapse, arrange the reinforcement in two layers with each layer complying with bar spacing rules for slabs (given above).
For early thermal and shrinkage cracking see Appendix CW.

Staircases (§3.9.2)

As for beams and slabs.

Bases (§3.10.6)

Satisfy bar spacing rules for beams, but reinforcement need not be provided in the side faces of bases to control cracking.

See also: Appendixes CU and CX and RC text.

APPENDIX CL

PRACTICAL GUIDES FOR DESIGN – LIGHTWEIGHT AGGREGATE CONCRETE

(Extracts from CP 110:1972, §3.12)

Lightweight aggregate concrete of Grade 15 or above can generally be designed as normal concrete with the modifications as given below. Refer to Appendix CU for concretes below Grade 15.

Modulus of elasticity (§App. D)

The properties of a particular lightweight aggregate are best obtained from the aggregate producer concerned. Alternatively, estimate short-term concrete modulus by multiplying the values from normal weight concrete by $(D_c/2300)^2$, i.e.

$$E_c = 1.72\, D_c^{\,2}\, (\sqrt[3]{f_{cu}} \times 10^{-3}) \text{ (MN/m}^2\text{ units)}$$

where D_c = density of concrete in kg/m³ at time of test and $1400 < D_c < 2300$.

Characteristic strength (§3.12.3)

See Appendix A for characteristic strengths and grades of lightweight aggregate concrete. Table A2 (or for Grade 15, the values in Table A2 for Grade 20 reduced by 25%) gives increases in strength with age for sintered pulverized fuel ash aggregate only; other lightweight aggregates may show little increase in strength after 28 days, especially for higher strengths.

Slenderness limits for columns and walls (§3.12.9, 10)

Some additional restrictions are necessary, due primarily to a reduced static tangent modulus of elasticity.

(a) *Short columns and walls:* $\ell_e/h \not> 10$ in given direction of bending

(b) *Slender columns and walls:* Increase additional moments (from §3.5.7, 3.8.6) by 1.75/1.2; e.g.

$$M_{tx} = M_{ix} + \beta_{cx}^{\,2}(1000 - 3.5\beta_{cx})Nh/1.2 \text{ etc.}$$

(c) *Plain slender walls:* Increase additional eccentricity (from §5.5.5) by 25/17 for plain slender wall (i.e. $e_a = \ell_e^2/1700h$)

Shear (§3.12.4,7)

As for normal concrete in both beams and slabs except that critical shear stress is given generally by 0.8 × value in Table CC1, i.e. by Table CL1.

Table CL1 Critical shear stress v_c for lightweight concrete (MN/m²)

$\frac{100 A_{st}}{bd}$	Characteristic cube strength (MN/m²)				
	15	20	25	30	≮40
0.25	0.15	0.28	0.28	0.28	0.28
0.50	0.20	0.36	0.40	0.44	0.44
1.00	0.20	0.48	0.52	0.56	0.60
2.00	0.25	0.64	0.68	0.72	0.76
3.00	0.30	0.68	0.72	0.76	0.80

In all cases $v \not> 0.6\sqrt{f_{cu}}$

i.e. for Grades 15 20 25 30 ≮40

$v \not>$ 2.30 2.68 3.00 3.28 and 3.80 respectively.

Torsion (§3.12.5)

As for normal concrete except that design torsion stresses ≯ 0.8 × values for dense aggregate concrete (see Table CL2).

Table CL2 Limiting design torsion stresses for lightweight concrete (MN/m²)

	Characteristic cube strength (MN/m²)				
	15	20	25	30	≮40
v_{tmin}	0.15	0.24	0.26	0.30	0.34
v_{tu}	2.00	2.68	3.00	3.28	3.80

<u>Cover, durability and fire resistance</u> (§3.12.2)

Nominal cover to reinforcement.

$\ngtr$ Grade 15 and internal non-corrosive conditions:	25 mm
$\nless$ Grade 20 and mild conditions:	as for normal concrete
$\nless$ Grade 20 and moderate or severe conditions:	10 mm greater than for normal concrete
Minimum cement contents:	as for normal concrete

<u>Bond and anchorage</u> (§3.12.11)

Use design bond stresses for normal concrete (Tables CD1 and CD2) modified as follows.

Grade 20 and above:	$\ngtr$ 0.5 × plain bar values
and	$\ngtr$ 0.8 × deformed bar values
Grade 15:	$\ngtr$ 0.8 × Grade 20 value for lightweight concrete

For bars horizontal during casting with foamed slag or similar aggregates reduce bond stresses even further.

<u>Bearing</u> (§3.12.12)

Local bearing stress inside bends $\ngtr f_{cu}/(1 + 2\Phi/a_b)$

<u>Limit state of deflection</u> (§3.12.6,8)

Deflection may be calculated by using appropriate value of E_c.
Alternatively, use additional multiplier ζ_4 (with Tables CE1 to CE3) of 0.85.
For slabs only, if characteristic imposed load $\ngtr 4\ kN/m^2$ use $\zeta_4 = 1.0$.

APPENDIX CU

PRACTICAL GUIDES FOR DESIGN – PLAIN CONCRETE WALLS

(Extracts from CP 110:1972, §5.5. Cross-references to code are shown in brackets.)

The definitions for reinforced walls (i.e. of 'walls', 'short', 'slender', 'braced', 'unbraced', etc.) apply generally to plain walls also. Disregard any reinforcement in plain walls (i.e. steel ratios < 0.4% as given in Appendix CD for strength considerations. Consider plain columns (i.e. where ratio: wall length to thickness < 4) as plain walls provided that the stresses are suitably reduced (see Table CU1).

Effective height of plain walls (§5.5.1.2)

The effective height, ℓ_e, is given by:

ξ × mean height (or length) between centres of lateral supports and ψ × mean height (or length) between lateral support and free edge. If the top of the wall is shaped like a gable, measure to midheight (not centroid) of gable.

(a) *Unbraced wall*

$\xi = 1\frac{1}{2}$ and $\psi = 2$

(b) *Braced wall*

If lateral supports resist:

(i) Lateral movement only, $\xi = 1$ and $\psi = 2\frac{1}{2}$

(ii) Both rotation and lateral movement, $\xi = \frac{3}{4}$ and $\psi = 2$

Lateral supports (§5.5.1.3)

Lateral supports for braced walls may be horizontal (e.g. floors) or vertical (e.g. other walls). The lateral supports attached to a braced wall should be capable of transmitting the sum of the following assumed forces to the elements of construction which provide lateral stability to the structure as a whole.

(i) The simple static reactions to the total applied horizontal forces at the point of lateral support,

(ii) $2\frac{1}{2}$% of total ultimate vertical design load acting on wall or pier at the point of lateral support.

Ignore other design loads when considering (i) and (ii) above.

Lateral supports provide resistance to rotation if:

(i) Lateral support and braced wall are concrete walls adequately detailed to provide bending restraint, or

(ii) Concrete floors have a bearing on the wall (along or across span) of $\nless \frac{2}{3} \times$ wall thickness, or connection is adequately detailed to provide bending restraint, or

(iii) the supported floors (or roofs) of any form of construction occur at same level on both sides of the wall.

Slenderness limit for plain walls (§5.5.1.2)

All cases. $\ell_e/h \ngtr 30$

Analysis of structure (§5.5.2,3,4)

Generally, as for reinforced concrete walls (see Appendix CB)

(a) *Eccentricity in plane at right angles to wall*

Assume line of action of load from concrete floor or roof acts at one-third the depth of the bearing area from the loaded face. However, for an in situ concrete floor on either side of the wall the common bearing area can be assumed to be shared equally by each floor. Eccentricities > ½ wall thickness may be applied through special fittings (e.g. joist hangers) if proven by suitable tests.

Resultant eccentricity of the total load at any level due to the eccentricity of all vertical loads and overturning moments of lateral forces above that level may be reduced as follows.

(i) *Unbraced:* No reductions.

(ii) *Braced:* Resultant eccentricity immediately above a lateral restraint is zero.

In all cases eccentricity $\nless$ 0.05 (as for RC walls).

(b) *Eccentricity in plane of wall*

As for RC walls (see Appendix CB), provided that when distributing horizontal forces between several walls, resultant eccentricity for any individual wall $\ngtr \frac{1}{3} \times$ wall length. If eccentricity of load in plane of wall is zero, assume uniform axial load per unit length of wall, n_w.

Appendix CU

Analysis of section (§5.5.5)

The maximum axial load per unit length of wall due to ultimate loads, n_w, at every section is given by

$$n_w \ngtr \beta_w \lambda_w \{h-2(e_x + e_a)\} f_{cu}$$

where h = thickness of wall

e_x = eccentricity of load (at right angles to plane of wall) due to primary moments ($\nless$ 0.05h) at level considered

e_a = additional eccentricity (see below)

f_{cu} = characteristic cube strength of the concrete

β_w = coefficient dependent on length to thickness ratio; given by 1.0, 0.9 and 0.8 if length is $\nless$ 4h, $2\frac{1}{2}$h, and h respectively.

λ_w = coefficient dependent on wall dimensions, grade and type of concrete as given in Table CU1.

Table CU1 Values of coefficient λ_w for walls

Type and grade of concrete	Clear height between supports		
	$\nless$1.5b*	b	$\ngtr$0.5b
Fully compacted $f_{cu} \nless$ 25 MN/m²	0.4	0.45	0.5
Fully compacted 10, 15, & 20 MN/m²	0.35	0.375	0.4
No-fines (aggregate to BS 882 or BS 1047)	0.3	0.325	0.35

* where b = length of wall.

Take values of e_a as follows

(a) *Short braced wall:* 0

(b) *Slender braced wall:*

(i) $\ell_e^2/2500h - 0.4e_x$

(ii) 0

(c) *Unbraced wall:*

(i) At top of wall: 0

(ii) At bottom of wall $\ell_e^2/2500h$

For (c) (ii) Substitute e_{x_2} for e_x where e_{x_2} is the resultant eccentricity calculated at the bottom of the wall.

Shear strengths for plain walls (§5.5.6)

Average horizontal shear strengths at collapse for forces in plane of wall assumed to be larger of either

(a) ¼ × mean vertical stress or

(b) 0.3 MN/m^2 (no-fines and Grades 10, 15, 20) and 0.45 MN/m^2($\nless$ Grade 25)

Shear strength for forces normal to wall satisfactory if

$\ell_e/h > 6$

Bearing stresses in plain walls (§5.5.7)

Local bearing stresses at collapse (e.g. at girder bearings)

$\ngtr 0.5\ f_{cu}$ (generally) and $\ngtr 0.6\ f_{cu}$($\nless$ Grade 25)

LIMIT STATES OF LOCAL DAMAGE (§5.5.9)

See Appendix CD and if necessary Appendix CW — thermal and shrinkage cracking,for minimum reinforcement requirements.

Cracks due to bending will generally not occur. Minimize cracks at the junction between the ends of simply supported slabs and the supporting walls, due to deflection of the slab under load, by complying with span/depth ratio limits given in Appendix CE.

LIMIT STATES OF DEFLECTION (§5.5.8)

Satisfy general requirements for plain walls.

For a cantilever shear wall: total height $\ngtr$ 10 × length.

APPENDIX CW

PRACTICAL GUIDES FOR DESIGN — WATER-RETAINING STRUCTURES

(Extracts from BS 5337:1976. Cross-references to code are shown in brackets.)

GENERAL DESIGN CONSIDERATIONS (§4)

Use *either* Limit state design based on CP 110:1972

or Alternative design based on CP 114 and CP 115.

Design floors and walls to resist the appropriate pressure and consider the exposure of external faces as Class A or Class B. Also prevent flotation. The build-up of external water pressure may be reduced by providing (i) drainage and (ii) relief valves discharging into the vessel (if contamination can be allowed).

Design load arrangements (§4.2)

Design all structures containing liquids for both full and empty conditions and for the most critical arrangement of loads. Refer to CP 2004 for guidance on ground pressure distribution and soil characteristics and allow for any active soil pressure. Ignore any passive soil pressure assisting a wall to contain liquid.

Analysis of walls (§4.3)

Design plane walls to resist liquid pressure by a combination of bending moments in vertical and horizontal planes, and direct tension in adjacent walls. The horizontal bending moments at all corners where the walls are rigidly joined require additional horizontal reinforcement. Cylindrical tanks may be constructed with a fixed, pinned or sliding joint between the walls and the foundation slab, and the vertical bending moments and hoop tensions estimated accordingly.

Control of cracking (§4.4)

Control cracking due to thermal and moisture movements as follows.

(a) Limit the extent of the thermal and moisture changes.
(b) Limit restraints by providing movement joints and by a planned construction sequence.
(c) Limit the effects of different conditions within the thickness of a member.
(d) Provide adequate reinforcement or prestress.

Control cracking due to minor uneven settlement by providing movement joints and by reinforcement or prestress (see §4.5). For cracking due to loading see §8 and §9. Restrict cracking by slow first filling of water-retaining structure at an approximately uniform rate of rise of water level $\ngtr$ 1m/day.

Settlement, subsidence and ground movement (§4.5)

Cracking due to differential settlement, subsidence or other movement can be excessive. Where sites affected by geological faults, mining, ground movement or subsoils of varying degrees of compressibility cannot be avoided, subdivide the structure into smaller compartments and provide joints suitable for the estimated movement if necessary. Also provide flexible joints to outlet pipes and other fittings.

Injurious soils (§4.6)

Guidance on the resistance of special cements to sulphates in soil and ground water is given in CP 110 itself (§6.3.3 and Table 49). In other and more serious conditions provide an isolating bituminous (or other suitable) membrane.

Durability (§4.7)

The code recommendations generally provide adequate durability of the concrete, reinforcement or prestressing steel, jointing materials and linings for the classes of exposure specified (in §4.9). Increase the cement content or grade of concrete and the cover to the steel if necessary where waters containing dissolved free carbon dioxide, natural acids or salts may be aggressive. If the concrete may be exposed to severe attack then provide either an impermeable protective lining or, if less severe, increased resistance with selected type of cement. Process liquids and sewage can also be aggressive, and biological attack, especially on the jointing material, should also be avoided.

Impermeability of the concrete (§4.8)

Thorough compaction of the concrete, without segregation, is essential. An adequately impermeable concrete is generally given by the concrete mixes, aggregates, minimum cement contents and strength, and curing recommended in the Code. Low impermeability directly affects leakage and influences the durability, resistance to leaching, chemical attack, erosion and frost damage of the concrete and the protection of embedded steel.

Appendix CW

Degrees of exposure (§4.9)

The three classes of exposure are defined as follows.

Class A Exposed to a moist or corrosive atmosphere or subject to alternate wetting and drying.

Class B Exposed to continuous or almost continuous contact with liquids.

Class C Not exposed to liquid nor to moist or corrosive conditions.

A member 225 mm or less thick and exposed to Class A or B conditions on one face, should be regarded as being similarly exposed on both faces. Roofs will generally come within Class A.

See also: reinforcement.

LIMIT STATE REQUIREMENTS (§5)

Design in accordance with CP 110 (§2) except as modified below.

Collapse Limit State (§5.2)

Take design imposed load due to contained liquid in load combination (1) as given by partial safety factor for load, γ_f of 1.6.

Deflection limit state (§5.3.1)

Generally as for non-liquid retaining structures.

Crack width limit states (§5.3.2)

The maximum calculated surface crack width should be as follows.

Reinforced Concrete:

Class A $\not>$ 0.1 mm (direct tension or flexure)

Class B $\not>$ 0.2 mm (direct tension or flexure)

Class C As CP 110.

Prestressed Concrete:

Class A No tensile stresses (i.e. as CP 110 Class 1)

Class B Tensile stresses $\ngtr$ permissible values in Table CW6 (i.e. similar to CP 110 Class 2)

Class C As Class 3 structures of CP 110 (§2.2.3.2)

ALTERNATIVE DESIGN REQUIREMENTS (§6)

Design in accordance with CP 114 and CP 115 for Class C exposure or as modified below for Class A or Class B exposures where design of members is based on two considerations as follows.

Reinforced Concrete:

(i) Adequate strength and (ii) adequate resistance to cracking (see §9.2 and §9.3).

Prestressed Concrete:

(i) Adequate strength and (ii) avoidance of cracking (see §12.1).

REINFORCEMENT

Cover (§4.10)

Concrete cover $\nless$ 40 mm in all cases (See also §4.7).

Minimum steel areas (§4.11)

(a) *Continuous walls*—Class A or Class B exposure

Minimum total steel ratio based on gross area, ρ_c, in each of two directions at right angles

$\nless$ 0.5% (mild steel)

or $\nless$ 0.3% (deformed high yield bars)

—placed as near concrete surface as possible consistent with cover, etc.

(i) Walls < 200 mm thick-may be placed in one layer of $\nless$ 0.5% (or 0.3% for deformed high yield bars) in each direction.

(ii) Walls 200 mm to 600 mm thick $\nless$ 0.25% (or 0.15% for deformed high yield bars) in each face and in each direction.

(b) *Floor slabs*— As for wall slabs of similar thicknesses in all cases.

Note: If calculations for limit state requirements are carried out then lower steel ratios can often be used, especially for walls with joints (§15), sections over 600 mm thick and ground slabs (§B3).

Bar spacings (§15.1)

Bar spacings $\ngtr$ 300 mm generally. See also §8.2.

LIMIT STATE DESIGN FOR REINFORCED CONCRETE (§8)

Design in accordance with §5 requirements and with CP 110 (§3) as modified below.

Basis of design (§8.1)

References in CP 110 §3.0 to CP 110 §2.0 should be replaced by reference to BS 5337 Section 2 (§4 and §5).

References to CP 110 §3.1.4.2 and CP 110 Table 2 are limited to Grades 2 and 30 for reinforced concrete.

CP 110 §3.11.1.4, 3.11.2 and 3.11.4 are superseded by BS 5337 Section (§13 to §19), 4.10 and 4.11 respectively.

CP 110 §3.2.2.3 does not apply.

Collapse limit state

CP 110 (§3.11.6.2) applies except that the anchorage (average) bond stresses for horizontal bars which are in sections in direct tension should be $\ngtr$ 0.7 × values given (see Table CD1).

Crack width limit states (§8.2)

CP 110 (§3.1.2.1) reference to allowing reinforcement spacing rules as an alternative to calculating crack widths does not apply for Class A or Class B exposure. Estimated maximum crack widths $\ngtr$ values given in §5.3.2. Alternatively, crack widths in the mature concrete may be deemed to be satisfactory if the appropriate steel stress and bar spacing requirements are satisfied as below for direct tension in mature and immature concrete and flexural tension in mature concrete.

Direct tension in mature concrete (§8.2)

Bar spacing conforms with CP 110 (§3.11.8.2) (and generally ≯ 300 mm) and steel stress ≯ appropriate value in Table CW1.

Table CW1 Allowable steel stresses in direct or flexural tension for serviceability limit states

Class of exposure	Allowable stress (MN/m^2)		
	A	B	C
Plain bars	85	115	125
Deformed bars	100	130	140

Direct tension in immature concrete (§App.B)

If $\rho_c \nless \rho_{crit} = f_{ct}/f_{yt}$

then the estimated maximum spacing of cracks, s_{max}, due to early thermal contraction (i.e. heat of hydration) and shrinkage in restrained reinforced members is given by:

$$s_{max} = \frac{f_{ct}}{f_b}\frac{\Phi}{2\rho_c} \qquad \text{(CW.1)}$$

where f_{ct} = direct tensile strength of the immature concrete taken at an age of 3 days as 1.15 and 1.3 MN/m^2 for Grades 25 and 30 respectively.

f_{yt} = characteristic strength of reinforcement as in CP 110 (see Table CD1)

f_b = average bond strength between immature concrete and steel

Φ = reinforcing bar size

f_{ct}/f_b = 1 for plain round bars; $\frac{4}{5}$ for deformed (Type 1) bars and $\frac{2}{3}$ for deformed (Type 2) bars.

Design of distribution reinforcement (§App.B)

Equation (CW.1) can be rearranged for design purposes as

Table CW2 Design options for control of early thermal contraction and shrinkage

Option	Method of control	Movement joint spacing	Minimum steel ratio† %	Comments*
1	Continuous construction	None	ρ_{crit}	Use small size bars at close spacing to avoid high steel ratios well in excess of ρ_{crit} (see §App.B)
2	Semi-continuous construction	Partial contraction joints at 7.5 m or complete contraction joints at 15 m	ρ_{crit}	Use small size bars but less steel than in 1 (see §App.B)
3	Moderate joint spacing with controlled cracking	$\not> 2s_{max}$ (see §15.1) and $\not> \dfrac{4w}{\varepsilon_{cs} + \varepsilon_{te} - 100 \times 10^{-6}}$	ρ_{crit}	Avoid small size bars which restrict the joint spacing
4	Close joint spacing with controlled cracking	$\not> 1.25s_{max}$ (see §15.1) and $\not> s_{max} + \dfrac{w}{\varepsilon_{cs} + \varepsilon_{te} - 100 \times 10^{-6}}$	$\frac{2}{3}\rho_{crit}$ and $\not< 0.2\%$	Use minimum steel ratio and bar spacing close to 300 mm
5	Close joint spacing and designed for no cracking	$\not> s_{max}$ (see §15.1)	As for 4	As for 4

w is the (design) maximum crack width, i.e. 0.1 mm for Class A or 0.2 mm for Class B exposure.

† In slabs over 200 mm thick $\not< \frac{1}{3}\rho_{crit}$ should be placed near each face.

* In all cases these comments assume that higher steel ratios for structural loading (direct or flexural tension) are not required.

$$n_b \Phi \nless \frac{f_{ct} 2bh}{f_b \pi s_{max}} \qquad \text{(CW.2)}$$

where n_b is number of bars in cross-section bh and
s_{max} is as given by Eqn (CW.3) or (CW.4) below for required w_{max}.

Design options for joint spacings (§15.1)

The design options for the spacing of movement joints to control early thermal contraction and shrinkage are summarized in Table CW2.

(a) *Continuous and semi-continuous thin walls and slabs (options 1 and 2)* (§App.B)

The estimated maximum width of fully developed cracks, w_{max}, due to early thermal contraction and shrinkage in restrained continuous walls and slabs is given by

$$w_{max} = s_{max} \{\varepsilon_{cs} + \varepsilon_{te} - (100 \times 10^{-6})\} \qquad \text{(CW.3)}$$

where ε_{cs} = estimated shrinkage strain

ε_{te} = estimated total thermal contraction after peak temperature due to heat of hydration.

Take the effective coefficient of thermal contraction as one half of the value for mature concrete (due to the high creep strain in immature concrete). For walls exposed to normal climatic conditions take ε_{cs} as less than 100×10^{-6}. Hence for cooling from hydration peak temperature to ambient

$$w_{max} = s_{max} \frac{\alpha}{2} T_1$$

where α = coefficient of thermal expansion of mature concrete for temperature fall from hydration peak to ambient

Take $T_1 \nless 30^{\circ}C$ for summer concreting (ambient $\geq 20^{\circ}C$) and
$\nless 20^{\circ}C$ for winter concreting (ambient $\leq 10^{\circ}C$)

Increase T_1 if cement content > 340 kg/m^3, if wall thickness > 400 mm or if timber shutters are used in summer. Any subsequent temperature fall T_2 due to seasonal variations occurs in mature concrete, for which the crack spacing is greatly reduced and for which restraints along the base and elsewhere tend to be more uniform. Hence take w_{max} as given by

$$w_{max} = s_{max} \frac{\alpha}{2}(T_1 + T_2) \qquad \text{(CW.4)}$$

Ignore T_2 if partial movement joints occur at $\ngtr$ 7.5 m centres, or complete movement joints occur at $\ngtr$ 15 m centres.

(b) *Sections with moderate or close movement joints (options 3, 4 and 5)* (§15)

Option 3 Movement joints provide a partially restrained system during contraction of the immature concrete. Thus if joint spacing $\ngtr 2s_{max}$ and complete movement joint spacing $\ngtr 4s_{max}$, assume that 50% of the total contraction occurs at the joints (i.e. actual w between joints is half w_{max} given by Eqn (CW.3)–hence the coefficient 4 in Table CW.2).

Options 4 and 5 If joint spacing $\ngtr 1.25s_{max}$ and complete movement joint spacing $\ngtr 2.5s_{max}$ most of the contraction can occur at the joint and the full critical reinforcement ratio is not required because the steel does not need to distribute cracks along lengths of wall of s_{max} (or more) but along only $0.25s_{max}$ of wall which exceeds s_{min} from a contraction joint. The reinforcement of $\frac{2}{3}\rho_{crit}$ is required only to assist the concrete of length s_{min} at either end to contract away from the joints and overcome the friction along the base. If the joint spacing is s_{max} (i.e. $2s_{min}$) both ends can contract towards the centre and no cracks should occur. If joint spacing is $1.25s_{max}$ then the central $0.25s_{max}$ only is effectively fully restrained and the crack width w for this short length is given by Eqn (CW.2).

Notice that the joint spacing (i.e. s_{max}) can be increased by increasing the bar diameter (subject to spacing generally $\ngtr$ 300 mm).

(c) *Thick sections* (§App.B)

The rates of cooling near the surfaces of thick sections are significantly faster than the core, and hence the requirement for (the entire thickness of) a thin member need only apply to the 'surface zones' of thick members.

Consider 'surface zone' for each exposed face as $\nless$ 250 mm thick if the centroid of the reinforcing bars $\ngtr$ 125 mm from the surface, i.e. when wall thickness > 500 mm, base ρ_{crit} on wall thickness of only 500 mm. However use temperature rise T_1 for the core of $\nless 10^{\circ}C$ higher than mean value for entire section.

(d) *Ground slabs* (§App.B)

For ground slabs over 500 mm thick assume top surface zone is 250 mm thick and bottom surface zone is only 100 mm thick. For ground slabs between 500 and 200 mm thick assume top surface zone is half the slab thickness and that the bottom surface zone is half the slab thickness less 100 mm, or 100 mm, whichever is the less.*

For ground slabs under 200 mm thick, assume top surface zone is 100 mm thick (maximum cover to reinforcement of 50 mm) and ignore bottom zone.

Flexural tension in mature concrete (§8.2 and §App.C)

Satisfactory if bar spacing conforms with CP 110 (§3.11.8.2) (and generally $\ngtr$ 300 mm) and if steel stress under service conditions $\ngtr$ appropriate value in Table CW1.

Alternatively the estimated maximum crack width may be calculated as given below provided that steel strain $\ngtr 0.8f_y/E_s$ and concrete stress $\ngtr 0.45f_{cu}$

$$w_{max} = 4.5a_{cr}\varepsilon_m/K_1$$

where $K_1 = 1 + 2.5(a_{cr} - c_{min})/(h - d_n)$

a_{cr} = distance from point considered to the surface of the nearest longitudinal bar

c_{min} = minimum cover to the tension steel

h = overall depth of member

d_n = depth of the neutral axis from elastic analysis as for ε_1

$\varepsilon_m = \varepsilon_1 - K_2$

ε_1 = average strain at level of point considered calculated from elastic analysis ignoring the stiffening effect of the concrete in the tension zone (i.e. ignoring factor K_2)

$K_2 = (0.7 \times 10^{-3})b_t hy/A_s(h - d_n)f_s$

b_t = width of the section at centroid of tension steel

y = distance from neutral axis to level of point considered

A_s = area of tension reinforcement

f_s = service stress in the reinforcement

In assessing the strains, take concrete modulus as half the instantaneous value. If shrinkage strain ε_{cs} exceeds 0.06% increase ε_m by adding $0.5\varepsilon_{cs}$.

* Not as BS 5337.

Appendix CW

ALTERNATIVE DESIGN FOR REINFORCED CONCRETE (§9)

Basis of design (§9.1)

Base design of members on strength and resistance to cracking (see §6) and on the following assumptions:

(i) Plane sections remain plane after bending.

(ii) Both steel and concrete are perfectly elastic with a modular ratio of 15.

(iii) For resistance to cracking in both flexure and direct tension, the concrete can sustain limited tensile stress. Consider the whole section of concrete, including the cover, together with the reinforcement.

(iv) For strength calculations, the concrete has no tensile strength.

Permissible concrete stresses (§9.2)

(a) For resistance to cracking use permissible concrete stresses shown in Table CW3. The total shear stress, V/bz, (where V = total shear, b = breadth and z = lever arm) should in no case exceed the values shown in Table CW3.

Table CW3 Permissible concrete stresses (MN/m^2) for resistance to cracking of reinforced concrete for Class A or Class B exposure

Concrete	Tension		Shear
Grade	Direct	Flexural	V/bz
25	1.31	1.84	1.94
30	1.44	2.02	2.19

(b) For strength calculations use permissible concrete stresses shown in Table CW4. Where the calculated shear stress (i.e. equal to the diagonal tension stress) in the concrete alone exceeds the permissible, provide secondary reinforcement to take the whole of the transverse component of the diagonal tension.

Table CW4 Permissible concrete stresses (MN/m^2) for strength of reinforced concrete for Class A or Class B exposure

Concrete Grade	Compression Direct	Compression Flexural	Shear	Bond* Average†	Bond* Local
25	6.95	9.15	0.77	0.9	1.36
30	8.37	11.0	0.87	1.0	1.49

* Bond stresses may be increased 40 per cent for deformed bars.

† Average (anchorage) bond stresses for horizontal bars in direct tension should be decreased by 30 per cent.

Permissible steel stresses (§9.3)

(a) For resistance to cracking the tensile steel stress is necessarily limited by the tensile stress in the concrete, e.g. steel stress $\ngtr$ 19.6 MN/m^2 if permissible direct tensile stress in concrete is 1.31 MN/m^2.

(b) For strength calculations the permissible steel stresses are as shown in Table CW5.

Table CW5 Permissible steel stresses in strength calculations

Condition	Class of exposure	Permissible stress (MN/m^2) Plain bars*	Permissible stress (MN/m^2) Deformed†
Direct or Flexural Tension or Shear	A	85	100
	B	115	130
Compression	A and B	125	140

* Including mild steel bars to BS 4449.

† Deformed high yield bars to BS 4449 or BS 4461 and high yield wire fabric to BS 4483.

Appendix CX
Stresses due to moisture or thermal changes (§9.4)

Allow for movement due to moisture and temperature changes as follows.

(i) Provide minimum reinforcement to §4.11.

(ii) If calculated to §4.11.1, *either* provide movement joints and suitable sliding layer beneath the construction (to §15.5) *or* design to be fully continuous.

(iii) Take precautions to avoid cracking of the concrete before and until the reservoir is put into use.

If concrete will never dry out and water or aqueous liquid is stored at or near ambient temperature then no further precautions are required. See also: BS 5337 (§App.A) itself.

APPENDIX CX
PRACTICAL GUIDES FOR DESIGN - JOINTS

(Extracts from CP 110:1972, §App. B and BS 5337:1976, Section 5 (§13 to 19). Cross-references to codes are shown in brackets.)

The cracking of concrete structures can generally be controlled by suitable provision of adequate reinforcement. In elevated structures where restraint is small, and in underground structures where the range of temperature is small, movement joints may not be required. However the most appropriate, or only, control measure may sometimes be the additional provision of movement joints if unacceptable cracking is to be avoided.

Closely spaced movement joints are generally necessary for structures subjected to large temperature variations-such as reservoirs or tanks storing warm liquids, or uninsulated roof slabs. For moderate service temperature fluctuations, a sliding layer can avoid cracking in a mature wall founded at or below ground level if the tensile strength of the wall section can overcome the resistance to sliding of one-half of the length of the wall between movement joints. However, a closer joint spacing may be desirable due to cracking in the immature concrete (see BS 5337 §15).

Movement joints should permit relative movement of adjacent parts without impairing the functional integrity of the member or structure as a whole. Thus in water-retaining structures watertightness should be maintained during the subsequent movement of the joint. Joints can prevent the build-up of harmful stresses by allowing controlled movement to occur and they can also act as the connecting link between the several parts of a member or structure.

Movement joints should, where possible, be located where cracks would otherwise develop (e.g. at changes in section). The location of all movement joints should be decided by the engineer and be indicated on the drawings.

JOINTS TO BS 5337:1976 (Section 5)

Construction joints and the various types of movement joints are described below. See BS 5337 (§App.D) itself for any illustrations of joints and discussion of joint materials. Joints to BS 5337 should generally be restricted to a single type.

Construction joints (§14)

Construction joints in the concrete are introduced for convenience in construction. Waterstops are not usually required, but the joint may be sealed on the water-face if necessary. There is no provision for subsequent relative movement and special measures are taken to achieve subsequent continuity. If complete continuity cannot be achieved treat as a movement joint.

Movement joints (§15)

Movement joints accommodate relative movement between adjoining parts of a structure and maintain watertightness. They include contraction joints, expansion joints and sliding joints. Reduce cracking due to thermal and shrinkage effects by limiting the thermal and humidity fluctuations (see BS 5337 §App.A). Align movement joints in walls, floors and wall footings with each other and reduce restraints to free expansion or contraction where possible. Sliding layers may be necessary for long walls or slabs founded at or below ground level. Piled foundations can introduce special restraints. To calculate spacings of movement joints see Appendix CW (§15.1)

Appendix CX

Design of movement joints (§15.2)

Movement joints with complete discontinuity of both concrete and steel (i.e. complete contraction joint or an expansion joint) should allow for repeated movement without loss of watertightness and without entry of grit and debris. Design joints to suit the characteristics of the material available. See BS 5337 (§App.D).

Partial contraction joints (§15.3.1)

Movement joints with discontinuity of the concrete and intended to confine contraction to predetermined lines. The steel is continuous and transfers nominal shear across the joint. Provide a waterstop and/or a sealing compound if necessary.

Complete contraction joints (§15.3.2)

Movement joints with deliberate discontinuity of both concrete and reinforcement but no initial gap, to accommodate substantial contraction of the concrete. Joggles in the face of vertical joints transfer shear. Provide sealing compound on one or both faces. A waterstop is essential.

Expansion joints (§15.4)

Movement joints with deliberate discontinuity of both concrete and reinforcement to accommodate both expansion and contraction. An initial gap is usual, and sliding surfaces may be incorporated if desired. Waterstops, joint sealing compound and joint filler are essential.

Sliding joints and sliding layers (§15.5)

Movement joints with complete discontinuity of both concrete and reinforcement and with special provision to faciliate movement in the plane of the joint. The concrete surface on the lower component should be flat, smooth and finished with a steel float; also provide an approved layer to prevent bond between the two faces to allow movement to take place. Provide a waterstop and sealing compound if subject to water pressure, e.g. for a wall/floor joint. A sliding or separating layer is a special category of sliding joint and permits sliding over a considerable area, e.g. between floor and sub-floor or blinding layer.

Joints in floors (§16)

A separating layer of thick polythene or thick bituminous material between floor slab and blinding concrete minimizes restraint where complete movement joints are provided.
Alternatively the floor may be designed for full restraint and cast direct on to the blinding concrete, and partial contraction joints provided if desired.

Joints in walls (§17)

Walls may be designed with movement joints at spacings as in §15.1 (see Appendix CW). These joints will be mainly of the partial (up to 7.5 m spacing) or complete (up to 15 m spacing) contraction type but include expansion joints where deemed necessary in appropriate circumstances.
Alternatively design the walls for full restraint against thermal and shrinkage contraction.

Joints in roof (§19)

Roofs will generally be subjected to greater thermal movement than floors, but to less restraint except at the corners of the reservoir. Ensure correspondence between movement joints where roof is monolithic with walls. If sliding joint allows for movement between the two, complete correspondence is desirable but not essential.

In all reservoirs > 60 mm long (and smaller reservoirs, if necessary,) provide a sliding joint between roof and wall which may be temporary until the roof is insulated, or permanent if roof is not insulated. Adequate reinforcement to limit cracking is particularly necessary at the corners of walls in rectangular reservoirs.

Appendix CX

Construction sequence (§17)

The construction sequence should be such as to minimize restraint to early thermal and shrinkage movement. For example, consecutive bay construction is preferred in walls, since construction in alternate panels results in greater restraint in the intermediate panels cast subsequently, but infill bays $\ngtr$ 1 m long may be used.

Pipes through walls and floors (§18)

Pipes should not coincide with a joint and are preferably cast into the panel when it is concreted. If necessary to box out, then the sides of the opening should be treated as construction joints.

JOINTS TO CP 110:1972 (§App.B)

Joints to CP 110 include the types described for BS 5337 as well as the following, and it may be necessary to design a joint to fulfil more than one requirement.

Hinged joint (§B.3)

Special provision for relative rotation of the members at the joint. A hinged joint can prevent the occurrence of undesirable reverse moments or restraints.

Settlement joint (§B.3)

Complete discontinuity in both steel and concrete permitting adjacent members or structures to settle or deflect relative to each other. Relative movements may be large: for example, between foundations of buildings due to mining subsidence.

Appendix D
Design stress-strain curves

(Extracts from CP 110:1972 §2.4.2.1)

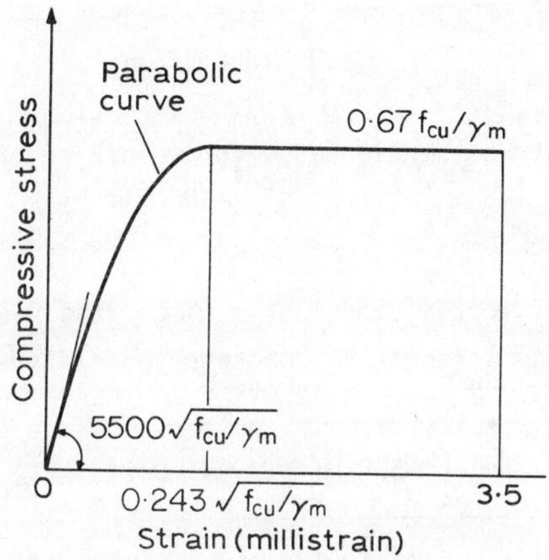

Fig.D1 Design stress-strain curve for concrete (in MN/m^2 units).

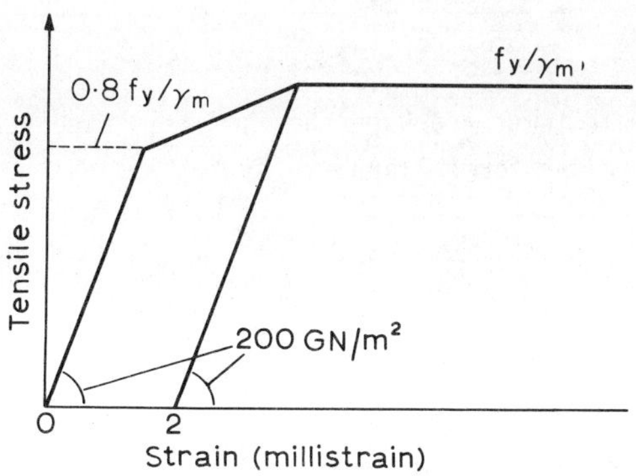

Fig.D2 Design stress-strain curve for reinforcement in tension.

Note: for serviceability limit states or for excessive loads γ_m for steel = 1.

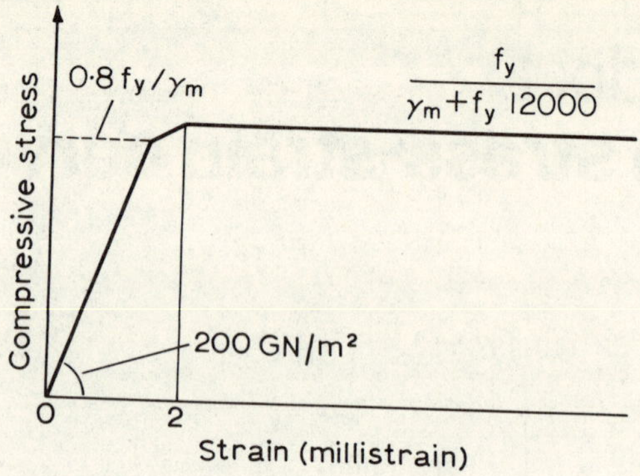

Fig.D3 Design stress-strain curve for reinforcement in compression (f_y in MN/m² units).

Table D1 Values of initial modulus, minimum strain at design strength and design strength of concrete for different limit states

Limit states	γ_m	$5500\sqrt{(f_{cu}/\gamma_m)}$	$0.243\sqrt{(f_{cu}/\gamma_m)}$	$0.67\ f_{cu}/\gamma_m$
Collapse 1,2,3	1.5	$4491\sqrt{(f_{cu})}$	$0.198\sqrt{(f_{cu})}$	0.45
Cracking *or* excessive loads	1.3	$4824\sqrt{(f_{cu})}$	$0.213\sqrt{(f_{cu})}$	0.52
Deflection	1.0	$5500\sqrt{(f_{cu})}$	$0.243\sqrt{(f_{cu})}$	0.67

Table D2 Values of lower and upper design yield strengths for typical steel grades for collapse (γ_m = 1.15)

Reinforcement	Lower $*f_y$	Upper $*f_{yc}$	Upper $*f_{yt}$	f_y
Hot rolled mild steel	174	196	217	250
Hot rolled high yield	285	303	357	410
Cold worked high yield				
$>$ 16 mm	296	312	370	425
$\not>$ 16 mm	320	333	400	460
Hard drawn wire				
$\not>$ 12 mm	337	348	422	485

Appendix E
Design charts for collapse

Fig. E1 General design chart for singly reinforced rectangular concrete beams

For f_y = 250 use upper full line for each value of f_{cu}

For f_y = 410 use chain line for each value of f_{cu}

For f_y = 485 use lower full line for each value of f_{cu}

Fig. E2 Rectangular reinforced concrete beams. f_{cu} = 20 MN/m^2, f_y = 410-425 MN/m^2, d'/d = 0.10

For f_y = 410 use lower lines for values of ρ' and x/d

For f_y = 425 use upper lines for values of ρ' and x/d

For f_y = 420 use interpolation

Fig. E2A Rectangular reinforced concrete beams. f_{cu} = 25 MN/m^2, f_y = 410-425 MN/m^2, d'/d = 0.10

For f_y = 410 use lower lines for values of ρ' and x/d

For f_y = 425 use upper lines for values of ρ' and x/d

For f_y = 420 use interpolation

Fig. E3 Rectangular reinforced concrete beams. f_{cu} = 30 MN/m^2, f_y = 410-425 MN/m^2, d'/d = 0.10

For f_y = 410 use lower lines for values of ρ' and x/d

For f_y = 425 use upper lines for values of ρ' and x/d

For f_y = 420 use interpolation

Fig. E4 Symmetrically reinforced rectangular columns. All steel grades; f_{cu} = 30 MN/m^2, d/h = 0.90

For f_y = 250 use outer full lines

For f_y = 410 use broken lines

For f_y = 485 use inner broken lines

For other values use interpolation.

Fig. E5 Symmetrically reinforced rectangular columns. All concrete grades; f_y = 250 MN/m², d/h = 0.90

For f_{cu} = 25 MN/m² use outer full lines

For f_{cu} = 50 MN/m² use inner broken lines

For other f_{cu} values use interpolation

$\rho_c f_y / f_{cu}$ = 0 to 1.2 in 0.1 steps.

Fig. E6 Symmetrically reinforced rectangular columns. All concrete grades; f_y = 410 MN/m²; d/h = 0.90

For f_{cu} = 25 MN/m² use outer full lines

For f_{cu} = 50 MN/m² use inner broken lines

For other f_{cu} values use interpolation

$\rho_c f_y / f_{cu}$ = 0 to 1.2 in 0.1 steps.

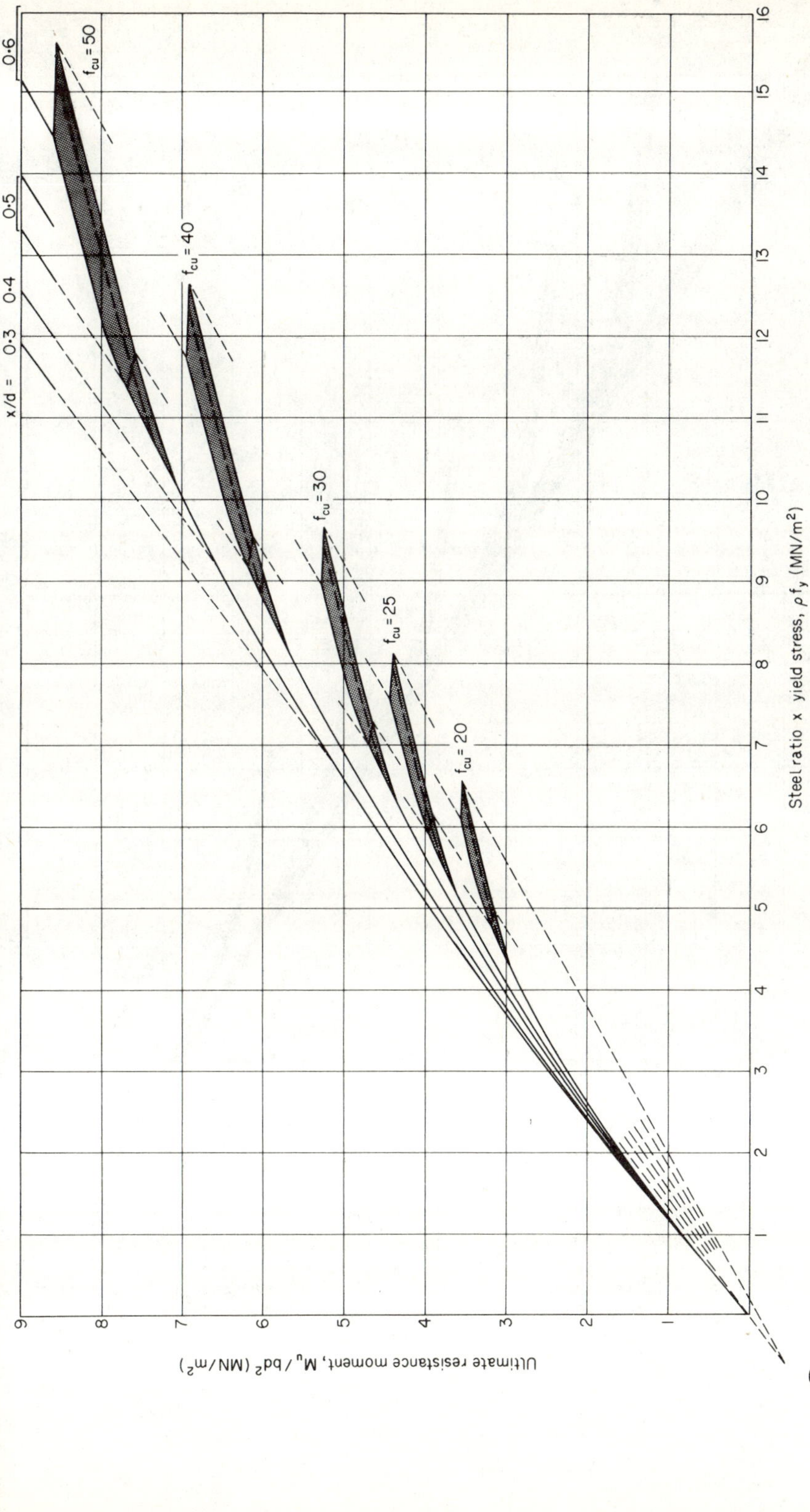

x/d = 0·3 0·4 0·5 0·6
f_{cu} = 50
f_{cu} = 40
f_{cu} = 30
f_{cu} = 25
f_{cu} = 20
Steel ratio x yield stress, ρf_y (MN/m²)
Ultimate resistance moment, M_u/bd^2 (MN/m²)

Fig. E1

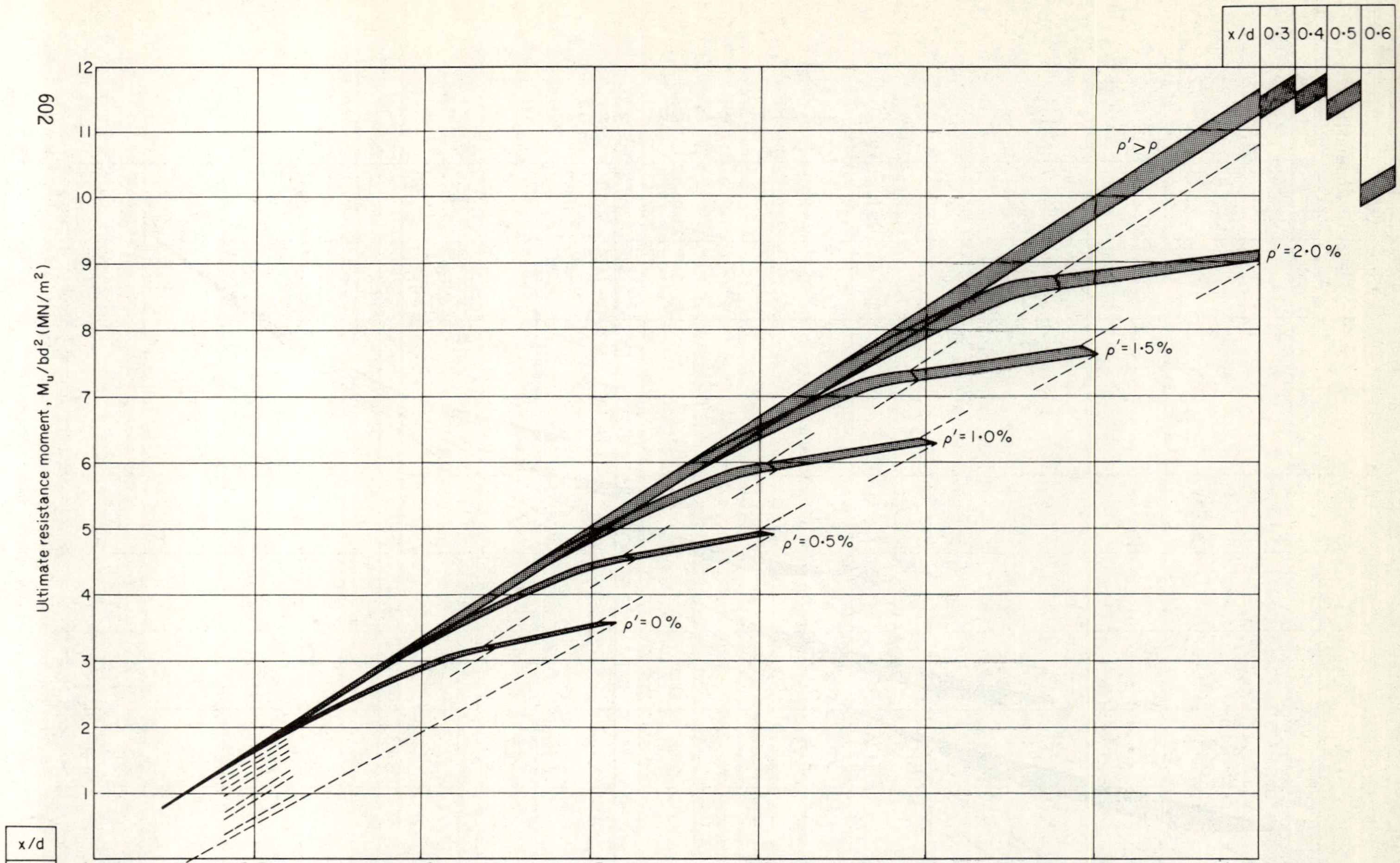
Ultimate resistance moment, M_u/bd^2 (MN/m^2)
x/d
0·3
0·4
0·5
0·6
ρ' > ρ
ρ' = 2·0 %
ρ' = 1·5 %
ρ' = 1·0 %
ρ' = 0·5 %
ρ' = 0 %
0·5
1·0
1·5
2·0
2·5
3·0
3·5
1
2
3
4
5
6
7
8
9
10
11
12

Fig. E2

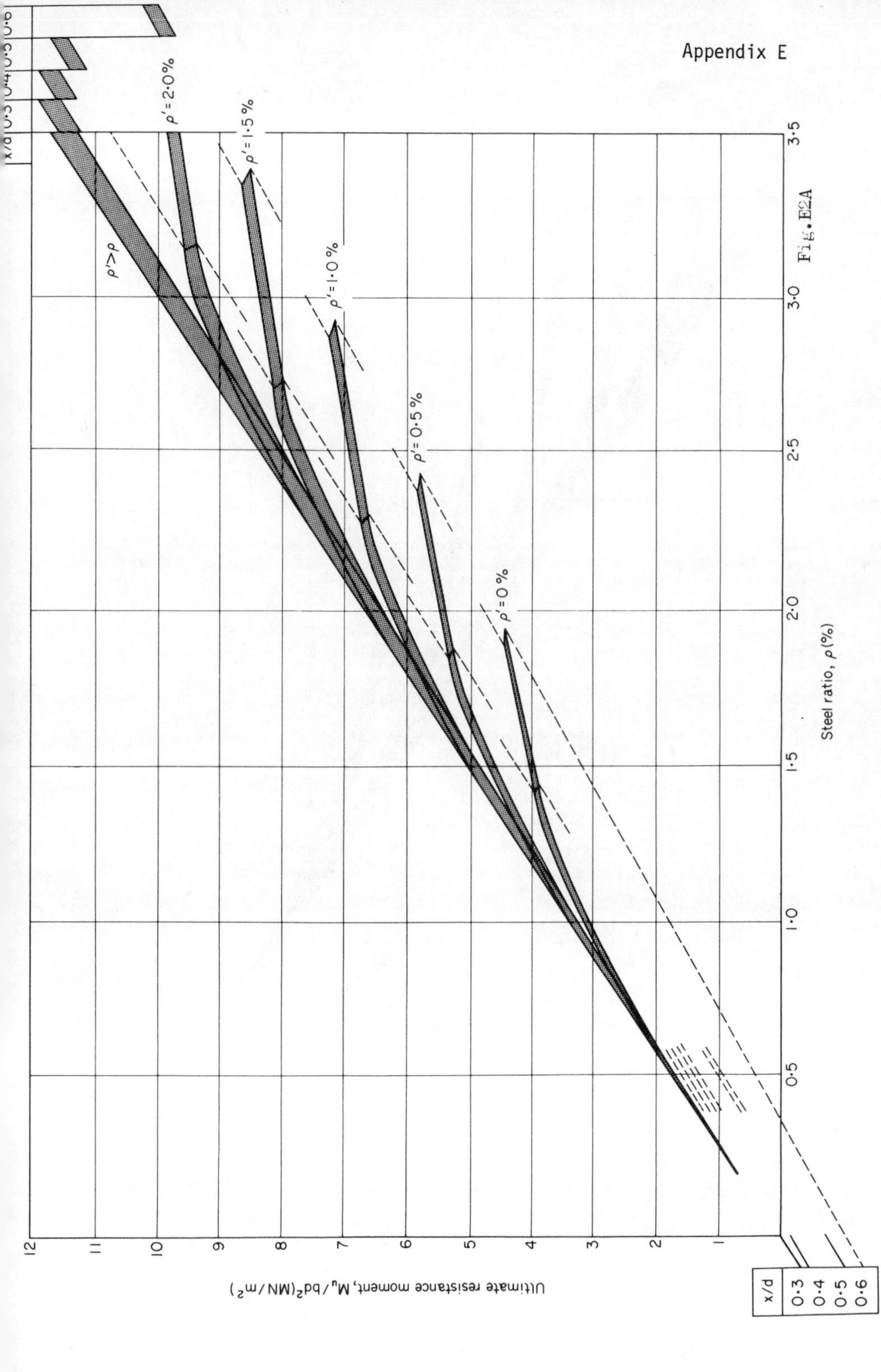

Ultimate resistance moment, M_u/bd^2 (MN/m²)
Steel ratio, ρ (%)
x/d
0·3
0·4
0·5
0·6
ρ' > ρ
ρ' = 2·0 %
ρ' = 1·5 %
ρ' = 1·0 %
ρ' = 0·5 %
ρ' = 0 %

Fig. E2A

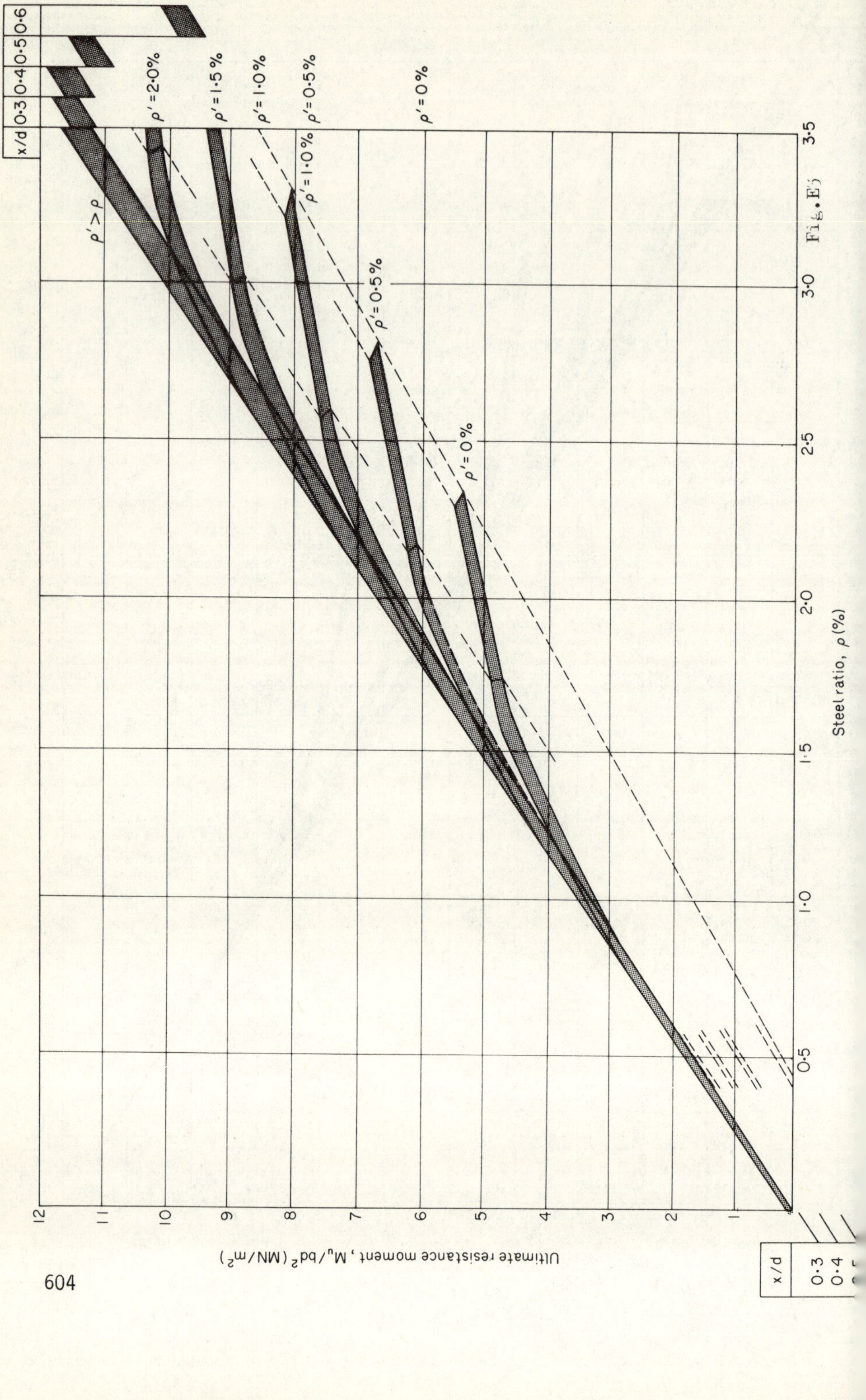
x/d
0·3
0·4
0·5
0·6
ρ' > ρ
ρ' = 2·0%
ρ' = 1·5%
ρ' = 1·0%
ρ' = 0·5%
ρ' = 0%
Ultimate resistance moment, M_u/bd² (MN/m²)
Steel ratio, ρ (%)
0·5
1·0
1·5
2·0
2·5
3·0
3·5
1
2
3
4
5
6
7
8
9
10
11
12

Fig. E5

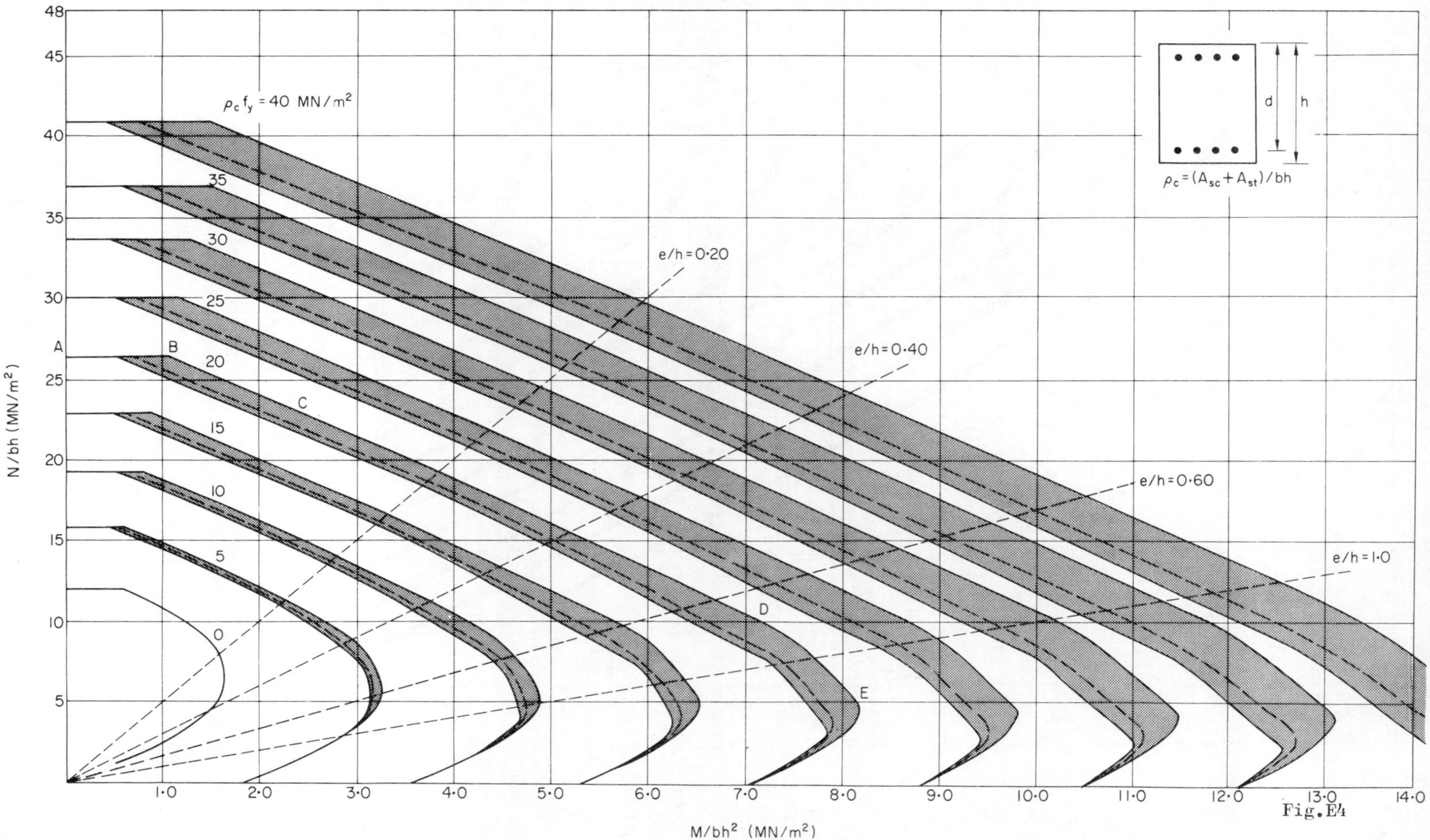

$\rho_c f_y = 40$ MN/m²
35
30
25
20
15
10
5
0
A
B
C
D
E
e/h = 0·20
e/h = 0·40
e/h = 0·60
e/h = 1·0
d
h
$\rho_c = (A_{sc} + A_{st})/bh$
N/bh (MN/m²)
M/bh² (MN/m²)
48
45
40
35
30
25
20
15
10
5
1·0
2·0
3·0
4·0
5·0
6·0
7·0
8·0
9·0
10·0
11·0
12·0
13·0
14·0

Fig. E4

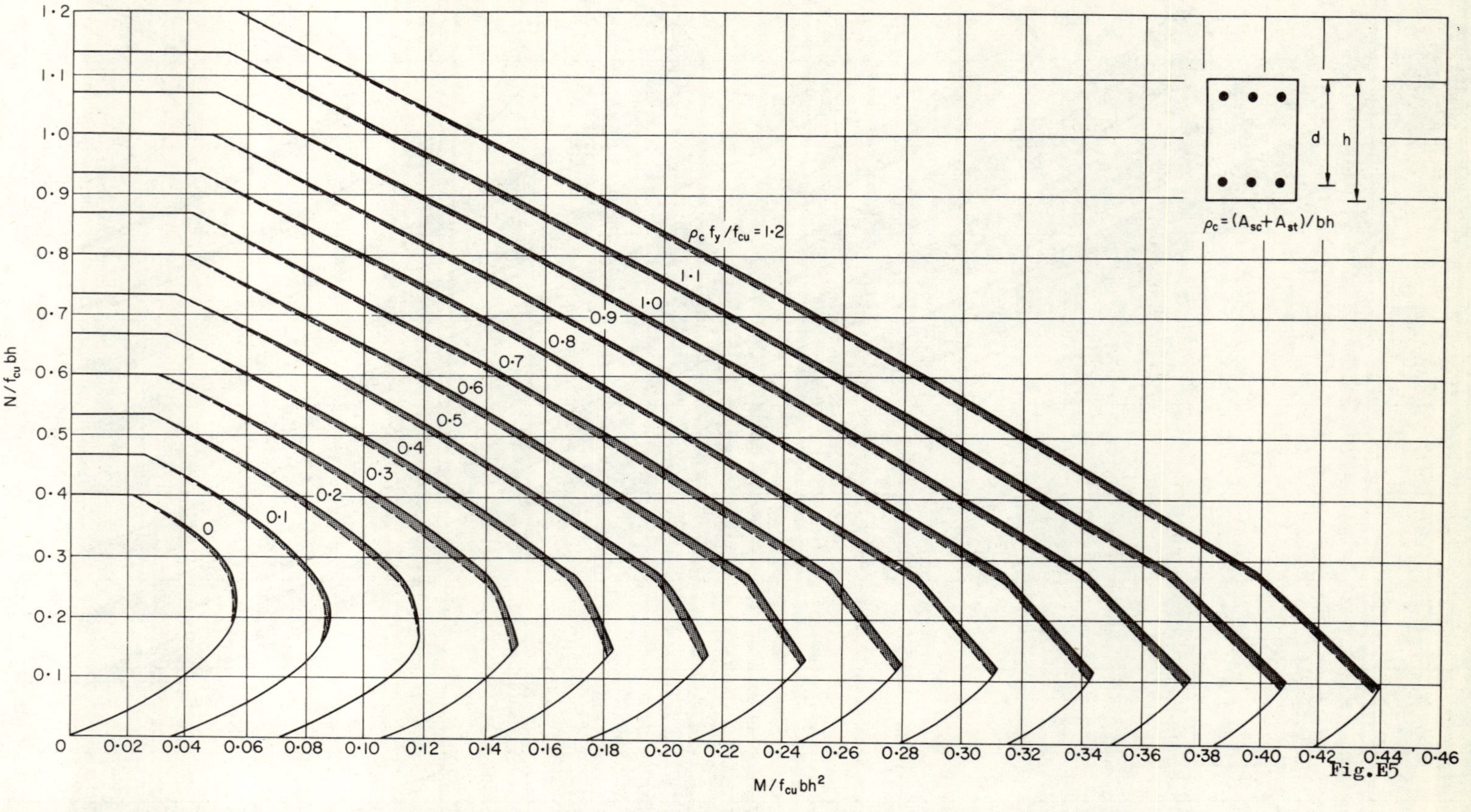

N/f_{cu}bh
M/f_{cu}bh^2
$\rho_c f_y/f_{cu}$ = 1·2
1·1
1·0
0·9
0·8
0·7
0·6
0·5
0·4
0·3
0·2
0·1
0
d
h
$\rho_c = (A_{sc} + A_{st})/bh$

Fig. E5

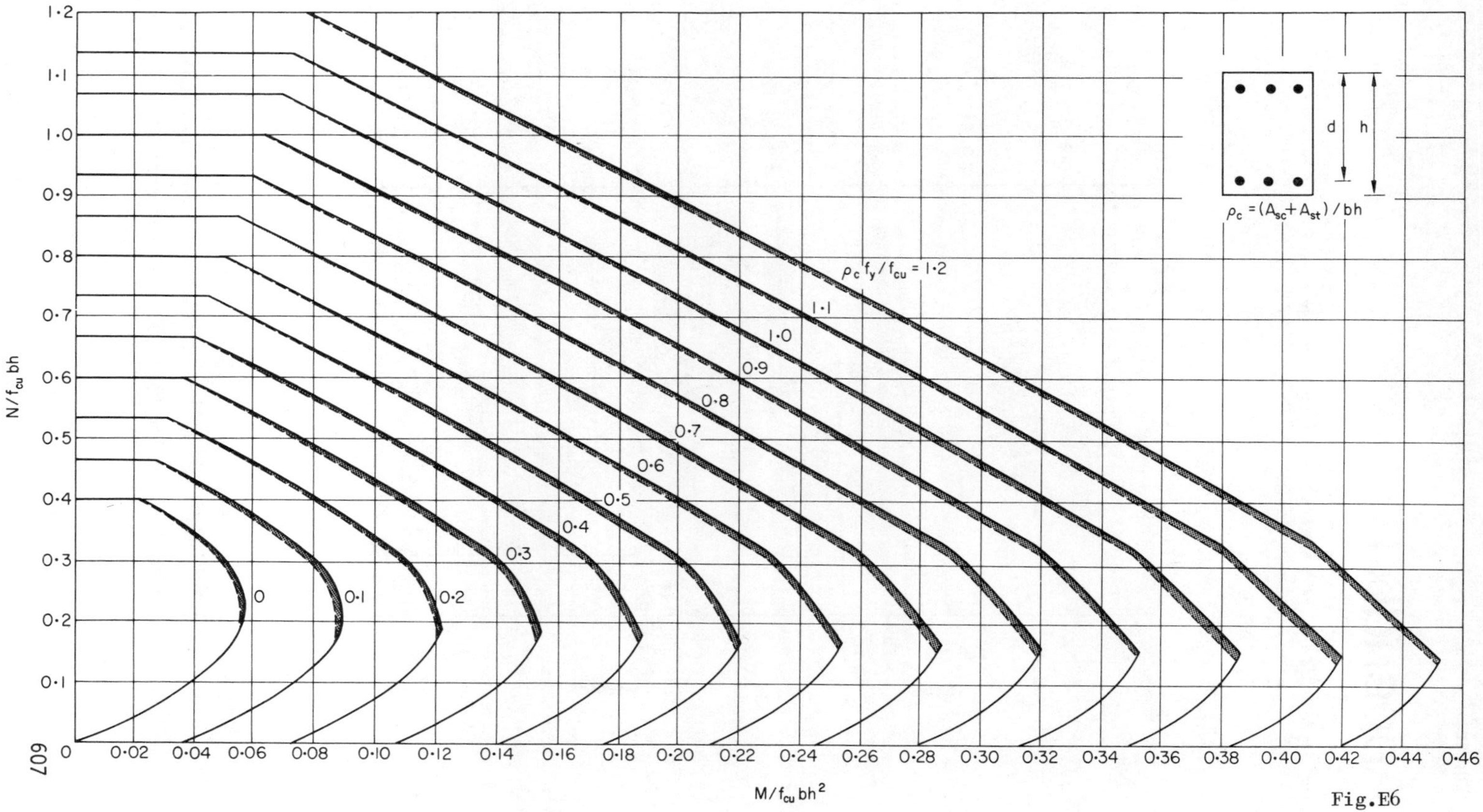

N/f_cu bh
M/f_cu bh²
ρ_c f_y / f_cu = 1·2
1·1
1·0
0·9
0·8
0·7
0·6
0·5
0·4
0·3
0·2
0·1
0
ρ_c = (A_sc + A_st) / bh
d
h

Fig.E6

Appendix F
Design charts for serviceability

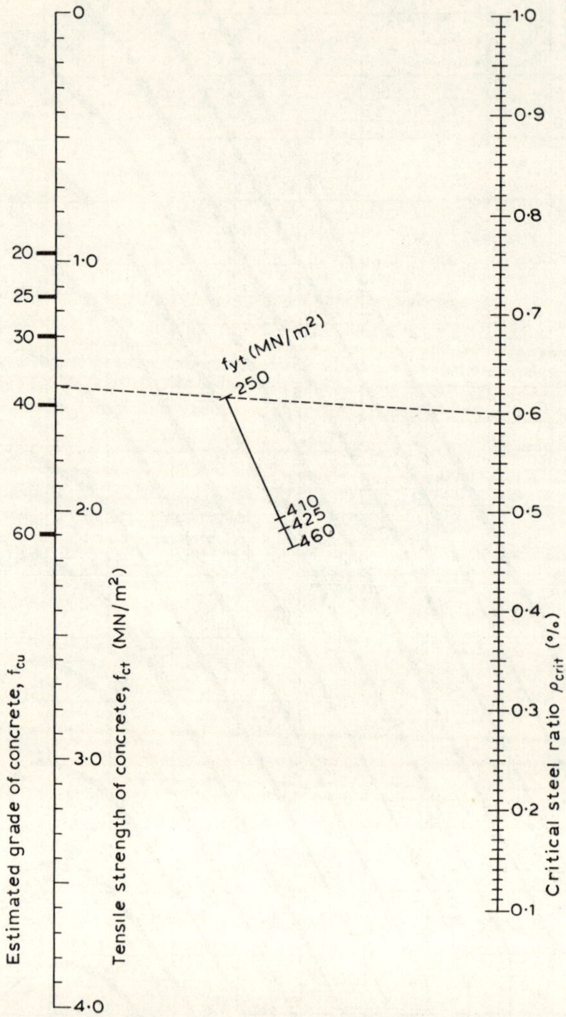

Fig.F1 Critical steel ratios for various grades of concrete in reinforced walls, etc.

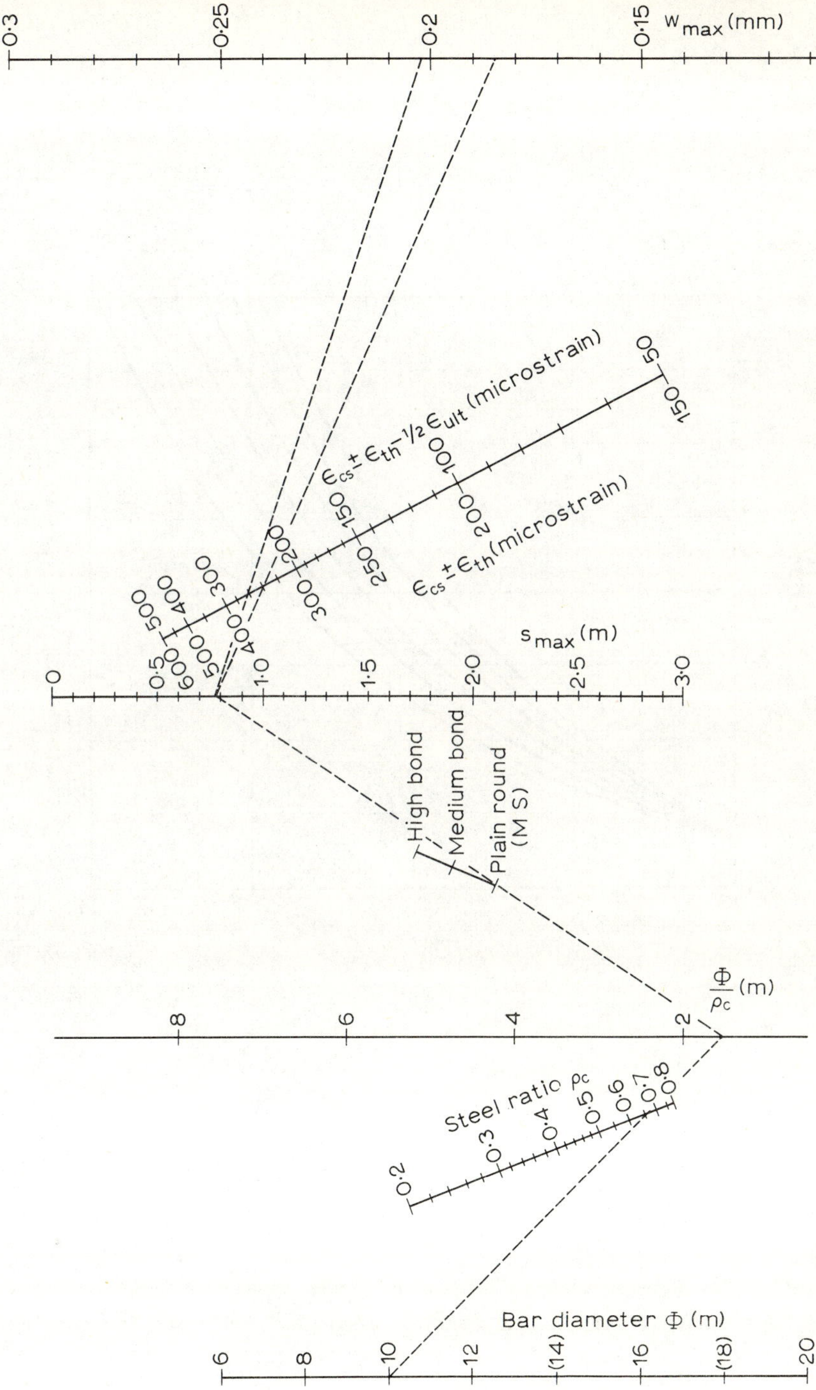

w_{max} (mm)
0·3
0·25
0·2
0·15
0·1
$\epsilon_{cs} + \epsilon_{th} - \frac{1}{2}\epsilon_{ult}$ (microstrain)
500
400
300
200
150
100
50
$\epsilon_{cs} + \epsilon_{th}$ (microstrain)
600
500
400
300
250
200
150
s_{max} (m)
0
0·5
1·0
1·5
2·0
2·5
3·0
High bond
Medium bond
Plain round (M S)
$\frac{\Phi}{\rho_c}$ (m)
8
6
4
2
Steel ratio ρ_c
0·2
0·3
0·4
0·5
0·6
0·7
0·8
Bar diameter Φ (m)
6
8
10
12
(14)
16
(18)
20

Fig.F2

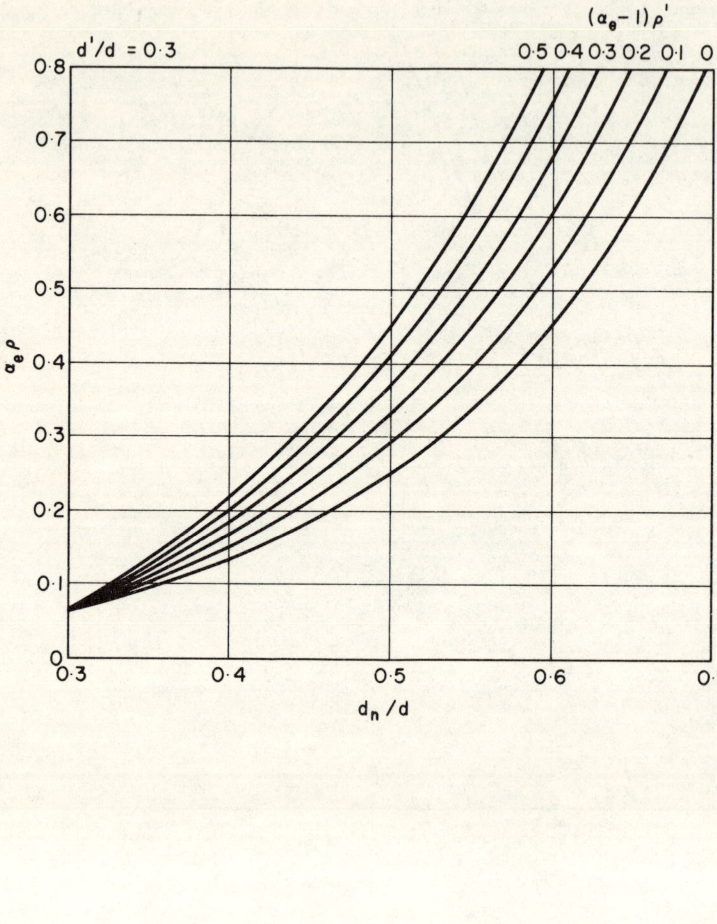

Fig. F3 Neutral axis depths (elastic theory) for rectangular sections

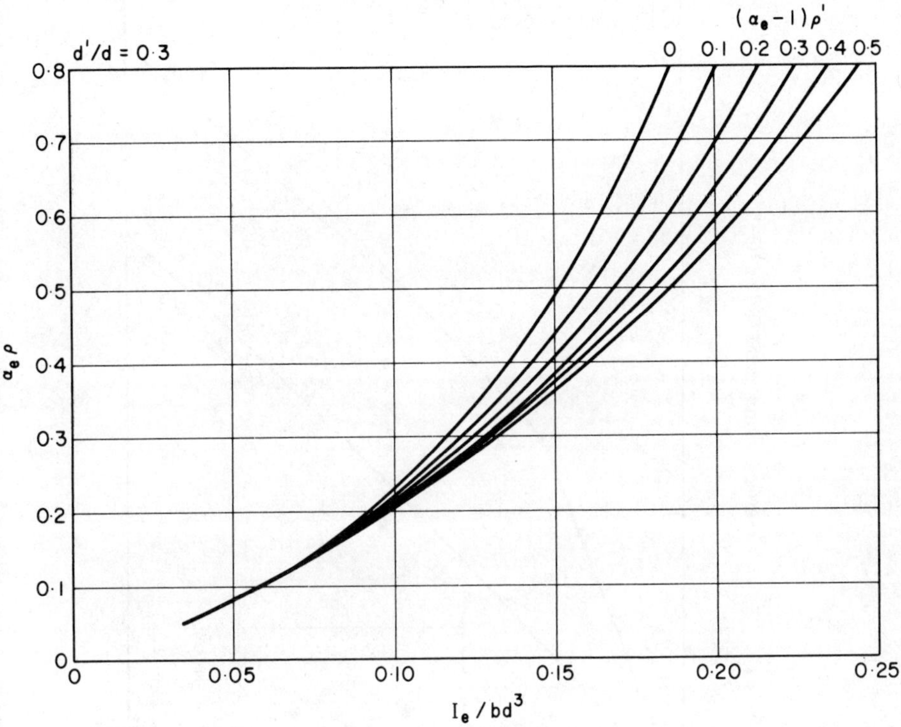

F4 Second moments of area of transformed (i.e. equivalent cracked) rectangular sections

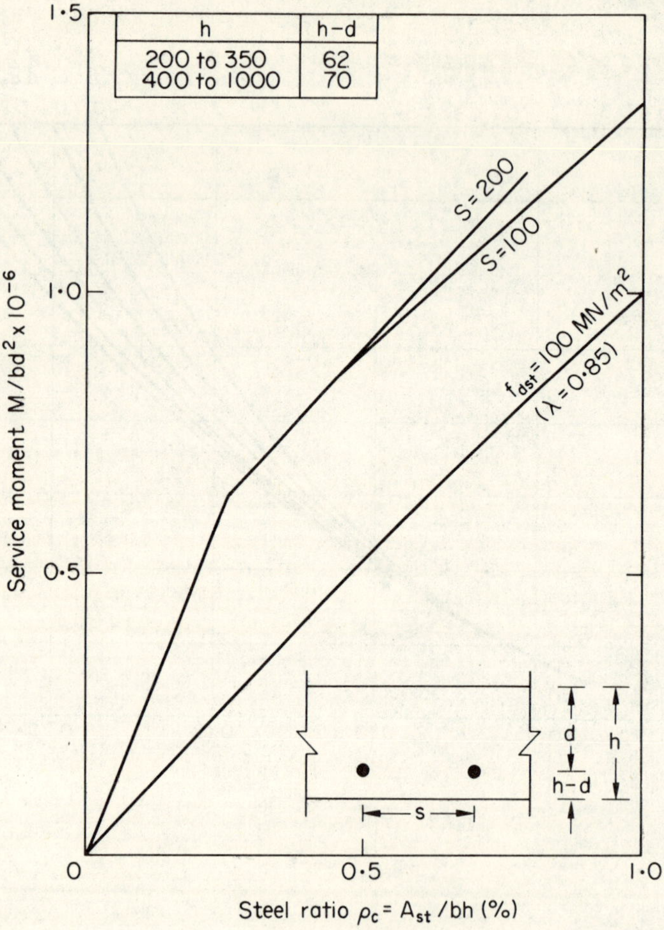

F5 Service moments for Class A exposure f_{cu} = 25 MN/m^2 f_y = 425 MN/m^2, 20 mm bars, 52 mm (h ≯ 350 mm) or 60 mm (h ≮ 400 mm) cover

Appendix G
Areas and weights of steel reinforcement

Table G1 Cross-sectional areas of groups of bars (mm^2)

No. of bars	Bar diameter (mm) 6	8	10	12	16	20	25	32	40
1	28.3	50.3	78.5	113	201	314	491	804	1260
2	56.6	101	157	226	402	628	982	1610	2510
3	84.9	151	236	339	603	943	1470	2410	3770
4	113	201	314	452	804	1260	1960	3220	5030
5	142	252	393	566	1010	1570	2450	4020	6280
6	170	302	471	679	1210	1890	2950	4830	7540
7	198	352	550	792	1410	2200	3440	5630	8800
8	226	402	628	905	1610	2510	3930	6430	10100
9	255	452	707	1020	1810	2830	4420	7240	11300
10	283	503	785	1130	2010	3140	4910	8040	12600

Table G2 Cross-sectional areas of bars per metre width for various spacings $(mm)^2$

Bar spacing (mm)	Bar diameter (mm) 6	8	10	12	16	20	25	32	40
50	566	1010	1570	2260	4020	6280	9820	16100	25100
75	377	670	1050	1510	2680	4190	6550	10700	16800
100	283	503	785	1130	2010	3140	4910	8040	12600
125	226	402	628	905	1610	2510	3930	6430	10100
150	189	335	523	754	1340	2090	3270	5360	8380
175	162	287	449	646	1150	1800	2810	4600	7180
200	142	252	393	566	1010	1570	2450	4020	6280
250	113	201	314	452	804	1260	1960	3220	5030
300	94.3	168	262	377	670	1050	1640	2680	4190

Table G3 Mass per unit length of round bars in kg/m (or weight per unit length in kgf/m)

Bar size (mm)	6	8	10	12	16	20	25	32	40
'Weight'	0.222	0.395	0.616	0.888	1.579	2.466	3.854	6.313	9.864

Appendix H
Guides for detailing reinforced concrete structures

(According to Technical Report TRCS2 (revised version, 1970). Standard method of detailing reinforced concrete, Joint Committee of the Concrete Society and the Institution of Structural Engineers; and BS 4466:1969 Bending dimensions and scheduling of bars for the reinforcement of concrete.)

General Principles

Drawings are prepared so that the designer can communicate his requirements to the contractor's workmen in a clear and unambiguous manner. Each drawing should be complete in itself and give all the information necessary for the construction of the portion of the work shown, omitting other irrelevant detail. Lettering should be clear and not too small so that it remains legible after the drawing has been used under adverse site conditions. Written instructions should be clear and concise, kept to a minimum and include special details of cambers, sequence of construction etc. Details should be simple and bars should have as few bends as possible. It is preferable to use the largest size bars consistent with the design requirements. The spacing of bars must be considered in relation to the placing of the concrete and the possible use of immersion vibrators. Allow for double the quantity of reinforcement at positions where bars are lapped.

Reinforcement sizes

Use the following preferred sizes of round area bars: 6, 8, 10, 12, 16, 20, 25, 32, 40 mm. Deformed bars of non-circular cross section are defined by the diameter of a circular bar of equivalent cross-sectional area. Reinforcing bars are available in lengths up to 12 m but 8 or 9 m is usually convenient and 6 m for bars less than 10 mm in diameter.

Appendix H

Identification of bars

The label attached to a particular group or set of bars should give both the bar mark number and the schedule reference number. This combination contains the essential information and ensures a unique reference.

Notation on drawings

Each bar or group of identical bars is defined by the code number or 'bar mark'. When referring to reinforcement on drawings the sequence of description should be as follows: Number of bars, type of steel, size, mark, spacing, and comment on location within the section. Thus the full description of a group of bars might be:

20R08—63—150T

which means 20 round mild steel bars, 8 mm diameter, bar mark 63, spaced at 150 mm centres in the top face of the concrete. This full description should appear on a drawing once only either in plan or elevation. Wherever else the group of bars appears on the drawing it should be defined by the bar mark only.

H.C. POSTLETHWAITE AND PARTNERS

Bar schedule reference 0401

Site reference Wigan Public Baths

Date: 1-8-68

Member	*Bar mark*	*Type & size*	*No of members*	*No in each*	*Total No*	*Length of each bar*	*Shape code*	*A*	*B*	*C*	*D*	*E/R*
Beam in	6	R10	6	33	198	1760	60	230	535			
Fig. H3	9	R10	6	10	60	1525	60	125	535			
			6									
	2	Y12	6	2	12	4040	20	4040				
			6									
	1	Y16	6	2	12	1830	37	305				
			6									
	10	R20	6	4	24	280	20	280				
	5	Y20	6	4	24	2490	20	2490				
	3	Y25	6	2	12	4060	20	4060				
	4	Y25	6	2	12	3120	20	3120				
	7	Y25	6	2	12	7700	20	7700				
	8	Y25	6	2	12	5080	20	5080				

Fig.H1 Typical bar schedule

Bar bending schedules

A bar bending schedule lists the location, mark, type, size, number off, length and bending details of each bar or sheet of welded fabric required for a structure. The bars should be grouped together for each structural element (e.g. beam, column, etc.) as shown by Fig. H1. Separate schedules are normally prepared for fabric reinforcement. The form of the bar schedule and the shapes of bar used should be in accordance with BS 4466 and the more commonly used shapes are given in Fig. H2. A reinforcement supplier often prepares cutting and bending lists from the schedules.

Bar schedule reference number

Schedules should be set out on A4 size paper and each sheet should be complete in itself. They should have simple consecutive reference numbers not exceeding six characters, and should be cross-referenced to the relevant drawing number. The first three characters can refer to the drawing number, the next two can refer to the schedule number and the last character can be used for revision letters. For example, 2908B refers to drawing 29, schedule 8, revision B. Thus each schedule has a different reference number and only refers to one drawing.

Drawings

BS 1192:1969 Drawing Office Practice for Architects and Builders lays down standards which are generally applicable to reinforced concrete detail drawings and are summarized (with any necessary variations) in the report. The report also recommends various abbreviations.

Sizes of drawings

The recommended normal sizes (with measurements in millimetres) are A4(210 × 297), A3(297 × 420), A2(420 × 594) and A1(594 × 841).

Types of drawing

The following types of drawings are recommended.

Shape code	Method of measurement of bending dimensions	Total length of bar	Sketch and dimensions to be given in schedule
20	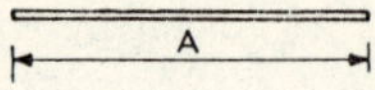	A	Straight
32	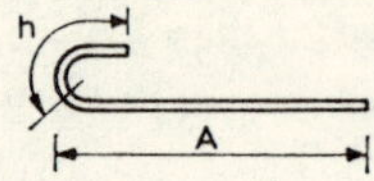	$A+h$	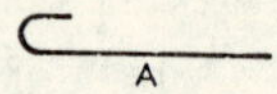
33	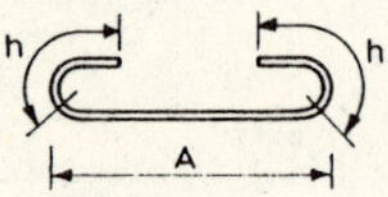	$A+2h$	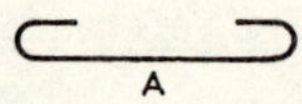
34	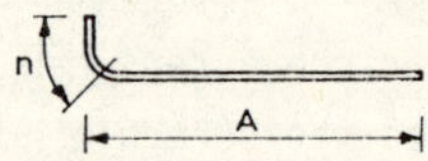	$A+n$	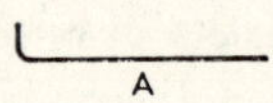
35	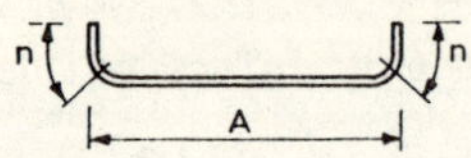	$A+2n$	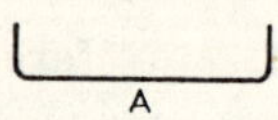
37	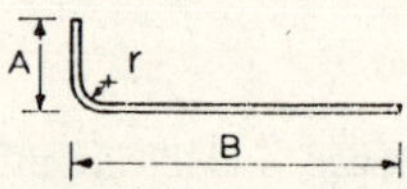	$A+B-\frac{1}{2}r-\Phi$	
38	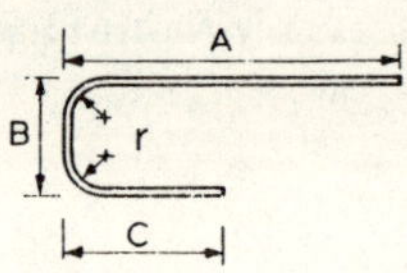	$A+B+C-r-2\Phi$	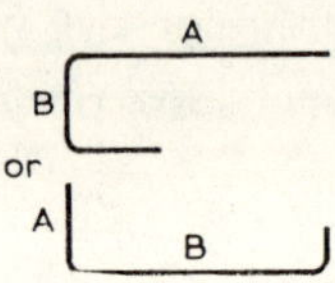
41	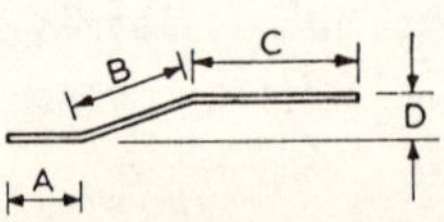	$A+B+C$	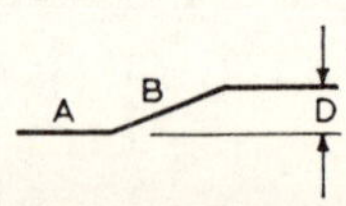

Note: all sketches on this page use outside dimensions (except *A* and *B* for shape code 41, which are along centre line).

Fig. H2 Some commonly used shape codes, from BS4466:1969.

Shape code	Method of measurement of bending dimensions	Total length of bar	Sketch and dimensions to be given in schedule
42*	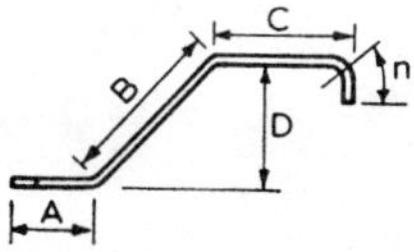	If angle with horizontal is 45° or less: $A+B+C+n$	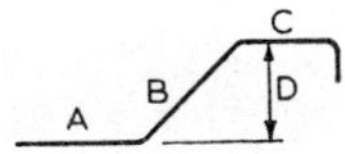
43	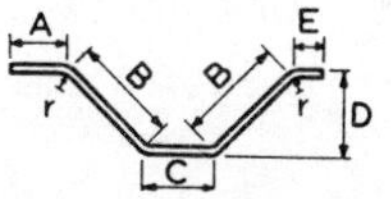	If angle with horizontal is 45° or less: $A+2B+C+E$	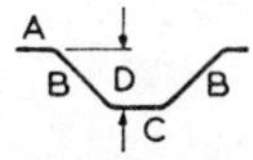
60	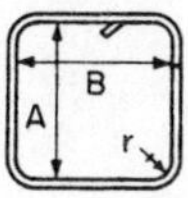	$2(A+B)+20\Phi$	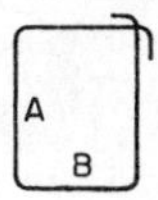
62	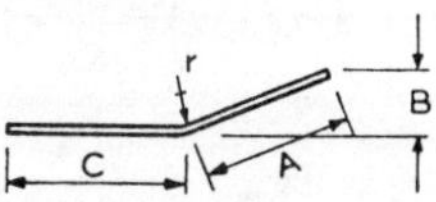	If angle with horizontal is 45° or less $A+C$	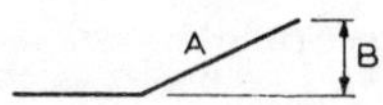
65*	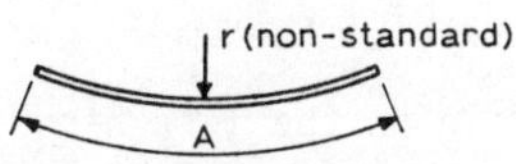	A	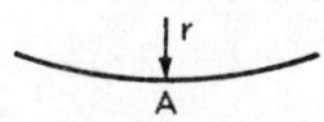
72*	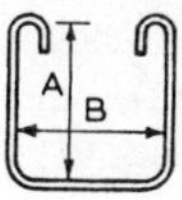	$2A+B+20\Phi$	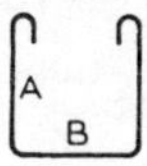
73*	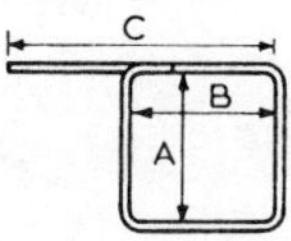	$2A+B+C+10\Phi$	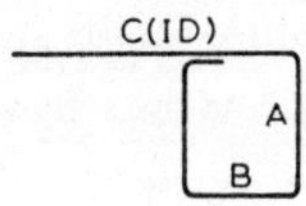
81	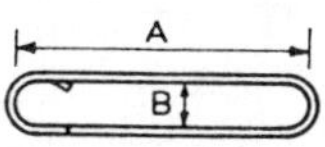	$2A+3B+22\Phi$	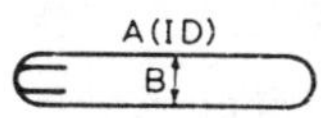
83	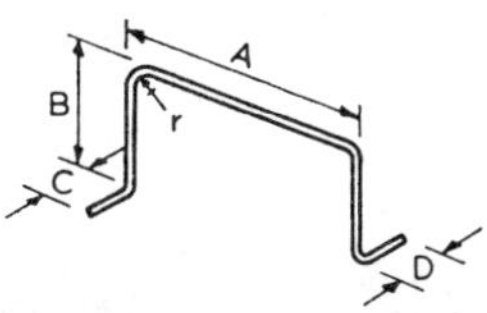	$A+2B+C+D-2r-4\Phi$	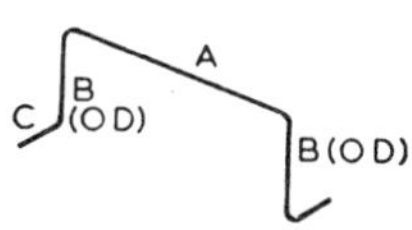

An asterisk indicates that the shape is not designated in the British Standard as a preferred shape. (I.D. = inside dimension, O.D. = outside dimension).

Fig. H2 Some commonly used shape codes, from BS 4466:1969. (continued) (See also Fig. 5·6·7)

(a) General arrangement drawings showing layout of all reinforced concrete work, setting out dimensions and sizes of all members to a small scale. Where beam sizes are given on plan the first figure should refer to the total depth, including the slab.

(b) Detail drawings

(i) Outline or profile drawings showing dimensions of elements of a structure (e.g. beams, columns, etc.) to a large scale, sizes and positions of all holes and fixings, etc., and intended to be read in conjunction with the general arrangement drawings.

(ii) Reinforcement drawings showing the position of all the steel in relation to the outline of the structure. A typical example is shown in Fig. H3.

These two types of drawing may be combined for a relatively simple structure.

Margins and panels

All drawings should have a 24-mm filing margin on the left hand side. A 10-mm margin should be provided on the other three sides for all sizes larger than A4. The 'title panel' should be 180 mm by 60 mm and should include all the following information: job title, drawing title, job number, drawing number and name of firm. The 'additional information panel' may be developed vertically or horizontally from the title panel and may include such additional information as scales, 'drawn by', 'checked by' date, notes and revisions.

Location and orientation

If a portion of the work has to be divided into several drawings it is convenient to have a small diagrammatic key on each drawing with the portion covered by that drawing clearly defined.

The direction of the North point should be clearly shown on all site plans. All other drawings relating to a particular building or major subdivision of a job should have consistent orientation with each other and also with the site plan orientation if possible.

Thickness of lines, lettering and spelling

Lines should generally be about 0.4 mm thick except for main reinforcement (0.6 mm) and dimension and centre lines (0.2 mm). Reinforcement is shown approximately to scale only on cross-sections of bars and in very large scale details. Lettering should be bold and clear, and capital letters should be used for titles, lower case for notes. The spelling of words should be in accordance with the Little Oxford Dictionary, using the first spelling where alternatives are given.

Standard abbreviations

Recommended abbreviations are as follows:

(a) General

RC	reinforced concrete	FFL	finished floor level
bwk	brickwork	SFL	structural floor level
drg	drawing	EL	existing level
FS	full size	hor	horizontal
NTS	not to scale	vert	vertical
		dia	diameter

(b) Type of reinforcement

R Round mild steel bars

Y High-yield, high bond-strength bars in the metric round area range

X A general abbreviation to cover any other types. X must be suitably described on drawings and schedules and in the specification.

(c) Comments relating to reinforcement

EW	each way	NF	near face
EF	each face	B	bottom
FF	far face	T	top

Dimensions

The general-arrangement drawing should show all setting-out dimensions and sizes of members. Reinforcement drawings should contain only those dimensions which are necessary for the correct location of the

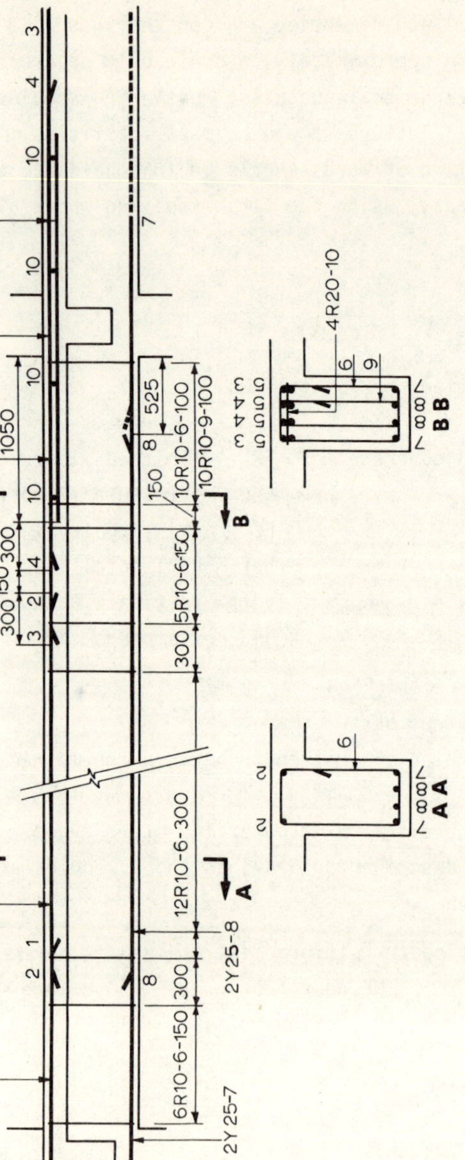

Fig.H3 Typical beam detail

reinforcement (see Fig. H3). Dimensions should be readable if viewed from either the bottom or the right-hand side of the drawing. For site layout and general-arrangement drawings the recommended unit is the metre, whilst for detail drawings the millimetre is usually more convenient. Values in millimetres should be expressed to an accuracy of 5 mm and it is not necessary to add 'mm' after the figure, e.g.

5
415

Values in metres should include 'm' after the figure or have 3 decimal places, e.g.

2 m or 2.000

Levels

A suitable fixed point should be selected as job datum such that all other levels are positive and are related to it as miltiples of 5 or 10 mm as appropriate. Finished floor levels, structural floor levels and existing levels should be indicated on plans thus:

FFL [12.335] SFL [12.000] EL [11.445]

Levels on section and elevation should be projected beyond the drawing with an arrowhead indicating the appropriate line thus:

▽SFL [12.000]

If the FFL is referred to, the thickness of the finish should be clearly shown. The job datum must be clearly defined.

Scales

Recommended scales are as follows:

Site layout and simple general arrangement	1:200
General arrangement	1:100 or 1:50
Simple wall and slab details	1:50
Beam and column elevations	1:20
Beam and column sections	1:20 or 1:10
Details of connections, hinges, etc	1:5, 1:2 or FS

Appendix H

Plans

Plans shall be drawn as a horizontal section through all structural members immediately above the portion of structure under consideration. Dimension lines should be well clear of the structural details.

Elevations

Elevations shall normally be drawn as a vertical section immediately adjacent to the element under consideration. Structural members cut by the section shall be shown in full lines and members behind the section in broken lines.

Sections

Sections shall show only the material cut by the section plane. Details behind the section are generally omitted. The reinforcement should be indicated by the bar mark only. The directions of sections shall be to the left for beams and downwards for columns. The section shall be drawn as close as possible to the plan or elevation to which it refers and should be described by letters, readable from the bottom of the drawing as shown in Fig. H3. Sections through concrete on drawings may be shown as lightly shaded areas.

Grid lines

Grid lines provide a reference system for beams and columns. It is convenient to commence from the top left corner of the drawing using letters for one direction and numbers for the other. Grid lines do not necessarily have to be straight or at right angles to each other, or actually pass through beams or columns (but should be sufficiently close for the reference to be obvious). Examples:

column D4 (column at or near intersection of line D and line 4)

beam 4a (beam along or near line 4 and at or adjacent to intersection with line A)

beam a4 (beam along or near line A and at or adjacent to intersection with line 4).

Notice that beams are denoted by lower-case letters although the lines themselves are denoted by upper-case letters. Additional grid lines can be added to provide subdivision of the main grid and a hyphened prefix can be used if location in elevation is also required. e.g. B-AA99 signifies level B on or near intersection of grid lines AA and 99.

Detailing Reinforcement

All bars shall be fully described once only as indicated under 'Notation on drawings' and elsewhere defined by their mark number only. Where the position of the bar is not clear the ends may be indicated by a tick and the mark number. For example, all main bars in a layer of a beam may be indicated by a single line in the elevation and ticks, suitably labelled with the mark of the bar concerned, should be used to show the ends of the individual bars (see Fig. H3). The position of bars should be established by dimensions to the faces of existing concrete or to the formwork. In beams all main bars are shown in full, but in columns only one bar of each type need be shown in full, the others being indicated by short lines. All bars that need to be fixed in a certain part before it can be concreted must be detailed with that part and then shown in broken lines in succeeding portions. The shape of a crank in main compressive steel should not normally exceed 1 in 10. Looped bars as in Fig. H4(a) should be avoided. Better arrangements are shown in Fig. H4(b) and H4(c).

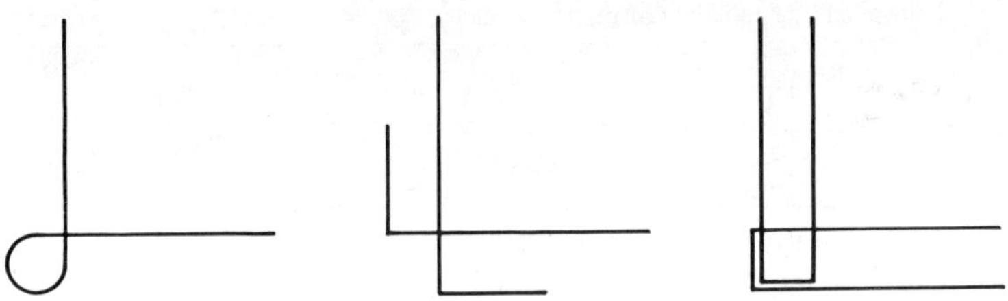

(a) (b) (c)

Fig.H4 Angle bars

Appendix H

Prefabrication

Detailing should be carried out with the possibility of prefabrication in mind. Reinforcement units should not be so heavy as to exceed the likely capacity of any lifting devices on the particular site and, in the case of cages, the detailing should prevent distortion during handling.

Cantilevers

Cantilevers are particularly susceptible to failures due to displacement of the steel during concreting. Main bars must therefore be held firmly and accurately at the support and, preferably, at the free end as well. Cantilevers are also very susceptible to thermal and shrinkage cracks: additional distribution reinforcement at the free edge, closer centres for contraction joints than in the body of the structure, and in large cantilevers the incorporation of distribution bars or fabric in the bottom of the slab, should be considered.

Tolerances on reinforcement

Cover to reinforcement is liable to variation on account of small errors in the dimensions of shuttering, and in cutting, bending and fixing the steel. All reinforcement shall be fixed to the nominal cover shown on the drawings. The term 'nominal cover' implies a permissible negative tolerance of 5 mm, i.e. the actual cover may be

Table H1 Deduction for tolerances in position of bent steel and at ends of straight bars

Overall concrete dimension (in direction of tolerance)	Deduction to determine bending dimension	Tolerance on member size
(a) Link and other bent bars:		
0–1000 mm	10 mm	$\ngtr$ 5 mm
1000–2000 mm	15 mm	$\ngtr$ 5 mm
over 2000 mm	20 mm	$\ngtr$ 10 mm
(b) At ends of straight bars:		
Any length of bar	40 mm	$\ngtr$ 10 mm

up to 5 mm less than the nominal cover. When a reinforcing bar is to fit between two concrete faces, (e.g. a rectangular stirrup in a beam) the dimensions on the bar schedule should be determined as the nominal dimension of the concrete, less the nominal cover on each face, and less an allowance for all other errors as in Table H1. If the tolerance on member size exceeds the values shown then either the deduction or the nominal cover should be increased.

Bar intersections

Special consideration is necessary at beam-to-column, beam-to-beam and slab-to-beam intersections to ensure that the different sets of bars avoid each other whilst still maintaining the correct cover. A system for achieving this end should be established for projects where several detailers may be working simultaneously on adjacent elements of the structure.

Slabs and walls

Slabs should generally be detailed in plan. Stairs and certain slabs may be more conveniently detailed in section. Walls should be detailed in elevation. In slabs and walls one bar of each type in each group is shown in full, the extent of the group being indicated by short lines. Beam reinforcement should not be shown on plans of slabs, but it should be indicated in the sections. Full lines should be used for all bars detailed on the drawing (e.g. for both top and bottom bars in slabs): only bars detailed on another drawing are shown by broken lines. Top and bottom bars can be detailed on separate plans if necessary but the symbols T and B alone usually provide adequate clarity. Vertical bars in walls are normally in the outside layer on each face (to assist the concrete placing) unless stresses require otherwise. In walls less than 120 mm thick only one layer of horizontal and one layer of vertical bars, in the centre of the wall, need be provided if stresses permit.

Holes up to 250 mm diameter or 150 mm square need not be indicated on the reinforcement drawings, but a suitable note should be added referring to the appropriate general arrangement drawing. It is often adequate if reinforcing bars are merely adjusted on site to

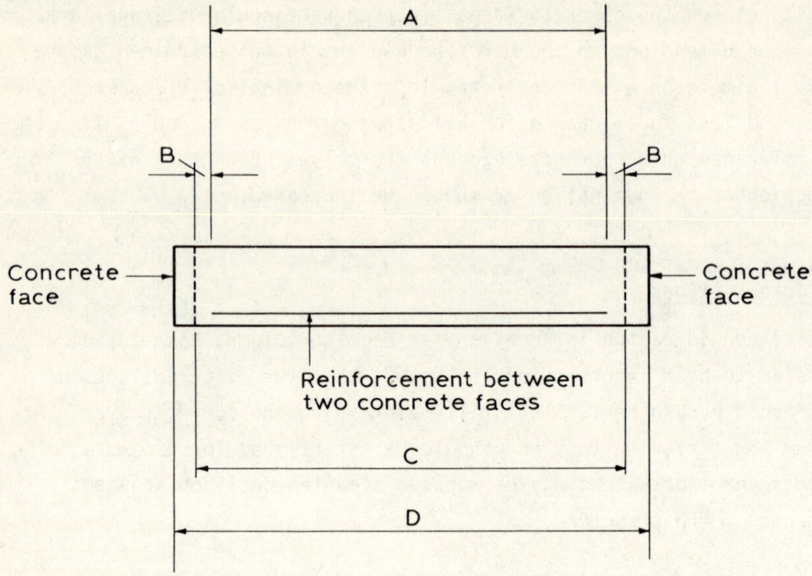
A
B
B
Concrete face
Concrete face
Reinforcement between two concrete faces
C
D

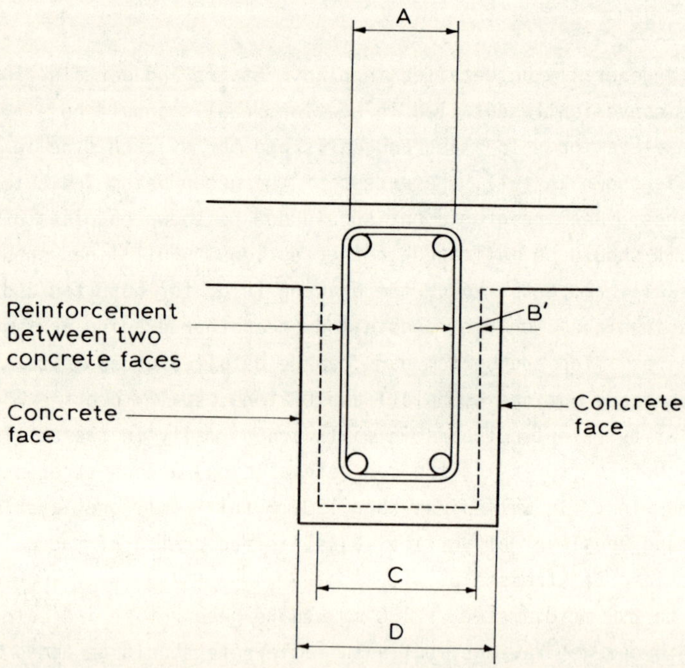
A
Reinforcement between two concrete faces
B'
Concrete face
Concrete face
C
D

Fig.H5

miss the holes. Larger holes should be indicated on the reinforcement drawing together with all trimming bars.

If a range of bars has to be detailed with progressively changing length (e.g. a wall tapering in height) it is often convenient to arrange the bars in groups and include a suitable lap so that the taper within a group can be accommodated by increasing the lap above minimum design requirement.

Links or stirrups

Only one stirrup or set of stirrups in each beam, and one link or set of links in each column, need to be shown in full: short lines are used to indicate the ends of each group at the same spacing. The semi-diameter of the main bars rarely equals the internal radius of the bend at the corner of the stirrup or link containing them (and therefore do not fit snugly in position). This must be allowed for when calculating tolerances so that the main bars can be fixed in their correct positions. Compression bars in beams (as distinct from nominal 'hanger' bars of, say, 12 mm diameter) may require an additional type of stirrup to provide the necessary lateral restraint to these bars. All bars in columns must be laterally restrained by links but the centre of the column should be left as free as possible from crossing steel. Transverse ties and nominal 'face' bars at intermediate heights are advisable for stiffening the stirrups in very deep beams.

Chairs and spacers

Chairs to shape code 83 (see Fig. H2), or welded continuous chairs, may be used to support steel in horizontal members and must be included in the bar schedules. Chairs must support the main steel and the weight of workmen during construction, e.g. for slabs the diameter must not be less than 10 mm and the spacing not more than 1.25 m in either direction at right angles. U-type spacers should be used to separate two layers of vertical steel in walls having at least one horizontal line of spacers of not less than 6 mm diameter for each lift of concrete. If straight lateral spacer bars are shown between layers of main bars in beams, they should be scheduled.

Appendix H

Welded fabric detailing

Steel fabric in flat sheets is sometimes used in slabs and walls and is detailed by drawing the outline of each sheet as a rectangle, with a diagonal line to indicate its extent. The direction of the diagonal indicates whether the sheet is in the top or bottom of the slab. The reference mark of the sheet is written along the diagonal. The amount of side or end lap should be shown on the drawing. Dimensions to show the positions of the sheets should be given on sections where possible, and the fabric in sections is indicated by heavy dashed lines —for oblong mesh, long or short dashes according to whether the section is parallel or at right angles to the main wires, respectively.

Standard reinforced concrete details

Guidance concerning five frequently occurring details:

(a) Shear reinforcement
(b) Column splices
(c) Beam-column intersections
(d) Wall details
(e) Holes in slabs

is given in the recommendations of the working party report: Standard Reinforced Concrete Details, Concrete Society, 1973, 28 pp.

Appendix I
Bending moment coefficients for continuous beams

M = Coefficient × Total load × Span (All spans loaded)

UNIFORMLY DISTRIBUTED LOAD

133(100)*88*
125

Δ 70 Δ 70 Δ
75(80)*85* **75**(80)*85*

90(80)*70* **90**(80)*70*
100 100

Δ 80 Δ 25 Δ 80 Δ
84(88)*92* **35**(45)*55* **84**(88)*92*

96(86)*75* **64**(57)*50* **96**(86)*75*
107 71 107

Δ 77 Δ 36 Δ 36 Δ 77 Δ
81(86)*90* **45**(54)*62* **45**(54)*62* **81**(86)*90*
95(84)*74* **71**(63)*55* **71**(63)*55* **95**(84)*74*
105 79 79 105

Δ 78 Δ 33 Δ 46 Δ 33 Δ 78 Δ
82(86)*91* **42**(51)*61* **54**(62)*70* **42**(51)*61* **82**(86)*91*

LOAD AT CENTRE POINT

169(150)*131*
188

Δ 156 Δ 156 Δ
166(175)*185* **166**(175)*185*

135(120)*105* **135**(120)*105*
150 150

Δ 175 Δ 100 Δ 175 Δ
182(190)*197* **115**(130)*145* **182**(190)*197*

	Support	Span	Support	Span	Support	Span	Support	Span	Support
Support moment	Δ		**145**(129)*112*		**96**(86)*75*		**145**(129)*112*		Δ
Support moment (elastic)			161		107		161		
Span moment (elastic)		170		116		116		170	
Span moment		**178**(186)*194*		**129**(143)*156*		**129**(143)*156*		**178**(186)*194*	

	Support	Span	Support	Span	Support	Span	Support	Span	Support	Span	Support
Support moment	Δ		**142**(126)*111*		**107**(95)*83*		**107**(95)*83*		**142**(126)*111*		Δ
Support moment (elastic)			158		118		118		158		
Span moment (elastic)		171		112		132		112		171	
Span moment		**179**(187)*195*		**126**(139)*153*		**143**(155)*167*		**126**(139)*153*		**179**(187)*195*	

LOAD AT THIRD POINTS

	Support	Span	Support	Span	Support
Support moment	Δ		**150**(133)*117*		Δ
Support moment (elastic)			167		
Span moment (elastic)		111		111	
Span moment		**177**(122)*128*		**177**(122)*128*	

	Support	Span	Support	Span	Support	Span	Support
Support moment	Δ		**120**(107)*93*		**120**(107)*93*		Δ
Support moment (elastic)			133		133		
Span moment (elastic)		120		33		120	
Span moment		**125**(129)*134*		**47**(60)*73*		**125**(129)*134*	

	Support	Span	Support	Span	Support	Span	Support	Span	Support
Support moment	Δ		**129**(114)*100*		**86**(76)*67*		**129**(114)*100*		Δ
Support moment (elastic)			143		95		143		
Span moment (elastic)		117		55		55		117	
Span moment		**122**(127)*132*		**66**(77)*88*		**66**(77)*88*		**122**(127)*132*	

	Support	Span	Support	Span	Support	Span	Support	Span	Support	Span	Support
Support moment	Δ		**126**(112)*98*		**95**(84)*74*		**95**(84)*74*		**126**(112)*98*		Δ
Support moment (elastic)			140		105		105		140		
Span moment (elastic)		117		49		61		49		117	
Span moment		**122**(127)*132*		**61**(72)*84*		**72**(82)*93*		**61**(72)*84*		**122**(127)*132*	

Values in heavy type, in brackets and in italics = 1000 × coefficients for moment redistributions of 10%, 20% and 30% respectively.

Table I1 Bending moment coefficients for dead loads

M = Coefficient × Total load × Span

UNIFORMLY DISTRIBUTED LOAD

133(100)*88*
125

96 Δ 96 Δ
86(80)*85* **86**(80)*85*

105(93)*82* **105**(93)*82*
117 117

101 Δ 75 Δ 101 Δ
91(83)*87* **68**(63)*69* **91**(83)*87*

104(93)*81* **96**(86)*75* **104**(93)*81*
116 107 116

100 Δ 80 Δ 80 Δ 100 Δ
90(83)*88* **72**(66)*70* **72**(66)*70* **90**(83)*88*

104(93)*81* **96**(85)*75* **96**(85)*75* **104**(93)*81*
116 106 106 116

100 Δ 79 Δ 86 Δ 79 Δ 100 Δ
90(83)*88* **71**(66)*70* **77**(68)*75* **71**(66)*70* **90**(83)*88*

LOAD AT CENTRE POINT

169(150)*131*
188

203 Δ 203 Δ
183(175)*185* **183**(175)*185*

157(140)*122* **157**(140)*122*
175 175

212 Δ 175 Δ 212 Δ
191(180)*189* **157**(155)*164* **191**(180)*189*

157(139)*122* **145**(129)*112* **157**(139)*122*
174 161 174

Δ 210 Δ 183 Δ 183 Δ 210 Δ
189(180)*189* **165**(159)*167* **165**(159)*167* **189**(180)*189*

157(139)*122* **144**(128)*112* **144**(128)*112* **157**(139)*122*
174 160 160 174

Δ 211 Δ 181 Δ 191 Δ 181 Δ 211 Δ
189(180)*189* **163**(159)*167* **172**(162)*173* **163**(159)*167* **189**(180)*189*

150(133)*117*
167

Δ 139 Δ 139 Δ
125(122)*128* **125**(122)*128*

LOAD AT THIRD POINTS

140(124)*109* **140**(124)*109*
155 155

Δ 143 Δ 100 Δ 143 Δ
129(123)*128* **90**(94)*100* **129**(123)*128*

139(124)*108* **129**(114)*100* **139**(124)*108*
155 143 155

Δ 142 Δ 111 Δ 111 Δ 142 Δ
127(123)*129* **100**(96)*101* **100**(96)*101* **127**(123)*129*

139(124)*108* **128**(114)*99* **128**(114)*99* **139**(124)*108*
155 142 142 155

Δ 142 Δ 108 Δ 114 Δ 108 Δ 142 Δ
128(123)*129* **97**(96)*101* **103**(99)*108* **97**(96)*101* **128**(123)*129*

Values in heavy type, in brackets and in italics = 1000 × coefficients for moment redistribution of 10%, 20% and 30% respectively.

Table I2 Bending moment coefficients for variable loads

Table I3 Bending moment coefficients for fixed end moments

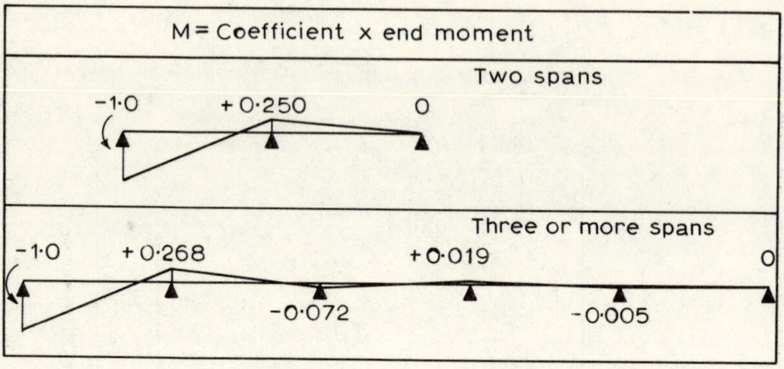

Table I4 Simplified CP 110 moments and shear forces in continuous beams if $Q_k \ngtr G_k$

Position	Moment	Shear
At outer support	0	0.45F
Near middle of end span	$F\ell/11$	—
At first interior support	$-F\ell/9$	0.6F
At middle of interior spans	$F\ell/14$	—
At internal supports	$-F\ell/10$	0.55F

Substantially uniformly distributed loads over 3 or more spans which differ by ≯ 15% of the longest span.

No redistribution of moments

F = total design load at collapse, Case I ($1.4G_k + 1.6Q_k$)

Note: see also Table 6.3.2 for simplified shear forces.

Appendix J
Curtailment of tensile bars

Sections at which bottom bars are no longer required, for a uniformly distributed load, are indicated in Fig. J1 and Table J1 in terms of the maximum distance from centre of support to point of bending up. If bar is not bent-up, sufficient bond distance must be provided from point of maximum stress in bar to point of curtailment. That is, suitable end anchorage must be provided beyond point where bar is no longer required (theoretically), from bending moment diagrams, as indicated in Fig. J1.

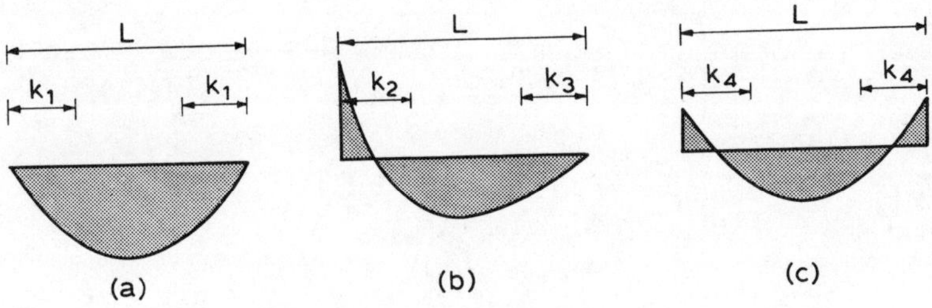

Fig.J1 Theoretical positions for 'bending up' bars. (a) Simple supports. (b) End span. (c) Interior span

Table J1 Order of 'bending up' bars

No. of bars at midspan	k_1 1st	2nd	3rd	4th	5th	6th	k_4 1st	2nd	3rd	4th	5th	6th
1	0						.09					
2	.15	0					.21	.09				
3	.21	.09	0				.27	.16	.09			
4	.25	.15	.07	0			.30	.21	.15	.09		
5	.27	.19	.12	.05	0		.31	.24	.18	.13	.09	
6	.30	.21	.15	.09	.04	0	.33	.27	.21	.16	.12	.09
7	.31	.23	.17	.12	.08	.04	.34	.29	.23	.19	.15	.12
8	.32	.25	.19	.15	.10	.07	.35	.30	.25	.21	.18	.15

Table J1 Order of 'bending up' bars (continued)

No. of bars at midspan	k_2 1st	2nd	3rd	4th	5th	6th	k_3 1st	2nd	3rd	4th	5th	6th
1	.11						0					
2	.24	.11					.13	0				
3	.30	.19	.11				.18	.08	0			
4	.33	.24	.17	.11			.22	.13	.05	0		
5	.35	.28	.21	.16	.11		.24	.15	.09	.04	0	
6	.36	.30	.24	.19	.14	.11	.25	.18	.13	.08	.03	0
7	.39	.32	.26	.22	.18	.14	.29	.20	.15	.11	.07	.03
8	.40	.33	.27	.24	.20	.17	.30	.22	.17	.13	.09	.05

Appendix K
Central deflections of beams

Table K1 Maximum bending moments and K-values for central deflections of beams

Loading	BMD	Maximum moment	Value of K
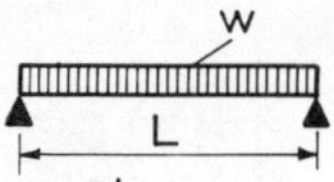		$\frac{wL^2}{8}$	$\frac{5}{48}$
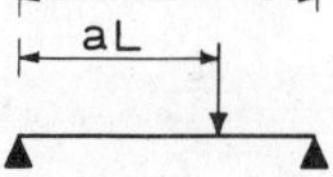	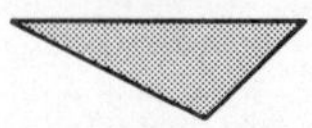	$Wa(1-a)L$	$a \nless \frac{1}{2}$; $\frac{4a^2-3a+1}{-48a}$
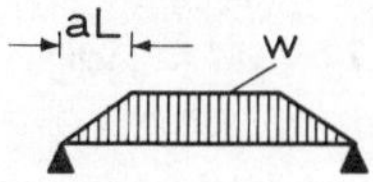		$\frac{(3-4a^2)wL^2}{24}$	$\frac{(5-4a^2)^2}{80(3-4a^2)}$
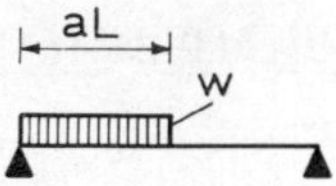		$\frac{a^2(2-a)^2wL^2}{8}$	$a \ngtr \frac{1}{2}$; $\frac{3-2a}{12(2-a)^2}$ $a > \frac{1}{2}$; $\frac{4a^4-10a^3+9a^2-2a+\frac{1}{4}}{12a^2(a-2)^2}$
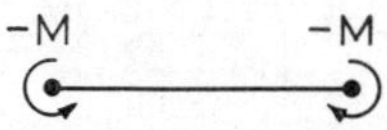		$-M$	$-\frac{1}{8}$
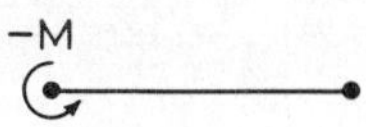	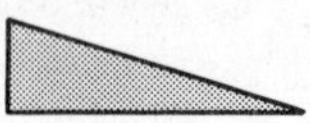	$-M$	$-\frac{1}{16}$
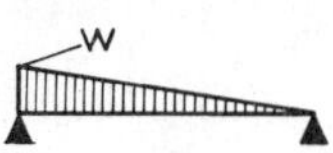	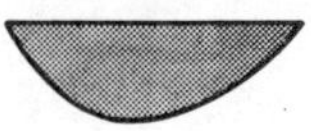	$\frac{wL^2}{7.81}$	$\frac{5}{48}$
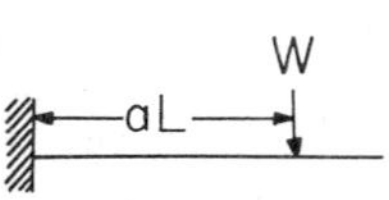		WaL	$\frac{a(3-a)}{6}^*$
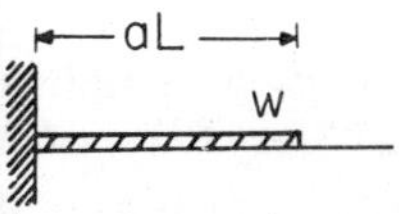	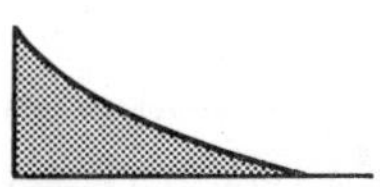	$\frac{wa^2L^2}{2}$	$\frac{a(4-a)}{12}^*$

Note: Midspan (or end*) deflection = $KL^2r_b^{-1}$ where r_b^{-1} is the maximum curvature.

*End deflection for cantilever.

Appendix L
Fixed end moments

Table L1 End moments for beams fixed at one or both ends

	One end fixed	Both ends fixed	
Type of loading	$-M_A$	$-M_A$	$-M_B$
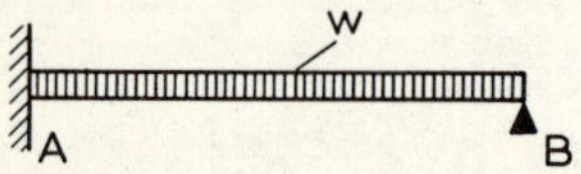	$\frac{wL^2}{8}$	$\frac{wL^2}{12}$	$\frac{wL^2}{12}$
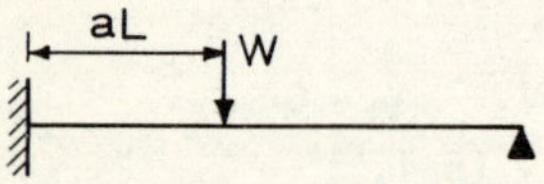	$a(1-a)(1-\frac{1}{2}a)WL$	$a(1-a)^2WL$	$a^2(1-a)WL$
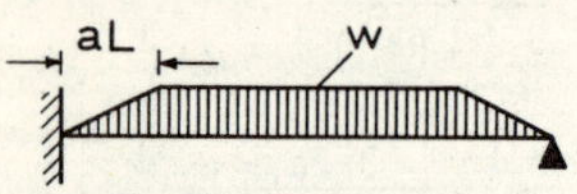	$\frac{(1-2a^2+a^3)wL^2}{8}$	$\frac{(1-2a^2+a^3)wL^2}{12}$	$\frac{(1-2a^2+a^3)wL^2}{12}$
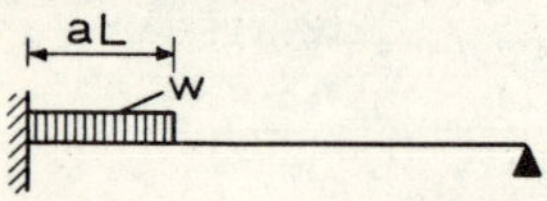	$\frac{a^2(4-4a+a^2)wL^2}{8}$	$\frac{a^2(6-8a+3a^2)wL^2}{12}$	$\frac{a^3(4-3a)wL^2}{12}$
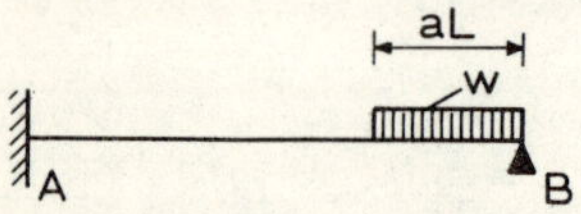	$\frac{a^2(2-a^2)wL^2}{8}$	$\frac{a^3(4-3a)wL^2}{12}$	$\frac{a^2(6-8a+3a^2)wL^2}{12}$
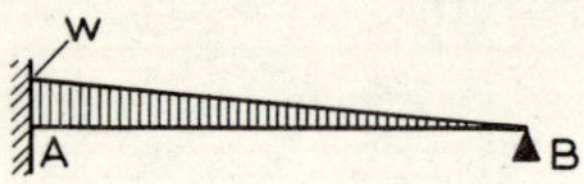	$\frac{wL^2}{15}$	$\frac{wL^2}{20}$	$\frac{wL^2}{30}$
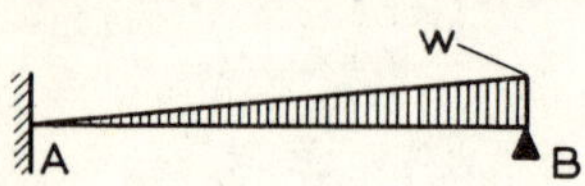	$\frac{7wL^2}{120}$	$\frac{wL^2}{30}$	$\frac{wL^2}{20}$

Note: with one end fixed stiffness should be taken as $\frac{3}{4}I/L$.

Appendix M
Limit states of fire resistance

It is a statutory requirement that all buildings are designed according to the requirements of the Building Regulations, and that a specified period of fire resistance is provided. The object is to save lives by preventing the spread of fire and to ensure that the structure does not collapse before it has been safely evacuated. The period of fire resistance to be specified is defined in the Regulations and varies from zero to 4 hours according to the use of the building and its location. The Regulations do not consider the prevention of damage to the building or its contents although the designer may wish to consider these matters, particularly in relation to the cost of fire insurance. A full discussion of these points and the general background to the subject is contained in references [1.2] and [1.3].

Fire tests

Basic information about the resistance to fire of various forms of concrete construction is obtained by testing specimens in a furnace under the appropriate design working load. The furnace temperature is raised according to a standard time/temperature curve and readings of deflection are taken. To pass the test, the structural element must sustain the working load without collapse or disintegration and must not deflect more than a specified amount.

It should be noted that the furnaces available only test structural elements and that the effects of continuity cannot at present be predicted on the basis of test information.

Effects of fire on a structure

Concrete as a material is resistant to fire but tends to crack, and in a reinforced section the cover to the reinforcement will fall away or spall after prolonged exposure. Steel loses strength fairly rapidly as heat is applied and at a temperature of 550°C the yield strength of normal high yield reinforcing steel is about 50% of the strength at normal temperature.

The Building Regulations require that the steel reinforcement is provided with sufficient thickness of concrete cover and, for longer periods of fire resistance, that supplementary reinforcement is provided in the cover itself to hold it in place and prevent spalling.

A fire normally starts in one limited area and the effects are initially local. As the fire grows, considerable thermal stresses are set up as the floors try to expand, and it is possible for stiff columns to fail in shear due to these horizontal forces. The object of the designer should be to ensure that the structure survives any fire which breaks out, contains the fire locally, and needs a minimum of repair after the fire has been extinguished.

Design procedures

The statutory requirements in the Building Regulations and Section 10 of CP 110 are not set out in limit state terms, but as rules which are expressed in terms of minimum member sizes and minimum thickness of concrete cover. It should also be noted that in respect of fire resistance, CP 110 is not deemed to satisfy Section E of the Building Regulations. A design can be prepared in accordance with Section A3 of the regulations in limit state terms if the designer can put forward appropriate information for consideration.

Limit states in fire

There are two possible forms of failure of a structure due to exposure to fire:

(1) Extreme deformation or collapse, due to reduction in strength.
(2) Failure to prevent the spread of fire due to lack of integrity or insufficient thermal insulation.

From a consideration of these two forms of failure of the structure it is possible to define two limit states:

(1) The ultimate limit state in fire.
(2) The limit state of fire spread.

The second limit state only applies to elements that form a part of compartment walls and floors.

Ultimate limit state in fire

When the local bearing capacity of the structure has been reduced by fire action to the equivalent of the design characteristic load the limit state has been reached.

A consideration of the strength of a structure under the action of fire involves the estimation of heat transfer through the various structural elements and an estimate of the temperature rise in the crucial elements of reinforcement. This is possible for simple flexural and compression members and an example of the calculation involved is contained in reference [1.2].

In a normal structure, the various slab, beam, and column elements are not arranged in a statically determinate manner but are redundant. If the tensile reinforcement at midspan in the bottom of a beam loses strength due to heating, the beam does not necessarily collapse. The reinforcement resisting the support moments is in the top of the beam sections, remote from the fire affecting the positive moment reinforcement and is at an appreciably lower temperature. The strength of the beam will then depend on the quantity of support reinforcement and the degree of continuity with the adjacent spans.

The limit state of fire spread

Buildings of any size are separated into fire compartments by walls and floors which are designed to have a specified resistance to the passage of fire. There are two criteria:

(a) The temperature rise on the side of the element which is remote from the fire must be limited so that fire is not spread by ignition from the warm surface.

(b) The element must retain its integrity so that cracks do not form and allow hot gases and flames to pass through.

Heat transfer through concrete may be estimated by reference to thickness and type of concrete, together with known thermal properties and the amount of moisture present. The thermal properties will vary according to the actual temperature of the element.

The retention of integrity is largely a matter of good detailing particularly of joints and holes for services and ducts. Particular attention must be given to fire-stopping all clearances around pipes and cables and ducts.

PROCEDURES TO JOINT COMMITTEE REPORT

The report of the Joint Committee on the Fire Resistance of Concrete Structures includes graphs which enable the fire resistance of a proposed section to be estimated by calculation. It should be emphasized that the information given in these graphs is incomplete and based on tests not conducted in this country. However, the following example indicates the general method of approach.

Example M1

Design a beam to carry a uniformly distributed load over a simply supported effective span of 8 metres, and have fire resistance of 4 hours.

Uniformly distributed imposed load = 15 kN/m
Uniformly distributed dead load = 10 kN/m

<u>Solution</u>

Assume concrete cover = 40 mm
From Fig. 58 of Joint Report [1.3], required loading factor = 0.5
Applied moment = $1/8 \times 25 \times 8^2 = 200$ kN/m
Required ultimate moment $M_u = 200/0.5 = 400$ kN/m
For normal design to CP 110, the required ultimate moment is

$M_u = 1/8\ (1.4 \times 10 + 1.6 \times 15) \times 8^2 = 304$ kN/m

Design to value $M_u = 400$ kN m to provide necessary fire resistance.
Assuming a section 600 × 400 and

$f_{cu} = 25$ MN/m², $f_{yt} = 425$ MN/m²

Then $A_{st} = 2526$ mm². Provide 2-Y25 and 2-Y32 (2592 mm²)
(CP 110 design required $A_{st} = 1903$ mm², 4-Y25 (1960 mm²)
and cover of 65 mm with supplementary reinforcement.)
If the temperature of the reinforcement is raised during the fire so that the yield strength is reduced to 185 MN/m², the ultimate moment of resistance of the beam will then equal the applied characteristic

moment and the beam will be on the point of failure.

To determine the temperature at which the load factor becomes unity then

$$\frac{\text{strength } T^{o}C}{\text{strength } 20^{o}C} = \frac{185}{425} = 0.43$$

and hence from Fig. 9 of the Joint Report [1.3] the critical temperature = 600°C.

As a check it is possible to refer to Fig. 57 of the Joint Report [1.3] where at an exposure of 4 hours, and 40 mm cover, the reinforcement temperature is approximately 650°C.

Appendix N
Statistical data for normal distributions

Table N1 Area under normal probability curve from z = 0 to z = z ($\times 10^3$)

z	0	1	2	3	4	5	6	7	8	9
0.0	000.0	004.0	008.0	012.0	015.9	019.9	023.9	027.9	031.9	035.9
0.1	039.8	043.8	047.8	051.7	055.7	059.6	063.6	067.5	071.4	075.3
0.2	079.3	083.2	087.1	091.0	094.8	098.7	102.6	106.4	110.3	114.1
0.3	117.9	121.7	125.5	129.3	133.1	136.8	140.6	144.3	148.0	151.7
0.4	155.5	159.1	162.8	166.4	170.0	173.6	177.2	180.8	184.4	187.9
0.5	191.5	195.0	198.5	201.9	204.5	208.8	212.3	215.7	219.0	222.4
0.6	225.7	229.1	232.4	235.7	238.9	242.2	245.4	248.6	251.8	254.9
0.7	258.0	261.1	264.2	267.3	270.4	273.4	276.4	279.4	282.3	285.2
0.8	288.1	291.0	293.9	296.7	299.5	302.3	305.1	307.8	310.6	313.3
0.9	315.9	318.6	321.2	323.8	326.4	328.9	331.5	334.0	336.5	338.9
1.0	341.3	343.8	346.1	348.5	350.8	353.1	355.4	357.7	359.9	362.1
1.1	364.3	366.5	368.6	370.8	372.9	374.9	377.0	379.0	381.0	383.0
1.2	384.9	386.9	388.8	390.7	392.5	394.4	396.2	398.0	399.7	401.5
1.3	403.2	404.9	406.6	408.2	409.9	411.5	413.1	414.7	416.2	417.7
1.4	419.2	420.7	422.2	423.6	425.1	426.5	427.9	429.2	430.6	431.9
1.5	433.2	434.5	435.7	437.0	438.2	439.4	440.6	441.8	443.0	444.1
1.6	445.2	446.3	447.4	448.5	449.5	450.5	451.5	452.5	453.5	454.5
1.7	455.4	456.4	457.3	458.2	459.1	459.9	460.8	461.6	462.5	463.3
1.8	464.1	464.9	465.6	466.4	467.1	467.8	468.6	469.3	469.9	470.6
1.9	471.3	471.9	472.6	473.2	473.8	474.4	475.0	475.6	476.2	476.7
2.0	477.2	477.8	478.3	478.8	479.3	479.8	480.3	480.8	481.2	481.7
2.1	482.1	482.6	483.0	483.4	483.8	484.2	484.6	485.0	485.4	485.7
2.2	486.1	486.5	486.8	487.1	487.5	487.8	488.1	488.4	488.7	489.0
2.3	489.3	489.6	489.8	490.1	490.4	490.6	490.9	491.1	491.3	491.6
2.4	491.8	492.0	492.2	492.5	492.7	492.9	493.1	493.2	493.4	493.6
2.5	493.8	494.0	494.1	494.3	494.5	494.6	494.8	494.9	495.1	495.2
2.6	495.3	495.5	495.6	495.7	495.9	496.0	496.1	496.2	496.3	496.4
2.7	496.5	496.6	496.7	496.8	496.9	497.0	497.1	497.2	497.3	497.4
2.8	497.4	497.5	497.6	497.7	497.7	497.8	497.9	498.0	498.0	498.1
2.9	498.1	498.2	498.3	498.3	498.4	498.4	498.5	498.5	498.6	498.6

z	3.0	3.1	3.2	3.3	3.4	3.5	4.0	4.5	5.0
area	498.7	499.0	499.3	499.5	499.7	499.77	499.968	499.9966	499.999

Example N1

Determine the value of z corresponding to 5% defectives for a normal distribution.

Solution

The proportion of defectives is given by the area from $z = -\infty$ to $z = -z$

i.e. area (from $z = -\infty$ to $z = -z$) = 0.05

Now area (from $z = -\infty$ to $z = 0$) = 0.5

and area (from $z = -z$ to $z = 0$) = 0.5 - 0.05 = 0.45

Hence from Table N1 for area (from $z = -z$ to $z = 0$) of 450×10^{-3}

$z = 1.645$ (Answer)

Appendix V
Wind loadings

(Extracts from the British Standard Code of Practice CP 3'Basic Data for the Design of Buildings', Chapter V. Loading: Part 2:1972, Wind loads.)

This part of the code applies to the design of buildings, structures and their components. It does not include buildings of unusual shape or location, or those subjected to serious wind-excited oscillation. The symbols used do not necessarily conform to the international recommendations and are therefore listed here separately with the minimum of amendment.

A	element of surface area
A_e	effective frontal area
b	breadth
C_f	force coefficient
$C_{f'}$	frictional drag coefficient
C_p	pressure coefficient
C_{pe}	external pressure coefficient
C_{pi}	internal pressure coefficient
d	depth
F	force
F'	frictional force
h	height to eaves of building
h_o	height above ground (overall height)
k	a constant
ℓ	length
p	pressure on surface
p_e	external pressure
p_i	internal pressure
q	dynamic pressure of wind (stagnation pressure)
S_1	topography factor

S_2	ground roughness, building size and height above ground factor
S_3	a statistical factor
V	basic wind speed
V_s	design wind speed
w	width of building
α	wind angle (from a given axis)

Definitions

Depth and breadth The dimensions of the building measured in the direction of the wind, and normal to the wind, respectively.

Height The height of a building above the ground adjacent to that building.

Length and width The greater and lesser horizontal dimensions of a building above the ground adjacent to that building.

Height above ground The dimension above the general level of the ground to windward.

Element of surface area The area of surface over which the pressure coefficient is taken to be constant.

Effective frontal area The area normal to the direction of the wind or 'shadow area'.

Dynamic pressure of wind The free stream dynamic pressure resultant from the design wind speed.

Pressure coefficient The ratio of the pressure acting at a point on a surface to the dynamic pressure of the incident wind.

Force coefficient A non-dimensional coefficient such that the total wind force on a body is the product of the force coefficient times the dynamic pressure of the incident wind times the effective frontal area.

Topography The nature of the earth's surface as influenced by the hill and valley configurations.

Ground roughness The nature of the earth's surface as influenced by small-scale obstructions such as trees and buildings (as distinct from topography).

Appendix V

Outline of procedure (§4)

(i) Assess the basic wind speed, V, appropriate to district from map of UK in CP 3—chapter V. Typical values in m/s from this map for a 3-second gust likely to be exceeded once in 50 years, from any horizontal direction, are:

London	38	Glasgow	51
Birmingham	44	Londonderry	52
Swansea	48	Western Isles	56

(ii) Obtain the design wind speed, V_s, from

$$V_s = V S_1 S_2 S_3$$

(iii) Convert design wind speed to dynamic pressure, q, using

$q = kV_s^2$ where

$k = 0.613$ for SI units (N/m^2 and m/s)

(iv) Obtain the pressure, p, exerted at any point on the surface of a building from

$$p = C_p q$$

C_p negative denotes suction as distinct from positive pressure.

Pressure coefficients apply to a surface or part of a building. The load F acting in a direction normal to the element equals

$$(C_{pe} - C_{pi})qA$$

F positive indicates an inwards acting force and F negative an outwards acting force.

Obtain the total wind load on a structure by vectorial summation of all the loads on all the surfaces. Alternatively obtain the total wind load on the building as a whole, F, from

$$F = C_f\, q\, A_e$$

if coefficients C_f are available.

Values of S_1

S_1	Topography
0.9	Steep-sided, enclosed, sheltered valleys
1.1	Very exposed hill slopes and crest. Valleys with funnel shape
1.0	Elsewhere

Values of S_2

S_2 values depend upon:

(a) Height above ground in metres h_o

(b) Ground roughness:

(1) Open country—no obstructions

(2) Open country—scattered wind-breaks

(3) Country with many windbreaks; small towns; outskirts of cities

(4) Large and frequent obstructions—city centres

(c) Class of building:

(A) Units of cladding, glazing, roofing.

(B) Buildings where neither maximum horizontal nor maximum vertical dimension exceeds 50 m.

(C) As (B) for either maximum dimension exceeding 50 m.

Typical values of S_2 are given in Table V1.

Table V1 Values of S_2 for ground roughness, class of building and height above ground

h_o		$\ngtr 3$	5	10	20	30	40	60	100	140	200
(1)	(A)	0.83	0.88	1.00	1.06	1.09	1.12	1.15	1.20	1.24	1.27
	(B)	0.78	0.83	0.95	1.01	1.05	1.08	1.12	1.17	1.20	1.24
	(C)	0.73	0.78	0.90	0.96	1.00	1.03	1.08	1.13	1.17	1.21
(2)	(A)	0.72	0.79	0.93	1.03	1.07	1.10	1.14	1.19	1.22	1.26
	(B)	0.67	0.74	0.88	0.98	1.03	1.06	1.10	1.16	1.19	1.24
	(C)	0.63	0.70	0.83	0.94	0.98	1.01	1.06	1.12	1.16	1.21
(3)	(A)	0.64	0.70	0.78	0.95	1.01	1.05	1.10	1.16	1.20	1.24
	(B)	0.60	0.65	0.74	0.90	0.97	1.01	1.06	1.12	1.17	1.21
	(C)	0.55	0.60	0.69	0.85	0.92	0.96	1.02	1.09	1.13	1.18
(4)	(A)	0.56	0.60	0.67	0.79	0.90	0.97	1.05	1.13	1.17	1.22
	(B)	0.52	0.55	0.62	0.75	0.85	0.93	1.02	1.10	1.15	1.21
	(C)	0.47	0.50	0.58	0.70	0.79	0.89	0.98	1.07	1.12	1.18

Appendix V

Values of S_3 (§App.C)

S_3 is based on statistical concepts and its value varies with the period of exposure and the probability level. See Code itself for explanation. For other than special cases use the following values of S_3 associated with a probability level of 0.63:

Period of exposure (years)

	2	5	10	20	50	100
S_3	0.77	0.83	0.88	0.93	1.00	1.05

Values of C_{pe} for rectangular clad buildings

(i) Sides of building

Table V2 Values of C_{pe} for rectangular clad buildings

Building height ratio h/w	Building plan ratio ℓ/w	Wind Angle α	C_{pe} for surface A	B	C	D	C_p local on E
$\frac{h}{w} \le \frac{1}{2}$	1 to 1½	0^0	+0.7	-0.2	-0.5	-0.5	-0.8
		90^0	-0.5	-0.5	+0.7	-0.2	
	$1\frac{1}{2} < \frac{\ell}{w} < 4$	0^0	+0.7	-0.25	-0.6	-0.6	-1.0
		90^0	-0.5	-0.5	+0.7	-0.1	
$\frac{1}{2} < \frac{h}{w} \le 1\frac{1}{2}$	1 to 1½	0^0	+0.7	-0.25	-0.6	-0.6	-1.1
		90^0	-0.6	-0.6	+0.7	-0.25	
	$1\frac{1}{2} < \frac{\ell}{w} < 4$	0^0	+0.7	-0.3	-0.7	-0.7	
		90^0	-0.5	-0.5	+0.7	-0.1	
$1\frac{1}{2} < \frac{h}{w} < 6$	1 to 1½	0^0	+0.8	-0.25	-0.8	-0.8	-1.2
		90^0	-0.8	-0.8	+0.8	-0.25	
	$1\frac{1}{2} < \frac{\ell}{w} < 4$	0^0	+0.7	-0.4	-0.7	-0.7	
		90^0	-0.5	-0.5	+0.8	-0.1	

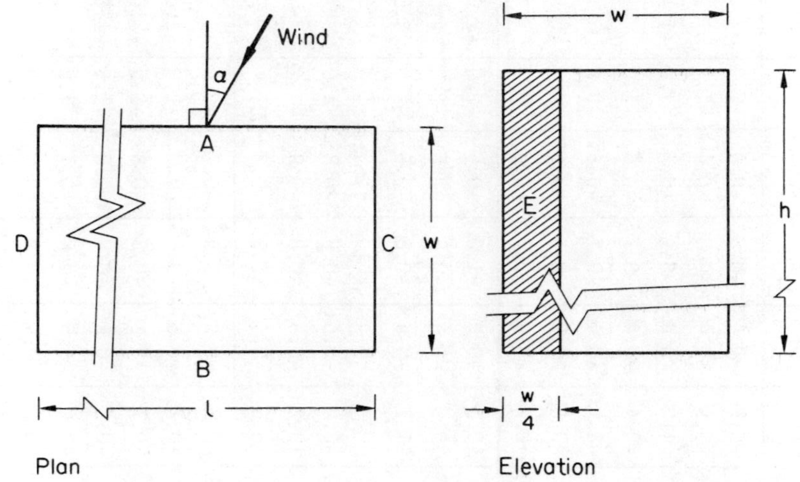

Fig.V1 Key to symbols for Table V2 (See also 'symbols' and 'definitions')

Appendix V

(ii) Roofs of buildings

Table V3 Values of C_{pe} for roofs of buildings

Building height ratio $\frac{h}{w}$	Roof angle degrees	Wind angle $\alpha = 0^o$		Wind angle $\alpha = 90^o$		Local coefficients on area			
		area EF	area GH	area EG	area FH	W	X	Y	Z
	0	-0.8	-0.4	-0.8	-0.4	-2.0	-2.0	-2.0	-
	5	-0.9	-0.4	-0.8	-0.4	-1.4	-1.2	-1.2	-1.0
	10	-1.2	-0.4	-0.8	-0.6	-1.4	-1.4		-1.2
0 to $\frac{1}{2}$	20	-0.4	-0.4	-0.7	-0.6	-1.0			-1.2
	30	0	-0.4	-0.7	-0.6	-0.8			-1.1
	45	+0.3	-0.5	-0.7	-0.6				-1.1
	60	+0.7	-0.6	-0.7	-0.6				-1.1
	0	-0.8	-0.6	-1.0	-0.6	-2.0	-2.0	-2.0	-
	5	-0.9	-0.6	-0.9	-0.6	-2.0	-2.0	-1.5	-1.0
	10	-1.1	-0.6	-0.8	-0.6	-2.0	-2.0	-1.5	-1.2
$\frac{1}{2}$ to $\frac{3}{2}$	20	-0.7	-0.5	-0.8	-0.6	-1.5	-1.5	-1.5	-1.0
	30	-0.2	-0.5	-0.8	-0.8	-1.0			-1.0
	45	+0.2	-0.5	-0.8	-0.8				
	60	+0.6	-0.5	-0.8	-0.8				
	0	-0.7	-0.6	-0.9	-0.7	-2.0	-2.0	-2.0	-
	5	-0.7	-0.6	-0.8	-0.8	-2.0	-2.0	-1.5	-1.0
	10	-0.7	-0.6	-0.8	-0.8	-2.0	-2.0	-1.5	-1.2
	20	-0.8	-0.6	-0.8	-0.8	-1.5	-1.5	-1.5	-1.2
$\frac{3}{2}$ to 6	30	-1.0	-0.5	-0.8	-0.7	-1.5			
	40	-0.2	-0.5	-0.8	-0.7	-1.0			
	50	+0.2	-0.5	-0.8	-0.7				
	60	+0.5	-0.5	-0.8	-0.7				

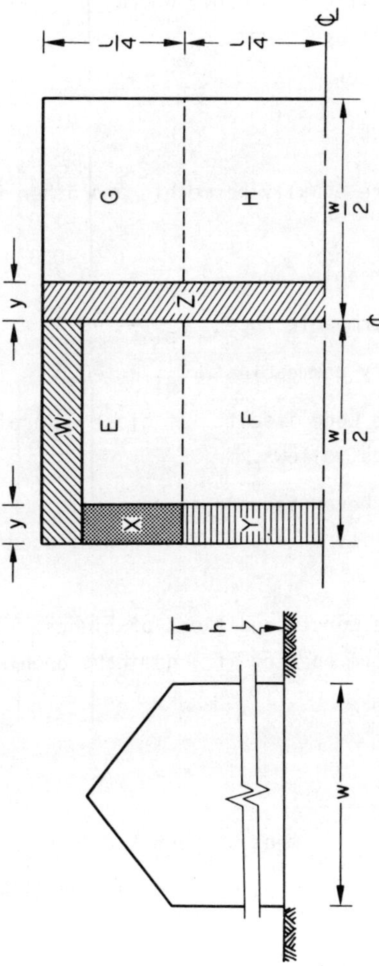

Fig.V2 Key to symbols to Table V3

Notes: (1) $w < \ell$

(2) $y = 0.15w$

(3) Pressure coefficient on underside or roof overhang equals that on adjoining wall surface.

(4) Overall coefficients apply where no local coefficients are given.

Estimates for C_{pi} (§App.E)

If two opposite faces are equally permeable and other faces are impermeable

(a) wind normal to permeable face $C_{pi} = +0.2$

(b) wind normal to impermeable face $C_{pi} = -0.3$

If four faces are equally permeable $C_{pi} = -0.3$

For greater detail see Code itself. If close estimate of C_{pi} is not possible, estimate as follows.

(a) +0.2 and -0.3, whichever is more onerous, if there is only negligible probability of a dominant opening occurring during a severe storm.

(b) 75% of C_{pe} (for the worst conditions of highest local external pressure) outside the opening if a dominant opening is likely to occur.

Estimates for force coefficients

Note: (1) b = breath = dimension normal to wind

(2) d = depth = dimension in direction of wind

(3) h = height of building

Table V4 Values of force coefficient C_f for rectangular clad buildings

$\frac{b}{d}$	C_f					
	Height/breadth (h/b) ratio					
	up to ½	1	2	4	6	10
4	1.2	1.3	1.4	1.5	1.6	
3	1.1	1.2	1.25	1.35	1.4	
2	1.0	1.05	1.1	1.15	1.2	
1½	0.95	1.0	1.05	1.1	1.15	
1	0.9	0.95	1.0	1.05	1.1	1.2
$\frac{2}{3}$	0.8	0.85	0.9	0.95	1.0	
½	0.75	0.75	0.8	0.85	0.9	
$\frac{1}{3}$	0.7	0.75	0.75	0.75	0.8	
¼	0.7	0.7	0.75	0.75	0.75	

Frictional drag

An additional force due to frictional drag is needed for buildings of special shape. For rectangular clad buildings this addition is only necessary where either d/h or d/b is greater than 4 and is given by:

$$F' = C_{f'} qb(d - 4z) + C_{f'} q2h(d - 4z)$$

where z = b or h whichever is least.

Use different values of $C_{f'}$ and q for roof (first term) and for walls (second term) as necessary.

$C_{f'}$ = 0.01 for smooth surfaces without ribs or corrugations across the wind direction

= 0.02 (or 0.04) for surfaces with corrugations (or ribs) across the wind direction.

Further information

See Code itself for additional information including the following.

(a) Pitched and saw tooth roofs of multi-span buildings and canopy roofs.

(b) Force coefficients for clad buildings of uniform section (e.g.

circular, triangular, curved corners, etc.),
(c) Pressure distribution around cylindrical structures,
(d) Force coefficients for unclad structures,
(e) Estimation of internal pressure coefficients,
(f) Ice formation on structures,
(g) Wind forces on circular sections,
(h) Addresses of advisory offices of the Meteorological Office.

APPROXIMATE METHODS OF ANALYSIS FOR LATERAL LOADING

Approximate methods of assessing the forces in a multi-storey building include the 'portal', 'cantilever' and 'continuous portal' methods, all of which assume that:

(a) Wind loads are applied at the panel points, i.e. the connections between beams and columns,
(b) A point of contraflexure exists at the midheight of each column,
(c) The total horizontal shear at any level is resisted by the columns at the points of contraflexure immediately below that level—the distribution between the columns varying with the actual method used.

Consider, for simplicity, a symmetrical single bay portal GABH, with the free body diagrams PABQ and RABS as shown in Fig. V3. The horizontal forces can be divided between each pair of columns as shown. The vertical forces are found by taking moments about the appropriate points of contraflexure. The shear force in beam AB is simply $W_1h_1/2\ell$ while the shear force in beam CD is the difference between the vertical forces in the columns

$$\text{i.e.}\quad \frac{W_1(h_1 + \tfrac{1}{2}h_2)}{\ell} + \frac{W_2h_2}{2\ell} - \frac{W_1h_1}{2\ell} = \frac{W_1h_1}{2\ell} + \frac{(W_1 + W_2)h_2}{2\ell}$$

Hence etc the bending moments in the beams and columns.
The following example illustrates the 'portal' method.

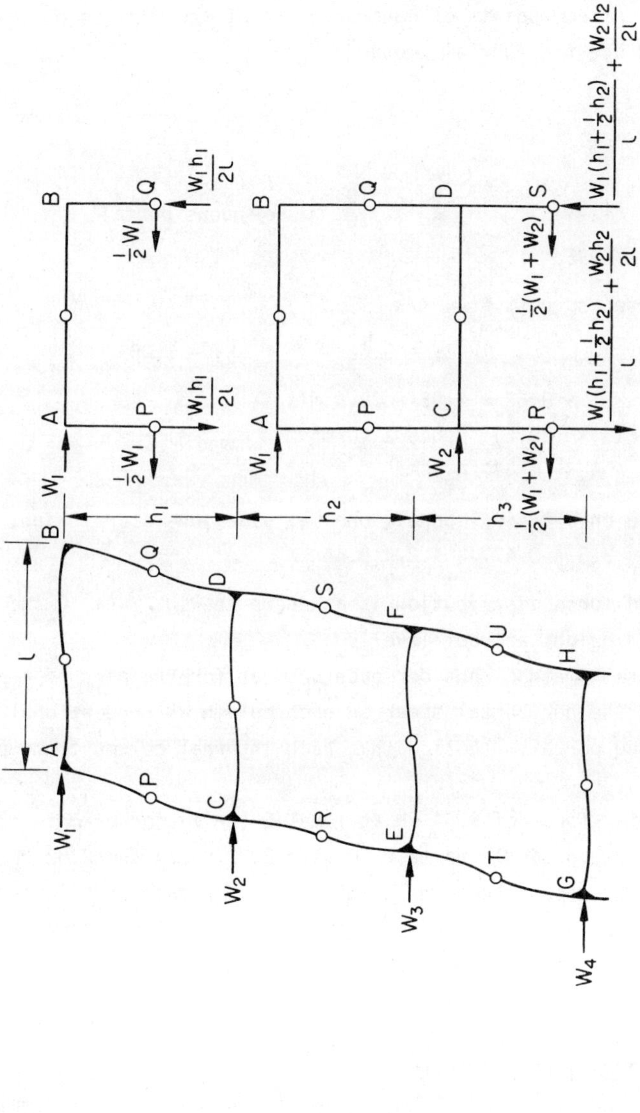

Fig.V3 Approximate analysis for lateral loading

Appendix V

Example V1

Fig. 9.6.1 shows the cross-section of a rectangular clad building which is required in the City of Birmingham for 100 years. There are 6 bays, each 5 m wide and 10 m overall height. Determine the distribution of bending moments along EFGH due to wind (γ_f = 1.2) for the limit state of collapse. Assume points of contraflexure at the centres of columns and beams and use the 'portal' method.

Solution

$$\frac{b}{d} = \frac{30}{20} = 1.5; \qquad \frac{h}{b} = \frac{10}{30} = 0.3$$

Therefore C_f = 0.95

Birmingham 3-second gust = 44 m/s

Take S_1 = 1.0; S_2 = 0.62; S_3 = 1.05

Hence $V_s = 44 \times 1 \times 0.62 \times 1.05 = 28.6$ m/s

$q = 0.613 \times (28.6)^2 = 0.502$ kN/m²

and $F = C_f\, q\, A_e = 0.477\, A_e$

The wind force on a typical panel, one bay wide, one storey high,

$= 5 \times 3.3 \times 0.477 \times 1.2 = 9.44$ kN

Hence the wind force distribution is as shown in Fig. V4(a) (wind force acts on 1.75 m height at roof level).

In the 'portal' method consider each beam as forming part of a portal sub-frame and the horizontal shear in each column as proportional to the span of the portal. (Notice that each internal column forms part of two portals.)

The horizontal force of 9.44 kN at level E (or I) for example is distributed between single portals to give 2.82, 3.78 and 2.82 kN; that is, 1.41 kN each on columns E and F, 1.89 kN each on columns F and G and 1.41 kN each on columns G and H. The total horizontal shear at column midheight above EFGH is 5.01 + 9.44 = 14.45 kN which can be distributed as shown in Fig.V4(b).

The moments resisted by beam EF at E

$= (2.17 + 3.58)1.65 = 9.49$ kN m

Similarly for the other moments which are shown in Fig.V4(c). The vertical forces at the points of contraflexure below E and F for portal MN are given by considering the horizontal forces shown in

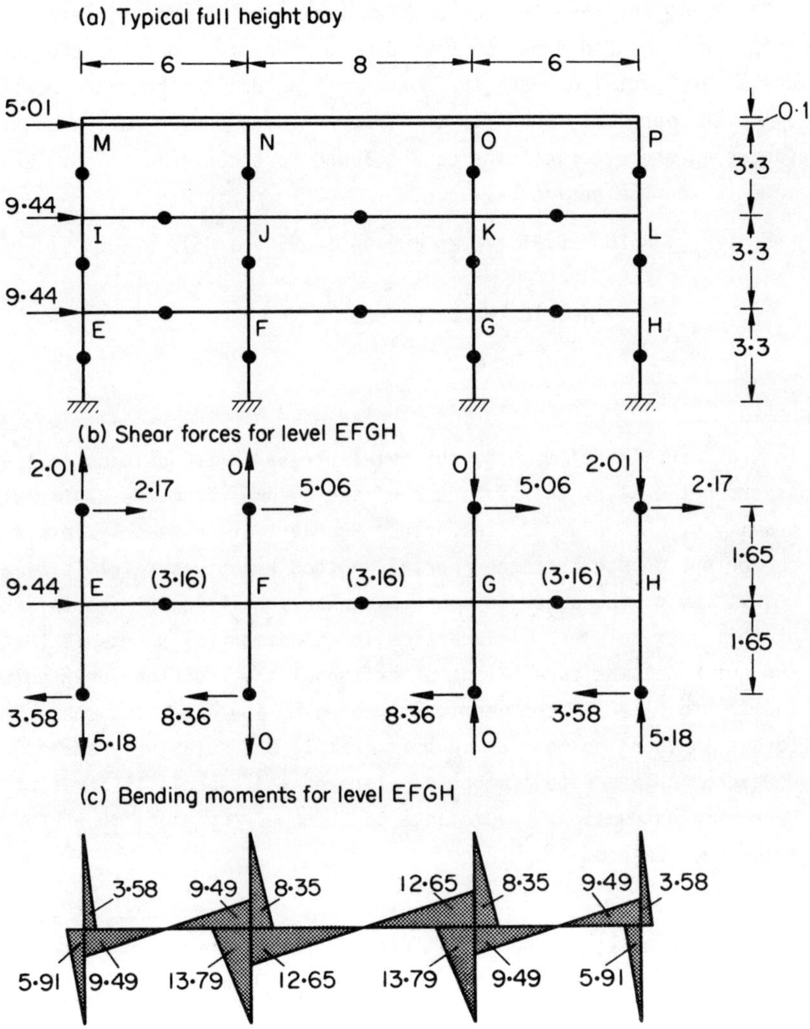

Fig.V4 Example V1

Fig. V4(b) and taking moments about the right-hand point of contraflexure (i.e. right-hand foot of the portal) as follows

$$(2.17 + 2.17)3.3 + (1.41 + 1.41)1.65 = 6 \times (F_{E(lower)} - 2.01)$$

or $F_{E(lower)} = 5.18$ kN

Similarly for the other columns as shown in Fig. V4(b), remembering that since each internal column forms part of two portals, the uplift due to one portal opposes the downward load due to the other portal. Since the vertical forces in the inner columns cancel out, then the forces in the outer columns can be found more conveniently by taking moments for the entire bay, e.g.

$$F_{E(lower)} \times 20 = 5.01 \times 8.25 + 9.44\ (4.95 + 1.65)$$

or

$$F_{E(lower)} = 5.18 \text{ kN} \quad \text{as before}$$

Notes

In the 'cantilever' method, the axial stress in the columns is taken as proportional to the distances of the columns from the centre of gravity of the columns as a whole. However, it is usually less convenient to use than the 'portal' method and gives fairly similar results, e.g. the outer columns take nearly half the horizontal shear of the inner columns. In practice it is common (a) to assume that the outer columns take half the horizontal shear of the inner columns and (b) to take all the vertical loading in the outer columns. The forces and bending moments in this example are relatively small. However, for a greater number of storeys, and especially for flat slab type of construction, the moments can become very significant for the members concerned.

Continuous portal method

This method, which is not in accordance with CP 110 and known as the continuous portal method, makes the following assumptions.

(a) There is a point of contraflexure at the midheight of each column.

(b) The axial stress in the columns is proportional to the distances of the columns from the centroid of the columns as a whole.

(c) The total horizontal shear is divided between the columns in proportion to their stiffnesses in the direction of the wind force.

The disadvantages of this method are the extra computations for unequally spaced columns of unequal area, and the calculated positions of the points of contraflexure for the outer beams. For a three-bay frame they are close to the inner columns, while for a five-bay frame of equal spans there is no point of contraflexure in the outer bays and it is necessary to obtain an approximate solution by halving the stiffness of the outer columns.

Appendix W
Dead loadings

The dead load is the load assumed to be produced by the static weight of all walls, partitions, floors, roofs, and finishes, including all other permanent construction. The dead load allowance for partitions and tanks as given in CP3 chapter V 'Dead and imposed loads', are included in Table W1 and the unit weights of some common materials are included in Table W2.

Table W1 Dead load allowances for partitions and tanks

Partitions shown on plans	Actual weight
Provision for partitions	Uniformly distributed load per metre:
	$\nless \frac{1}{3}$ × weight per metre run of finished portion
	and $\nless$ 1 kN/m^2 (offices only)
Tanks and other receptacles	Actual weights, full and empty

Table W2 Typical unit weights of materials

Density (kg/m^3)	Material	Weight (kN/m^3)
800-960	Aerated concrete (typical)	7.85–9.42
2550	Aluminium (cast)	25.0
640	Ashes (dry)	6.28
2080	Asphalt—paving	20.4
1760	Ballast—gravel and sand mixed	17.3
1920	—undisturbed	18.8
1350	Bitumen (prepared)	13.2
	Bricks (common burnt clay)	
1600–1920	—stacked	15.7–18.8
1200	—uncrushed ballast	11.8
1800	—crushed hardcore	17.7
1920	Brickwork—common burnt clay	18.8
2240	—heavy pressed brick	22.0
1200–1360	Cement (Portland)—loose	11.8–13.3
1120–1280	—bags	11.0–12.6
1280–1440	—bulk	12.6–14.1
1040	Clay—(dry fill, lumps)	10.2
2080	—(solid i.e. wet, compact)	20.4

Table W2 Typical unit weights of materials (*continued*)

Density (kg/m²)	Material	Weight (kN/m³)
800	Clinker, or ash (loose)	7.85
	Concrete (plain OPC)	
1840	—brick aggregate (typical)	18.1
1440	—clinker aggregate (typical)	14.1
2240	—stone aggregate (typical)	22.0
2370	Concrete (reinforced)—1% steel	23.2
2420	—2% steel	23.7
2580	—5% steel	25.3
8760	Copper (cast)	85.9
1280	Earth (loamy or dry, loose)	12.6
1760–1840	Earth (moist, compact)	17.3–18.1
1600	Gravel—loose	15.7
1920–2160	—undisturbed	18.8–21.2
1920	Hardcore (well consolidated)	18.8
910	Ice	8.93
7200	Iron (cast)	70.6
2100	Macadam	20.6
1920–2080	Mortar (set)—cement	18.8–20.4
2300	—cement screeds	22.6
1600–1760	—lime	15.7–17.3
1280	Sand—fill, damp when filled	12.6
1600	—dry	15.7
2000	—undisturbed, saturated	19.6
1010–1040	Sea water	9.91–10.20
100	Snow—fresh and dry	0.98
320	—wet compact	3.14
7830	Steel (cast or mild)	76.8
2080	Stone (solid)—Bath	20.4
2640	—granite	25.9
2560	—limestone	25.1
2240	—Portland	22.0
	Stone (crushed uncompacted)	
1600	—granite	15.7
800–880	Timber (seasoned)—English oak	7.85–8.63
650	—Pitch pine	6.38
480	—Softwoods	4.71

Appendix X
Imposed floor loadings

(Extracts from the British Standard Code of Practice CP3 'Code of Basic Data for the Design of Buildings', chapter V part 1:1967 'Dead and imposed loads').

The imposed load is the load assumed to be produced by the intended occupancy or use, including distributed, concentrated, impact, inertia and snow loads, but excluding wind loads. CP3 chapter V part 1:1967 requires that all beams shall be designed to carry the distributed load which is appropriate for the intended use.

Table X1 Distributed loads (kN/m^2)

Intended occupancy	Load intensity
Domestic and college bedrooms. Houses. Flats. Dormitories. Stairs (3 storeys or less).	1.5
Hotel bedrooms. Lounges and dining rooms. Hospital wards. Toilet rooms.* Billiard rooms.	2.0
Book stores per metre of storage height. (6.5kN/m^2 minimum)	2.4
Reading rooms without book storage. Light workrooms. Offices. Car parking for light cars and vans only.	2.5
Chapels and churches. Class rooms. Banking halls.* Minimum for stairs (over 3 storeys). Laundries. Kitchens. Laboratories.	3.0
Offices with computing equipment.	3.5
Public halls, clubs, colleges and theatres with fixed seating.* Shop floors. Libraries. Museums (minimum). Corridors subject to crowd loading. Stationery stores per metre of storage height.	4.0
Public and drill halls, clubs, college and theatres without fixed seating. Hotel bars and vestibules.* Gymnasia. Maximum for stairs (over 3 storeys). Cold storage per metre of storage height (15.0 kN/m^2 minimum). Office filing and storage spaces. Minimum for garages generally.	5.0
Boiler rooms. Theatre stages.	7.5
Minimum for foundries.*	20.0
Balconies: same as rooms to which they give access.	

* No concentrated load need be considered for slabs.

All floor slabs, or beams, ribs and joists spaced at centres not exceeding 1 metre, shall be designed to carry the appropriate distributed or concentrated imposed load, whichever produces the greater stresses in the part of the floor slab under consideration. The concentrated loads shall be considered to be applied in the positions which produce the maximum stresses and, where deflection is the design criteria, in the positions which produce maximum deflections. The concentrated imposed load need not be considered where the floor slabs are capable of effective lateral distribution of this load.

Concentrated loads (§4.1)

Table X2 Concentrated loads (kN), applied over an area 300 mm square

Load	Use
1.4	Domestic bedrooms. Houses. Flats.
1.5	Balconies per metre run concentrated at the edge.
1.8	Dormitories. Hotel and college bedrooms. Hospital wards. Stairs (3 storeys or less).
2.7	Lounges and dining rooms. Class rooms. Billiard rooms. Chapels and churches.
3.6	Public halls, clubs, colleges and theatres without fixed seating. Gymnasia. Shop floors.
4.5	Reading rooms with or without book storage. Corridors subject to crowd loading. Laundries. Minimum for kitchens and laboratories.
9.0	Minimum for drill halls, vehicle ramps and garages.

Ceilings and skylights (§4.2)

The supports of ceilings, ribs of skylights, etc., which may need to support the weight of a man, shall be designed for a concentrated load of 0.9 kN over an area 125 mm square, in addition to 0.25 kN/m^2 uniformly distributed over the whole area or area supported. If any member will in no circumstances need to support the weight of a man, the concentrated load may be neglected.

Allowable reductions in imposed floor loads (§5)

Reductions in the total imposed floor loads assumed for the beams may be taken as shown in Table X3 in designing columns, piers, walls, their supports and foundations. A roof may be regarded as a floor.

Table X3 Reduction in total distributed floor loads

No. of floors	1	2	3	4	5-10	over 10
Reduction (%)	0	10	20	30	40	50

Where a single span of a beam or girder supports not less than 46 m^2 of floor at one general level the imposed load for the beam may be reduced by 5% for each 46 m^2 supported subject to a maximum reduction of 25%. This reduction, or that given in Table X3, whichever is greater, may be assumed for the columns or other members supporting such a beam. The reductions given by Table X3 for factories and workshops designed for 5 kN/m^2 or more are subject to the limitation that the mean value assumed for all floors should not become less than 5 kN/m^2.

<u>Imposed roof loads</u> (§6)

Table X4 Imposed roof loads other than wind loads

	Distributed load (kN/m^2)	Concentrated load (kN)*
Flat roofs (slope $\not> 10^0$) with ready access	1.5	1.8*
Flat or sloping roofs (slope $\not> 30^0$) without ready access	0.75	0.9*
Sloping roofs, slope $\not< 75^0$	0.0	0.0
Sloping roofs, 30^0 < slope < 75^0	Use linear interpolation	

* applied over an area 300 mm square.

Roofs should be designed for the appropriate load (i.e. distributed or concentrated) in Table X4 which produces the greater stress. The distributed loads include for snow.

Dynamic loading (§7)

Table X5 Dynamic loadings, for crane gantry construction for single-crane operations

	Electric overhead crane	Hand-operated crane
Vertical loads	25%†	10%†
Horizontal force transverse to rails +	10%*	5%*
Horizontal force along the rails +	5%†	5%†

+ at rail level.

† % of maximum static wheel load.

* % of combined weight of crab and load lifted.

For forklift trucks allow 100% for impact loading $\overset{\S}{\nabla}$

Parapets and balustrades (§8)

Table X6 Minimum horizontal forces (N/m run) acting at handrail or coping level

Light access stairs $\not>$ 0.6 m wide	220
Light access stairs > 0.6 m wide Domestic landings and balconies	360
All other stairways and parapets	740
Panic barriers	3000

$\overset{\S}{\nabla}$ Not as CP3 chapter V.

See code itself concerning vehicle barriers for car parks.

Appendix Y
Sign conventions

CONVENTION (1): used for reinforced or prestressed concrete when considering the direct forces in the steel and the concrete separately.

(a) Steel It is normally assumed that steel is used in tension, thus tensile stresses and forces are positive unless specially indicated otherwise with the subscript c for compression:
i.e. F_s or F_{st} are both tension positive (+), and F_{sc} is compression positive (+) and tension negative (-).

(b) Concrete It is normally assumed that concrete is used in compression, thus compressive stresses and forces are positive unless specially indicated otherwise with the subscript t for tension:
i.e. F_c or F_{cc} are both compression positive (+), and F_{ct} is tension positive (+).

Thus in a column, for example, with steel bars in both tension and compression, but with resultant compression, then ΣF_s is negative whereas ΣF_{sc} is positive.

This rigid sign convention can conform with the natural and logical use of the materials and therefore avoids any anomalies. For example, the prestress force in prestressed concrete is consistently positive both for the steel (which is always tensile) and the concrete (which is always compressive) since the object of prestressing is to produce compression in the concrete (which is weak in tension).

CONVENTION (2): used for all sections *when considered in their entirety*.

Tensile stresses and forces are positive.

Angles are measured clockwise positive from x-direction.

y is always positive when in direction at 90° clockwise from positive x-direction (i.e. y-axis positive downwards for beams).

A shear stress τ_{xy} is positive if it acts in the positive y-direction on the x-plane whose outward drawn normal is the positive x-direction (hence $\tau_{xy} = + \tau_{yz}$).

(Notation for plates and slabs: first suffix indicates plane, second suffix indicates direction; for example τ_{xy} is shear stress on plane normal to x-axis, acting in y-direction).

Note: 'Construction of symbols', item (7) in Appendix F.1 of CP 110:1972 clearly states that the sign of a computed stress is given by positive (+) for tension and negative (-) for compression. Thus the symbol 'f' implies tension positive. This is not reversed by the subscript 's' for steel since 'f_s' also implies tension positive. The internationally agreed subscript for compression is 'c' as given in Table 70. Thus if the signs are to be reversed then it is essential to specify compression: i.e. use 'f_{sc}' for steel stress in compression as indicated in Table 73(with no abbreviation possible) In Eqns (28) and (29) of section 3.5.5 in CP 110:1972 it is clear that sign convention (1) above is intended for 'f_{s2}' and hence this should read 'f_{sc2}' in order to conform, because an abbreviated form is not acceptable in this case. Furthermore, for clarity and consistency the same subscripts should be used for the corresponding steel areas. Also, if signs are to be consistently reversed only with the preferred subscript "c" then "A'_s" should also be in the form "A_{sc}". The consequental changes which are necessary to conform with the international recommendations are therefore as shown under (I) in Table Y1.

The only other changes in the CP 110:1972 symbols which are necessary concern SI units (see Appendix Z) and these are shown under (II). In this connection it can be noted that the 'pascal' has been accepted by SI as a unit of pressure and is commonly used in some countries. Some countries are also considering using it as a unit for stress. If the pascal is so adopted then the change from the present recommended SI form is of little consequence (it merely involves the adoption of a special name for stress), whereas the change from the present CP 110 form also involves the change to the preferred single prefix in the numerator.

Table Y1 Alternative symbols and names for units

CP 110 symbols	Meaning	Preferred symbols
(I)		
A'_{s1}	Area of compression reinforcement in the highly compressed face	A'_{sc1}
A_{s2}	Area of reinforcement in other face	A_{sc2}
A_{sc}	Total area of compression reinforcement in columns	A_{sc}
A'_{s}	Total area of compression reinforcement	A_{sc}
A_{s}	Total area of tension reinforcement	A_{st}
$A_{s} + A'_{s}$	Total area of reinforcement	A_{s}
(II)		
kN/mm^2	10^9 N/m^2 or 10^9 Pa	GN/m^2 or GPa
N/mm^2	10^6 N/m^2 or 10^6 Pa	MN/m^2 or MPa

Appendix Z
SI units

The Système International d'Unités (SI) is essentially an extension and refinement of the traditional metric system. It embodies features which make it logically superior to any other system as well as practically more convenient: it is rational, coherent and comprehensive.

The metric system spread to several countries following the aftermath of the French Revolution and began to be adopted in scientific work in English-speaking countries towards the end of the nineteenth century. Over the years much thought was given to its extension and improvement until finally, in 1960, the Conférence Générale des Poids et Mesures, the body responsible for maintaining standards of measurements (of which the UK is an active participant), formally approved SI. The metrication of the construction industry in the UK was programmed to be effectively complete by December 1972. Many countries have already decided to make SI the only legally accepted system and it is clearly destined to become the universal currency of science and commerce.

Main features of SI

The main features of SI can be summarized as follows:

1. There are seven basic units, the metre and kilogramme taking the place of the centimetre and gramme of the old metric system.
2. The unit of force, the newton (kgm/s^2), is independent of the Earth's gravitation and the often confusing introduction, in some branches of science and technology, of g into equations is no longer necessary.
3. The unit of energy in all forms is the joule (newton × metre), and of power, the joule per second (watt); thus the variously defined calories, together with the kilowatt hour, the Btu and the horsepower, are all superseded.

4. 'Electrostatic' and 'electromagnetic' units are replaced by SI electrical units.
5. Multiples of units are normally to be restricted to steps of a thousand and, similarly, fractions to steps of a thousandth. A prefix should normally be used in the numerator only.

Basic SI units

There are in all seven basic SI units: metre, kilogramme, second ampere, kelvin, mole and candela. The abbreviations are m, kg, s, A, K, mol and cd respectively. Derived units with special names include newton, joule, watt, volt, ohm, degree Celsius. (N,J,W,V,Ω,°C). Derived units without special names include stress*(N/m^2), moment (Nm), etc. Selected conversion factors are given on the attached list. The term Celsius (°C) is retained for use when quoting values of the Celsius temperature.

* but see page 671

Prefixes

The names and symbols of the prefixes representing numerical factors are as follows:

k	kilo	=	10^3	m	milli	=	10^{-3}
M	mega	=	10^6	μ	micro	=	10^{-6}
G	giga	=	10^9	n	nana	=	10^{-9}
T	tera	=	10^{12}	p	pico	=	10^{-12}

These prefixes are both convenient in obviating the need to write large numbers of zeros (or high powers of ten) and also helpful in establishing familiarity with the numerical framework of modern science. Notice that the recommended prefixes are limited to $10^{\pm 3n}$. Notice also that the use of a prefix constitutes a new unit: thus mm^2, for example, $= (10^{-3}m)^2 = 10^{-6}m^2$ and not $10^{-3}(m)^2$.

Duplicated prefixes

SI recommends that wherever possible any numerical prefix should appear in the numerator of an expression and that two prefixes should not be used. For example, the units of stress and pressure should be expressed as N/m^2, kN/m^2, MN/m^2 etc. There are two reasons for this. First, there are fewer chances of making errors and, second, there are fewer demands on the memory. For example, consider a typical value of E for steel of 200 GN/m^2. This value can be remembered by two items only: the number 200 and the prefix giga, since one is referring them to a constant reference datum consisting of basic units. Compare this with the three chances of error when writing down 200 kN/mm^2 in terms of basic SI units as $200 \times 10^3/(10^{-3})^2$.

Additional prefixes

SI allows the use of the following additional prefixes for units involving cubes or higher powers of the basic unit:

d	deci	=	10^{-1}	da	deka	=	10
c	centi	=	10^{-2}	h	hecto	=	10^2

There is a strongly felt need in such cases since the intervals between units involving cubes, for example, become as great as $10^{\pm 9}$, and with higher powers the intervals between powers of the basic units are even greater. Thus second moments of area may be expressed in terms of dm^4 or cm^4 as well as m^4 and mm^4.

Units of time

The common units of time will persist. It is clear that many human activities are related to daily and yearly cycles and that it would be irrational to express all units in terms of ks, Ms, etc. Years, days and possibly hours should be retained.

Appendix Z

Table Z1 SI units-selected conversion factors

	British	SI		Reciprocal
Length	1 in	25.4×10^{-3}m		39.37
	1 ft	0.3048 m		3.281
Area	1 in^2	$645.16\times10^{-6}m^2$	(1)	1550
	1 ft^2	$92.90\times10^{-3}m^2$		10.76
Volume	1 ft^3	$28.32\times10^{-3}m^3$	(2)	35.31
	1 yd^3	0.7646 m^3		1.308
	1 gal	$4.546\times10^{-3}m^3$		220.0
2nd Moment of area	1 in^4	41.62 cm^4		2.403×10^{-2}
Mass	1 lb	0.4536 kg		2.205
	1 slug	14.59 kg		68.52×10^{-3}
Density	1 lb/ft^3	16.02 kg/m^3		62.43×10^{-3}
Force	1 lbf	4.448 N		0.2248
	1 tonf	9964 N		100.4×10^{-6}
Force/unit length	1 lbf/ft	14.59 N/m		68.52×10^{-3}
Pressure, stress	1 lbf/ft^2	47.88 N/m^2	(3)	20.88×10^{-3}
	1 lbf/in^2	6895 N/m^2		145.0×10^{-6}
	1 tonf/in	$15.444\times10^6N/m^2$		64.75×10^{-9}
Moment, torque	1 lbfin	0.1130 Nm		8.850
	1 tonfft	3037 Nm		0.3293×10^{-3}
Work, Energy	1 ftlbf	1.356 J	(4)	0.7376
Heat	1 Btu	1055 J		947.9×10^{-6}
Power	1 HP	745.7 W	(5)	0.001341
Temperature	1 ^oF	5/9 K (or ^oC)		9/5

	Old metric	SI	Reciprocal
Area	1 cm^2	$10^{-4}m^2$	10^4
Force	1 kgf	9.807 N	0.1020
Pressure, stress	1 kgf/cm^2	98070 N/m^2	10.20×10^{-6}
	1 bar	10^5 N/m^2	10^{-5}
Heat	1 cal	4.187 J	0.2388
Dynamic viscosity	1 poise	$10^{-1}Ns/m^2$	10

Notes

(1) $10^{-6}m^2$ = 1 mm^2

(2) $10^{-3}m^3$ = 1 litre

(3) 1 MN/m^2 = 1 N/mm^2 = 1MPa

(4) 1 joule = 1 mN. Note that dimensionally a moment of 1 Nm equals 1J

(5) 1 watt = 1 J/s

Answers to problems

Chapter 1

1 68.6 (exact value, 68.54); 4.74

Chapter 3

1 116 kN m

2 (a) 102.5 kNm (d_n = 0.242 and M_{dc} = 159 kN m)
(b) 118 kN m (x = 0.114, M_u = 189 kN m)

Chapter 4

1 150.3 kN m; 1.43%

2 434 mm^2 compression; 1930 + 360 mm^2 tension

Chapter 5

1 (a) (V = 36.6 kN;) 977 kN/m^2; 3 m; No, must satisfy $\ngtr \frac{3}{4}d$ (i.e. $\ngtr$ 188 mm) and nominal links requirement ($\simeq$ 186 mm)
(b) 0.47 MN/m^2)

2 (V = 380 kN); 3.38 MN/m^2; 195 mm(196); 40.3 mm; 426 mm

3 (a) If h = 625·mm and d $\simeq$ 588 mm, A_{st} = 1570 mm^2
(b) No; 340 mm spacing required, even allowing for offset of d from support.

Chapter 6

1 (a) -277 (support) and +204 kN m (3.66 m from A).

Chapter 7

1 1.6 × 10^{6}lbf in (181 kN m); under-reinforced beam

2 (a) 522 kN m (b) 432 kN m
Differences due primarily to partial safety factors for material strength but also to differences in stress-strain curves.

Chapter 8

1 (a) Yes, M_{ut} = 39500 lbf ft (>M = 39050 lbf ft)
(b) No, M_{uc} = 44500 lbf ft ($\ngtr$M = 46100 lbf ft)

2 See chapter 3, 2(b)

3 262 and -78.8 kN m; -20 and -252 kN m; 475 mm, 505 mm, 1983 mm^2

Chapter 9

1 (b) (i) 972 kN (ii) 1048 kN

2 b = a/3; h = 2a/3

3 2.11 MN (x = 0.245); 6%

4 1.515 MN (x = 0.26); 11%

5 (ρ_c = 5.5% for maximum axial load and only 2.2% for maximum moment;) 3430 mm^2; 4 No. 40 mm diameter bars

6 324 kips (1.44 MN)

Chapter 10

1 0.030 in (0.77 mm); 925 lbf/ft run

Chapter 11

1 (ii) 0.15 mm

2 (a) 0.26 mm (b) wide (uncontrolled); Yes: 1.48 m. Note: the actual crack width for predicted 1.48 m spacing would be greater at the soffit (due to eccentric effects) although stirrups could act as fine crack inducers and so relieve contraction at main cracks.

3 (a) $\rho_{crit} > \rho_c > \frac{2}{3}\rho_{crit}$ hence alternate complete contraction joints are necessary at 3.58 m (or up to 1¼ × 3.58m) spacing.
(b) Stability of wall against overturning = 1.065 if water seeps through.

4 0.31 mm—almost satisfies requirement of 0.3 mm.

Chapter 13

1 254 lbf/ft^2(12.2 kN/m^2)

2 $M = wl^2/72$

3 ℓ = 2.95 m; m = 35.3 kN m/m

4 143 lbf/ft^2(6.85 kN/m^2)

5 +27.3%; 0%

6 m'_x = 31.1 kN m/m at E; m_x = 89 and m_y = 11.1 kN m/m at F
Yes, ratio = 1.95 for 15% reduction in m'_x at E

7 $\alpha = A\beta,\ \beta = \dfrac{-B \pm \sqrt{|B^2 + 24(m + m')m/\lambda^2|}}{8(m + m')/\lambda^2}$

where $A = \sqrt{(1 + m'/m)}$; $B = (m + m')/2A + m(1 + \frac{1}{2}A)$

$\beta = 0.211$; $\alpha = 0.299$; $w = 47.9$ kN/m^2

Chapter 15

1 34.1 MN/m^2; 3.07 to 2.88 MN/m^2 (depending on method used)

(a) 24.1 to 24.6 MN/m^2; (b) 25.1 MN/m^2

(i) (a) Just unsatisfactory and (b) satisfactory

(ii) Satisfactory—all seven results exceed 29.5 MN/m^2

5.18 MN/m^2

2 1067, 783, 364, 188 kg/m^3; 0.795

Index